AF864216

Radiobiology and Environmental Security

NATO Science for Peace and Security Series

This Series presents the results of scientific meetings supported under the NATO Programme: Science for Peace and Security (SPS).

The NATO SPS Programme supports meetings in the following Key Priority areas: (1) Defence Against Terrorism; (2) Countering other Threats to Security and (3) NATO, Partner and Mediterranean Dialogue Country Priorities. The types of meeting supported are generally "Advanced Study Institutes" and "Advanced Research Workshops". The NATO SPS Series collects together the results of these meetings. The meetings are co-organized by scientists from NATO countries and scientists from NATO's "Partner" or "Mediterranean Dialogue" countries. The observations and recommendations made at the meetings, as well as the contents of the volumes in the Series, reflect those of participants and contributors only; they should not necessarily be regarded as reflecting NATO views or policy.

Advanced Study Institutes (ASI) are high-level tutorial courses to convey the latest developments in a subject to an advanced-level audience

Advanced Research Workshops (ARW) are expert meetings where an intense but informal exchange of views at the frontiers of a subject aims at identifying directions for future action

Following a transformation of the programme in 2006 the Series has been re-named and re-organised. Recent volumes on topics not related to security, which result from meetings supported under the programme earlier, may be found in the NATO Science Series.

The Series is published by IOS Press, Amsterdam, and Springer, Dordrecht, in conjunction with the NATO Emerging Security Challenges Division.

Sub-Series

A.	Chemistry and Biology	Springer
B.	Physics and Biophysics	Springer
C.	Environmental Security	Springer
D.	Information and Communication Security	IOS Press
E.	Human and Societal Dynamics	IOS Press

http://www.nato.int/science
http://www.springer.com
http://www.iospress.nl

Series C: Environmental Security

Radiobiology and Environmental Security

edited by

Carmel E. Mothersill
Department of Medical Physics and Applied Radiation Sciences
McMaster University
Hamilton, ON, Canada

Victoria Korogodina
Joint Institute for Nuclear Research
Dubna, Russia

and

Colin B. Seymour
Department of Medical Physics and Applied Radiation Sciences
McMaster University
Hamilton, ON, Canada

Springer

Published in Cooperation with NATO Emerging Security Challenges Division

Proceedings of the NATO Advanced Research Workshop on
Radiological Issues Pertaining to Environmental Security and Ecoterrorism
Alushta, Ukraina
9–14 October 2010

Library of Congress Control Number: 2011935038

ISBN 978-94-007-1999-6 (PB)
ISBN 978-94-007-1938-5 (HB)
ISBN 978-94-007-1939-2 (e-book)
DOI 10.1007/978-94-007-1939-2

Published by Springer,
P.O. Box 17, 3300 AA Dordrecht, The Netherlands.

www.springer.com

Printed on acid-free paper

All Rights Reserved
© Springer Science+Business Media B.V. 2011
No part of this work may be reproduced, stored in a retrieval system, or transmitted in any form or by any means, electronic, mechanical, photocopying, microfilming, recording or otherwise, without written permission from the Publisher, with the exception of any material supplied specifically for the purpose of being entered and executed on a computer system, for exclusive use by the purchaser of the work.

Foreword

This book comes from a NATO Advanced Research Workshop held in Alushta, Ukraine in October 2010. The Workshop was held in parallel with the Third International conference "Modern problems of genetics, radiobiology, radioecology, and evolution" dedicated to the famous Russian scientists N. W. Timofeeff-Ressovsky, V. I. Korogodin and Vladimir A. Shevchenko. Nikolay W. Timofeeff-Ressovsky was the first to initiate investigations of radiation effects in ecology, his pupil Vladimir I. Korogodin was a pioneer researcher of postradiation recovery and "cascade mutagenesis" in cells, and Vladimir A. Shevchenko was the investigator of radiation-polluted areas over the world.

The subject of the workshop was to consider fundamental science which contributes to our understanding of the potential risks from ecological terrorism – i.e. dirty bombs, atomic explosions, intentional release of radionuclides into water or air.

None of these have happened yet but there is grave concern about radioactive sources which appear to have gone missing and about the expansion of nuclear programs in politically unstable areas.

The workshop considered how radiation acts to cause health effects in humans and in ecosystems. There was a particular focus on low dose effects and on the new science which is suggesting that DNA is not the only important target for radiation effects but that systemic effects operating via signaling mechanisms are highly important and probably dominate the response in environmentally relevant dose ranges. The last 15 years have seen a major paradigm shift in radiation biology. Several discoveries challenge the DNA centric view which holds that DNA damage is the critical effect of radiation irrespective of dose. This theory leads to the assumption that dose and effect are simply linked – the more energy deposition, the more DNA damage and the greater the biological effect. This is embodied in radiation protection (RP) regulations as the linear-non-threshold (LNT) model. However the science underlying the LNT model is being challenged particularly in relation to the environment because it is now clear that at low doses of concern in RP, cells, tissues and organisms respond to radiation by inducing responses which are not predictable by dose. These include adaptive responses, bystander effects, genomic instability and low dose hypersensitivity and are commonly described as *stress* responses, while recognizing that "stress" can be good as well as bad. The phenomena contribute to observed radiation responses and appear to be influenced by genetic, epigenetic and environmental factors, meaning that dose and response are not simply related. The question is whether our discovery of these phenomena means that we need to re-evaluate RP approaches. The so called "non-targeted" mechanisms mean that low dose radiobiology is very complex and supra linear or hormetic responses are equally probable but their occurrence is unpredictable for a given individual. Issues which may need consideration are synergistic or antagonistic effects of other pollutants because RP at present only looks at radiation dose but the new radiobiology means that chemical or physical pollutants which interfere with tissue responses to low doses of radiation could critically modulate the predicted risk. Similarly, the "health" of the organism could determine the effect of a given low dose by enabling or disabling a critical response.

The contributions in the book cover genetics and radiation genetics, radiobiology, radioecology, radiation epidemiology and risk assessment. The evolution regularities are also considered. The contributions represent state-of-the-art contributions from leading scientists in the low dose field who were invited to attend.

In addition to the major support from the NATO Peace and Security section, we wish to acknowledge support from the Joint Institute for Nuclear Research, International Union of Radioecology, Genetic Society of America, Max Delbrük Center, Russian Academy of Sciences, McMaster University, Canada, National Academy of Sciences of Armenia, National Academy of Sciences of Belarus, National Academy of Sciences of Ukraine, Russian Medicine Radiology Center of Russian Academy of Medicine Sciences, Russian Institute of Agricultural Radiology and Agroecology of Russian Agriculture Academy of Sciences, Scientific Society "Biosphere and Humanity" after N.W. Timofeeff-Ressovsky, Russian Vavilov Society of geneticists and selectionists.

кафе "ДУБНА"

Contents

Chapter 1
Hypermutability Associated with Double-Strand Break Repair

Dmitry A. Gordenin

Abstract Double-strand breaks (DSBs) are the most toxic kind of DNA damage caused by ionizing radiation as well as by a number of other environmental factors and drugs. DSBs lead to gross chromosome rearrangements, genetic disease, and cancer or cell death. However cells can be programmed to generate DSBs in their own DNA. Programmed DSBs are a key element of many biological functions such as meiotic recombination and segregation, adaptive immunity, regulation switches and viral life cycles. Either damage-induced or programmed DSBs should be repaired in order to retain cell viability. Over the last years it has been established that DSB repair can be associated with up to 10,000-fold increase in frequency of base substitutions and small insertions/deletions (indels). This *localized* hypermutability represents additional genotoxic threat as well as a potential for generating rare multiple mutant alleles with high fitness without overloading the rest of the genome with mutations.

Keywords DNA repair • Mutagenesis • Double-strand breaks • Multiple mutations

1.1 A Challenge of Multiple Mutations in a Single ORF

Even a single mutation in DNA can alter biological functions either to a benefit or to a detriment of a cell or an organism. Thus important balance must be maintained between limiting mutation frequency to avoid harmful changes on one hand and on the other hand allowing the level of mutagenesis which can generate sufficient amount of

D.A. Gordenin (✉)
Laboratory of Molecular Genetics, National Institute of Environmental Health Sciences, NIH, DHHS, Research Triangle Park, NC 27709, USA
e-mail: Gordenin@niehs.nih.gov

C.E. Mothersill et al. (eds.), *Radiobiology and Environmental Security*, NATO Science for Peace and Security Series C: Environmental Security, DOI 10.1007/978-94-007-1939-2_1, © Springer Science+Business Media B.V. 2012

rare adaptive changes fueling evolution. In various species the rates of spontaneous mutations on a genome scale are kept down by a firewall of replication fidelity and repair systems [15, 25]. Based on studies with mutation reporters mutation rate per genome in various species is low; no more than one out of a hundred new cells would carry a single new mutation in the entire genome [10, 11]. Thus accumulation of multiple mutations in a genome over just few generations would be extremely unlikely. Even less likely would be the incidence of simultaneous changes in a single gene; however multiple mutations in a gene would carry the strongest biological effects, i.e. reduction in gene function, increase in gene function or even creation of a novel function. The stronger potential of multiple mutations is evident for the case of gene inactivation, since the majority of base pair substitutions (bps) and even some of small insertions and deletions (indels) would leave the gene function on a biologically sufficient level. Multiple mutations are also more likely to generate changes increasing fitness. Limited number of studies aimed to generate enzymes with enhanced or even new activity established that this can be achieved primarily by multiple mutations [4, 47]. Importantly, multiple mutations showed *sign epistasis*, i.e. a condition when individual changes within a multiple mutant are neutral or even deleterious [24, 56]. On the evolution scale it translates into a requirement of multiple mutations to avoid *fitness valleys* (i.e. steps with reduced fitness within the succession chain of mutations resulting in alleles with high fitness) by following *fitness ridges* involving only the succession of mutation steps that do not lead to fitness reduction [9, 58]. As established by comparisons across wide range of taxons sign epistatsis as well as fitness ridges and valleys are the present characteristics of protein evolution [43]. However, fitness valleys are not a concern if multiple mutations of the advantageous alleles occur by simultaneous or closely timed events. Multiple mutations that appear to be simultaneous or coordinated in time (*chronocoordinate*) were also detected in normal mouse and human tissues [7, 55] and in tumors [5, 6, 23]. While the fraction of such mutations is small they may play bigger role only in some categories of cancers especially in cancers associated with high density of DNA damage (see below).

Since beneficial mutations represent only a tiny subset of all possible changes, only very small group of multiple mutations can be expected to bring high fitness. In order to get a specific set of simultaneous specific mutations in a single ORF with a realistic probability, individual changes should have a very high chance. Based on the maximal estimate of spontaneous mutation rate for as 10^{-9} per nt [10, 11] even the simultaneous mutations of three specific bases become practically impossible (10^{-27}), because it would require unrealistic amounts of biological material (for example 10^{27} yeast cells would weigh 100 billion tons). Several orders of magnitude greater mutation rates are required to make simultaneous multiple mutations realistic. For example, the rate of 10^{-4} per nt would allow a simultaneous mutation of three specified nucleotides to be found among 10^{12} cells which corresponds to just 100 g of yeast or around 1 kg of human cells. Such high mutation probabilities are impossible on a genome-wide scale even for a single cell generation. A minimal approximation for 40,000 one-kilobase ORFs (minimal estimate for a diploid human genome) gives around 4,000 mutations, which would create intolerable mutation load by coincidence of allelic recessive lethals and by inactivation of haplo-insufficient genes.

1.2 Pathway Generating Multiple Mutant Alleles

The genome-wide mutation overload can be avoided if high mutation density would be generated only in a small section of a genome – a phenomenon we define as *localized hypermutability* (LHM). Some regions of a genome could be permanently more prone to mutations than others. The presence of at-risk motifs capable of forming DNA structures that are poor substrates for DNA repair and mutation avoidance systems [18, 62] and chromatin structure and modification [21] are among possible causes. Studies of adaptive mutagenesis in *E. coli* by Rosenberg et al. [20, 48] indicated that double-strand break (DSB) repair could be mutagenic. Soon after that Jeff Strathern and colleagues have established in the model yeast system using a defined site-specific DSB repaired by homologous recombination that the repair in fact is associated with up to several hundred fold increase in mutation frequency in the area around a DSB [44, 53]. Thus there is a potential for LHM to emerge in any given place of the genome, if there is a DSB followed by DSB-repair. Later the hypermutability of DNA adjacent to a site-specific DSB has been shown also in *E. coli* thus establishing the generality of the phenomenon across microbial taxons [16, 42]. Recently we confirmed one of the sources of LHM proposed by the Strathern group – hypermutability of long single-strand DNA (ssDNA) formed by strand-biased $5' \rightarrow 3'$ DNA degradation (resection) [59]. The second proposed source – error-prone DNA synthesis creating two new strands in the course of filling a double-strand gap was demonstrated recently by the Haber lab [22]. In all systems the rates of mutation per nt in the absence of applied DNA damage ("spontaneous") was close to 10^{-6} per nt initially estimated by the Strathern group [45] (see also Table 1.1). Not surprisingly this level of hypermutability produced only single mutations.

The rates of spontaneous LHM associated with DSB repair vary around the estimated value of *in vivo* error rates in yeast cells carrying double defect in DNA polymerase proofreading and in post-replicative mismatch repair (MMR) (1.5×10^{-6} per nt in the *URA3* gene [35]). Importantly, combination of proofreading and MMR defects did not produce multiple *URA3* mutations even through growing yeast cells for several generations. There are no indications so far that significantly higher *in vivo* rate of errors capable of producing simultaneous multiple mutations within a single ORF could be achieved during synthesis on long undamaged templates. However, very high mutation density can be achieved if LHM is associated with DNA damage. The striking case of *programmed* damage-induced increase of mutation frequency by about million-fold as compared to genome-wide rate has been well established for a small region of the Ig-locus in genomes of immune B cells [29, 38] and Table 1.1). The increase in somatic mutation (somatic hypermutability –SHM) is confined to a small region within the Ig locus. SHM is driven by a specialized enzyme – activation-induced deaminase (AID), which converts a part of cytosines into uracils in the SHM region. Since this region is involved in determination of the affinity to an antigen, SHM results in a very fast accumulation of multiple mutant alleles providing sufficient material for selecting cells producing antibodies with several orders of magnitude greater affinity to the antigen. Because of the

Table 1.1 Mutation frequencies associated with various kinds of localized hypermutability

Cause of hypermutability (species)	Mutation frequency per megabase	References
DSB-repair in F′-episome (E. coli)[a]	19	Gonzalez et al. [17]
DSB-repair, gap filling (yeast)[b]	0.7	Hicks et al. [22]
DNA adjacent to repaired DSB (yeast)[c]	17	Rattray et al. [45]
ssDNA adjacent to DSB repair (yeast)[d]	13	Yang et al. [59]
Damage-induced LHM (yeast)[e]	300	Yang et al. [59, 60]
SHM, adaptive immunity (humans)[f]	1,000	Odegard and Schatz [38], Liu and Schatz [29]

[a] *E. coli* mutations selected in F′ cod A, B genes with mutation target size 848 bp use for calculations as determined in the referenced work
[b] Mutations in the *URA3* gene of Sacharomyces cerevisiae occurring during a double-stranded (ds) gap repair (minimal estimate of mutation target of 125 bp as determined by [26])
[c] Frequency of spontaneous mutations in the vicinity of a DSB repaired by homologous recombination. Site-specific DSB was induced next to chromosomal *CAN1* gene and repaired by homologous recombination with a truncated copy of *CAN1* in the same chromosome. This system did not allow distinguishing between hypermutability in transiently formed ssDNA (as in), or hypermutability during repair of a ds-gap (as in Hicks et al. [22])
[d] In this system *CAN1* reporter gene was placed in the vicinity of a DSB that was repaired by a short oligonucleotide, so *CAN1* sequence did not participate in a recombination act therefore the most likely hypermutable intermediate was a stretch of transient ssDNA next to a DSB
[e] Average frequencies of mutations induced by UV-C (45 J/m^2) and MMS (30 min in 11.8 mM (0.1%) MMS) in yeast ssDNA around DSB or next to uncapped telomere (see also Fig. 1.1)
[f] Approximate frequency of mutations associated with somatic hypermutation in the Ig-genes

specially organized cell division control the very same cells also have proliferation advantage over the cells producing antibodies with lower affinity. LHM can bring not only high fitness alleles with improved or new function. It also increases the frequency of gene inactivation. Albeit with much lower efficiency, AID expressed in immune cells is mutagenic for several other genomic regions, which makes them prone for undesired changes [29, 30, 54]. However, since LHM is mostly confined to a small region within the Ig-locus, it can generate multiple mutant alleles with high fitness without overloading genome with additional mutations and thus avoiding accumulation of lethal or low fitness alleles in other genes.

Can damage-induced LHM operate in a *non-programmed* mode? The existence of multiple powerful repair systems enables living cells to repair vast number of damages through a single cell cycle. Even the number of endogenous damages in normal human cells is estimated in tens of thousands [28]. The tolerable number of damages that can be caused by exogenous sources can be orders of magnitude greater reaching a density of one damage per several thousand nt [39, 46, 49, 57]. Unrepaired damages often lead to mutations, if copied by error-prone translesion synthesis (TLS) DNA polymerase [14, 40]. A cell with high density of DNA damage will inevitably die if it lacks DNA repair throughout the genome. However, the lack of repair in a small section of the genome could be tolerated. In this situation error-prone TLS during replication of the damaged section can produce a stretch of multiple mutations. Most of DNA repair systems operate in double-strand DNA (dsDNA). However, damage in rare stretches of single strand DNA (ssDNA) would

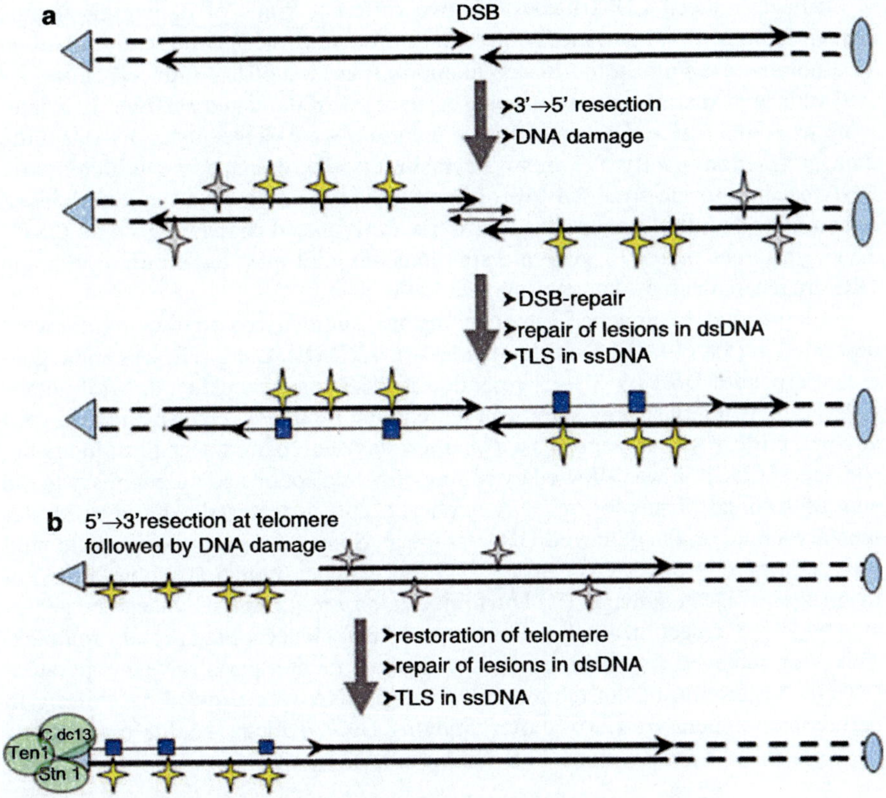

Fig. 1.1 Damage-induced localized hypermutability associated with transient regions of single-strand DNA generated at double-strand breaks and uncapped telomeres

be often left unrepaired and thus lead to mutation. Coincidence in formation of large stretches of ssDNA with DNA-damage could, in principle, be a source of LHM, if damaged ssDNA is capable of recovery to ds-state. However, if long ssDNA with multiple damages is lost because of degradation or cell death triggered by checkpoint reaction, the opportunity for multiple mutations will be lost.

We sought to determine if long stretches of ssDNA formed around DSB or at uncapped telomeres can recover with multiple mutations [59]. For that purpose we developed special genetic systems in model eukaryote, yeast *S. cerevisiae* where stretches of long ssDNA can be formed around inducible site-specific double strand break or uncapped telomeres formed by holding a telomere-capping *cdc13-1* mutant in non-permissive temperature (Fig. 1.1, Table 1.1). In these model systems we observed the frequencies of damage-induced LHM comparable to those observed in programmed SHM within Ig-locus (Table 1.1). Importantly, we observed up to six simultaneous mutations in a single ORF, while mutagenesis in other genomic locations was barely detectable.

Damage-induced LHM caused by two different kinds of damaging agents, ultraviolet light (UV) and methylmethane sulfonate (MMS) completely relied on TLS polymerase Pol. Strand-biased mutation spectrum of UV-induced mutations [59] indicated that mutations are caused by TLS in damaged ssDNA. In a later work we found that in the case of MMS mutations were caused by ssDNA-specific damage (predominantly N-3-methyl cytosine), so the damage was inflicted after DNA became single-stranded [60]. Importantly, we observed a large number of strand-biased multiple mutations (up to 6 widely spaced changes in a 4 kb ORF). Thus, yeast cell is set to generate simultaneous multiple mutations in a single ORF by mechanism of damage-induced LHM.

Experimental approached and conclusions summarized in this figure were described in [59, 60]. **(A) Double-strand break (DSB).** Long ssDNA can be generated around a DSB by $5' \rightarrow 3'$ resection if DSB repair is delayed. While DNA damage in ds regions (grey stars) can be repaired by major repair pathways, such as BER, NER, PRR, damage in ssDNA often stays unrepaired. Repair of inducible site-specific DSB in was allowed by adding oligonucleotides complementary to the ends of the break. Trans-lesion DNA synthesis (TLS) is required to build a complementary strand on the damaged DNA template. Some TLS events will create wild type sequence, however error-prone TLS can generate mutations (blue boxes) at many sites of DNA damage. **(B) Uncapped telomere.** Long ssDNA can be generated by $5' \rightarrow 3'$ resection at telomeres that transiently lost capping protein complex. This was achieved by shifting *cdc13-1* mutant to non-permissive temperature (37°C). Restoration of the telomere cap and dsDNA was allowed by shifting to permissive temperature (23°C) after applying DNA damage. Multiple mutations were generated by error-prone TLS similar to (A).

1.3 Mechanisms of Damage-Induced LHM

In our experiments LHM was observed after restoration of damaged ssDNA formed at unprotected DNA ends such as DSBs and uncapped telomeres. Long ssDNA was well documented at unprotected DNA ends in pro- and eukaryotic microbes, however resection tracts may be less long in mammalian systems [1, 34, 50, 51]. It is worth note that unlike in microbial systems mammalian resection studies mostly rely on detection of ssDNA-interacting proteins rather than direct monitoring of ssDNA formation. The prevailing current view that the resection machinery is conserved through eukaryotes but in normal conditions end-resection capacity in mammalian cells is limited. Shorter resection tracts were also suggested based on high capacity of mammalian cells to non-homologous end-joining (NHEJ) or to microhomology mediated end-joining (MMEJ). These pathways efficiently operate with blunt or minimally degraded DNA ends and thus quickly eliminate the substrates of end-resection. An additional factor reducing the size and number of resection tracts could be the ssDNA binding proteins inhibiting resection. Even if normal resection tracts in mammalian cells are shorter than in yeast all conserved proteins participating in resection are there so infrequent long resection tracts are possible especially when inhibitors of resection would not act.

Another potential source of long ssDNA is uncoupling between unwinding and DNA synthesis in leading and/or lagging strands of replication fork that can be formed by infrequent spontaneous miscoordination and/or in response to DNA damage blocking DNA polymerase [19, 31]. However, the frequency of uncoupling and sizes of ssDNA regions are not well documented so far. There were also communications about very long stretches of ssDNA formed in cultured cancer cells [3]. The origin of and mechanism producing this form of ssDNA are unknown. However, if it can restore to ds state this could be one more source of damage-induced LHM.

In principle damage-induced LHM should not be necessarily associated with ssDNA. It can originate from any cause that would inhibit DNA repair from the time of damage trough the next DNA replication. While several factors such as chromatin type, nucleosome position or transcription status can affect efficiency of DNA repair and/or mutation frequency there were no demonstration of really strong, multi-fold effects of these factors leading to clusters of simultaneously occurring multiple mutations. We expect that with accumulation of whole-genome mutagenesis data (see e.g., [36, 37, 61]) will reveal important information about pathways and molecular mechanisms of damage-induced LHM.

1.4 Biological Context and Evolution Implications for Damage-Induced LHM

Examples of wide spread clusters of multiple mutations (*mutation showers*) have been detected among mutation spectra in mice and humans [7, 55] however the mechanisms generating these clusters were not addressed. Hypermutability and mutation clusters in our experiments [59, 60] were caused by damaging artificially formed ssDNA around inducible site specific DSB or in the vicinity of uncapped telomeres in G2-arrested *cdc13-1* mutant yeast. However, ssDNA can be formed by resection at unprotected ends of spontaneous or damage-induced DSBs. Importantly, the vast number of DNA damaging agents can induce both DSBs and mutagenic base or nucleotide damage [15]. For example we demonstrated that base alkylation by methyl methanesulfonate (MMS) can results in DSBs via faulty repair of closely opposed damages [32, 33] as well as in high frequency of base substitutions in artificially formed ssDNA around site-specific DSB [60]. However, there were no or very small fraction of multiple mutations found within spectra of spontaneous or damage-induced forward mutations in common mutation reporters. Mutation reporters targeted to detect low frequencies of clustered multiple mutations are under development in our lab.

Mutations have generated all current variety of DNA sequences over the time of biological evolution. On the evolution timescale localized increases in the numbers of mutation accumulated during human evolution from a common ancestor with chimpanzees have been associated with meiotic DNA breaks [2, 12, 13, 27, 41].

These studies have identified a number of regions (*human accelerated regions* – HARs) in which over the past 10 million years of primate evolution much more mutations have accumulated than over preceding 100 million years of mammalian evolution. Interestingly, association was detected between hotspots of meiotic recombination in human males and HARs. Another distinct feature of HARs is a mutation bias of A-T or T-A pairs changing into G-C or C-G pairs. One explanation is that there is biased gene conversion (BGC) in which G-T and C-A mismatches are more frequently corrected towards G-C and C-G as compared with correction towards A-T and T-A. This would lead to increased fixation of G-C and C-G mutant base pairs. However HARs could also reflect the increased mutability around meiotic DSBs which can be further enhanced by endogenous damage to ssDNA formed around DSBs triggering male meiotic recombination. Recently, based on analysis of the vast amount of human sequencing data it was concluded that the increased rates of base substitutions on an evolution, a population and even on a single tumor or cell line timescales are associated with rearrangement breakpoints which could at least in part be caused by hypermutability of break-associated ssDNA ([8] and references therein). Another study associated increased mutation rates in the human evolution line with late replicating regions of genome which could be also associated with higher frequency of breakage during mitotic divisions in the germline or/and with increase in ssDNA formation [52]. These observations together with >1,000-fold increase of mutation frequency in damaged ssDNA as compared with dsDNA regions demonstrated in our experiments makes us to speculate that significant fraction of mutations in nature may come from error-prone translesion synthesis in ssDNA. Future studies integrating model experiments with genotoxic factors and whole genome mutation analyses on various timescales will shed light on the role of damage induced LHM in evolution, biology of species as well as in human health.

1.5 Funding Note

This work was supported by intramural research funds from NIEHS to Chromosome Stability Group under project Z-01-ES06507.

References

1. Amundsen SK, Smith GR (2003) Interchangeable parts of the Escherichia coli recombination machinery. Cell 112(6):741–744
2. Berglund J, Pollard KS, Webster MT (2009) Hotspots of biased nucleotide substitutions in human genes. PLoS Biol 7(1):e26
3. Bjursell G, Gussander E, Lindahl T (1979) Long regions of single-stranded DNA in human cells. Nature 280(5721):420–423
4. Camps M, Herman A, Loh E, Loeb LA (2007) Genetic constraints on protein evolution. Crit Rev Biochem Mol Biol 42(5):313–326

5. Chen Z, Feng J, Buzin CH, Sommer SS (2008) Epidemiology of doublet/multiplet mutations in lung cancers: evidence that a subset arises by chronocoordinate events. PLoS One 3(11):e3714
6. Chen Z, Feng J, Saldivar JS, Gu D, Bockholt A, Sommer SS (2008) EGFR somatic doublets in lung cancer are frequent and generally arise from a pair of driver mutations uncommonly seen as singlet mutations: one-third of doublets occur at five pairs of amino acids. Oncogene 27(31):4336–4343
7. Colgin LM, Hackmann AF, Emond MJ, Monnat RJ Jr (2002) The unexpected landscape of *in vivo* somatic mutation in a human epithelial cell lineage. Proc Natl Acad Sci USA 99(3):1437–1442
8. De S, Babu MM (2010) A time-invariant principle of genome evolution. Proc Natl Acad Sci USA 107(29):13004–13009
9. DePristo MA, Weinreich DM, Hartl DL (2005) Missense meanderings in sequence space: a biophysical view of protein evolution. Nat Rev Genet 6(9):678–687
10. Drake JW (1999) The distribution of rates of spontaneous mutation over viruses, prokaryotes, and eukaryotes. Ann N Y Acad Sci 870:100–107
11. Drake JW, Holland JJ (1999) Mutation rates among RNA viruses. Proc Natl Acad Sci USA 96(24):13910–13913
12. Dreszer TR, Wall GD, Haussler D, Pollard KS (2007) Biased clustered substitutions in the human genome: the footprints of male-driven biased gene conversion. Genome Res 17(10):1420–1430
13. Duret L, Arndt PF (2008) The impact of recombination on nucleotide substitutions in the human genome. PLoS Genet 4(5):e1000071
14. Friedberg EC (2005) Suffering in silence: the tolerance of DNA damage. Nat Rev Mol Cell Biol 6(12):943–953
15. Friedberg EC, Walker GC, Siede W, Wood RD, Schultz RA, Ellenberger T (2006) DNA repair and mutagenesis. ASM Press, Washington, DC
16. Galhardo RS, Hastings PJ, Rosenberg SM (2007) Mutation as a stress response and the regulation of evolvability. Crit Rev Biochem Mol Biol 42(5):399–435
17. Gonzalez C, Hadany L, Ponder RG, Price M, Hastings PJ, Rosenberg SM (2008) Mutability and importance of a hypermutable cell subpopulation that produces stress-induced mutants in Escherichia coli. PLoS Genet 4(10):e1000208
18. Gordenin DA, Resnick MA (1998) Yeast ARMs (DNA at-risk motifs) can reveal sources of genome instability. Mutat Res 400(1–2):45–58
19. Hamdan SM, van Oijen AM (2010) Timing, coordination, and rhythm: acrobatics at the DNA replication fork. J Biol Chem 285(25):18979–18983
20. Harris RS, Longerich S, Rosenberg SM (1994) Recombination in adaptive mutation. Science 264(5156):258–260
21. Hassa PO, Hottiger MO (2005) An epigenetic code for DNA damage repair pathways? Biochem Cell Biol 83(3):270–285
22. Hicks WM, Kim M, Haber JE (2010) Increased mutagenesis and unique mutation signature associated with mitotic gene conversion. Science 329(5987):82–85
23. Kashuba VI, Pavlova TV, Grigorieva EV, Kutsenko A, Yenamandra SP, Li J, Wang F, Protopopov AI, Zabarovska VI, Senchenko V, Haraldson K, Eshchenko T, Kobliakova J, Vorontsova O, Kuzmin I, Braga E, Blinov VM, Kisselev LL, Zeng YX, Ernberg I, Lerman MI, Klein G, Zabarovsky ER (2009) High mutability of the tumor suppressor genes RASSF1 and RBSP3 (CTDSPL) in cancer. PLoS One 4(5):e5231
24. Kogenaru M, de Vos MG, Tans SJ (2009) Revealing evolutionary pathways by fitness landscape reconstruction. Crit Rev Biochem Mol Biol 44(4):169–174
25. Kunkel TA, Bebenek K (2000) DNA replication fidelity. Annu Rev Biochem 69:497–529
26. Lang GI, Murray AW (2008) Estimating the per-base-pair mutation rate in the yeast Saccharomyces cerevisiae. Genetics 178(1):67–82
27. Lercher MJ, Hurst LD (2002) Human SNP variability and mutation rate are higher in regions of high recombination. Trends Genet 18(7):337–340
28. Lindahl T, Barnes DE (2000) Repair of endogenous DNA damage. Cold Spring Harb Symp Quant Biol 65:127–133

29. Liu M, Schatz DG (2009) Balancing AID and DNA repair during somatic hypermutation. Trends Immunol 30(4):173–181
30. Liu M, Duke JL, Richter DJ, Vinuesa CG, Goodnow CC, Kleinstein SH, Schatz DG (2008) Two levels of protection for the B cell genome during somatic hypermutation. Nature 451(7180):841–845
31. Lopes M, Foiani M, Sogo JM (2006) Multiple mechanisms control chromosome integrity after replication fork uncoupling and restart at irreparable UV lesions. Mol Cell 21(1):15–27
32. Ma W, Resnick MA, Gordenin DA (2008) Apn1 and Apn2 endonucleases prevent accumulation of repair-associated DNA breaks in budding yeast as revealed by direct chromosomal analysis. Nucleic Acids Res 36(6):1836–1846
33. Ma W, Panduri V, Sterling JF, Van Houten B, Gordenin DA, Resnick MA (2009) The transition of closely opposed lesions to double-strand breaks during long-patch base excision repair is prevented by the coordinated action of DNA polymerase delta and Rad27/Fen1. Mol Cell Biol 29(5):1212–1221
34. Mimitou EP, Symington LS (2009) DNA end resection: many nucleases make light work. DNA Repair (Amst) 8(9):983–995
35. Morrison A, Johnson AL, Johnston LH, Sugino A (1993) Pathway correcting DNA replication errors in Saccharomyces cerevisiae. EMBO J 12(4):1467–1473
36. Nishant KT, Singh ND, Alani E (2009) Genomic mutation rates: what high-throughput methods can tell us. Bioessays 31(9):912–920
37. Nishant KT, Wei W, Mancera E, Argueso JL, Schlattl A, Delhomme N, Ma X, Bustamante CD, Korbel JO, Gu Z, Steinmetz LM, Alani E (2010) The baker's yeast diploid genome is remarkably stable in vegetative growth and meiosis. PLoS Genet 6(9):e1001109
38. Odegard VH, Schatz DG (2006) Targeting of somatic hypermutation. Nat Rev Immunol 6(8):573–583
39. Pachkowski BF, Tano K, Afonin V, Elder RH, Takeda S, Watanabe M, Swenberg JA, Nakamura J (2009) Cells deficient in PARP-1 show an accelerated accumulation of DNA single strand breaks, but not AP sites, over the PARP-1-proficient cells exposed to MMS. Mutat Res 671(1–2):93–99
40. Plosky BS, Woodgate R (2004) Switching from high-fidelity replicases to low-fidelity lesion-bypass polymerases. Curr Opin Genet Dev 14(2):113–119
41. Pollard KS, Salama SR, King B, Kern AD, Dreszer T, Katzman S, Siepel A, Pedersen JS, Bejerano G, Baertsch R, Rosenbloom KR, Kent J, Haussler D (2006) Forces shaping the fastest evolving regions in the human genome. PLoS Genet 2(10):e168
42. Ponder RG, Fonville NC, Rosenberg SM (2005) A switch from high-fidelity to error-prone DNA double-strand break repair underlies stress-induced mutation. Mol Cell 19(6):791–804
43. Povolotskaya IS, Kondrashov FA (2010) Sequence space and the ongoing expansion of the protein universe. Nature 465(7300):922–926
44. Rattray AJ, Strathern JN (2003) Error-prone DNA polymerases: when making a mistake is the only way to get ahead. Annu Rev Genet 37:31–66
45. Rattray AJ, Shafer BK, McGill CB, Strathern JN (2002) The roles of *REV3* and *RAD57* in double-strand-break-repair-induced mutagenesis of Saccharomyces cerevisiae. Genetics 162(3):1063–1077
46. Resnick MA, Setlow JK (1972) Repair of pyrimidine dimer damage induced in yeast by ultraviolet light. J Bacteriol 109(3):979–986
47. Romero PA, Arnold FH (2009) Exploring protein fitness landscapes by directed evolution. Nat Rev Mol Cell Biol 10(12):866–876
48. Rosenberg SM, Longerich S, Gee P, Harris RS (1994) Adaptive mutation by deletions in small mononucleotide repeats. Science 265(5170):405–407
49. Santos JH, Meyer JN, Mandavilli BS, Van Houten B (2006) Quantitative PCR-based measurement of nuclear and mitochondrial DNA damage and repair in mammalian cells. Methods Mol Biol 314:183–199
50. Sartori AA, Lukas C, Coates J, Mistrik M, Fu S, Bartek J, Baer R, Lukas J, Jackson SP (2007) Human CtIP promotes DNA end resection. Nature 450(7169):509–514

51. Shrivastav M, De Haro LP, Nickoloff JA (2008) Regulation of DNA double-strand break repair pathway choice. Cell Res 18(1):134–147
52. Stamatoyannopoulos JA, Adzhubei I, Thurman RE, Kryukov GV, Mirkin SM, Sunyaev SR (2009) Human mutation rate associated with DNA replication timing. Nat Genet 41(4):393–395
53. Strathern JN, Shafer BK, McGill CB (1995) DNA synthesis errors associated with double-strand-break repair. Genetics 140(3):965–972
54. Unniraman S, Schatz DG (2007) Strand-biased spreading of mutations during somatic hypermutation. Science 317(5842):1227–1230
55. Wang J, Gonzalez KD, Scaringe WA, Tsai K, Liu N, Gu D, Li W, Hill KA, Sommer SS (2007) Evidence for mutation showers. Proc Natl Acad Sci USA 104(20):8403–8408
56. Weinreich DM, Watson RA, Chao L (2005) Perspective: sign epistasis and genetic constraint on evolutionary trajectories. Evolution 59(6):1165–1174
57. Wheatcroft R, Cox BS, Haynes RH (1975) Repair of UV-induced DNA damage and survival in yeast. I. Dimer excision. Mutat Res 30(2):209–218
58. Wright S (1932) The roles of mutation, inbreeding, crossbreeding and selection in evolution. In: Proceedings of the sixth international congress on genetics, Ithaca, New York, 1932
59. Yang Y, Sterling J, Storici F, Resnick MA, Gordenin DA (2008) Hypermutability of damaged single-strand DNA formed at double-strand breaks and uncapped telomeres in yeast Saccharomyces cerevisiae. PLoS Genet 4(11):e1000264
60. Yang Y, Gordenin DA, Resnick MA (2010) A single-strand specific lesion drives MMS-induced hyper-mutability at a double-strand break in yeast. DNA Repair (Amst) 9(8):914–921
61. Zanders S, Ma X, Roychoudhury A, Hernandez RD, Demogines A, Barker B, Gu Z, Bustamante CD, Alani E (2010) Detection of heterozygous mutations in the genome of mismatch repair defective diploid yeast using a Bayesian approach. Genetics 186:493–503
62. Zhao J, Bacolla A, Wang G, Vasquez KM (2010) Non-B DNA structure-induced genetic instability and evolution. Cell Mol Life Sci 67(1):43–62

Chapter 2
A Test of Kimura's Mutation-Rate Conjecture

John W. Drake

Abstract Mutation is the first of the four great engines of evolution and the prime source of heritable variation in all living and semi-living organisms and their molecular parasites. As a result, the processes and products of mutation have been deeply explored at both the phenomenological and the mechanistic levels for more than a century. There has been notable progress at both levels over the last half century, with the result that the phylogeny of mutation has become fairly well codified, especially the rates and kinds of mutation in as diverse organisms as could be probed. These explorations began to coalesce in the 1990s and soon comprised a small number of broad generalizations. These include very high genomic mutation rates in riboviruses and some higher eukaryotes, and strikingly similar genomic mutation rates in most DNA microbes. However, highly informative exceptions have recently been observed.

Keywords Mutation • Mutation rate • Ribovirus • Thermophile

2.1 Introduction

My scientific hobby for the last 40 years has been the exploration of the phylogeny of mutation, specifically the rates of mutation in as many organisms as could be probed and the kinds of mutations that result. These explorations began to coalesce in the 1990s and soon comprised some broad generalizations. More recently, I and several colleagues sought exceptions to these generalizations in order to test particular hypotheses, with sometimes surprising results.

J.W. Drake (✉)
Laboratory of Molecular Genetics, National Institute of Environmental Health Sciences, Research Triangle Park, NC 27709-2233, USA
e-mail: drake@niehs.nih.gov

C.E. Mothersill et al. (eds.), *Radiobiology and Environmental Security*,
NATO Science for Peace and Security Series C: Environmental Security,
DOI 10.1007/978-94-007-1939-2_2, © Springer Science+Business Media B.V. 2012

At the beginning, it is important to stress that mutation has long been accepted to be a highly evolved process shaped by natural selection [6, 10, 16]. For instance, most mutations are deleterious and thus reduce the average fitness of progeny, providing a basis for adaptive evolution based on mutation down-modifiers. As such, the mutational target of an organism is its whole genome, or at least that fraction which is subject to purifying or adaptive selection. Thus, the most interesting parameter is usually the *genomic* mutation rate μ_g, although it is sometimes subdivided in various ways. However, the only part of the genomic mutation rate that is subject to selection is that fraction that can generate deleterious or adaptive mutations, so that spacer DNA that shows no traces of purifying selection is of a very limited significance in this respect; accordingly, I will sometimes refer to the fraction of the genome in which new mutations are subject to selection as the *effective* genome and the corresponding genomic mutations rate as μ_{eg}. This is close to the total genome in most microbes but may be a small fraction of the genome in higher eukaryotes.

Mutation rates are most often measured starting with a specific mutation-reporter locus. The mutation frequency f is determine by counting mutants among total organisms screened. The frequency is then converted into a rate μ using equations that reflect the mode of chromosome replication, for instance, iterative copying (the stamping-machine model) or exponential replication. Although complex equations are sometimes required, there exist very simple equations that suffice for many situations. Thus, for riboviruses, which replicate by a stamping-machine mechanism, $\mu = f/2c$ where c is the number of sequential infection cycles that generated the test population [3]. For geometrically replicating organisms (such as numerous microbes), provided that the population is initiated with a number of organisms small enough to contain no mutants and then grows until it has experienced many (more than 30) mutations, $\mu = f/\ln(N\mu)$ where N is the final population size [1, 14]; the median rate from five or more cultures is the best measure.

The rate per locus is then converted into a rate per average base pair, which is finally converted into a genomic rate. It is important that the mutation-reporter locus be free of selective forces and certain other behaviors that bias mutation frequencies, and that it be large enough to mirror the genome as a whole. The main difficulty with mutation-reporters, which are usually protein-encoding sequences, is that they report base-pair substitutions (BPSs) inefficiently because most BPSs do not produce a mutant phenotype. Because of the efficiency and low cost of DNA sequencing, this constraint can now sometimes be overcome by calculating BPS rates using only BPSs that create chain-terminating mutations, which are expected to be scored efficiently.

Other modes of measuring mutation are sometimes used. One involves massive genomic sequencing. This has the advantage of scoring mutations without regard to their phenotypes but may have the disadvantages of young technologies.

2.2 Mutation Rates Codified

Table 2.1 presents an overview of rates of spontaneous mutation across major groups of organisms. They fall into three categories.

Table 2.1 Rates of mutation in diverse categories of organisms

Group	μ_{eg} per genome replication	μ_{eg} per "generation"
Riboviruses	0.1–1	0.2–2
Retroelements	0.03–0.43	0.08–1.3
Most DNA microbes	0.003–0.005	(same)
Some animals	0.004–0.014	0.04–1.6

Ribovirus rates are from Drake and Holland [3] and Malpica et al. [12]. Retrovirus rates are from Drake et al. [4]. DNA-microbe rates are from Drake [2]. Animal rates are from Drake et al. [4]

Riboviruses and retroelements (including retroviruses and retrotransposons) have the highest rates per genome replication, and when these are presented per "generation" (two replications for riboviruses, three for retrotransposons), the rates are close to the maximum that can be tolerated without extinguishing the population (variously called "error catastrophe" or "mutational meltdown"). Most of the published values are based on very short reporter sequences and thus provide uncertain values when extrapolated to the whole genome, which may account for the wide range of values.

Most DNA-based microbes have rates far lower than do the RNA-based microbes, in the neighborhood of 0.004 per genome replication. For the DNA-based microbes, a genome replication usually occurs once per cell division.

Several animals (*Caenorhabditis elegans, Drosophila melanogaster, Mus musculus, Homo sapiens*) also have rates (per *effective* haploid genome) in the neighborhood of 0.004 per genome replication in the germline, but animals often experience large numbers of germline replications per sexual generation (mostly in the male) and therefore have very high rates per effective genome per sexual generation. As a result, therefore, animals to some extent resemble the RNA world. The highest rate, 1.6 mutations per haploid genome per generation, belongs to humans, and suggests that any long-extended increase in this rate might extinguish the species.

All but one of the mutation rates considered thus far were determined using strains well adapted to the laboratory growing under generally benign conditions and should be viewed as *basal* mutation rates. For most of these organisms, the basal rates are so large that even small increases could be disastrous. However, because of their much lower basal mutation rates, the DNA microbes can tolerate sharply increased rates, at least for a while. However, under diverse stresses, bacterial mutation rates often increase transiently, or at least for as long as the stress persists (*e.g.*, Petrosino et al. [13], Gibson et al. [8]). More generally, a considerable fraction of mutation spectra from all categories of life contain one or more mutants bearing two or more mutations, where the "multiples" are usually far more frequent than anticipated from a random distribution of mutations and are likely to arise from bouts of transient hypermutability such as might result from errors of transcription, translation, or folding in proteins involved in diverse DNA transactions (Drake et al. [5]). In addition, closely spaced complex mutations are sometimes produced by DNA primer extension using ectopic templates, but these can usually be recognized by finding the ectopic donor sequences [15]. All of these kinds of supra-basal mutation rates may well contribute to evolution, but to an unknown extent.

2.3 Informative Exceptions

The single most remarkable aspect of comparative mutation is the very narrow range of genomic rates among the DNA microbes, and perhaps also among animals if the estimates in Table 2.1 are correct. Because genome sizes vary by about 6,000-fold among the microbes that make up that entry in the table, average mutation rates per base pair necessarily vary inversely by about the same factor, an observation that is about two decades old [1] but remains unexplained: while much is known about fidelity mechanisms in many of these model organisms, it is not known why "0.004" is such a strongly preferred value.

One of the most frequently evoked hypotheses has its origins in the work of Motoo Kimura, who pointed out that down-modifiers of mutation rates should be adaptive because most mutations are deleterious, at least within the effective genome (the latter concept not being anticipated at that time because of the lack of DNA sequences to reveal the large fractions of many genomes that consists of DNA not subject to strong purifying selection). As he appreciated, however, there will always be a cost associated with further reducing the error frequency [10]. (He also noted that up-modifiers of mutation rates would be inefficiently selected through the adaptive mutations they might produce because the favorable alleles would soon be separated from the mutator allele by recombination, an efficient process in frequently sexual organisms and not necessarily sufficiently absent from bacteria to prevent such selection, although the issue is complex.) In any case, the Kimura conjecture implies an equilibrium mutation rate, but does not imply that the equilibrium would be the same in such a diverse set of organism with so sharply differing life histories and molecular mechanisms for maintaining replication fidelity, especially when it encompasses a 6,000-fold variation in average mutation rate per base pair.

An experimental test of the Kimura conjecture might consist of artificially increasing the mutation rate with a physical or chemical mutagen and then determining how the overall rate adjusts during subsequent evolution in the laboratory, but there are many pitfalls in this approach, one being uncertainly about the time required to reach equilibrium and another being about the limited set of environments that can be explored in the laboratory. Our first test of the conjecture involved characterizing mutation in the hyperthermophile *Sulfolobus acidocaldarius* because we wondered if growth at extreme temperatures might introduce so much DNA damage as to force the mutation rate upwards. However, the result was just the opposite: mutation rates were lower in *S. acidocaldarius* than in the mesophilic microbes examined to date [9]. We then considered the impact of temperature on the average base pair in a protein-encoding gene, knowing that one of the easiest kind of conditional mutation to isolate historically in many microbes was a temperature-sensitive allele. It seemed reasonable to surmise that many neutral missense mutations in mesophiles would become deleterious at higher temperatures, in which case purifying selection against nonsynonymous mutations would be expected to be more intense in thermophiles than in mesophiles. This turned out to be correct: the estimator dN/dS (where N refers to nonsynonymous mutations and

Table 2.2 Rates of spontaneous mutation in DNA microbes

Organism	Genome (nt)	μ_b	μ_g	μ_g(BPS)	μ_g(indel)
Phage M13	6.4×10^3	7.5×10^{-7}	0.0048	0.0038	0.0010
Phage λ	4.9×10^4	5.3×10^{-8}	0.0026	0.0022	0.0004
Herpes simplex virus	1.5×10^5	1.8×10^{-8}	0.0043	0.0035	0.0008
Phage T4	1.7×10^5	1.8×10^{-8}	0.0038	0.0030	0.0008
Escherichia coli	4.6×10^6	6.4×10^{-10}	0.0030	0.0025	0.0004
Saccharomyces cerevisiae	1.2×10^7	3.6×10^{-10}	0.0044	0.0041	0.0003
Schizosaccharomyces pombe	1.3×10^7	2.4×10^{-10}	0.0030	0.0026	0.0004
Mesophile mean			**0.0038**	**0.0032**	**0.0006**
Thermus thermophilus	2.1×10^6	3.2×10^{-10}	0.00067	0.00054	0.00013
Sulfolobus acidocaldarius	2.2×10^6	1.7×10^{-10}	0.00037	0.00011	0.00026
Thermophile mean			**0.00052**	**0.00033**	**0.00019**

Values are from Drake [2]. The value for phage λ is for its rates when growing lytically. Where a mutation reporter was impacted by an extreme indel hotspot, the posted values are for the rates without the hotspot; the overall profile is only modestly impacted by this procedure. The total genomic rate μ_g is fractionated into its base-pair substitution and indel components, μ_g(BPS) and μ_g(indel), respectively

S to synonymous mutations) averaged to about 0.14 in mesophilic DNA microbes but fell to about 0.09 in thermophiles [7]. Also, because *S. acidocaldarius* was the first archaeon to be examined for its mutational propensities, we repeated the analysis in a bacterial thermophile, *Thermus thermophilus*, and observed again that the mutation rate was lower than seen in mesophiles [11]. The more reliable of the mesophile and thermophile rates are listed in Table 2.2 and were calculated using only chain-terminating mutations to estimate BPS rates. Not only are total genomic rates seen to be reduced in thermophiles (by roughly sevenfold), but the main cause of this decrease is a tenfold decrease in BPS rates. These results provide a striking confirmation of the concept of a reciprocal relationship between the impact of mutations and the genomic mutation rate, as expected from the Kimura conjecture.

Acknowledgements Research conducted in the author's laboratory was supported by funds allocated to project number ES065016 of the Intramural Research Program of the NIH, National Institute of Environmental Health Sciences.

References

1. Drake JW (1991) A constant rate of spontaneous mutation in DNA-based microbes. Proc Natl Acad Sci USA 88:7160–7164
2. Drake JW (2009) Avoiding dangerous missense: thermophiles display especially low mutation rates. PLoS Genet 5:e1000520
3. Drake JW, Holland JJ (1999) Mutation rates among RNA viruses. Proc Natl Acad Sci USA 96:13910–13913

4. Drake JW, Charlesworth B, Charlesworth D, Crow JF (1998) Rates of spontaneous mutation. Genetics 148:1667–1686
5. Drake JW, Bebenek A, Kissling GE, Peddada S (2005) Clusters of mutations from transient hypermutability. Proc Natl Acad Sci USA 102:12849–12854
6. Fisher RA (1930) The genetical theory of natural selection. Oxford University Press, Oxford
7. Friedman R, Drake JW, Hughes AL (2004) Genome-wide patterns of nucleotide substitution reveal stringent functional constraints on the protein sequences of thermophiles. Genetics 167:1507–1512
8. Gibson JL, Lombardo MJ, Thornton PC, Hu KH, Galhardo RS, Beadle B, Habib A, Magner DB, Frost LS, Herman C, Hastings PJ, Rosenberg SM (2010) The sigma(E) stress response is required for stress-induced mutation and amplification in *Escherichia coli*. Mol Microbiol 77:415–430
9. Grogan DW, Carver GT, Drake JW (2001) Genetic fidelity under harsh conditions: analysis of spontaneous mutation in the thermoacidophilic archaeon *Sulfolobus acidocaldarius*. Proc Natl Acad Sci USA 98:7928–7933
10. Kimura M (1967) On the evolutionary adjustment of spontaneous mutation rates. Genet Res 9:23–34
11. Mackwan RR, Carver GT, Kissling GE, Drake JW, Grogan DW (2008) The rate and character of spontaneous mutation in *Thermus thermophilus*. Genetics 180:17–25
12. Malpica JM, Fraile A, Moreno I, Obies CI, Drake JW, García-Arenal F (2002) The rate and character of spontaneous mutation in an RNA virus. Genetics 162:1505–1511
13. Petrosino JF, Galhardo RS, Morales LD, Rosenberg SM (2009) Stress-induced beta-lactam antibiotic resistance mutation and sequences of stationary-phase mutations in the *Escherichia coli* chromosome. J Bacteriol 191:5881–5889
14. Rosche WA, Foster PL (2000) Determining mutation rates in bacterial populations. Methods 20:4–17
15. Schultz GE Jr, Drake JW (2008) Templated mutagenesis in bacteriophage T4 involving imperfect direct or indirect sequence repeats. Genetics 178:661–673
16. Sturtevant AH (1937) Essays on evolution. I. On the effects of selection on mutation rate. Quart Rev Biol 12:464–467

Chapter 3
Micro-algae as a Model System for Studying of Genotype Resistance to Oxidative Stress and Adaptive Response

Stefka G. Chankova and Nadezhda Yurina

Abstract Here we discuss the possible contribution of DSB DNA repair and chaperone systems for the formation of genotype resistance to oxidative stress and the possible correlation between cells' genotype resistance and the magnitude of adaptive response (AR). Mutant strains of *Chlamydomonas reinhardtii* showing different levels of gamma- rays, paraquat- and zeocin-resistance as well as species *Chlorella* isolated from habitats with extreme environmental conditions are used as a model system. On the basis of results obtained it could be speculated that genotype resistance to oxidative stress may not always correlate with the initial level of DSB induced, but rather with higher DSB repair capacity and higher content of constitutive HSP70B. Our results concerning the relationship between genotype resistance to oxidative stress and adaptive response confirm our previous suggestion that up-regulated DSB DNA rejoining could be considered as one of the mechanisms involved in the formation of AR in this organism. New data are provided that strains with a relatively lower genotype resistance demonstrate a stronger AR. On the other hand the higher genotype resistance of strains did not abrogate their competence to adapt. A relationship between DNA repair capacity estimated on the basis of single dose treatment and the DNA repair capacity estimated on the basis of the AR is obtained.

Keywords Genotype resistance • Adaptive response • DSB DNA induction • Repair capacity • Chloroplast chaperones • Paraquat • Zeocin • Heat shock

---

S.G. Chankova (✉)
Institute of Biodiversity and Ecosystem Research, BAS, 1113 Sofia, Bulgaria
e-mail: stephanie.chankova@yahoo.com

N. Yurina
A.N. Bach Institute of Biochemistry, Russian Academy of Sciences,
119071 Moscow, Russia

C.E. Mothersill et al. (eds.), *Radiobiology and Environmental Security*,
NATO Science for Peace and Security Series C: Environmental Security,
DOI 10.1007/978-94-007-1939-2_3, © Springer Science+Business Media B.V. 2012

3.1 Introduction

Organisms are continuously subjected to naturally occurring changes and additionally to disturbances originated by human activity. The impact of these changes on whole ecosystem depends on the response of single organism. During the evolution many defense strategies have been developed in organisms to minimize/overcome the potential harmful action of environmental stress and to increase cell tolerance or adaptation to oxidative stress.

The term adaptive response (AR) generally refers to the phenomenon that low-dose exposure to a genotoxic agent can protect cells from a subsequent damaging high-dose exposure [1, 20, 24, 26, 27, 39, 57, 58]. Accumulating data in the literature strongly suggest that some defense mechanisms probably underlying the AR could involve acceleration of DNA repair, *de novo* synthesized proteins; activation and/or partial contribution of the antioxidant system, chaperone accumulation, more efficient detoxification of free radicals etc. [17, 20, 29, 34, 35]. A well known hypothesis is that the AR may be triggered by some low level of primary induced DNA double-strand breaks (DSB) which could probably serve as a signal for activation of DSB repair [3, 16, 32, 57].

For example, it has been previously shown by us that model *C. reinhardtii* strains could develop strong radiation-, zeocin (Zeo)- or paraquat (PQ)-induced AR involving accelerated rejoining of DSB [8, 13, 15, 20, 21]. Intriguingly, heat pretreatment (37–42°C) was not sufficient to induce a statistically significant increasing of DSB in *C. reinhardtii* strain CW15 and did not induce AR to test dose gamma rays irradiation in the form of accelerated DSB rejoining [16].

At present, however, there is limited evidence whether there could be a correlation between genotype resistance (i.e. single cell radio-/chemo-resistance) and the capability to develop AR. The published data are still insufficient and the relationships between cell resistance and the ability to develop an AR are difficult to predict. For example, it has been proposed that certain tumors which are difficult to cure by X-rays may exhibit a strong induced radio-resistance [56]. Some radio-sensitive tumor lines and cells of ataxia telangiectasia patients do not show an AR in the form of induced radio-resistance [28]. On the other hand Zasukhina et al. [60] reported that the level of radiation-induced AR was the same in blood cells of patients with Bloom syndrome (human autosomal recessive disorder with chromosomal instability and increased risk of malignancy at an early age) as that in cells from healthy donors. For better understanding of some mechanisms involved in the formation of genotype resistance and adaptive response (AR) in plants, mutant strains of *C. reinhardtii* with different level of radio- and chemo-resistance as well as algae from habitats with extreme environmental conditions have been used as a model system.

Our hypothesis is that resistant to oxidative stress mutant strains and algal species isolated from habitats with extreme environmental conditions would have similar and/or more efficient cellular defense mechanisms.

We address two main questions: What is the possible contribution of DSB DNA repair and chaperone systems for the formation of genotype resistance to oxidative stress? Is there a relationship between genotype resistance to oxidative stress and the magnitude of adaptive response (AR) in micro- algae?

3.2 Why Micro-algae Have Been Chosen as a Model System?

Unicellular green algae have been developed into a "good" test-system in modern eco-toxicology and for studying of genotype resistance and AR for several reasons: cell/organism, with a simple life cycle, relatively inexpensive experimental conditions and methods for cultivation, a growing array of tools and techniques for molecular genetic studies, quick methods with good resolution, sensitivity, potential for extrapolation of results obtained to higher plants, easily isolated mutants with different levels of radio- and chemo-resistance etc. [13–15, 19–21].

On the other hand an understanding of the molecular mechanisms of genotype resistance to oxidative stress and adaptive response exhibited by mutant strains of *Chlamydomonas reinhardtii* may lead us to a deeper understanding of plant cells response to oxidative stress in general. This organism is particularly interesting from the point of view of its powerful adaptive response to radiation [3, 4–6, 8], and in addition it promises to provide indirectly further insight into the adaptive response of normal and malignant human cells at low radiation doses [26, 27]. Up to now, most studies of repair on *C. reinhardtii* have focused on nuclear and extra nuclear DNA using mutant repair-deficient strains [33, 40, 48, 52–54]. However, radio resistant mutants are scarce [11–13]. We have been interested in the question about whether radio resistant and chemo- resistant strains of this organism would, in addition to its resistance in single-dose experiments, show an adaptive response to oxidative stress in split-dose experiments. The main characteristics of *C. reinhardtii* strains and *Chlorella* species are presented in Table 3.1.

3.3 Genotype Resistance to Oxidative Stress

Previously it was described by us that radio resistant genotype of *Chlorella vulgaris* and *Chlamydomonas reinhardtii* would be characterized with increased SSB DNA repair efficiency [10, 11], higher levels of constitutive SOD, SH-group, pigments, especially carotenoids and chl "a" [9, 11], and stability of the ultrastructural cell components or/and of the presence of cell wall [11]. Furthermore we aimed to throw more light on the contribution of initial level of DSB DNA, repair efficiency and constitutive levels of HSPs for the formation of genotype resistance.

In order to test our hypothesis that algal species isolated from habitats with extreme environmental conditions could be used as a model for the investigation of genotype resistance to oxidative stress, the responses of three *Chlorella* species to

Table 3.1 Characterization of strains and species of micro-algae

C. reinhardtii strains	Characteristics
137C	Wild type (WT), St Petersburg collection
AK-9-9	Obtained by chemical mutagenesis ([9], St Petersburg University), *Can res,* radioresistant, cross-resistant to Zeo and PQ [11–14, 20]
CW15	Culture Collection of Algae and Protozoa (CCAP 11/32CW15+), Ambleside, UK; cell-wall-less with WT radio resistance [12, 18]
H-3	Hybrid strain, obtained by Chankova and Bryant [11, 13, 14], radio resistant, cross – resistant to Zeo and PQ [20, 21].
Chlorella species	
Chlorella kessleri	Mesophilic, isolated from USA freshwater, axenic, mainly used for fundamental research, from Czech collection
Chlorella sp.	Antarctic – psychrophilic, isolated from soil on the island of Livingston, the South Shetland archipelago
Chlorella vulgaris strain 8/1	Thermophilic strain, isolated in 1968 from the thermal spring (t=40–50° C) near by Petrich (Bulgaria); during 20 years laboratory cultivated at room temperature

UV-B irradiation, Zeo and heat treatment are compared. A large range of doses of UV-B irradiation, λ=312 nm (BLX, Life Technology, UV crosslinker), Zeo concentrations and temperatures are applied. As a first step cells' response of *Chlorella sp* (antarctic), *Chlorella vulgaris 8/1* (thermophilic) and *Chlorella kesslery* (mesophilic) is examined on the basis of various endpoints – spot-test, micro-colonies assay and growth rate. Species resistance to oxidative stress is evaluated on the basis of LD_{20}, LD_{50} and LD_{90} [13, 14]. Our pilot results show that *Chlorella* species differ in their response. The most pronounced growth rate and cell survival are calculated for *Chlorella sp.* It is also found that *Chlorella* species are photo reactivation proficient and differ in their capacity to overcome harmful effect of UV-B. Species resistance to UV-B induced stress could be arranged in the following order depending on post irradiation conditions: *Ch.sp*>*Ch.vulgaris*8/1=*Ch.kesslery* (samples with photo-reactivation) and *Ch.sp*>*Ch.vulgaris*8/1>*Ch.kesslery* (samples without photo-reactivation).

3.3.1 On the Role of Initial Level of DSB DNA for the Formation of Genotype Resistance

In our paper [20] we describe that the level of DSB DNA induced by single PQ treatment is higher in strain CW15 comparing with that for strain 137C. No statistically significant increasing in the level of DSB is found for the radio-resistant strains H-3

Table 3.2 Repair capacity of *C. reinhardtii* strains after single dose treatment with Zeo and PQ. Mean data from at least three independent experiments

Treatment	ZEO	PQ	ZEO	PQ	ZEO	PQ	ZEO	PQ
Recovery time	1 h		2 h		3 h		4 h	
137C	1.37	1.44	1.35	1.46	1.44	2.11	1.61	2.74
CW15	1.10	1.30	1.14	1.96	1.25	2.49	1.30	2.56
H-3	2.42		3.70		5.54		10.42	
AK-9-9	1.81		2.23		2.91		2.96	

and AK-9-9 even at high concentrations. For the same strains an increasing of DSB induced by single Zeo treatment in a concentration depended manner up to about 100 μg ml^{-1} is shown [21]. Above this concentration the dose response curves levelled out to a plateau. As a whole, no statistically significant differences are observed between DSB values in strains 137C, AK-9-9 and H-3. One way ANOVA with Tukey multiple comparison test confirm that these three strains have a similar Zeo resistance in terms of DSB induction in the range of 10–300 μg ml^{-1}. The level of DSB is higher for strain CW15 [15, 21].

3.3.2 *On the Role of DSB DNA Repair Efficiency for the Formation of Genotype Resistance*

To clarify the contribution of DNA repair factor in the formation of genotype resistance to oxidative stress, repair capacity of strains is analyzed. Our results indicate that differences in Zeo resistance not always correlate with the initial level of DSB induced, but rather with higher DSB repair capacity (Table 3.2).

These results are in agreement with the findings of other authors [23, 41]. Moreover, DNA repair has been pointed out as probably the most crucial mechanism employed by cells to avert bleomycin-induced genotoxicity [42]. However, it is likely that, apart from DNA repair, other factors such as chromatin structure, mutations, cell cycle, altered DNA damage checkpoint response, defective apoptosis, or cell wall maintenance, etc. could also play a critical role toward determining cell resistance or sensitivity [2, 7, 49].

3.3.3 *On the Role of Chloroplast HSP70B/Chaperone for the Formation of Genotype Resistance*

Heat stresses proteins (HSPs) as a key components contributing to cellular homeostasis take a special place among defense systems [25, 38, 51, 55]. A special role in protecting cells from oxidative and heat stress belongs to HSP70 proteins, that together with co-chaperones proteins take part in the formation of chloroplast "foldosome" [44]. There are evidences of the involvement of chloroplast HSP70B/chaperons in

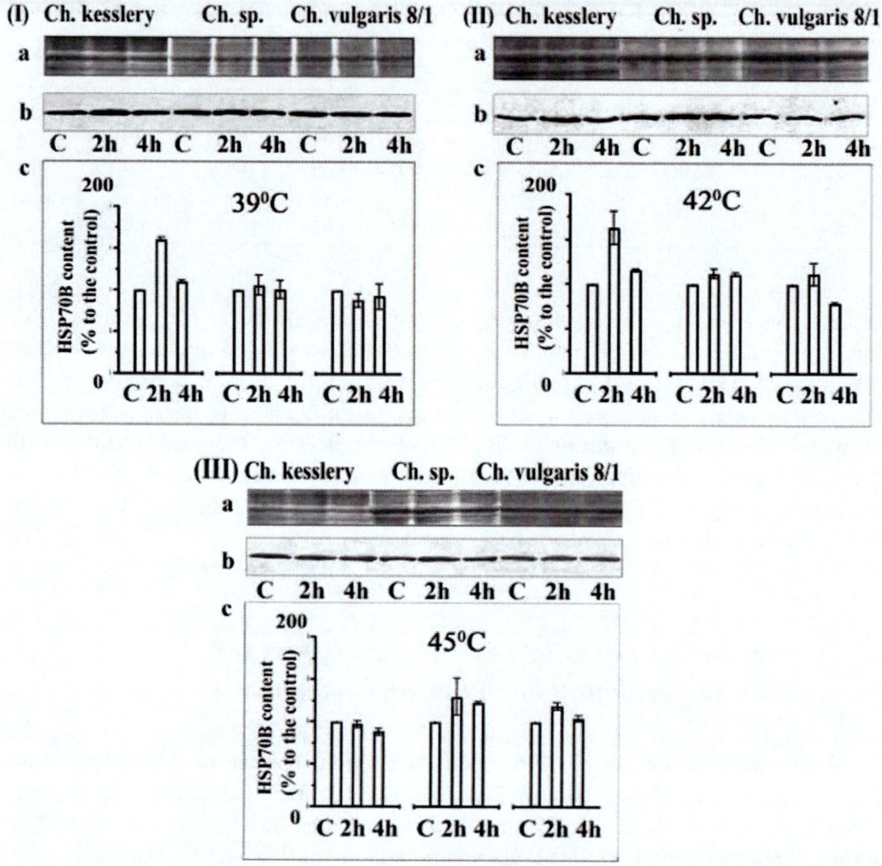

Fig. 3.1 Comparison of HSP70B level in *Chlorella* species after heat stress: (*I*) – 39°C/30 min, (*II*) 42°C/5 min and (*III*) 45°C/5 min; (**a**) Coomassie R-250 staining. The same amount of total protein at the gel was applied. Equal application of the protein on the gel was controlled by Coomassie R-250 staining; (**b**) Western blot analysis of HSP70B level in *Chlorella* species incubated at heat stress; (**c**) results of densitometry of HSP70B contents

refolding of denatured proteins [22], in the biogenesis of thylakoid membranes, the synthesis or assembly of components of the new reaction centers and in the protection and repair of photosystem II during and after photoinhibition [43, 45].

To throw more light on the possible role of HSP70B protein in the formation of cell resistance to temperature induced stress we addressed two questions: whether *Chlorella* species that differ in their response to oxidative stress would differ in their content of constitutive HSP70B and whether *Chlorella* species would differ in their response to heat shock (Figs. 3.1 and 3.2). Western blot analysis is used to measure HSP70B content 2 and 4 h after HS treatment [16, 59].

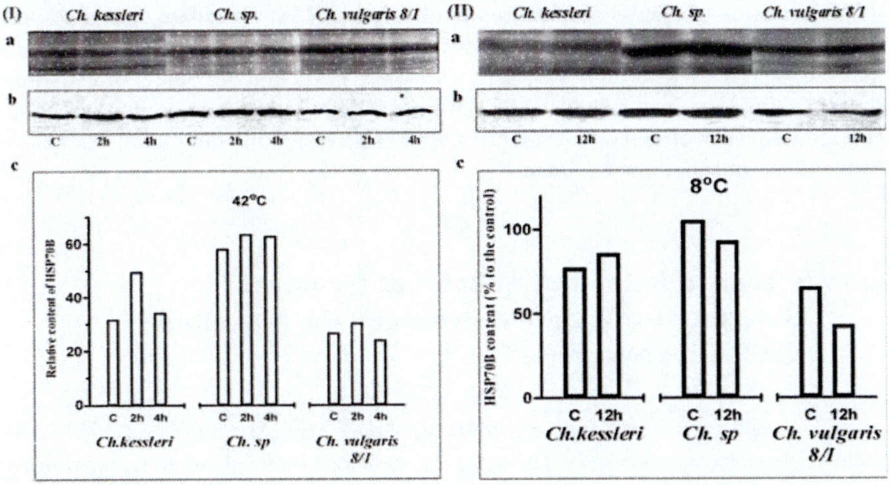

Fig. 3.2 Effect of higher (*I* – 42°C/5 min) and lower (*II* – 8°C/12 h) temperature treatment on HSP70B level in *Chlorella* species. (**a**) Coomassie R-250 staining; (**b**) Western blot analysis of HSP70B level in *Chlorella* species incubated with heat stress; (**c**) results of densitometry of HSP70B contents

Results presented in Fig. 3.1 show different stress response of algal species.

About 60% higher content of HSP70B is found for *Ch. kessleri* after the treatment with 39°C /30 min and 42°C/5 min and no statistically significant changes are obtained after the treatment at 45°C/5 min. The lack of induction of protein at 45°C corresponds well with our survival experiments (data not shown) where about 50% cell death is determined. Suppressed protein synthesis is possible also at this temperature. The only temperature that can induce heat stress in both species isolated from habitats with extreme environmental conditions *Ch. sp* (antarctic), and *Ch. vulgaris* 8/1 (thermophilic) is 45°C/5 min. The thermophilic strain 8/1 shows a slight HSP70B induction, but psychrophilic *Chlorella sp* demonstrates the increased HSP70B induction, indicating the occurrence of stress state in these cells. Small changes of HSP70B content in thermophilic *Ch. vulgaris* 8/1 may indicate participation of other defense factors. In Fig. 3.2 (I) the relative contents of constitutive HSP70B are presented.

Markedly higher content of HSP70B is measured for *Ch. Sp* (Antarctic) comparing with other two species at different experimental conditions. The difference between HSP70B contents for *Ch. kessleri* and *Ch. vulgaris* 8/1 is found to be insignificant. Incubation of *Ch. sp.* (antarctic) at low temperature did not lead to a decrease in HSP70B level (Fig. 3.2, II). This result speaks in favor of the assumption that increased HSP70B content in Antarctic algae is evolutionary fixed sign promoting to the survival of cells under extreme conditions. It is interesting that *Ch. sp.* (antarctic) is also characterized by increased activity of superoxidedismutase and catalase at high temperatures, and increased efficiency of antioxidant system than other strains [31, 37].

On the basis of results obtained it could be possible to speculate that genotype resistance to oxidative stress may not always correlate with the initial level of DSB induced, but rather with higher DSB repair capacity and higher content of constitutive HSP70B. To show whether algae from habitats with extreme environmental conditions have evolutionary developed more effective cellular defense mechanisms further experiments must be done.

3.4 Is There a Relationship Between Genotype Resistance to Oxidative Stress and the Magnitude of Adaptive Response?

Little is currently known about the correlation between cells' genotype resistance and their response to radiation [47]. The magnitudes of Zeo induced AR in *C. reinhardtii* strains presented in Fig. 3.3 show that all strains have the capacity to develop AR.

A stronger AR is demonstrated in strains with a relatively lower genotype resistance. Interestingly, it was obtained that the higher genotype resistance of strains H-3 and AK-9-9 did not abrogate their competence to adapt. Our observation that the magnitude of induced AR in *C. reinhardtii* depends on the cell genotype is in agreement with the suggestion that the development of AR in human lymphocytes probably depends on genetic factors and not only on physiological ones [36, 50].

Quite different is the picture presented in Fig. 3.4.

Strong PQ induced AR is manifested in strains with WT resistance – 137C and CW15 and no/or very slight PQ-induced AR in resistant strain AK-9-9 and H-3 (about 3%). No AR has been also induced when heat pretreatment has been used several hours before test treatment with gamma-rays or heat shock [16].

Looking for the reason we have analyzed the level of DSB initially induced by pretreatment with low conditioning doses of PQ and heat. The absence of statistically significant higher level of DSB DNA induced by priming (conditioning)

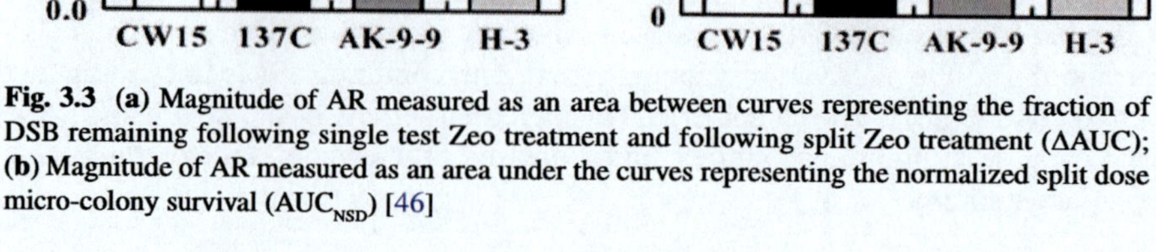

Fig. 3.3 (**a**) Magnitude of AR measured as an area between curves representing the fraction of DSB remaining following single test Zeo treatment and following split Zeo treatment (ΔAUC); (**b**) Magnitude of AR measured as an area under the curves representing the normalized split dose micro-colony survival (AUC_{NSD}) [46]

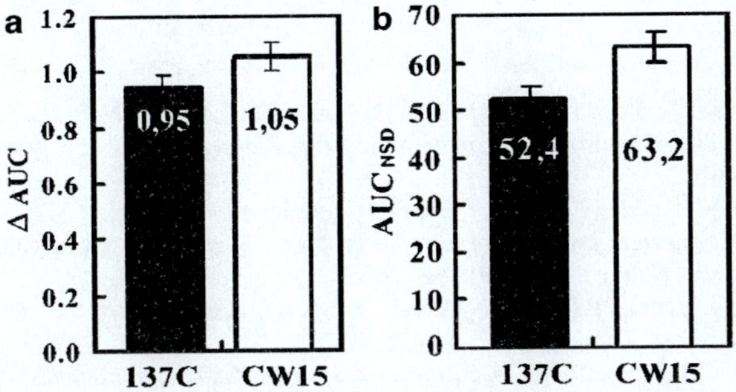

Fig. 3.4 (**a**) Magnitude of AR measured as an area between curves representing the fraction of DSB remaining following single test PQ treatment and following split PQ treatment (ΔAUC); (**b**) Magnitude of AR measured as an area under the curves representing the normalized split dose micro-colony survival (AUC_{NSD}). [46]

treatment is obtained for both cases. Our data add evidence to the hypothesis that a certain primary level of DNA damage could serve as a triggering event signalling the activation of DNA repair systems [3, 16, 20, 30, 32, 57]. The analysis of our results could suggest that the level of DSB DNA that can trigger an AR in *C. reinhardtii* strains should be at least 1.5-fold higher than those in control non-treated cells. Comparing repair capacity of strains, a relationship between DNA repair capacity estimated on the basis of single dose treatment and the DNA repair capacity estimated on the basis of the AR is obtained.

3.5 Conclusion

Our results concerning the relationship between genotype resistance to oxidative stress and adaptive response, confirm our previous suggestion that up-regulated DSB DNA rejoining could be considered as one of mechanisms involved in the formation of AR in this organism. New data are provided that strains with a relatively lower genotype resistance demonstrate a stronger AR. On the other hand the higher genotype resistance of strains did not abrogate their competence to adapt. A relationship between DNA repair capacity estimated on the basis of single dose treatment and the DNA repair capacity estimated on the basis of the AR is obtained.

Acknowledgements This work was funded under the projects: D-002-317 "Antarctic algae – model system for oxidative stress resistance", Bulgarian Ministry of Education, Young and Science; "Molecular mechanisms of induced resistance in plants to oxidative stress", agreement between RAS and BAS; "WETLANET", FP7 CSA – SUPPORT ACTION, contract No. 229802, 2008; Russian Foundation for Basic Research (project № 06-04-48923a), Program MCB RAS.

We thank Prof. Michael Schroda (Albert-Ludwigs-Universität Freiburg, Institut füz BiologieII, Germany) for the antiserum against HSP70B.

References

1. Asad NR, Asad LMBO, de Almeida CEB, Felzenszwalb I, Cabral-Neto JB, Leitão AC (2004) Several pathways of hydrogen peroxide action that damage the E. coli genome. Genet Mol Biol 27:291–303
2. Bao S, Wu Q, McLendon RE, Hao Y, Shi Q, Hjelmeland AB, Dewhirst MW, Bigner DD, Rich JN (2006) Glioma stem cells promote radioresistance by preferential activation of the DNA damage response. Nature 444(7120):756–760
3. Boreham DR, Mitchel RE (1991) DNA lesions that signal the induction of radioresistance and DNA repair in yeast. Radiat Res 128:19–28
4. Bryant PE (1968) Survival after fractionated doses of radiation: modification by anoxia of the response of Chlamydomonas. Nature 219:75–77
5. Bryant PE (1975) Decrease in sensitivity of cells after split-dose recovery: evidence for the involvement of protein synthesis. Int J Radiat Biol 27:95–102
6. Bryant PE (1976) Absence of an oxygen effect for induction of resistance to ionizing radiation. Nature 261:588–590
7. Chalmers AJ (2007) Radioresistant glioma stem cells – therapeutic obstacle or promising target? DNA Repair (Amst) 6(9):1391–1394
8. Chankova SG, Bryant PE (2002) Acceleration of DNA-double strand rejoining during the adaptive response of *Chlamydomonas reinhardtii.* Radiats Biol Radioecol 42:600–603
9. Chankova SG, Vinarova KM, Nikolov SH, Mehandjiev AD, Sergeeva SA, Ptitzina SN, Semov AB, Shevchenko VA (1990) Changes in *Chlorella vulgaris* B. populations after chronic influence with chemical and physical mutagenic factors. Biol Plant 32(1):35–41
10. Chankova SG, Mehandjiev AD, Blagoeva ED, Ptitsina SN, Sergeeva SA, Shevchenko VA (1994) Plant radioresistance and DNA repair efficiency in Chlamydomonas reinhardtii and Pisum sativum. Biol Plant 36:583–589
11. Chankova SG, Kapchina-Toteva VM, Stoyanova DP (2000) Some aspects of plant radioresistance. Radiat Biol Radioecol 40:536–544
12. Chankova SG, Kapchina-Toteva VM, Frolova O, Stoilov LM, Blagoeva ED (2001) Characterization of new radioresistant strains of *Chlamydomonas reinhardtii.* Compt Rend Acad Bull Sci 54:51–54
13. Chankova SG, Matos JA, Simões F, Bryant PE (2005) Adaptive response of a new radioresistant strain of *Chlamydomonas reinhardtii* and correlation with increased DNA double-strand break rejoining. Int J Radiat Biol 81:509–514
14. Chankova S, Stoyanova-Koleva D, Kapchina-Toteva V, Bryant P (2005) Characterisation of new radioresistant strains of *Chlamydomonas reinhardtii.* J Environ Prot Ecol 6:660–668
15. Chankova SG, Dimova E, Dimitrova M, Bryant PE (2007) Induction of DNA double-strand breaks by zeocin in *Chlamydomonas reinhardtii* and the role of increased DNA double-strand breaks rejoining in the formation of an adaptive response. Radiat Environ Biophys 46:409–416
16. Chankova SG, Yurina NP, Dimova EG, Ermohina OV, Oleskina YP, Dimitrova MT, Bryant PE (2009) Pretreatment with heat does not affect double-strand breaks DNA rejoining in *Chlamydomonas reinhardtii.* J Therm Biol 34:332–336
17. Coleman MA, Yin E, Peterson LE, Nelson D, Sorensen K, Tuckera JD, Wyrobeka AJ (2005) Low-dose irradiation alters the transcript profiles of human lymphoblastoid cells including genes associated with cytogenetic radioadaptive response. Rad Res 164:369–382
18. Darling D (1993) Investigation of induced repair in a cell wall-less mutant of Chlamydomonas reinhardtii exposed to low doses of gamma-rays, using a clonal assay. BSc thesis, University of St Andrews, St Andrews
19. Dimitrova M, Dimova E, Mitrovska Z, Kapchina-Toteva V, Chankova S (2007) Testing of polluted soil samples for genotoxic potential using *Chlamydomonas reinhardtii.* Algol Stud 123:111–121
20. Dimova EG, Bryant PE, Chankova SG (2008) "Adaptive response" – some underlying mechanisms and open questions. Genet Mol Biol 31:396–408

21. Dimova E, Dimitrova M, Miteva D, Mitrovska Z, Yurina NP, Bryant PE, Chankova S (2009) Does single-dose cell resistance to the radio-mimetic zeocin correlate with a zeocin-induced adaptive response in *Chlamydomonas reinhardtii* strains? Radiat Environ Biophys 48:77–84
22. Drzymalla C, Schroda M, Beck CF (1996) Light-inducible gene HSP70B encodes a chloroplast-localized heat shock protein in *Chlamydomonas reinhardtii*. Plant Mol Biol 31:1185–1194
23. Eguchi-Kasai K, Kosaka T, Sato K, Kaneko I (1991) Reparability of DNA double-strand breaks and radiation sensitivity in five mammalian cell lines. Int J Radiat Biol 59(1):97–104
24. Girigoswami BK, Ghosh R (2005) Response to gamma-irradiation in V79 cells conditioned by repeated treatment with low doses of hydrogen peroxide. Radiat Environ Biophys 44:131–137
25. Huang B, Xu C (2008) Identification and characterization of proteins associated with plant tolerance to heat stress. J Integr Plant Biol 50:1230–1237
26. Joiner MC, Lambin P, Malaise EP, Robson T, Arrand JE, Skov KA, Marples B (1996) Hypersensitivity to very-low single radiation doses: Its relationship to the adaptive response and induced radioresistance. Mutat Res 358:171–183
27. Joiner MC, Lambin P, Marples B (1999) Adaptive response and induced resistance. CR Acad Sci III 322:167–175
28. Lambin P, Fertil B, Malaise EP, Joiner MC (1994) Multiphasic survival curves for cells of human tumor cell lines: induced repair or hypersensitive subpopulation? Radiat Res 138:S32–S36
29. Lanza V, Pretazzoli V, Olivieri G, Pascarella G, Panconesi A, Negri R (2005) Transcriptional response of human umbilical vein endothelial cells to low doses of ionizing radiation. J Radiat Res 46:265–276
30. Leonard BE (2007) Adaptive response and human benefit: part I.A microdosimetry dose – dependent model. Int J Radiat Biol 83:115–131
31. Malanga G, Gonzalez PM, Estevez MS, Abele D, Puntarulo S (2008) Oxidative stress in Antarctic algae and mollusks. Ber Polarforsch Meeresforsch 571:208–215
32. Matsumoto H, Takahashi A, Ohinishi T (2004) Radiation-induced adaptive response and bystander effects. Biol Sci Space 18:247–254
33. Miadokova E, Podstavkova S, Cervenak Z, Vlcek D (1994) Different responses of repair-deficient strains of Chlamydomonas reinhardtii to UV and MNNG treatments. Biol Plant 49:633–637
34. Miura Y (2004) Oxidative stress, radiation – adaptive responses, and aging. J Radiat Res 45:357–372
35. Miura Y, Tamao E (2010) Survival responses to oxidative stress and aging. Geriatr Gerontol Int 10(Suppl1):S1–S9
36. Mortazavi SMJ, Cameron R, Niroomand-rad A (2003) Is the adaptive response an efficient protection against the detrimental effects of space radiation. In: Proceedings of 28th international cosmic ray conference. 31 Jul–7 Aug 2003, Tsukuba. University Academic Press Inc., Japan, pp 4299–4302
37. Nedeva D, Puneva I (2009) Changes in electrophoretic profiles of proteins and some antioxidant enzymes in Antarctic alga *Choricystis minor* and *Chlorella* sp. as affected by temperature and oxidative stress. Biotechnol Biotech Eq 23:233–236
38. Niu P, Liu L, Gong Z, Tan H, Wang F, Yuan J, Feng Y, Wei Q, Tanguay RM, Wu T (2006) Overexpressed heat shock protein 70 protects cells against DNA damage caused by ultraviolet C in a dose-dependent manner. Cell Stress Chaperones 11:162–169
39. Patra J, Sahoo MK, Panda BB (2003) Persistence and prevention of aluminium- and paraquat-induced adaptive response to methyl mercuric chloride in plant cells *in vivo*. Mutat Res 538:51–61
40. Podstavkova S, Vlcek D, Miadokova E (1992) New DNA repair deficient mutants of Chlamydomonas reinhardtii. Mut Res 293:65–69
41. Ramotar D, Wang H (2003) Protective mechanisms against the anti-tumor agent bleomycin: lessons from Saccharomyces cerevisiae. Curr Genet 43:213–224
42. Schaue D, McBride WH (2005) Counteracting tumor radioresistance by targeting DNA repair. Mol Cancer Ther 4:1548–1550

43. Schroda M (2004) The genome reveals its secrets: chaperone genes and the potential role of their gene products in the chloroplast. Photosynth Res 82:221–240
44. Schroda M, Muehlhaus T (2009) A "foldosome" in the chloroplast? Plant Signal Behav 4:301–303
45. Schroda M, Vallon O, Wollman FA, Beck CF (1999) A chloroplast-targeted heat shock protein 70 (HSP70) contributes to the photoprotection and repair of photosystem II during and after photoinhibition. Plant Cell 11:1165–1178
46. Serafin AM, Akudugu JM, Böhm L (2003) Studies on the influence of DNA repair on radiosensitivity in prostate cell lines. Urol Res 31:227–231
47. Silva P, Homer JJ, Slevin NJ, Musgrove BT, Sloan P, Price P, West CML (2007) Clinical and biological factors affecting response to radiotherapy in patients with head and neck cancer: a review. Clin Otolaryngol 32:337–345
48. Small GD (1987) Repair systems for nuclear and chloroplast DNA in Chlamydomonas reinhardtii. Mut Res 181:31–35
49. Strasser H, Grabenbauer GG, Sprung CN, Sauer R, Distel LVR (2007) DNA double-strand break induction and repair in irradiated lymphoblastoid, fibroblast cell lines and white blood cells from ATM, NBS and radiosensitive patients. Strahlenther Onkol 183(8):447–453
50. Tapio S, Jacob V (2007) Radioadaptive response revisited. Radiat Environ Biophys 46:1–12
51. Timperio AM, Egidi MG, Zolla L (2008) Proteomics applied on plant abiotic stresses: role of heat shock proteins (HSP). J Proteomics 7:391–411
52. Vlcek D, Podstavkova S, Miadokova E, Adams GMW, Small GD (1987) General characteristics, molecular and genetic analysis of two new UV-sensitive mutants of Chlamydomonas reinhardtii. Mut Res 183:169–175
53. Vlcek D, Podstavkova S, Miadokova E (1995) Interaction between photolyase and dark repair processes in Chlamydomonas reinhardtii. Mut Res 336:251–256
54. Vlcek D, Slivkova A, Podstavkova S, Miadokova E (1997) A Chlamydomonas reinhardtii UV-sensitive mutant uvs15 is impaired in a gene involved in several repair pathways. Mut Res 385:243–249
55. Wang W, Vinocur B, Shoseyov O, Altman A (2004) Role of plant heat-shock proteins and molecular chaperones in the abiotic stress response. Trends Plant Sci 9:244–252
56. Ward JF (1988) DNA damage produced by ionizing radiation in mammalian cells: identities, mechanisms of formation, and reparability. Prog Nucleic Acid Res Mol Biol 35:95–125
57. Wolff S (1998) The adaptive response in radiobiology: evolving insights and implications. Environ Health Perspect 106:277–283
58. Yan G, Hua Z, Du G, Chen J (2006) Adaptive response of Bacillus sp. F26 to hydrogen peroxide and menadione. Curr Microbiol 52:238–242
59. Yurina NP, Kloppstech K (2001) Accumulation of plastid protein precursors under norflurazon-induced carotenoid deficiency and oxidative stress in barley. Plant Physiol Biochem 39:807–814
60. Zasukhina GD, Vasyleva IM, Semyachkina AN (2000) Independence of DNA repair after gamma irradiation and radioadaptive response in lymphocytes of patients with Bloom syndrome. Radiats Biol Radioecol 40:513–515 (In Russian)

Chapter 4
Flavonoid Biosynthesis Genes in Wheat and Wheat-Alien Hybrids: Studies into Gene Regulation in Plants with Complex Genomes

Elena K. Khlestkina, Olesya Tereshchenko, and Elena Salina

Abstract Pigmentation by flavonoid pigments is the oldest trait employed for studies in genetics. In the current chapter, we describe application of the flavonoid biosynthesis (FB) genes as a model in the study of regulatory-target gene relationships in allopolyploid wheat genome (*Triticum aestivum*, AABBDD, $2n = 6x = 42$) and gene functioning in a foreign background in wheat-alien hybrids. Investigation of this multicomponent gene system showed that FB gene regulation cuts across genomes of allopolyploid wheat, the regulatory FB genes contribute more to the functional divergence between the diploid genomes of allopolyploid wheat than do the structural genes, and a good cooperation of the wheat and alien FB gene systems is observed in wheat-alien hybrids.

Keywords Allopolyploids • Wheat • Flavonoid biosynthesis • Multicomponent gene system • Genetic mapping • Gene regulation

4.1 Introduction

Bread wheat (*Triticum aestivum* L., $2n=6x=42$) is derived from a complex hybridization process involving three diploid species carrying the three homoeologous genomes A, B and D [49, 50]. The progenitors of the A, B and D genomes (*T. urartu*, *Aegilops speltoides* and *Ae. tauschii*, respectively) are closely related, and the gene content in these three genomes is largely conserved. Thus, there are commonly three representatives of each single copy gene present in bread wheat. They are referred as homoeologous genes. Due to allopolyploid nature of wheat

E.K. Khlestkina (✉) • O. Tereshchenko • E. Salina
Institute of Cytology and Genetics SB RAS, Lavrentjeva Ave. 10,
Novosibirsk 630090, Russia
e-mail: khlest@bionet.nsc.ru

C.E. Mothersill et al. (eds.), *Radiobiology and Environmental Security*,
NATO Science for Peace and Security Series C: Environmental Security,
DOI 10.1007/978-94-007-1939-2_4, © Springer Science+Business Media B.V. 2012

chromosome engineering technologies can be successfully applied to its improvement by introgression of alien genetic material from some other cereal species. It is of interest how genetic material derived from different species or genera manages to coexist in a common nucleus. The mechanisms underlying adaptation of different genomes in a common nucleus are known to be acting at the different levels of cell organization. Besides changes in genome size [77] and genome structure [37, 58, 76, 81, 87, 92], alteration of gene expression in allopolyploids in comparison with the diploid parents is observed. This chapter will focus mainly on the regulation of gene transcription in multicomponent flavonoid biosynthesis gene system in allopolyploid wheat and wheat-alien hybrids.

Flavonoid substances are divided into 12 classes. Eight of the 12 classes are known to be either plant pigments or precursors of the pigments. Different plant organs can be colored with flavonoid pigments. Mutations in the genes determining coloration are nonlethal and confer easily scorable phenotypes, therefore flavonoid biosynthesis (FB) gene system provides a colorful tool for studies in genetics and biotechnology. FB genes contributed to understanding and discovery of many genetic and epigenetic phenomena, such as basic underlying principles of heredity [64] and gene polymery [72, 73], transposable elements and epigenetic gene regulation [60], posttranscriptional gene silencing [71] and paramutations [11]. FB also has contributed to understanding of the organization of metabolic pathways in plants [85, 96]. Recently, FB genes have been used as a model to investigate how genes function in a foreign background in wide hybrids and to study interaction between structural genes and their homoeologous regulatory genes in allopolyploid wheat genome [39, 42, 44].

4.2 Biological Role of Flavonoids

Coloration of different plant organs (stems, flowers, seeds etc.) with flavonoid pigments is not a luxury. It plays multiple biological functions in plants. Flavonoids are known to be important for plant growth, development and reproduction, as well as for biotic and abiotic stress tolerance.

In 1883, Charles Darwin noted in his book that pigmented canes, grapes and onion are more resistant to diseases than white varieties of these species, and that red wheat is hardier in northern climates than white wheat [18]. Later, many researchers reported relationships between pigmentation intensity (or content of certain flavonoid substance and transcriptional level of FB structural genes) and level of resistance to different stressors (or extent of stress-induced damage) [4, 13, 15, 27, 33, 62, 79, 91]. In particular, pigmentation of plant organs with anthocyanins is important for plants defense against UV-light, cold, draught, salinity stress, heavy metals, irradiation exposure, nutrient deficiency and fungal diseases [13, 22, 35, 52, 70, 75, 82, 83, 86, 93, 94, 6].

Furthermore, flavonoids affect plant growth, development and reproduction, membrane permeability and seed germination, and prevent pre-harvest sprouting [16, 19, 24, 26, 31, 38, 66, 80, 88].

4.3 Genetic Bases of Flavonoid Biosynthesis in Wheat

Flavonoid biosynthesis genes are divided into the regulatory and the target genes. FB enzymes are encoded by the structural genes representing the target genes in the network. Regulatory genes control tissue-specific expression of the target genes. Genetic bases of flavonoid biosynthesis have been intensively studied in maize, rice, barley, *Arabidopsis*, petunia and morning glory [14, 67]. In wheat, until recently, information on flavonoid biosynthesis gene system was very scant. Different wheat plant organs may be colored with flavonoid pigments such as anthocyanins (grain pericarp, coleoptile, culm, leaf blade, leaf sheath, auricle, anther and glume; [43, 61]), proanthocyanidins (seed coat; [65]) and phlobaphenes (glume; [39]). In wheat genome, at least 70 loci were found, either determining pigmentation of different organs or encoding FB enzymes [61]. Some of the genes determining organ pigmentation may encode transcriptional factors regulating expression of the FB structural genes ([30, 47]; Khlestkina et al. unpublished).

4.3.1 Genes Determining Pigmentation of Different Organs

Three genes determining red coleoptiles coloration were reported in hexaploid wheat, localized on homoeologous chromosomes 7A [89], 7B [25] and 7D [36]. Microsatellite mapping of these genes showed them to be homoeoloci which were designated *Rc-A1*, *Rc-B1* and *Rc-D1*, respectively [40]. Two genes controlling purple culm were localized on chromosomes 7B [51] and 7D [59]. Later they and the third gene on chromosome 7A were mapped in homoeologous positions (loci designations: *Pc-A1*, *Pc-B1* and *Pc-D1*) in close linkage with the *Rc-1* loci [45, 47]. Furthermore, novel genes determining purple leaf blade color (*Plb-A1*, *Plb-B1* and *Plb-D1*) and purple leaf sheath color (*Pls-A1*, *Pls-B1* and *Pls-D1*) were mapped in close linkage with the *Rc-1* and *Pc-1* genes [45, 47]. Two genes determining purple anther (*Pan-A1* and *Pan-D1*) were mapped on chromosomes 7A [5] and 7D [45] in a short distance from *Rc-A1* and *Rc-D1*, respectively. Comparative mapping in wheat and maize demonstrates that the wheat *Rc/Pc/Plb/Pls/Pan* gene cluster locates in syntenic region in comparison with the maize genome region carrying locus *c1* encoding Myb-like regulatory protein and determining anthocyanin pigmentation in maize [20, 43]. Furthermore, the *c1* gene was used as a probe in Southern hybridization-based mapping in wheat, and its homolog was mapped to the chromosomes 7A, 7B and 7D in position highly comparable with that of the wheat *Rc/Pc/Plb/Pls/Pan* gene cluster [54].

Two complementary genes determining purple grain pericarp, *Pp1* and *Pp3*, were mapped to non-homoeologous chromosomes 7B and 2A, respectively in both hexaploid [21] and tetraploid wheats [48]. Comparative mapping in wheat, rice and maize demonstrates that the *Pp3* gene locates in syntenic region of the rice and maize genomes carrying loci *Pb* (*Ra*; [34, 95]) and *Lc* (*R*; [57]), respectively, encoding Myc-like regulatory protein determining anthocyanin pigmentation.

Three genes determining red auricles, *Ra1*, *Ra2*, *Ra3*, were assigned to non-homoeologous chromosomes 1D [28], 4B and 6B [63], respectively.

The pigmentation traits described above are due to anthocyanin biosynthesis. In wheat, seed coat, glumes and awns may be colored with proanthocyanidins or phlobaphenes. Three genes determining red seed coat color, *R-A1*, *R-B1* and *R-D1*, were mapped to the homoeologous chromosomes 3A, 3B and 3D [23]. Three homoeologous loci determining red, smokey-grey or black glume color, *Rg-A1*, *Rg-B1* and *Rg-D1*, were mapped to chromosomes 1A, 1B and 1D [41, 46]. Homeoloci determining black (*Bla*) or red (*Raw*) awn color are closely linked to the *Rg-1* genes [7, 78].

4.3.2 *Genes Encoding Flavonoid Biosynthesis Enzymes*

The genes for phenylalanine ammonia-lyase (*Pal*) were cloned [53, 55] and mapped [54] to chromosomes of homoeologous groups 3 and 6. The genes for chalcone synthase (*Chs*) were cloned [97] and mapped [54] to chromosomes of homoeologous groups 1 and 2. The genes encoding chalcone-flavanone isomerase (*Chi*) and anthocyanidin-3-glucoside rhamnosyltransferase (*3Rt*) were cloned and mapped to chromosomes of homoeologous group 5 ([44, 54]; Tereshchenko and Khlestkina, unpublished). The genes for flavanone 3-hydroxylase (*F3h*) and dihydroflavonol-4-reductase (*Dfr*) were cloned and mapped to chromosomes of homoeologous group 2 [42] and 3 [29, 69], respectively. The anthocyanidin synthase (*Ans*) genes were cloned and mapped to chromosomes of homoeologous group 6 [32].

4.4 Regulation in Flavonoid Biosynthesis Gene Systems in Wheat and Wheat-Alien Hybrids

Mapping the genes participating in pigmentation traits formation in wheat allowed to choose proper genetic models for the further investigation of FB gene regulation in allopolyploid wheat and wheat-alien hybrids.

4.4.1 *Transcription of Flavonoid Biosynthesis Structural Genes in Wheat*

4.4.1.1 Transcription of Flavonoid Biosynthesis Genes in Different Organs

Analysis of wheat near-isogenic lines showed that wheat *R* gene determining red seed coat color is a tissue-specific Myb-like transcriptional activator of the FB structural genes [30].

Using wheat 'Chinese Spring' with green coleoptile and chromosome substitution line 'Chinese Spring'('Hope' 7A) with red coleoptile, Ahmed et al. [2] showed that *Rc-A1* gene may be a transcriptional activator of the structural genes *Dfr*, *Ans* and *Ufgt* (UDPG flavonol 3-O-glucosyl transferase). The use of such models as substitution, recombinant and introgression lines demonstrated that *Rc-A1*, *Rc-B1* and *Rc-D1* activate transcription of the *F3h* gene in colored wheat coleoptiles [42, 47]. Transcriptional activating function of the *Rc-1* genes is in agreement with the comparative mapping in wheat and maize, suggesting *Rc* to be an orthologue of the maize *c1* gene encoding Myb-like regulatory protein determining anthocyanin pigmentation [20, 43]. The gene *Pc-1* which is closely linked to *Rc-1* was shown to activate transcription of the *Chs*, *F3h* and *Ans* genes in purple culm (Tereshchenko and Khlestkina, unpublished). Probably, the wheat *Rc*/*Pc*/*Plb*/*Pls*/*Pan* gene cluster in wheat chromosome 7 originates from the single ancestor Myb-like gene-orthologue of the maize gene *c1*.

Using wheat near-isogenic lines we demonstrated that the *Rg-1* genes activate *Chi* transcription in red and black glumes, whereas the *Pp3* gene activate the *Chs*, *F3h* and *Ans* genes in purple pericarp. Furthermore, analysis of expression of the candidate gene for *Pp3* in pericarp of the isogenic lines confirmed the suggestion resulted from comparative mapping in wheat, rice and maize that *Pp3* is a Myc-like transcriptional activator for the structural FB genes ([39]; Tereshchenko and Khlestkina, unpublished).

4.4.1.2 Transcription of Homoeologous Genes in Wheat

In 70–99% homoeologous gene series of allopolyploids, all homoeologues are reportedly co-expressed [8, 17, 37]. Analysis of transcriptional level of individual gene copies showed that expression levels of co-expressed homoeologues can sometimes be equal [42, 68, 90] or vary [3, 74, 90]. However until recently the question remained as to whether, in an allopolyploid, the interaction between regulatory and target is genome-specific, or whether regulation cuts across genomes. To clarify this, relationships between structural genes and their homoeologous regulatory genes were studied using the *Rc*-*F3h* gene pair as a regulatory-target model [42]. It was shown that each dominant *Rc-1* allele affects the expression of the three *F3h-1* homoeologues equally, but the level of *F3h* expression was dependent on the identity of the dominant *Rc-1* allele present. The lack of any genome-specific relationship between *F3h-1* and *Rc-1* implies an integrative evolutionary process among the three diploid genomes, following the formation of hexaploid wheat. Another conclusion made from this study was that the regulatory genes contribute more to the functional divergence between the wheat genomes than do the structural genes themselves [42].

4.4.2 Gene Regulation in Wheat-Alien Hybrids

Alien germplasm keeps stirring interest as a source of genes useful for crop plant species. Novel phenotypes can be obtained using transgenic technologies or chromosome engineering approaches. Finely coordinated work of the FB genes of different species [1, 10, 12, 56] or *vice versa* their inability to cooperate [9, 84] were shown by transgenic technologies.

Alternatively to transgenic plants, a wide range of the wheat-alien hybrids provides an important tool to investigate relationships between foreign regulatory and target genes. Using such wheat-rye chromosome substitution and addition lines, it was shown that the rye regulatory *Rc* gene can activate the wheat target gene *F3h* and *vice versa* wheat *Rc* induces expression of rye *F3h*. However, lower level of expression of rye *F3h* in comparison with that of the two wheat orthologues in the wheat-rye chromosome substitution line 2R(2D) was observed [44]. It was suggested that transcriptional dominance observed in the substitution line was a result of post-hybridization changing methylation patterns of rye *F3h* or was due to divergence between rye and wheat *F3h cis*-regulatory elements, which although did not prevent activation of rye *F3h* governed by the wheat *Rc* gene, affected specifically expression level of the gene [39]. Analysis of wheat *F3h* transcription in wheat-*Aegilops* chromosome addition and substitution lines, wheat-rye chromosome addition line and wheat-barley chromosome substitution line demonstrated that each of the *Aegilops*, rye or barley *Rc* genes is able to activate wheat *F3h-1* genes. However, the observed compensatory effect of the alien chromosomes was partial, and the bigger genetic distance between wheat and a donor species was, the lower transcriptional level of wheat *F3h-1* genes was observed [39].

4.5 Conclusions

Pigmentation traits formation in wheat is based on the functioning of the multicomponent gene system including the genes controlling biosynthesis of flavonoid pigments anthocyanins, proanthocyanidins and phlobaphenes in different plant organs and the genes encoding flavonoid biosynthesis enzymes (structural genes). The genes determining pigmentation of different organs are the tissue-specific regulatory genes activating transcription of the structural genes. During pigmentation traits formation in wheat, coordinated work of the regulatory and structural genes located in different diploid genomes composing allopolyploid wheat genome is observed. When some wheat components of this gene system are substituted by orthologous genes from *Aegilops*, rye or barley, the partial compensatory effect takes place.

References

1. Ahmed N, Maekawa M, Utsugi S, Himi E, Ablet H, Rikiishi K, Noda K (2003) Transient expression of anthocyanin in developing wheat coleoptile by maize *c1* and *B-peru* regulatory genes for anthocyanin synthesis. Breed Sci 52:29–43

2. Ahmed N, Maekawa M, Utsugi S, Rikiishi K, Ahmad A, Noda K (2006) The wheat *Rc* gene for red coleoptile colour codes for a transcriptional activator of late anthocyanin biosynthesis genes. J Cereal Sci 44:54–58
3. Appleford NE, Evans DJ, Lenton JR, Gaskin P, Croker SJ, Devos KM, Phillips AL, Hedden P (2006) Function and transcript analysis of gibberellin-biosynthetic enzymes in wheat. Planta 223:568–582
4. Bahler D, Steffen KL, Orzolek MD (1991) Morphological and biochemical comparison of a purple-leafed and a green-leafed pepper cultivar. Hortscience 26:736
5. Blanco A, Bellomo MP, Cenci A, De Giovanni C, D'Ovidio R, Iacono E, Laddomada B, Pagnotta MA, Porceddu E, Sciancalepore A, Simeone R, Tanzarella OA (1998) A genetic linkage map of durum wheat. Theor Appl Genet 97:721–728
6. Bogdanova ED, Sarbaev AT, Makhmudova KK (2002) Resistance of common wheat to bunt. In: Proceedings of the research conference on genetics. Moscow, Russia, pp 43–44
7. Börner A, Schumann E, Fürste A, Cöster H, Leithold B, Röder MS, Weber WE (2002) Mapping of quantitative trait loci determining agronomic important characters in hexaploid wheat (*Triticum aestivum* L.). Theor Appl Genet 105:921–936
8. Bottley A, Xia GM, Koebner RM (2006) Homoeologous gene silencing in hexaploid wheat. Plant J 47:897–906
9. Bovy A, de Vos R, Kemper M, Schijlen E, Almenar Pertejo M, Muir S, Collins G, Robinson S, Verhoeyen M, Hughes S, Santos-Buelga C, van Tunen A (2002) High-flavonol tomatoes resulting from the heterologous expression of the maize transcription factor genes *LC* and *C1*. Plant Cell 14:2509–2526
10. Bradley JM, Davies KM, Deroles SC, Bloor SJ, Lewis DH (1998) The maize *Lc* regulatory gene up-regulates the flavonoid biosynthetic pathway of petunia. Plant J 13:381–392
11. Brink RA (1956) A genetic change associated with the *R* locus in maize which is directed and potentially reversible. Genetics 41:872–889
12. Butelli E, Titta L, Giorgio M, Mock HP, Matros A, Peterek S, Schijlen EG, Hall RD, Bovy AG, Luo J, Martin C (2008) Enrichment of tomato fruit with health-promoting anthocyanins by expression of select transcription factors. Nat Biotechnol 26:1301–1308
13. Chalker-Scott L (1999) Environmental significance of anthocyanins in plant stress responses. Photochem Photobiol 70:1–9
14. Chopra S, Hoshino A, Boddu J, Iida S (2008) Flavonoid pigments as tools in molecular genetics. In: Grotewold E (ed) The science of flavonoids. Springer, New York, pp 147–173
15. Christensen AB, Gregersen PL, Olsen CE, Collinge DB (1998) A flavonoid 7-O-methyltransferase is expressed in barley leaves in response to pathogen attack. Plant Mol Biol 36:219–227
16. Clegg MT, Durbin ML (2000) Flower color variation: a model for the experimental study of evolution. Proc Natl Acad Sci USA 97:7016–7023
17. Comai L, Tyagi AP, Winter K, Holmes-Davis R, Reynolds SH, Stevens Y, Byers B (2000) Phenotypic instability and rapid gene silencing in newly formed arabidopsis allotetraploids. Plant Cell 12:1551–1568
18. Darwin C (1883) The variation of animals and plants under domestication. D. Appelton&Co, New York, pp 1–495
19. Debeaujon I, Léon-Kloosterziel KM, Koornneef M (2000) Influence of the testa on seed dormancy, germination, and longevity in *Arabidopsis*. Plant Physiol 122:403–414
20. Devos KM, Chap S, Li QY, Simonetti MC, Gale MD (1994) Relationship between chromosome 9 of maize and wheat homoeologous group 7 chromosomes. Genetics 138:1287–1292
21. Dobrovolskaya OB, Arbuzova VS, Lohwasser U, Röder MS, Börner A (2006) Microsatellite mapping of complementary genes for purple grain colour in bread wheat (*Triticum aestivum* L.). Euphytica 150:355–364
22. Farrant JM (2000) A comparison of mechanisms of desiccation tolerance among three angiosperm resurrection plant species. Plant Ecol 151:29–39
23. Flintham JE, Gale MD (1995) Dormancy gene maps in homoeologous cereal genomes. In: Proceedings of 7th international symposium on pre-harvest sprouting in cereals, Osaka, Japan, pp 143–149

24. Freed RD, Everson EH, Ringlund K, Gullord M (1976) Seed coat color in wheat and the relationship to seed dormancy at maturity. Cereal Res Commun 4:147–149
25. Gale MD, Flavell RB (1971) The genetic control of anthocyanin biosynthesis by homoeologous chromosomes in wheat. Genet Res Camb 18:237–244
26. Gfeller F, Svejda F (1960) Inheritance of post-harvest seed dormancy and kernel colour in spring wheat lines. Can J Plant Sci 40:1–6
27. Giovanini MP, Puthoff DP, Nemacheck JA, Mittapalli O, Saltzmann KD, Ohm HW, Shukle RH, Williams CE (2006) Gene-for-gene defense of wheat against the Hessian fly lacks a classical oxidative burst. Mol Plant Microbe Interact 19:1023–1033
28. Gulyaeva ZB (1984) Localization of the genes for pubescence of the glumes and coloration of the auricles in the leaf sheath in winter wheat variety Ul'yanovka. Bull Appl Bot Genet Plant Breed 85:95–96
29. Himi E, Noda K (2004) Isolation and location of three homoeologous dihydroflavonol-4-reductase (DFR) genes of wheat and their tissue-dependent expression. J Exp Bot 55:365–375
30. Himi E, Noda K (2005) Red grain colour gene (*R*) of wheat is a Myb-type transcription factor. Euphytica 143:239–242
31. Himi E, Mares DJ, Yanagisawa A, Noda K (2002) Effect of grain colour gene (*R*) on grain dormancy and sensitivity of the embryo to abscisic acid (ABA) in wheat. J Exp Bot 53:1569–1574
32. Himi E, Osaka T, Noda K (2006) Isolation and characterization of wheat ANS genes. GenBank2006.http://www.ncbi.nlm.nih.gov/sites/entrez?term=himi%20osaka%20noda&cmd=Search&db=nuccore&QueryKey=4
33. Hoch WA, Singsaas EL, McCown BH (2003) Resorption protection. Anthocyanins facilitate nutrient recovery in autumn by shielding leaves from potentially damaging light levels. Plant Physiol 133:1296–1305
34. Hu J, Anderson B, Wessler R (1996) Isolation and characterization of rice *R* genes: evidence for distinct evolutionary paths in rice and maize. Genetics 142:1021–1031
35. Izdebski R (1992) Utilization of rye genetic resources – initial material selection. Hereditas 116:179–185
36. Jha KK (1964) The association of a gene for purple coleoptile with chromosome 7D of common wheat. Can J Genet Cytol 6:370–372
37. Kashkush K, Feldman M, Levy AA (2002) Gene loss, silencing and activation in a newly synthesized wheat allotetraploid. Genetics 160:1651–1659
38. Khan M, Cavers PB, Kane M, Thompson K (1996) Role of the pigmented seed coat of proso millet (*Panicum miliaceum* L.) in imbibition, germination and seed persistence. Seed Sci Res 7:21–25
39. Khlestkina EK (2010) Regulatory-target gene relationships in allopolyploid and hybrid genomes. In: Urbano KV (ed) Advances in genetics research, vol 3. NOVA Science Publishers, Inc., New York, pp 311–328
40. Khlestkina EK, Pestsova EG, Röder MS, Börner A (2002) Molecular mapping, phenotypic expression and geographical distribution of genes determining anthocyanin pigmentation of coleoptiles in wheat (*Triticum aestivum* L.). Theor Appl Genet 104:632–637
41. Khlestkina EK, Pshenichnikova TA, Röder MS, Arbuzova VS, Salina EA, Börner A (2006) Comparative mapping of genes for glume colouration and pubescence in hexaploid wheat (*Triticum aestivum* L.). Theor Appl Genet 113:801–807
42. Khlestkina EK, Röder MS, Salina EA (2008) Relationship between homoeologous regulatory and structural genes in allopolyploid genome – a case study in bread wheat. BMC Plant Biol 8:88
43. Khlestkina EK, Röder MS, Pshenichnikova TA, Simonov AV, Salina EA, Börner A (2008) Genes for anthocyanin pigmentation in wheat: review and microsatellite-based mapping. In: Verrity JF, Abbington LE (eds) Chromosome mapping research developments. NOVA Science Publishers, Inc, New York, pp 155–175
44. Khlestkina EK, Tereshchenko OYU, Salina EA (2009) Anthocyanin biosynthesis genes location and expression in wheat-rye hybrids. Mol Genet Genomics 282:475–485

45. Khlestkina EK, Pshenichnikova TA, Röder MS, Börner A (2009) Clustering anthocyanin pigmentation genes in wheat group 7 chromosomes. Cereal Res Commun 37:391–398
46. Khlestkina EK, Salina EA, Pshenichnikova TA, Röder MS, Börner A (2009) Glume coloration in wheat: allelism test, consensus mapping and its association with specific microsatellite alleles. Cereal Res Commun 37:37–43
47. Khlestkina EK, Röder MS, Pshenichnikova TA, Börner A (2010) Functional diversity at *Rc* (red coleoptile) locus in wheat (*Triticum aestivum* L.). Mol Breed 25:125–132
48. Khlestkina EK, Röder MS, Börner A (2010) Mapping genes controlling anthocyanin pigmentation on the glume and pericarp in tetraploid wheat (*Triticum durum* L.). Euphytica 171:65–69
49. Kihara H (1944) Discovery of the DD-analyser, one of the ancestors of *Triticum vulgare*. Agric Hortic (Tokyo) 19:13–14
50. Kihara H (1954) Origin of wheat. Wheat Inf Serv 1:35–42
51. Kuspira J, Unrau J (1958) Determination of the number and dominance relationships of genes on substituted chromosomes in common wheat *Triticum aestivum* L. Can J Plant Sci 38:119–205
52. Lachman J, Dudjak J, Miholová D, Kolihová D, Pivec V (2005) Effect of cadmium on flavonoid content in young barley (*Hordeum sativum* L.) plants. Plant Soil Environ 51:513–516
53. Li HP, Liao YC (2003) Isolation and characterization of two closely linked phenylalanine ammonia-lyase genes from wheat. Yi Chuan Xue Bao 30:907–912
54. Li WL, Faris JD, Chittoor JM, Leach JE, Hulbert SH, Liu DJ, Chen PD, Gill BS (1999) Genomic mapping of defense response genes in wheat. Theor Appl Genet 98:226–233
55. Liao YC, Li HP, Kreuzaler F, Fischer R (1996) Nucleotide sequence of one of two tandem genes encoding phenylalanine ammonia-lyase in *Triticum aestivum*. Plant Physiol 112:1398–1398
56. Lloyd AM, Walbot V, Davis RW (1992) *Arabidopsis* and *Nicotiana* anthocyanin production activated by maize regulators *R* and *C1*. Science 258:1773–1775
57. Ludwig SR, Habera LF, Dellaporta SL, Wessler SR (1989) *Lc*, a member of the maize *R* gene family responsible for tissue-specific anthocyanin production, encodes a protein similar to transcriptional activators and contains the myc-homology region. Proc Natl Acad Sci USA 86:7092–7096
58. Ma XF, Gustafson JP (2008) Allopolyploidization-accommodated genomic sequence changes in triticale. Ann Bot (Lond) 101:825–832
59. Maystrenko OI (1992) The use of cytogenetic methods in ontogenesis study of common wheat. In: Ontogenetics of higher plants. Kishinev, Shtiintsa, pp 98–114
60. McClintock B (1956) Controlling elements and the gene. Cold Spring Harbor Symp Quant Biol 21:197–216
61. McIntosh RA, Yamazaki Y, Dubcovsky J, Rogers J, Morris C, Somers DJ, Appels R, Devos KM (2008) Catalogue of gene symbols for wheat. http://www.grs.nig.ac.jp/wheat/komugi/genes/
62. Mehdy MC, Lamb CJ (1987) Chalcone isomerase cDNA cloning and mRNA induction by fungal elicitor, wounding and infection. EMBO J 6:1527–1533
63. Melz G, Thiele V (1990) Chromosome locations of genes controlling 'purple leaf base' in rye and wheat. Euphytica 49:155–159
64. Mendel G (1865) Versuche uber Pflanzen-Hybriden. Verh Naturforsch Ver Brunn 4:3–47
65. Miyamoto T, Everson EH (1958) Biochemical and physiological studies of wheat seed pigmentation. Agron J 50:733–734
66. Miyamoto T, Tolbert NE, Everson EH (1961) Germination inhibitors related to dormancy in wheat seeds. Plant Physiol 36:739–746
67. Mol J, Grotewold E, Koes R (1998) How genes paint flowers and seeds. Trends Plant Sci 3:212–217
68. Morimoto R, Kosugi T, Nakamura C, Takumi S (2005) Intragenic diversity and functional conservation of the three homoeologous loci of the KN1-type homeobox gene *Wknox1* in common wheat. Plant Mol Biol 57:907–924

69. Munkvold JD, Greene RA, Bermudez-Kandianis CE, La Rota CM, Edwards H, Sorrells SF, Dake T, Benscher D, Kantety R, Linkiewicz AM, Dubcovsky J, Akhunov ED, Dvorak J, Miftahudin, Gustafson JP, Pathan MS, Nguyen HT, Matthews DE, Chao S, Lazo GR, Hummel DD, Anderson OD, Anderson JA, Gonzalez-Hernandez JL, Peng JH, Lapitan N, Qi LL, Echalier B, Gill BS, Hossain KG, Kalavacharla V, Kianian SF, Sandhu D, Erayman M, Gill KS, McGuire PE, Qualset CO, Sorrells ME (2004) Group 3 chromosome bin maps of wheat and their relationship to rice chromosome 1. Genetics 168:639–650
70. Nagata T, Todoriki S, Masumizu T, Suda I, Furuta S, Du Z, Kikuchi S (2003) Levels of active oxygen species are controlled by ascorbic acid and anthocyanin in *Arabidopsis*. J Agric Food Chem 51:2992–2999
71. Napoli C, Lemieux C, Jorgensen R (1990) Introduction of a chimeric chalcone synthase gene into petunia results in reversible co-suppression of homologous genes *in trans*. Plant Cell 2:279–289
72. Nilson-Ehle H (1914) Zur Kenntnis der mit der keimungsphysiologie des weizens in zusammenhang stehenden inneren faktoren. Z Pflanzenzüct 2:153–187
73. Nilsson-Ehle H (1909) Kreuzungsversuchungen an Hafer und Weizen. Lands Univ Aersskrift NF Afd 2:1–122
74. Nomura T, Ishihara A, Yanagita RC, Endo TR, Iwamura H (2005) Three genomes differentially contribute to the biosynthesis of benzoxazinones in hexaploid wheat. Proc Natl Acad Sci USA 102:16490–16495
75. Nozzolillo C, Isabelle P, Andersen OM, Abou-Zaid M (2002) Anthocyanins of jack pine *(Pinus banksiana)* seedlings. Can J Bot 80:796–801
76. Ozkan H, Levy AA, Feldman M (2001) Allopolyploidy-induced rapid genome evolution in the wheat (*Aegilops-Triticum*) group. Plant Cell 13:1735–1747
77. Ozkan H, Tuna M, Arumuganathan K (2003) Nonadditive changes in genome size during allopolyploidization in the wheat (*Aegilops-Triticum*) group. J Hered 94:260–264
78. Panin VM, Netsvetaev VP (1986) Genetic control of gliadins and some morphological characters of spike in durum winter wheats. Nauchno Tekhn Bull USGI Odessa (in Russian) 2:31–36
79. Parker J (1962) Relationships among cold hardiness, watersoluble protein, anthocyanins, and free sugars in *Hederu helix* L. Plant Physiol 37:809–813
80. Peer WA, Murphy AS (2008) Flavonoids as signal molecules. In: Grotewold E (ed) The science of flavonoids. Springer, New York, pp 239–268
81. Pestsova EG, Goncharov NP, Salina EA (1998) Elimination of a tandem repeat of telomeric heterochromatin during the evolution of wheat. Theor Appl Genet 97:1380–1386
82. Plaza BM, Jimenez S, Segura ML, Contreras JI, Lao MT (2009) Physiological stress caused by salinity in cordyline fruticosa and its indicators. Commun Soil Sci Plant Anal 40:473–484
83. Pozolotina VN, Molchanova IV, Karavaeva EN, Mihkaylovskaya LN, Antonova EV, Karimullina EM (2007) Analysis of current state of terrestrial ecosystems in the East-Ural Radioactive Trace. Iss Rad Safety (In Russian) Special Issue 'The East Ural Radioactive Trace Marks Its 50 Year Anniversary':32–44
84. Quattrocchio F, Wing JF, van der Woude K, Mol JN, Koes R (1998) Analysis of bHLH and MYB domain proteins: species-specific regulatory differences are caused by divergent evolution of target anthocyanin genes. Plant J 13:475–488
85. Quattrocchio F, Baudry A, Lepiniec L, Grotewold E (2008) The regulation of flavonoid biosynthesis. In: Grotewold E (ed) The science of flavonoids. Springer, New York, pp 97–122
86. Ryan KG, Swinny EE, Markham KR, Winefield C (2002) Flavonoid gene expression and UV photoprotection in transgenic and mutant *Petunia* leaves. Phytochemistry 59:23–32
87. Salina EA, Numerova OM, Ozkan H, Feldman M (2004) Alterations in subtelomeric tandem repeats during early stages of allopolyploidy in wheat. Genome 47:860–867
88. Sasaki K, Takahashi T (2002) A flavonoid from *Brassica rapa* flower as the UV-absorbing nectar guide. Phytochemistry 61:339–343

89. Sears ER (1954) The aneuploids of common wheat. Univ Missouri Agr Sta Res Bull 572:1–59
90. Shitsukawa N, Tahira C, Kassai K, Hirabayashi C, Shimizu T, Takumi S, Mochida K, Kawaura K, Ogihara Y, Murai K (2007) Genetic and epigenetic alteration among three homoeologous genes of a class E MADS box gene in hexaploid wheat. Plant Cell 19:1723–1737
91. Singh KB, Malhotra RS, Saxena MC (1995) Additional sources of tolerance to cold in cultivated and wild *Cicer* species. Crop Sci 35:1491–1497
92. Song KM, Lu P, Tang KL, Osborn TC (1995) Rapid genome changes in synthetic polyploids of *Brassica* and its implications for polyploidy evolution. Proc Natl Acad Sci USA 92:7719–7723
93. Suzuki T, Honda Y, Mukasa Y (2005) Effects of UV-B radiation, cold and desiccation stress on rutin concentration and rutin glucosidase activity in tartary buckwheat (*Fagopyrum tataricum.*) leaves. Plant Sci 168:1303–1307
94. Tereshchenko OYu, Khlestkina EK, Gordeeva EI, Arbuzova VS, Salina EA (2010) Biosynthesis of flavonoids under salinity stress in *Triticum aestivum* L. In: Abstract of international conference "plant genetics, genomics, biotechnology", Novosibirsk, Russia, 7–10 June, p 84
95. Wang C, Shu Q (2007) Fine Mapping and candidate gene analysis of purple pericarp gene *Pb* in rice (*Oryza sativa* L.). Chinese Sci Bull 52:3097–3104
96. Winkel BSJ (2008) The biosynthesis of flavonoids. In: Grotewold E (ed) The science of flavonoids. Springer, New York, pp 71–96
97. Yang G, Li B, Gao J, Liu J, Zhao X, Zheng Q, Tong Y, Li Z (2004) Cloning and expression of two chalcone synthase and a flavonoid 3′5′-Hydroxylase 3′-end cDNAs from developing seeds of blue-grained wheat involved in anthocyanin biosynthetic pathway. J Integr Plant Biol (Acta Bot Sin) 46:588–594

Chapter 5
Implications for Human and Environmental Health of Low Doses of Radiation

Carmel E. Mothersill and Colin B. Seymour

Abstract Recent advances in our understanding of the mechanisms underlying the biological effects of low dose effects of ionizing radiation have revealed that similar mechanisms can be induced by chemicals in the environment. Therefore interactions between radiation and chemicals are likely and that the outcomes following mixed exposures to radiation and chemicals may not be predictable for human health, by consideration of single agent effects. Our understanding of the biological effects of low dose exposure has undergone a major paradigm shift. We now possess technologies which can detect very subtle changes in cells due to small exposures to radiation or other pollutants. We also understand much more now about cell communication, systems biology and the need to consider effects of low dose exposure at different hierarchical levels of organization from molecules up to and including ecosystems. We also understand at least in part, some of the mechanisms which drive low dose effects and which perpetuate these not only in the exposed organism but also in its progeny and even its kin. This means that previously held views about safe doses or lack of harmful effects cannot be sustained. ICRP and all national radiation and environmental protection organisations have always accepted a theoretical risk and have applied the precautionary principle and the LNT (linear-non-threshold) model which basically says that there is no safe dose of radiation. Therefore even in the absence of visible effects, exposure of people to radiation is strictly limited. This review will consider the history of the new discoveries and will focus on evidence for emergent effects after mixed exposures to combined stressors which include ionizing radiation. The implications for regulation of low dose exposures to protect human health and environmental security will be discussed.

C.E. Mothersill (✉) • C.B. Seymour
Department of Medical Physics and Applied Radiation Sciences,
McMaster University, Hamilton L8S 4K1, ON, Canada
e-mail: mothers@mcmaster.ca

C.E. Mothersill et al. (eds), *Radiobiology and Environmental Security*,
NATO Science for Peace and Security Series C: Environmental Security,
DOI 10.1007/978-94-007-1939-2_5, © Springer Science+Business Media B.V. 2012

5.1 Biology of "Non-targeted" Effects

Within conventional radiobiology as accepted in the 1950s continuing through to the 1990s there was little consideration of epigenetic effects, because the traditional concept of radiobiology was based on target theory [53, 67]. For an effect to occur, radiation had to hit a defined target within the cell, assumed to be DNA. Assumptions about the number of targets hit could then be made from measurements of dose and dose rate [2, 15]. The evolution of non-targeted effect (NTE) radiobiology meant that at low doses the previous assumptions needed to be reconsidered in the light of the existence of non-DNA mechanisms [28, 44, 47]. The mechanisms underlying radiation effects are not constant with respect to dose and it would now be generally accepted that low dose effects are mechanistically different to high doses effects. This is not to say the mechanisms are necessarily mutually exclusive but it does mean that NTE's will contribute more to the overall outcome at low doses where targeted effects are small. Targeted effects will predominate at high doses and in situations where NTE's have been inhibited or otherwise prevented. In terms of the progression of radiobiological thinking in this field, disease caused by radiation no longer had to be exclusively genetically based, but radiation could promote or exacerbate systemic disease. This disease could have been caused for example by a chemical mutagen [4, 25, 58]. Equally, the radiation could facilitate a non-mutation based inflammatory type disease [33, 35, 36, 40]. These concepts, although largely accepted theoretically by the radiobiology community, have been difficult to prove epidemiologically because of what are generally called "confounding variables" such as smoking, drinking, age, gender, or concurrent past or future exposures to the same or a different pollutant [57, 64]. These factors actually reflect the futility of trying to assign causation, as defined in epidemiology, to one agent when the doses are low! Others argue that radiation and many chemical "pollutants" might actually boost the immune system and be good [5, 7, 59]. The hormetic argument has many interesting applications but is unproven with regard to multiple pollutants. This adds to the confusion and controversy surrounding low dose exposures. The essential point is that there will be huge individual variation due to involvement of epigenetic and non-targeted factors in the response [17, 55, 71]. At any one time we are as unique epigenetically as we are genetically. Epigenetic differences are linked to gender and lifestyle. In theory therefore a low dose of radiation could cause any number of effects ranging from beneficial to death-inducing disease depending on the context of the exposure and the interplay of factors such as cell communication, microenvironment, tissue infrastructure and a whole host of systemic variables which influence outcome from a cellular track of ionizing radiation [19, 70].

5.2 Is Radiation "Just Another Stressor" or Is It Unique?

Key developments leading to the current widespread acceptance of ionizing radiation as "just another stressor" include:

1. The development of sensitive techniques such as m-FISH, for detecting chromosomal abnormalities [14, 26, 54, 65].

2. Studies showing that delayed or persistent sub-optimal survival (reproductive death) could be seen in surviving progeny of irradiated cells [8, 46, 63, 66].
3. The emergence of genomic instability as a mechanism by which low doses of radiation could cause delayed or persistent damage to chromosomes [32, 34, 41, 56, 68].
4. The accumulation of knowledge of "bystander effects" whereby chromosome damage, death, DNA damage and various other consequences occur in cells receiving signals from cells irradiated with low doses of radiation [6, 27, 37, 38, 62, 69].
5. Criticism of the epidemiological research undertaken after the Hiroshima and Nagasaki bombs as ignoring the damage from residual radiation and fall out [45, 61].

The NTE paradigm emerged initially as a result of re examination of firmly held beliefs and some odd results in the laboratory which did not fit the DNA paradigm. Proof of the new hypotheses required the techniques such as molecular imaging, M-FISH, and SKY as well as the development of tissue culture techniques for human *normal* tissues which permitted functional studies to be performed [18]. Older studies tended to use high doses on a limited number of cell lines or highly inbred animal strains. These tended to thrive in the laboratory in the laboratory but were often unrepresentative of tissues in the outbred human or non-human [1, 16, 29].

5.2.1 *Important Consequence for Radiation Protection and Risk Assessment of NTE's*

5.2.1.1 The Concept of Hierarchical Levels

Hierarchical levels stretch from the individual to smaller units (organs – tissues – cells – organelles –genes) and to bigger units – populations (multiple individuals/ single species) and ecosystems (multiple individuals and multiple species). Confusion in the low dose exposure field (both radiation and chemical) arise from lack of consideration of this concept. Most of the arguments about whether radiation is good or bad for you fail due to lack of consideration of the hierarchical level at which the effects occur and because most of the arguments are anthropocentric. For example cell death is seen as a bad effect but if it removes a potentially carcinogenic cell from the population of cells in a tissue it could prevent cancer starting and could be seen as good. Survival of cells is seen sometimes as desirable but if they survive with unrepaired or mis-repaired damage, they could facilitate development of a cancer. Similarly in non-human populations – death of radiosensitive individuals which cannot adapt to the changed (now radioactive and or chemically polluted) environment, could be good at the population level in evolutionary terms depending on the life stage and reproductive status when the effects manifest, although death will be bad for the individual. It is only by considering responses in context, that any conclusions can be drawn about risk or harm.

5.2.1.2 Spatio-Temporal Concepts

There are two aspects to this – one is simply the age of the irradiated unit and the spatial deposition pattern of the ionizing energy. This concept is relevant across all hierarchical levels. Obvious considerations are the age or maturity of the cell, organ, life-form or ecosystem receiving the track, the density of the energy deposition, the lifetime of the unit and its importance in the context of functionality of the higher hierarchical levels. Young units tend to be less stable and thus more vulnerable (or more adaptive?) than old or mature units because of their faster metabolic rate, higher rate of growth/cell division and at the ecosystem level, because of their less strongly developed interdependencies. There is also (usually), more redundancy in young units, for example there are more available individuals, better reproductive rates and better viability from young progenitors , whether cells or individuals. The other aspect is that the delayed effects of radiation and bystander effects mean that radiation effects are not fixed in time or space to the energy deposition along ionizing track. The effects can persist and manifest at distant points in time and space. These concepts are also discussed elsewhere [4, 58].

5.2.1.3 The Importance of Mixed Exposure Analysis

Pollutants including radiation seldom occur in isolation. In fact most environmental radioactivity comes from radioisotopes which are chemical entities. This means that there is always a mixed exposure and that both the chemical and radioactive aspects need to be considered. Additive damage used to be an acceptable way to deal with mixed exposures (if any way were used!). The new field of non-targeted effects with the consequent realization that emergent properties can exist, which were not predictable from the individual agent dose response data, makes this no longer acceptable. The complexities of mixed pollutant scenarios call for a re-think of fundamental approaches to both epidemiological causation after low dose exposures to anything. They also question the need regulators have to regulate to a number (dose unit/exposure unit). Some of the issues concerning the latter position include the following:

- How to ensure compliance if there is no "safe" or legal limit?
- How to deal with multiple stressors especially if the interactions are not known
- How to correct for dose rate/time of exposure
- How to deal with mixed chronic and acute exposures
- How to factor in possible adaptive, hormetic or antagonistic effects
- How to regulate in pristine versus dirty environments

The issues of legal causation are highly relevant to the former point but outside the scope of this review. Discussion of these issues can be found elsewhere [22, 42, 43]. Ultimately, in order to resolve these issues, more data are needed for mixed exposure scenarios using relevant species. Systems biology approaches involving close interaction between experimental biologists and modelers are also required.

5.3 Data Concerning Low Dose Effects of Mixed Exposures

There are very little data where low dose exposures to multiple stressors/mixed contaminants involving radiation and a chemical are investigated. The field was reviewed by Mothersill et al [50] in 2006. Recent interest in non-targeted effects probably means more attention will be paid to this area in future. Gowans et al. [23] have data showing chemical induction of genomic instability. Data from the authors' own and other laboratories shows that heavy metals singly or in combination can cause genomic instability [3, 9–13, 20, 21, 24, 39, 48, 49, 52]. Delayed death and chromosome aberrations in human cells following nickel, titanium or cadmium exposure have been reported [9, 10, 20, 21]. Similar effects have been reported in fish cell lines [11–13, 39, 49, 52], and more recently in live fish exposed to very low doses of gamma radiation 4–75 m Gy over 48 h in the presence of heavy metals at levels just above background [51, 60].

Organic pesticides and detergents such as prochloras, nonoylphenol, nonoxynol and dichloroaniline have also been found to cause delayed lethal mutations in fish cells [11, 13, 48].

Chromium and vanadium used in implants and dentures lead to a variety of genetic and reproductive delayed effects in vivo and to multiple endpoints associated with non-targeted effects in vitro [9, 10, 20, 21].

5.4 Implications for Environmental Protection and Human Health

While many of the studies cited above are concerned with fish or in vitro cells rather than humans, the data show that non-targeted effects can be induced by low dose exposures to a number of environmental chemicals as well as ionizing radiation. This means that combined exposures to low doses of these agents cannot be regulated in isolation and that studies of potential mechanistic interactions are important. Radiation protection of humans could find use from the approaches which are being taken by the task groups within ICRP, IAEA and the US-DoE (see for example [30, 31]) who have to formulate policy to regulate exposure of non-human biota. Many of the issues involved such as dealing with non-cancer endpoints, mixed contaminants or chronic low dose exposure are real issues in human radiation protection.

5.5 Summary Thoughts and Recommendations

The challenge in the low dose exposure field is to tease out the "noise". Noise is the euphemistic term we use when the level of the disease which is un attributable to our favoured causative agent, is too high to prove causation formally in any strict scientific or legal sense. Perhaps we should accept that we cannot assign causation and instead view ionizing radiation as one among many agents which *together* contribute to cause disease. Before we can do this it is vital to understand the key mechanisms and in particular to find areas of mechanistic commonality suggesting common causation. Biomarkers may be useful to identify possible common mechanisms and to validate their relevance across different hierarchical levels. If this is achieved it should be possible to model links between *effects* at one level e.g. cellular or individual leading to *harm and risk* at higher levels – in this example the individual or the population. Biomarker studies do need to be interpreted cautiously however because they are often used as surrogates for risk when in fact they may merely be pointing to change in the system. Without the back-up modeling and multi-level analysis of their relevance they may lead to false conclusions and confusion about the true risk of an inducing agent.

The problem of establishing causation following mixed exposures still exists as does the issue of what constitutes "harm". In the non-human biota field, there is great concern about doing more harm than good, if action levels are enforced which might require "remediation" of a habitat – i.e. removal of contaminated vegetation and soil. This could cause much more harm to the ecosystem than the original stressor. In the realm of human protection against low dose stressors, issues might include the ethics of genetic screening to identify sensitive sub-populations. If a sensitivity marker were available, who should be tested and when? Should diagnostic screening be forbidden to these individuals because of their possible sensitivity to low doses of radiation? There are also issues regarding lifestyle choices and risk benefit analysis at the biological level. Evolutionary adaptation leads to a fitter population (of cells, individuals) by eliminating the weak units but how is that population changed? In dealing with concepts of adaptation to environmental stressors where "nature" sorts things out in the optimal way, is "nature's way" to Nietzian for Man?

It would be good to conclude this reflection with a "way forward" but as we are still in the very early stages of accepting that radiation doses effects at low doses are non-linear, that multiple stressors impact the final outcome, and that what appears to be bad (or good) may be good (or bad)– it is perhaps best to recommend caution and consideration of these points rather than a great new regulatory framework!

Acknowledgements I acknowledge continued support from the Canada Research Chairs program, NSERC, and the EU NOTE integrated project.

References

1. Alper T (1973) The relevance of experimental radiobiology to radiotherapy. Present limitations and future possibilities. Br Med Bull 29(1):3–6
2. Alper T (1979) Cellular radiobiology. Cambridge University Press, Cambridge
3. Bagwell CE, Milliken CE, Ghoshroy S, Blom DA (2008) Intracellular copper accumulation enhances the growth of Kineococcus radiotolerans during chronic irradiation. Appl Environ Microbiol 74(5):1376–1384
4. Baverstock K, Rönkkö M (2008) Epigenetic regulation of the mammalian cell. PLoS One 3(6):e2290

5. Boonstra R, Manzon RG, Mihok S, Helson JE (2005) Hormetic effects of gamma radiation on the stress axis of natural populations of meadow voles (Microtuspennsylvanicus). Environ Toxicol Chem 24(2):334–343
6. Brenner DJ et al (2001) The bystander effect in radiation oncogenesis. Radiat Res 155:402–408; 13
7. Calabrese EJ, Baldwin LA (2000) The effects of gamma rays on longevity. Biogerontology 1(4):309–319
8. Coates PJ, Lorimore SA, Wright EG (2005) Cell and tissue responses to genotoxic stress. J Pathol 205(2):221–235
9. Coen N, Mothersill C, Kadhim M, Wright EG (2001) Heavy metals of relevance to human health induce genomic instability. J Pathol 195(3):293–299
10. Coen N, Kadhim MA, Wright EG, Case CP, Mothersill CE (2003) Particulate debris from a titanium metal prosthesis induces genomic instability in primary human fibroblast cells. Br J Cancer 88(4):548–552
11. Dowling K, Mothersill C (1999) Use of rainbow trout primary epidermal cell cultures as an alternative to immortalized cell lines in toxicity assessment: a study with nonoxynol. Environ Toxicol Chem 18:2846–2850
12. Dowling K, Mothersill C (2001) The further development of rainbow trout primary epithelial cell cultures as a diagnostic tool in ecotoxicology risk assessment. Aquat Toxicol 53:279–289
13. Dowling K, Seymour C, Mothersill C (2005) Delayed cell death and bystander effects in the progeny of Chinook salmon embryo cells exposed to radiation and a range of aquatic pollutants. Int J Radiat Biol 81(1):89–96
14. Edwards AA et al (2005) Review of translocations detected by FISH for retrospective biological dosimetry applications. Radiat Prot Dos 113:396–402
15. Elkind MM (1967) Radiobiology of cultured mammalian cells. Gordon and Breach Science Publishers, New York
16. Elkind MM (1988) The initial part of the survival curve: does it predict the outcome of fractionated radiotherapy? Radiat Res 114(3):425–436
17. Fike JR, Rola R, Limoli CL (2007) Radiation response of neural precursor cells. Neurosurg Clin N Am 18(1):115–127
18. Freshney RI (2005) Culture of animal cells a manual of basic technique. Wiley, Liss
19. Gault N, Rigaud O, Poncy JL, Lefaix JL (2007) Biochemical alterations in human cells irradiated with alpha particles delivered by macro- or microbeams. Radiat Res 167(5):551–562
20. Glaviano A, Nayak V, Cabuy E, Baird DM, Yin Z, Newson R, Ladon D, Rubio MA, Slijepcevic P, Lyng F, Mothersill C, Case CP (2006) Effects of hTERT on metal ion-induced genomic instability. Oncogene 25(24):3424–3435
21. Glaviano A, Mothersill C, Case CP, Rubio MA, Newson R, Lyng F (2009) Effects of hTERT on genomic instability caused by either metal or radiation or combined exposure. Mutagenesis 24(1):25–33
22. González AJ, Lauriston S (2005) Taylor Lecture: radiation protection in the aftermath of a terrorist attack involving exposure to ionizing radiation. Health Phys 89(5):418–446
23. Gowans ID, Lorimore SA, McIlrath JM, Wright EG (2005) Genotype-dependent induction of transmissible chromosomal instability by gamma-radiation and the benzene metabolite hydroquinone. Cancer Res 65(9):3527–3530
24. Grygoryev D, Moskalenko O, Zimbrick JD (2008) Non-linear effects in the formation of DNA damage in medaka fish fibroblast cells caused by combined action of cadmium and ionizing radiation. Dose Response 6(3):283–298
25. Gundy S (2006) The role of chemical and physical factors in cancer development. Magy Onkol 50(1):5–18
26. Hande MP, Brenner DJ et al (2003) Past Exposure to densely ionising radiation leaves a unique permanent signature in the genome. Am J Hum Genet 72:1162–1170
27. Hei TK, Persaud R, Zhou H, Suzuki M (2004) Genotoxicity in the eyes of bystander cells. Mutat Res 568(1):111–120
28. Hei TK, Zhou H, Ivanov VN, Hong M, Lieberman HB, Brenner DJ, Amundson SA, Geard CR (2008) Mechanism of radiation-induced bystander effects: a unifying model. J Pharm Pharmacol 60(8):943–950

29. Hornsey S, Kutsutani Y, Field SB (1975) Damage to mouse lung with fractionated neutrons and x rays. Radiology 116(1):171–174
30. ICRP Publication 91 (2003) A framework for assessing the impact of ionising radiation on non-human species. Ann ICRP 33(3):201–270
31. ICRP Publication 103 (2009). New Recommendations of the ICRP. Ann ICRP 37(2–4):1–230, Elsevier BV.
32. Kadhim MA, Macdonald DA, Goodhead DT, Lorimore SA, Marsden SJ, Wright EG (1992) Transmission of chromosomal instability after plutonium alpha-particle irradiation. Nature 355(6362):738–740
33. Kusunoki Y, Hayashi T (2008) Long-lasting alterations of the immune system by ionizing radiation exposure: implications for disease development among atomic bomb survivors. Int J Radiat Biol 84(1):1–14
34. Little JB, Nagasawa H (1992) Induction of sister chromatid exchanges by extremely low doses alpha-particles. Cancer Res 52(22):6394–6396
35. Little MP, Tawn EJ, Tzoulaki I, Wakeford R, Hildebrandt G, Paris F, Tapio S, Elliott P (2008) A systematic review of epidemiological associations between low and moderate doses of ionizing radiation and late cardiovascular effects, and their possible mechanisms. Radiat Res 169(1):99–109
36. Lorimore SA, Wright EG (2003) Radiation-induced genomic instability and bystander effects: related inflammatory-type responses to radiation-induced stress and injury? Int J Radiat Biol 79(1):15–25
37. Lorimore SA, Kadhim MA, Pocock DA, Papworth D, Stevens DL, Goodhead DT, Wright EG (1998) Chromosomal instability in the descendents of unirradiated surviving cells after alpha-particle irradiation. Proc Natl Acad Sci USA 95(10):5730–5733
38. Lui Z, Mothersill CE, McNeill FE, Lyng FM, Byun SH, Seymour CB, Prestwich WV (2006) A dose threshold for a medium transfer bystander effect for a human skin cell line. Radiat Res 166(1 Pt 1):19–23
39. Lyng FM, Lyons-Alcantara M, Olwell P, Ní Shuilleabháin S, Seymour C, Cottell DC, Mothersill C (2004) Ionizing radiation induces a stress response in primary cultures of rainbow trout skin. Radiat Res 162(2):226–232
40. Manton KG, Volovik S, Kulminski A (2004) ROS effects on neurodegeneration in Alzheimer's disease and related disorders: on environmental stresses of ionizing radiation. Curr Alzheimer Res 1(4):277–293
41. Marder BA, Morgan WF (1993) Delayed chromosomal instability induced by DNA damage. Mol Cell Biol 13(11):6667–6677
42. Masse R (2000) Ionizing radiation. C R Acad Sci III 323(7):633–640
43. Miller C (2006) Causation in personal injury: legal or epidemiological common sense? Legal Studies 26(4):544–569
44. Morgan WF, Sowa MB (2006) Non-targeted bystander effects induced by ionizing radiation. Mutat Res 616(1–2):159–164, Epub
45. Mossman KL (2001) Deconstructing radiation hormesis. Health Phys 80(3):263–269
46. Mothersill C, Seymour C (1997) Lethal mutations and genomic instability. Int J Radiat Biol 71(6):751–758
47. Mothersill C, Seymour C (2006) Radiation-induced bystander and other non-targeted effects: novel intervention points in cancer therapy? Curr Cancer Drug Targets 6(5):447–454
48. Mothersill C, Crean M, Lyons M, McSweeney J, Mooney R, O'Reilly J, Seymour CB (1998) Expression of delayed toxicity and lethal mutations in the progeny of human cells surviving exposure to radiation and other environmental mutagens. Int J Radiat Biol 74(6):673–680
49. Mothersill C, Lyng F, Mulford A, Seymour C, Cottell D, Lyons M, Austin B (2001) Effect of low doses of ionizing radiation on cells cultured from the hematopoietic tissue of the Dublin Bay Prawn, Nephrops norvegicus. Radiat Res 156:241–250
50. Mothersill C, Mosse I, Seymour CB (2006) Proceedings of the NATO Advanced research workshop on multipollution exposure and risk assessment – a challenge for the future, Minsk, Belarus, 1-5. Multiple stressors: a challenge for the future 6(3):283–298, Epub 2007 ISBN: 978-1-4020-6334-3

51. Mothersill C, Salbu B, Heier LS, Teien HC, Denbeigh J, Oughton D, Rosseland BO, Seymour CB (2007) Multiple stressor effects of radiation and metals in salmon (Salmo salar). J Environ Radioact 96(1–3):20–31
52. Ní Shúilleabháin S, Mothersill C, Sheehan D, O'Brien NM, O'Halloran J, van Pelt FN, Kilemade M, Davoren M (2006) Cellular responses in primary epidermal cultures from rainbow trout exposed to zinc chloride. Ecotoxicol Environ Saf 65:332–341
53. Osborne JC, Jr M, Jay H, Kempner ES (2000) Molecular mass and volume in radiation target theory. Biophys J 78:1698–1702
54. Pinkel et al (1988) Fluorescence in situ hybridization with human chromosome-specific libraries: Detection of trisomy 21 and translocations of chromosome 4. Proc Natl Acad Sci USA 85:9138–9142
55. Pinto M, Howell RW (2007) Concomitant quantification of targeted drug delivery and biological response in individual cells. Biotechniques 43(1):64, 66–71, PMID: 17695254
56. Ponniaya B et al (1997) Radiation-induced chromosomal instability in BALB/c and C57BL/6 mice: the difference is as clear as black and white. Radiat Res 147(2):121–125
57. Prasad KN, Cole WC, Hasse GM (2004) Health risks of low dose ionizing radiation in humans: a review. Exp Biol Med 229(5):378–382
58. Preston RJ (2005) Bystander effects, genomic instability, adaptive response, and cancer risk assessment for radiation and chemical exposures. Toxicol Appl Pharmacol 207(2 Suppl):550–556
59. Sakai K (2006) Biological responses to low dose radiation–hormesis and adaptive responses. YakugakuZasshi 126(10):827–831
60. Salbu B, Denbeigh J, Smith RW, Heier LS, Teien HC, Rosseland BO, Oughton D, Seymour CB, Mothersill C (2008) Environmentally relevant mixed exposures to radiation and heavy metals induce measurable stress responses in Atlantic salmon. Environ Sci Technol 42(9):3441–3446
61. Sawada S (2007) Cover-up of the effects of internal exposure by residual radiation from the atomic bombing of Hiroshima and Nagasaki. Med Conf Surviv 23(1):58–74
62. Schettino G, Folkard M, Michael BD, Prise KM (2005) Low-dose binary behaviour of bystander cell killing after microbeam irradiation of a single cell with focused c(k) x-rays. Radiat Res 163(3):332–336
63. Seymour CB, Mothersill C, Alper T (1986) High yields of lethal mutations in somatic mammalian cells that survive ionizing radiation. Int J Radiat Biol Relat Stud Phys Chem Med 50(1):167–179
64. Sigurdson AJ, Ron E (2004) Cosmic radiation exposure and cancer risk among flight crew. Cancer Invest 22(5):743–761
65. Speicher et al (2006) Multiplex-fluorescence in situ hybridisation for chromosome karyotyping. Nat Protoc 1:1172–1184
66. Stamato T, Weinstein R, Peters B, Hu J, Doherty B, Giaccia A (1987) Delayed mutation in Chinese hamster cells. Somat Cell Mol Genet 13(1):57–65
67. Timofeeff-Ressovky NW, Zimmer KG, Delbrück M (1935) Über die Natur der Genmutation und der Genstruktur, Nachrichten von der Gesellschaft der Wissenschaften zu Göttingen. Mathematische-Physikalische Klasse, Fachgruppe VI, Biologie 1(13):189–245
68. Watson GE, Lorimore SA, Clutton SM, Kadhim MA, Wright EG (1997) Genetic factors influencing alpha-particle-induced chromosomal instability. Int J Radiat Biol 71(5):497–503
69. Weber TJ, Siegel RW, Markillie LM, Chrisler WB, Lei XC, Colburn NH (2005) A paracrine signal mediates the cell transformation response to low dose gamma radiation in JB6 cells. Mol Carcinog 43(1):31–37
70. Wright EG (2007) Microenvironmental and genetic factors in haemopoietic radiation responses. Int J Radiat Biol 83(11–12):813–818
71. Wright EG, Coates PJ (2006) Untargeted effects of ionizing radiation: implications for radiation pathology. Mutat Res 597(1–2):119–132

Chapter 6
Role of Epigenetic Changes in Radiation-Induced Genome Instability

Slava Ilnytskyy, Jody Filkowski, and Olga Kovalchuk

Abstract Ionizing radiation (IR) is an important diagnostic and treatment modality, yet it is also a potent genotoxic agent that causes genome instability and carcinogenesis. While modern cancer radiation therapy has led to increased patient survival rates, the risk of radiation treatment-related complications is becoming a growing problem as radiation poses a threat to the exposed individuals and their progeny. Radiation-induced genome instability, which manifests as an elevated mutation rate (both delayed and non-targeted), chromosomal aberrations and changes in gene expression, has been well-documented in directly exposed cells and organisms. However, it has also been observed in distant, naïve, out-of-field, 'bystander' cells and their progeny. Enigmatically, this increased instability is even observed in the pre-conceptually exposed progeny of animals, including humans. The mechanisms by which these distal effects arise remain obscure and, recently, have been proposed to be epigenetic in nature.

Epigenetic alterations which comprise mitotically and meiotically heritable changes in gene expression that are not caused by changes in the primary DNA sequence, are increasingly being recognized for their roles in health and disease. Three major areas of epigenetics—DNA methylation, histone modifications and small RNA-mediated silencing, are known to have profound effects on controlling gene expression. Yet, the exact nature of the epigenetic changes and their precise roles in IR responses and IR-induced genome instability still need to be delineated. Here we will focus on the nature of epigenetic changes in directly exposed and bystander tissues. We will also discuss the emerging evidence that support the role of epigenetic deregulation in transgenerational effects.

Keywords Radiation • Bystander effects • Genome instability • Transgenerational effects • Epigenetics • DNA methylation • Histone modifications • Small RNAs

S. Ilnytskyy • J. Filkowski • O. Kovalchuk (✉)
Department of Biological Sciences, University of Lethbridge, Lethbridge,
AB T1K3M4, Canada
e-mail: olga.kovalchuk@uleth.ca

C.E. Mothersill et al. (eds.), *Radiobiology and Environmental Security*,
NATO Science for Peace and Security Series C: Environmental Security,
DOI 10.1007/978-94-007-1939-2_6, © Springer Science+Business Media B.V. 2012

6.1 Epigenetics – A Brief Historical Overview of a Science Reborn

The term 'epigenetics' (outside of genetics) was introduced by developmental biologist Conrad Hall Waddington well before the discovery the molecular structure of DNA itself. Waddington's model proposed that epigenetics describes how genes within a multicellular organism interact with other genes and their environment to yield a certain phenotype [107]. In later works by Robin Holliday, epigenetics was characterized as the cellular "mechanisms of temporal and spatial control of gene activity during development" [44].

After the discovery of DNA as the genetic material, the field of epigenetics was overshadowed by studies focused on the role of changes in DNA sequence between various normal and pathological conditions. It was not until the near completion of the human, and several other genome projects, that an interest in epigenetics was 'reborn.' By this time, many researchers had come to understand that genome sequences failed to fully explain the complexity of cellular processes, regulation or, most importantly, the fine-tuning of cell-organism-environment interactions.

One can not discount the importance of the genetic code: it is indispensible for life. Besides containing the coding information for RNAs and proteins, the DNA sequence also harbours a variety of gene expression guidelines within regulatory sequences such as promoters, enhancers and other elements. Nevertheless, it has become clear that regulatory elements may function differently under certain conditions or cell states without requiring any DNA sequence alterations. However, it has been noted that these changes are usually accompanied by chromatin modifications. Thus, it has become necessary to focus on the non-sequential characteristics of the genome: those aspects not detailed in the genetic code but, rather outside of it, the 'epi-gentics' so to speak. And indeed, in practice the term "epigenetics" has been adopted to refer to the information contained in chromatin rather than in the actual DNA sequence [11, 49].

6.2 What Is Epigenetics?

Modern science defines epigenetics as meiotically and mitotically stable alterations in gene expression that are not based on DNA sequence changes and involve processes that impact chromatin structure such as DNA methylation, histone modifications and genomic imprinting [11, 112, 113]. More recently, RNA-mediated silencing was proposed as another epigenetic phenomenon [10].

6.2.1 DNA Methylation

Cytosine DNA methylation was the first epigenetic alteration identified and has become the most widely studied [34, 35]. It is known to be associated with inactive chromatin state and in most cases with the repressed gene expression activity

[49, 55, 112] and, thus, is crucially important for the regulation of gene expression, silencing of parasitic sequences, X-chromosome inactivation and overall, for the normal development, cell proliferation and proper maintenance of genome stability of a given organism [49, 90].n mammals, DNA methylation occurs predominantly in the context of CG dinucleotides which are methylated 60–80% of the time [112]. This methylation occurs specifically at cytosine residues and requires the activity of DNA methyltransferase (DNMT) enzymes to form 5-methylcytosine. In mammals, three DNA methyltransferases (DNMT1, DNMT3a and DNMT3b) are primarily responsible for establishing and maintaining DNA methylation patterns at CpG sites [39, 89, 94]. DNMT1 is the major enzyme involved in maintaining the existing pattern of DNA methylation following DNA replication [63]. It localizes to the replication fork, where it can directly modify nascent DNA immediately after replication [39, 52, 112]. In contrast, Dnmt3a and Dnmt3b are *de novo* methyltransferases that target unmethylated and hemi-methylated sites in response to intra- and inter- cellular signals that are not yet fully elucidated [39, 81, 112]. Deregulation of these proteins is frequently recognized in diseased states and likely is involved in the altered methylation patterns that are a hallmark of numerous diseases [39, 52, 112].

Altered global DNA methylation patterns are a well-known characteristic of cancer cells [113]. Frequently, cancer cells are characterized by global genome hypomethylation with concurrent hypermethylation of selected CpG islands within gene promoters [8, 9, 49, 113]. The global loss of DNA methylation at CpG dinucleotides was the first epigenetic abnormality identified in cancer cells, and it has been linked to the activation of transposable elements, elevated chromosome breakage, aneuploidy, increased mutation rates and, thus, to the phenomenon of a global genomic instability [89, 91, 112, 113].

6.2.2 Histone Modifications

Changes in DNA methylation do not appear to be isolated, independent events. Such changes accompany global chromatin deregulation and, sometimes, histone modifications [49, 51]. The term 'histone code' is used to describe the plethora of potential covalent modifications that may occur on several of the core histones. These modifications, many of which are still not fully understood, allow for an enormous amount of flexibility and complexity of the regulatory responsibilities of chromatin and include (not exclusively) acetylation, methylation, ribosylation, ubiquitination, sumolation and phosphorylation. Although the full scope of their effects remains mostly unexplored, many of these modifications are known to alter the spatial arrangement of the DNA strand, thus, changing the accessibility for numerous DNA associating proteins. Frequently, these spatial rearrangements impact transcriptional regulation. Examples of such modifications include histone acetylation, methylation and phosphorylation [51, 113].

Acetylated histone tails lose their positive charge reducing their affinity for the negatively charged DNA and, thus, lead to a more relaxed chromatin packaging. The increasingly relaxed acetylated loci experience increased transcriptional activation, while histone deacetylation has an opposite effect—transcriptional repression [51]. Histone methylation is not so straight forward. Methylation can result in different transcriptional consequences depending upon the residue affected [18, 95]. For example, methylation of lysine 9 of histone H3 is associated with chromatin compaction and gene silencing, while methylation of lysines 4 or 27 of histone H3 results in chromatin relaxation and transcriptional activation. Additionally, histone residues can be either mono-, di- or tri-methylated, and again, lead to different transcriptional profiles (both between the degree of methylation and the recipient residues) [18, 43, 95, 113]. In a practical example, it has been shown that tumours undergo a massive loss of tri-methylation at lysine 20 of histone H4 [37, 105]. This loss occurs along with DNA hypomethylation and is linked to chromatin relaxation, aberrant expression and has been suggested to be a universal marker for malignant transformation [37, 96].

Although influencing transcription and expression is considered the paramount purpose of the epigenome, it also has been implicated in identifying DNA lesions, recruiting repair complexes and facilitating in the repair process. Phosphorylation, another important facet of the histone code, [43] is best studied in the modification of histone H2AX (a member of the H2A family). The phosphorylation of H2AX is crucially important for the repair of DNA strand breaks and the maintenance of genome stability [15]. As H2AX becomes phosphorylated at serine 139 (γH2AX) it accumulates in foci at the break sites, possibly as one of the earliest cellular responses to double strand breaks (DSBs) [82, 93, 97]. A direct correlation exists between H2AX phosphorylation and the number of radiation induced DSBs [16]. Again, it appears that this mechanism does not act exclusively as γH2AX recruits histone acetyl transferase (HATs) to relax the neighbouring chromatin.

6.2.3 Small RNAs – New Regulators of Gene Expression

Another mechanism of epigenetic control is mediated through small regulatory RNAs [10] with a particular interest in microRNAs (miRNAs). miRNAs are abundant, small, single-stranded, non-coding RNAs that regulate gene expression and are conserved across species [47, 99]. To control the translation of the target mRNAs, miRNAs associate with the RNA-induced silencing complex (RISC) proteins and bind to the 3′UTR of mRNAs, thus, serving as translational suppressors that regulate protein synthesis [46]. Regulatory miRNAs impact cellular differentiation, proliferation, apoptosis and, possibly, even predisposition to cancer [30, 47]. Aberrant levels of miRNAs have been reported in a variety of human cancers [108, 114]. Furthermore, it has been suggested that small RNAs may be involved in the regulation of chromatin packaging [10, 41].

Another class of short, non-coding RNAs are 29–30 nucleotides long and form complexes with Piwi proteins (piRNAs) [5, 6, 14, 21, 32]. This novel class of small RNA molecules, discovered in 2007, are expressed in the mammalian germline and have been linked to silencing of retrotransposons and other germline genetic elements [6, 14, 21].

6.2.4 Epigenetics and Environment

Epigenetic regulation appears to be vitally important for all organisms—especially eukaryotes, and during this decade it has become increasingly apparent that epigenetic changes modulate the molecular, cellular and organism responses to a changing environment [49, 52, 71, 111, 113]. Through interaction with their surrounding environment, living organisms are constantly exposed to a variety of stressors including physical, chemical, biological and social. Although these interactions may be positive and negative, all of them may potentially re-shape not only the genome, but also the epigenome. The body of work supporting this hypothesis continues to grow as more effort is applied to investigating the molecular players involved in communicating the signals and effecting the epigenetic changes. The full potential of epigenetics and the role it plays in maintaining a balanced physiology cannot be fully appreciated until all the mechanisms are fully researched and understood: a daunting, albeit, critically important task.

6.3 Direct Effects of Radiation Exposure

Besides diagnostic and therapeutic medical radiation exposures, which usually represent finite doses, most exposure comes from more ubiquitous, unobvious sources that tend to supply chronic levels of radiation. These include backround radiation, cosmic rays, radioactive waste, radon decay, nuclear tests and accidents (e.g. Chernobyl and other nuclear power plants). It is important to note that all living organisms are exposed to ionizing radiation (IR) on a day-to-day basis (both purposeful and unavoidable) yet, the underlying mechanisms of such exposure is not fully understood. On one hand, IR is a well-known cancer-inducing agent. The carcinogenic potential of IR was recognized soon after its discovery, when the first radiation-induced tumor was reported in 1902 [17, 64, 65]. Over the past decade, intense research efforts have been made to elucidate the cellular and molecular mechanisms of radiation-induced carcinogenesis in mammalian cells. On the other hand, radiation is one of the primary clinical methods used for detecting and fighting human malignancies: medical practices that are responsible for prolonging countless lives. It can be difficult to justify the use of a known mutagen and carcinogen in such clinical methods without fully understanding the proximal, distal, immediate and persistent effects that are caused by radiation exposure, which makes investigations into the direct and indirect effects of IR much more imperative.

IR can disrupt a variety of processes in exposed cells including gene expression, mitochondrial processes, cell cycle arrest and apoptotic cell death [2–4, 19, 33, 48, 50, 85, 92, 106]. Perhaps its most important characteristic is that IR is a potent DNA damaging agent capable of producing lesions such as cross linking, nucleotide base damage and single and double strand breaks [45, 64]. The accumulation of DNA damage caused by IR in conjunction with the disrupted cellular regulation processes can lead to carcinogenesis [64].

6.4 Epigenetic Changes in the Directly Exposed Tissue

Direct radiation exposure strongly influences epigenetic effectors. DNA damaging agents including ionizing radiation have been reported to affect DNA methylation patterns [53, 60, 104]. Acute exposures to low LET x-rays or γ-rays were noted to result in global hypomethylation [53, 104]. Surprisingly, this effect was not a generic response as it was later shown that the IR exposure leads to a profound dose-dependent and sex- and tissue- specific global DNA hypomethylation [56, 68, 83, 84, 86]. Furthermore, the changes in methylation status did not appear to occur indiscriminately throughout the genome, but rather may be focused to particular loci as it was noted that IR exposure also affects methylation of the promoter of the p16 tumour suppressor in a sex- and tissue-specific manner [60]. These changes in methylation patterns correlated with the radiation-induced alterations in the expression of DNA methyltransferases, especially *de novo* methyltransferases DNMT3a and DNMT3b suggesting that the changes are part of purposeful radiation-induced mechanism and not just another immediate, direct radiation-induced structural change [84, 86]. Most importantly, the radiation-induced global genome DNA hypomethylation appeared to be linked to genome instability in the exposed tissue [57, 83, 84, 86].

DNA methylation is closely connected with other components of chromatin structure and although much attention has been given to the radiation-induced changes in DNA methylation, histones have been largely overlooked. Among the histone modifications that change upon radiation exposure, phosphorylation of histone H2AX has been studied most intensively. Recent studies have also indicated that radiation-induced global loss of DNA methylation may correlate with the changes in histone methylation, specifically with the loss of histone H4 lysine tri-methylation [84].

6.5 Epigenetic Determinants of the Indirect Radiation Effects: Bystander Effect

Despite a significant body of evidence that points towards the epigenetic nature of radiation-induced bystander and transgenerational effects, until recently few studies addressed the exact nature of epigenetic changes related to the indirect radiation response. The pioneering work of Kaup and colleagues has shown that DNA methylation is important for the maintenance of radiation-induced bystander effects in cultured cells. Using cultured human keratinocytes, they demonstrated that the dysregulation of DNA methylation profiles in naïve cells exposed to media from irradiated cells persists for 20 passages. Over a similar period of culture under similar conditions, these cells also exhibited an increased and persistent level of chromosome and chromatid aberrations, reproductive cell death, apoptosis and other signs of genome instability [54].

Epigenetic changes were also shown to be important in whole-tissue- and whole-organism-based bystander effect models. The reconstituted 3D human tissue model offers an excellent alternative to cell cultures. The recent study by Sedelnikova and colleagues examined bystander effects in two reconstructed human 3D tissue models—bronchial epithelial and full-thickness skin. Following microbeam irradiation of cells located in a thin plane through the tissue, a variety of biological endpoints were analyzed in distal bystander cells (up to 2.5 mm away from the irradiated cell plane) as a function of post-exposure time (0 h to 7 days). In bystander cells, they detected a significant increase in the levels of phosphorylated H2AX; apoptosis (persistent); micronuclei formation; loss of nuclear DNA methylation; growth arrest (persistent) and an increasing number of senescent cells. Of a special interest is the observed loss of DNA methylation in bystander cells as it may be indicative of an epigenetic nature of bystander effect in 3D human tissue models [98].

Further insight into the role of epigenetic changes in the bystander effect comes from animal-based studies, where irradiation was shown to induce DNA damage and modulate the epigenetic effectors in distant bystander tissues. The Kovalchuk and Engelward laboratories pioneered *in vivo* studies on the role of epigenetic changes in radiation-induced bystander effects. To analyze *in vivo* bystander effects, they developed a mouse model whereby half of an animal body was exposed to radiation, while the other half was protected by a medical grade shield [57]. This model was used to monitor the induction and repair of DNA strand breaks in the unexposed cutaneous tissue. In addition to this well-established endpoint, the authors also explored the possibility of epigenetic mechanisms (*i.e.* DNA methylation and alterations in DNA methyltransferases and methyl-binding proteins) in the generation and/or maintenance of a radiation-induced bystander effect in the unexposed cutaneous tissue. They have shown that radiation exposure to one half of the body leads to elevated levels of DNA strand breaks, and altered levels of key proteins that modulate methylation patterns and silencing in the bystander half of the body at least 0.7 cm from the irradiated tissue. These are some of the first data to clearly demonstrate that the epigenetically regulated bystander effects occur *in vivo* in distant tissues. Importantly, these epigenetic changes in bystander tissues are not due to the insufficient shielding or radiation scattering [57].

To be relevant for carcinogenesis, the epigenetic manifestations of bystander effects should accumulate and/or persist over a long period of time. To investigate the possibility that the localized X-ray irradiation induces persistent epigenetically modulated bystander effects in distant tissues, Koturbash and colleagues monitored the occurrence of epigenetic changes (i.e. DNA methylation, histone methylation and miRNA expression) in spleen tissue 7 months after the localized cranial irradiation. This analysis has revealed that the localized cranial radiation exposure leads to the decreased levels of global DNA methylation. It also alters the levels of key proteins that modulate methylation patterns and silencing (i.e. *de novo* methyltransferase DNMT3a and methyl-binding protein MeCP2) and contributes to the reactivation of the LINE1 retrotransposon in the bystander spleen, located at least 16 cm from the irradiation site. Importantly, it is the first evidence that down regulation of

DNMT3a and MeCP2 is probably triggered and maintained by higher activity of a small regulatory RNA, microRNA *miR-194*. These experiments have demonstrated that *miR-194* is up-regulated in the bystander rat spleen. These data have also clearly demonstrated that the bystander effect occurs *in vivo* in distant tissue, persists over a long period of time, and are epigenetically regulated [58].

The observed altered expression of *miR-194* in the bystander rat spleen was quite intriguing and promoted further studies of microRNAome changes in bystander tissues. Using the microRNA microarray platform, microRNAome patterns have been profiled in skin and spleen tissues of mice subjected to sham treatment, whole-body or head exposure. The radiation exposure led to significant alterations in the microRNA expression profiles in bystander skin and spleen [59] The pronounced microRNAome alterations can also be seen in the bystander tissues using the 3D model [61]. These data suggest that the microRNA expression changes really occur in bystander tissues. Their exact function in the bystander effect still has to be delineated. Furthermore, due to their small size and high stability, microRNAs may be plausible candidates for the bystander signal.

6.6 Transgenerational Effects

Targeted effects of radiation are caused by direct interaction of ionizing particles with genetic material. Direct germline mutations are transmitted to subsequent generations in according to the laws of classical genetics and may lead to a whole spectrum of deleterious effects including, but not limited to, dominant lethality and congenital malformations [38, 73].

A growing body of data shows that heritable effects of radiation exposure are not limited to those caused by targeted mutations. A number of studies have shown *de novo* mutations in the non-exposed progeny of an irradiated parent that resemble radiation induced genomic instability [13, 69, 70, 100–102]. Early work that focused on the study of radiation-induced, dominant, lethal mutations in mice, unexpectedly, showed an excess of intrauterine deaths of fetuses fathered by pre-conceptually irradiated (at spermatogonial stage) mice. This effect continued to the F2 generation [70]. Later, the discovery of hypermutable mini satellite loci (short tandem repeats in human and mouse genome) provided an invaluable biomarker for the study of radiation-induced germ line mutations [23]. Pioneering works by Dubrova et al. showed increase of germ line mutation rates in the progeny of fathers exposed to radiation as a result of the Chernobyl accident [24, 25]. The same methods were used to detect the increased frequency of germ line mutations in multi-generation families living close to the Semipalatinsk nuclear test site and among the Techa River population [1, 28, 29].

Animal studies conducted by the same group showed that germ line mutations at mini satellite loci were induced at a frequency that was an order of magnitude higher than is expected in the case of direct mutations [26, 27]. Importantly, the increased rate of mini satellite mutations was still observed in F2 progeny of irradiated male mice and this trait was exhibited by most of the offspring in violation of Mendelian laws [7]. Induction of transgenerational genomic instability was influenced by the dose, type of irradiation and stage of germ cell development [26, 31, 75].

Transgenerational genomic instability may be linked to transgenerational radiation induced carcinogenesis that manifests itself as an increase of cancer risk in the unexposed progeny of irradiated parents. Increased predisposition to tumour formation in the progeny of irradiated mice was first demonstrated by Nomura [76–78, 80]. This effect, although slightly carcinogenic, was transmitted to subsequent generations in a non-mendelian manner and was greatly enhanced upon application of secondary carcinogens [109, 110]. A number of experiments were designed to confirm and expand initial findings by Nomura. Daher et al. [20] reported marginally significant increase of leukemia incidence in the progeny of X-ray irradiated N5 mice, as well as earlier onset in leukemia development in the progeny of mice injected with tritium. Similarly, modulation in leukemia/lymphoma development was observed after in BDF1 and CBA mice after pre-conceptual treatment with plutonium 239, followed by secondary exposure to methylnitrosourea [67]. A lifetime study that included 4,279 mice showed that pre-conceptual treatment to a high, acute dose of X-Rays leads to increased hematopoietic malignancy in the female and bronchioloalveolar adenocarcinomas in male progeny. On the other hand the same study failed to show increased tumour induction by secondary exposure to urethane [72]. The exact manifestations of radiation induced transgenerational carcinogenesis were shown to depend on the dose, type of irradiation, stage of germ cell development and genetic background of the model organism [22, 66, 79].

To date, the exact molecular mechanism of radiation induced transgenerational effects remains unclear: the high frequency of induction and the fact that these effects are present in most, if not all, of the offspring are incompatible with the involvement of genetic mutations. Radiation induced free radicals could be considered, however, their short lived nature and negligible cytoplasmic component in sperm make them an unlikely candidate. Manifestations and the mode of transmission of this effect suggests that it has to be fixed across multiple loci in most if not all of the germ cells produced by affected gonad. It is tempting to hypothesize that these changes are memorialized in the germ cell through epigenetic mechanism. DNA methylation marks at imprinted loci and certain repeat elements were shown to be resistant to epigenetic reprogramming events in developing embryos [42, 62]. A series of recent studies described the ability of sperm-bound small RNA to trigger developmental changes in the embryo that could be transmitted to subsequent generations in non-mendelian manner [40, 87, 88]. No data is available on the ability of histone marks to trigger similar responses, however, histones were shown to be present in sperm covering genomic areas responsible for embryonic development [12].

A preliminary study conducted by Kovalchuk's group found significant loss of cytosine DNA methylation in the thymus of the progeny after paternal exposure to X-rays. The loss of DNA methylation was paralleled by a significant decrease in the levels of maintenance (DNMT1) and *de novo* methyltransferases DNMT3a, DNMT3b and methyl-CpG-binding protein (MECP2). They also noted significant accumulation of DNA strand breaks in the thymus [57]. Localized cranial irradiation of male rats leads to accumulation of DNA lesions and loss of global DNA methylation in mature sperm [103]. Loss of DNA methylation along with dysregulation of DNA methylation enzymes was also found in the progeny of cranially exposed male rats [103]. A microarray expression profiling revealed altered miRNA levels in the thymus of adult male mice pre-conceptually exposed to 2.5 Gy of X-Ray [36]. Changes in miRNAome were paralleled by dysregulation in miRNA processing machinery and decreased expression of LSH [36]—a chromatin remodelling factor implicated in the regulation of DNA methylation at repeatable elements [74]. DNA methylation profiling of repeats in the same samples revealed loss of CpG methylation at LINE1 and SINEB2 retrotransposable elements [36].

Cumulative evidence points to the epigenetic origins of radiation-induced transgenerational genomic instability and cancer predisposition, however, their exact mechanism remains unclear. Future research in this area has to rely on the use of microarray technology, next generation sequencing and bioinformatic approaches in order to extract functionally relevant, causal changes influencing epigenetic reprogramming and genomic stability across generations. Such research has both practical and fundamental value, as it may offer an understating of how genotoxic factors contribute to complex disease by altering our epigenome across generations.

References

1. Akleev AV, Dubrova Iu E et al (2007) The effects of chronic radiation exposure on the frequency of mutations at minisatellite DNA loci in residents of the Techa Riverside Villages. Radiats Biol Radioecol 47(5):558–566
2. Amundson SA, Fornace AJ Jr (2003) Monitoring human radiation exposure by gene expression profiling: possibilities and pitfalls. Health Phys 85(1):36–42
3. Amundson SA, Lee RA et al (2003) Differential responses of stress genes to low dose-rate gamma irradiation. Mol Cancer Res 1(6):445–452
4. Andreev SG, Eidelman YA et al (2006) Mechanistic modelling of genetic and epigenetic events in radiation carcinogenesis. Radiat Prot Dosimetry 122(1–4):335–339
5. Aravin AA, Sachidanandam R et al (2007) Developmentally regulated piRNA clusters implicate MILI in transposon control. Science 316(5825):744–747
6. Aravin AA, Sachidanandam R et al (2008) A piRNA pathway primed by individual transposons is linked to de novo DNA methylation in mice. Mol Cell 31(6):785–799
7. Barber R, Plumb MA et al (2002) Elevated mutation rates in the germ line of first- and second-generation offspring of irradiated male mice. Proc Natl Acad Sci USA 99(10):6877–6882
8. Baylin SB, Chen WY (2005) Aberrant gene silencing in tumor progression: implications for control of cancer. Cold Spring Harb Symp Quant Biol 70:427–433
9. Baylin SB, Ohm JE (2006) Epigenetic gene silencing in cancer–a mechanism for early oncogenic pathway addiction? Nat Rev Cancer 6(2):107–116
10. Bernstein E, Allis CD (2005) RNA meets chromatin. Genes Dev 19(14):1635–1655
11. Bird A (2007) Perceptions of epigenetics. Nature 447(7143):396–398
12. Brykczynska U, Hisano M et al (2010) Repressive and active histone methylation mark distinct promoters in human and mouse spermatozoa. Nat Struct Mol Biol 17(6):679–687
13. Carls N, Schiestl RH (1999) Effect of ionizing radiation on transgenerational appearance of p(un) reversions in mice. Carcinogenesis 20(12):2351–2354

14. Carmell MA, Girard A et al (2007) MIWI2 is essential for spermatogenesis and repression of transposons in the mouse male germline. Dev Cell 12(4):503–514
15. Celeste A, Difilippantonio S et al (2003) H2AX haploinsufficiency modifies genomic stability and tumor susceptibility. Cell 114(3):371–383
16. Celeste A, Fernandez-Capetillo O et al (2003) Histone H2AX phosphorylation is dispensable for the initial recognition of DNA breaks. Nat Cell Biol 5(7):675–679
17. Chauveinc L, Giraud P et al (1998) Radiotherapy-induced solid tumors: review of the literature and risk assessment. Cancer Radiother 2(1):12–18
18. Cheung P, Lau P (2005) Epigenetic regulation by histone methylation and histone variants. Mol Endocrinol 19(3):563–573
19. Criswell T, Leskov K et al (2003) Transcription factors activated in mammalian cells after clinically relevant doses of ionizing radiation. Oncogene 22(37):5813–5827
20. Daher A, Varin M et al (1998) Effect of pre-conceptional external or internal irradiation of N5 male mice and the risk of leukemia in their offspring. Carcinogenesis 19(9):1553–1558
21. Das PP, Bagijn MP et al (2008) Piwi and piRNAs act upstream of an endogenous siRNA pathway to suppress Tc3 transposon mobility in the Caenorhabditis elegans germline. Mol Cell 31(1):79–90
22. Draper GJ (1989) General overview of studies of multigeneration carcinogenesis in man, particularly in relation to exposure to chemicals. IARC Sci Publ 96:275–288
23. Dubrova YE, Jeffreys AJ et al (1993) Mouse minisatellite mutations induced by ionizing radiation. Nat Genet 5(1):92–94
24. Dubrova YE, Nesterov VN et al (1996) Human minisatellite mutation rate after the Chernobyl accident. Nature 380(6576):683–686
25. Dubrova YE, Nesterov VN et al (1997) Further evidence for elevated human minisatellite mutation rate in Belarus eight years after the Chernobyl accident. Mutat Res 381(2):267–278
26. Dubrova YE, Plumb M et al (1998) Stage specificity, dose response, and doubling dose for mouse minisatellite germ-line mutation induced by acute radiation. Proc Natl Acad Sci USA 95(11):6251–6255
27. Dubrova YE, Plumb M et al (2000) Transgenerational mutation by radiation. Nature 405(6782):37
28. Dubrova YE, Bersimbaev RI et al (2002) Nuclear weapons tests and human germline mutation rate. Science 295(5557):1037
29. Dubrova YE, Ploshchanskaya OG et al (2006) Minisatellite germline mutation rate in the Techa River population. Mutat Res 602(1–2):74–82
30. Fabbri M, Ivan M et al (2007) Regulatory mechanisms of microRNAs involvement in cancer. Expert Opin Biol Ther 7(7):1009–1019
31. Fan YJ, Wang Z et al (1995) Dose-response of a radiation induction of a germline mutation at a hypervariable mouse minisatellite locus. Int J Radiat Biol 68(2):177–183
32. Farazi TA, Juranek SA et al (2008) The growing catalog of small RNAs and their association with distinct Argonaute/Piwi family members. Development 135(7):1201–1214
33. Fei P, El-Deiry WS (2003) P53 and radiation responses. Oncogene 22(37):5774–5783
34. Feinberg AP (2004) The epigenetics of cancer etiology. Semin Cancer Biol 14(6):427–432
35. Feinberg AP, Vogelstein B (1983) Hypomethylation distinguishes genes of some human cancers from their normal counterparts. Nature 301(5895):89–92
36. Filkowski JN, Ilnytskyy Y et al (2010) Hypomethylation and genome instability in the germline of exposed parents and their progeny is associated with altered miRNA expression. Carcinogenesis 31(6):1110–1115
37. Fraga MF, Ballestar E et al (2005) Loss of acetylation at Lys16 and trimethylation at Lys20 of histone H4 is a common hallmark of human cancer. Nat Genet 37(4):391–400
38. Goldman M (1982) Ionizing radiation and its risks. West J Med 137(6):540–547
39. Goll MG, Bestor TH (2005) Eukaryotic cytosine methyltransferases. Annu Rev Biochem 74:481–514

40. Grandjean V, Gounon P et al (2009) The miR-124-Sox9 paramutation: RNA-mediated epigenetic control of embryonic and adult growth. Development 136(21):3647–3655
41. Grewal SI, Moazed D (2003) Heterochromatin and epigenetic control of gene expression. Science 301(5634):798–802
42. Hajkova P, Erhardt S et al (2002) Epigenetic reprogramming in mouse primordial germ cells. Mech Dev 117(1–2):15–23
43. He S, Dunn KL et al (2008) Chromatin organization and nuclear microenvironments in cancer cells. J Cell Biochem 104(6):2004–2015
44. Holliday R (1990) DNA methylation and epigenetic inheritance. Philos Trans R Soc Lond B Biol Sci 326(1235):329–338
45. Huang L, Snyder AR, Morgan WF (2003) Radiation-induced genomic instability and its implications for radiation carcinogenesis. Oncogene 22:5848–5854
46. Hutvagner G, Zamore PD (2002) A microRNA in a multiple-turnover RNAi enzyme complex. Science 297(5589):2056–2060
47. Hwang HW, Mendell JT (2006) MicroRNAs in cell proliferation, cell death, and tumorigenesis. Br J Cancer 94(6):776–780
48. Iliakis G, Wang Y et al (2003) DNA damage checkpoint control in cells exposed to ionizing radiation. Oncogene 22(37):5834–5847
49. Jaenisch R, Bird A (2003) Epigenetic regulation of gene expression: how the genome integrates intrinsic and environmental signals. Nat Genet 33(Suppl):245–254
50. Jeggo P, Lobrich M (2006) Radiation-induced DNA damage responses. Radiat Prot Dosimetry 122(1–4):124–127
51. Jenuwein T, Allis CD (2001) Translating the histone code. Science 293(5532):1074–1080
52. Jirtle RL, Skinner MK (2007) Environmental epigenomics and disease susceptibility. Nat Rev Genet 8(4):253–262
53. Kalinich JF, Catravas GN et al (1989) The effect of gamma radiation on DNA methylation. Radiat Res 117(2):185–197
54. Kaup S, Grandjean V et al (2006) Radiation-induced genomic instability is associated with DNA methylation changes in cultured human keratinocytes. Mutat Res 597(1–2):87–97
55. Klose RJ, Bird AP (2006) Genomic DNA methylation: the mark and its mediators. Trends Biochem Sci 31(2):89–97
56. Koturbash I, Pogribny I et al (2005) Stable loss of global DNA methylation in the radiation-target tissue–a possible mechanism contributing to radiation carcinogenesis? Biochem Biophys Res Commun 337(2):526–533
57. Koturbash I, Rugo RE et al (2006) Irradiation induces DNA damage and modulates epigenetic effectors in distant bystander tissue in vivo. Oncogene 25(31):4267–4275
58. Koturbash I, Boyko A et al (2007) Role of epigenetic effectors in maintenance of the long-term persistent bystander effect in spleen in vivo. Carcinogenesis 28(8):1831–1838
59. Koturbash I, Zemp FJ et al (2008) Sex-specific microRNAome deregulation in the shielded bystander spleen of cranially exposed mice. Cell Cycle 7(11):1658–1667
60. Kovalchuk O, Burke P et al (2004) Methylation changes in muscle and liver tissues of male and female mice exposed to acute and chronic low-dose X-ray-irradiation. Mutat Res 548(1–2):75–84
61. Kovalchuk O, Zemp FJ et al (2010) microRNAome changes in bystander three-dimensional human tissue models suggest priming of apoptotic pathways. Carcinogenesis 31(10):1882–1888
62. Lane N, Dean W et al (2003) Resistance of IAPs to methylation reprogramming may provide a mechanism for epigenetic inheritance in the mouse. Genesis 35(2):88–93
63. Liang G, Chan MF et al (2002) Cooperativity between DNA methyltransferases in the maintenance methylation of repetitive elements. Mol Cell Biol 22(2):480–491
64. Little JB (2000) Radiation carcinogenesis. Carcinogenesis 21(3):397–404
65. Little MP (2003) Risks associated with ionizing radiation. Br Med Bull 68:259–275
66. Lord BI (1999) Transgenerational susceptibility to leukaemia induction resulting from preconception, paternal irradiation. Int J Radiat Biol 75(7):801–810
67. Lord BI, Woolford LB et al (1998) Tumour induction by methyl-nitroso-urea following preconceptional paternal contamination with plutonium-239. Br J Cancer 78(3):301–311

68. Loree J, Koturbash I et al (2006) Radiation-induced molecular changes in rat mammary tissue: possible implications for radiation-induced carcinogenesis. Int J Radiat Biol 82(11):805–815
69. Luke GA, Riches AC et al (1997) Genomic instability in haematopoietic cells of F1 generation mice of irradiated male parents. Mutagenesis 12(3):147–152
70. Luning KG, Frolen H et al (1976) Genetic effects of 239Pu salt injections in male mice. Mutat Res 34(3):539–542
71. Minamoto T, Mai M et al (1999) Environmental factors as regulators and effectors of multistep carcinogenesis. Carcinogenesis 20(4):519–527
72. Mohr U, Dasenbrock C et al (1999) Possible carcinogenic effects of X-rays in a transgenerational study with CBA mice. Carcinogenesis 20(2):325–332
73. Mole RH (1979) Radiation effects on pre-natal development and their radiological significance. Br J Radiol 52(614):89–101
74. Muegge K (2005) Lsh, a guardian of heterochromatin at repeat elements. Biochem Cell Biol 83(4):548–554
75. Niwa O, Fan YJ et al (1996) Induction of a germline mutation at a hypervariable mouse minisatellite locus by 252Cf radiation. J Radiat Res (Tokyo) 37(3):217–224
76. Nomura T (1982) Parental exposure to x rays and chemicals induces heritable tumours and anomalies in mice. Nature 296(5857):575–577
77. Nomura T (1983) X-ray-induced germ-line mutation leading to tumors. Its manifestation in mice given urethane post-natally. Mutat Res 121(1):59–65
78. Nomura T (1989) Role of radiation-induced mutations in multigeneration carcinogenesis. IARC Sci Publ 96:375–387
79. Nomura T (2006) Transgenerational effects of radiation and chemicals in mice and humans. J Radiat Res (Tokyo) 47(Suppl B):B83–B97
80. Nomura T, Nakajima H et al (2004) Transgenerational transmission of radiation- and chemically induced tumors and congenital anomalies in mice: studies of their possible relationship to induced chromosomal and molecular changes. Cytogenet Genome Res 104(1–4):252–260
81. Okano M, Bell DW et al (1999) DNA methyltransferases Dnmt3a and Dnmt3b are essential for de novo methylation and mammalian development. Cell 99(3):247–257
82. Pilch DR, Sedelnikova OA et al (2003) Characteristics of gamma-H2AX foci at DNA double-strand breaks sites. Biochem Cell Biol 81(3):123–129
83. Pogribny I, Raiche J et al (2004) Dose-dependence, sex- and tissue-specificity, and persistence of radiation-induced genomic DNA methylation changes. Biochem Biophys Res Commun 320(4):1253–1261
84. Pogribny I, Koturbash I et al (2005) Fractionated low-dose radiation exposure leads to accumulation of DNA damage and profound alterations in DNA and histone methylation in the murine thymus. Mol Cancer Res 3(10):553–561
85. Powell SN, Kachnic LA (2003) Roles of BRCA1 and BRCA2 in homologous recombination, DNA replication fidelity and the cellular response to ionizing radiation. Oncogene 22(37):5784–5791
86. Raiche J, Rodriguez-Juarez R et al (2004) Sex- and tissue-specific expression of maintenance and de novo DNA methyltransferases upon low dose X-irradiation in mice. Biochem Biophys Res Commun 325(1):39–47
87. Rassoulzadegan M, Grandjean V et al (2006) RNA-mediated non-mendelian inheritance of an epigenetic change in the mouse. Nature 441(7092):469–474
88. Rassoulzadegan M, Grandjean V et al (2007) Inheritance of an epigenetic change in the mouse: a new role for RNA. Biochem Soc Trans 35(Pt 3):623–625
89. Robertson KD (2001) DNA methylation, methyltransferases, and cancer. Oncogene 20(24):3139–3155
90. Robertson KD (2002) DNA methylation and chromatin–unraveling the tangled web. Oncogene 21(35):5361–5379
91. Robertson KD, Wolffe AP (2000) DNA methylation in health and disease. Nat Rev Genet 1(1):11–19

92. Rodemann HP, Blaese MA (2007) Responses of normal cells to ionizing radiation. Semin Radiat Oncol 17(2):81–88
93. Rogakou EP, Pilch DR et al (1998) DNA double-stranded breaks induce histone H2AX phosphorylation on serine 139. J Biol Chem 273(10):5858–5868
94. Rountree MR, Bachman KE et al (2001) DNA methylation, chromatin inheritance, and cancer. Oncogene 20(24):3156–3165
95. Saha A, Wittmeyer J et al (2006) Chromatin remodelling: the industrial revolution of DNA around histones. Nat Rev Mol Cell Biol 7(6):437–447
96. Sanders SL, Portoso M et al (2004) Methylation of histone H4 lysine 20 controls recruitment of Crb2 to sites of DNA damage. Cell 119(5):603–614
97. Sedelnikova OA, Pilch DR et al (2003) Histone H2AX in DNA damage and repair. Cancer Biol Ther 2(3):233–235
98. Sedelnikova OA, Nakamura A et al (2007) DNA double-strand breaks form in bystander cells after microbeam irradiation of three-dimensional human tissue models. Cancer Res 67(9):4295–4302
99. Sevignani C, Calin GA et al (2006) Mammalian microRNAs: a small world for fine-tuning gene expression. Mamm Genome 17(3):189–202
100. Shiraishi K, Shimura T et al (2002) Persistent induction of somatic reversions of the pink-eyed unstable mutation in F1 mice born to fathers irradiated at the spermatozoa stage. Radiat Res 157(6):661–667
101. Slovinska L, Elbertova A et al (2004) Transmission of genome damage from irradiated male rats to their progeny. Mutat Res 559(1–2):29–37
102. Streffer C (2006) Transgenerational transmission of radiation damage: genomic instability and congenital malformation. J Radiat Res (Tokyo) 47(Suppl B):B19–B24
103. Tamminga J, Koturbash I et al (2008) Paternal cranial irradiation induces distant bystander DNA damage in the germline and leads to epigenetic alterations in the offspring. Cell Cycle 7(9):1238–1245
104. Tawa R, Kimura Y et al (1998) Effects of X-ray irradiation on genomic DNA methylation levels in mouse tissues. J Radiat Res (Tokyo) 39(4):271–278
105. Tryndyak VP, Kovalchuk O et al (2006) Loss of DNA methylation and histone H4 lysine 20 trimethylation in human breast cancer cells is associated with aberrant expression of DNA methyltransferase 1, Suv4-20 h2 histone methyltransferase and methyl-binding proteins. Cancer Biol Ther 5(1):65–70
106. Valerie K, Yacoub A et al (2007) Radiation-induced cell signaling: inside-out and outside-in. Mol Cancer Ther 6(3):789–801
107. Van Speybroeck L (2002) From Epigenesis to Epigenetics. Ann N Y Acad Sci 981(1):61–81
108. Volinia S, Calin GA et al (2006) A microRNA expression signature of human solid tumors defines cancer gene targets. Proc Natl Acad Sci USA 103(7):2257–2261
109. Vorobtsova IE, Kitaev EM (1988) Urethane-induced lung adenomas in the first-generation progeny of irradiated male mice. Carcinogenesis 9(11):1931–1934
110. Vorobtsova IE, Aliyakparova LM et al (1993) Promotion of skin tumors by 12-O-tetradecanoylphorbol-13-acetate in two generations of descendants of male mice exposed to X-ray irradiation. Mutat Res 287(2):207–216
111. Wade PA, Archer TK (2006) Epigenetics: environmental instructions for the genome. Environ Health Perspect 114(3):A140–A141
112. Weber M, Schubeler D (2007) Genomic patterns of DNA methylation: targets and function of an epigenetic mark. Curr Opin Cell Biol 19(3):273–280
113. Weidman JR, Dolinoy DC et al (2007) Cancer susceptibility: epigenetic manifestation of environmental exposures. Cancer J 13(1):9–16
114. Wiemer EA (2007) The role of microRNAs in cancer: no small matter. Eur J Cancer 43(10):1529–1544

Chapter 7
Radiation and Terror

Colin B. Seymour and Carmel E. Mothersill

The best description of terrorism I have encountered is in this poem by Yeats, an Irish writer

The second coming

Turning and turning in the widening gyre
The falcon cannot hear the falconer;
Things fall apart; the centre cannot hold;
Mere anarchy is loosed upon the world,
The blood – dimmed tide is loosed, and everywhere
The ceremony of innocence is drowned;
The best lack all conviction, while the worst
Are full of passionate intensity.

-William Butler Yeats

This paper is presented to examine certain radiological aspects of terrorism, and to demonstrate that the perception of any event, and the consequent reaction to that perception, is more important than the reality of the event. Radiation, as a terrorist weapon, has been considered for a number of years. A report from 1969 (classified for 15 years) examined radiation sickness or death caused by surreptitious administration of ionizing radiation to an individual [1]. It was initiated because of a suggestion that Alexander Dubcek, Czeckoslovak Communist Party Leader, had radiation sickness due to radiation exposure or consumption of radioisotopes during his August 1968 captivity in Moscow.

This was not confirmed or denied by the report, which concluded that surreptitious administration of a lethal or sublethal dose to an individual was technically possible. It would however be technically difficult to achieve an immediately lethal dose, or for the subsequent illness to be undiagnosable, and at sub lethal doses the

C.B. Seymour • C.E. Mothersill (✉)
Department of Medical Physics and Applied Radiation Sciences,
McMaster University, Hamilton L8S 4K1, ON, Canada
e-mail: mothers@mcmaster.ca

C.E. Mothersill et al. (eds.), *Radiobiology and Environmental Security*,
NATO Science for Peace and Security Series C: Environmental Security,
DOI 10.1007/978-94-007-1939-2_7, © Springer Science+Business Media B.V. 2012

outcome would be uncertain. The final conclusion was "In view of the existence of other tried and true methods of political assassination, the use of ionizing radiation for this purpose would seem unnecessarily cumbersome unless the enemy were to expect some compelling advantage peculiar to the administration of radiation". A striking example of this was the recent poisoning of the Russian security agent Litvinenko with Polonium 210 which served as a graphic and very public demonstration of the misfortune that could befall dissenters.

The object of terrorism is to strike terror into a general population, but the motives for terrorism may vary considerably. Simplistic terrorism, in historical terms, was primarily intended to cause regime change. Often the terrorism was an expression of alienation from the state apparatus of power, with disenfranchised intellectuals and middle classes giving some tacit support.

The success of terrorism in it's primary objective, that of inducing a state of terror, means that it is useful to both revolutionaries, who wish to change the established order, and those who wish to consolidate power and enable more draconian powers for the existing government. There is also the possibility of a "third player" who may wish to manipulate events for financial or other reasons that may not be immediately obvious.

In terms of striking terror into a general population, a "dirty bomb" that ensures widespread radiological contamination is obviously the weapon of choice, and is aided by the fear of radiation that pervades most of the population. The bomb would have minor health effects, but would cause major panic, disruption and chaos. The panic caused would depend on the objectivity or sensationalism of the reporting, and the level of trust in the population of the government. The style of reporting would also to some extent depend on the trust of the media of the government, and so the argument could be made that the effect of terrorism in this context could depend entirely on the trust of the population in their government and institutions. There is no doubt that trust in government has eroded considerably over recent years. Consider the following quote from Lord Denning, a senior English judge.

Before his Lordship's court were the six men who had been convicted of the Birmingham pub bombings. They alleged that the West Midlands police had beaten them up, and were suing for damages. Denning realized that accepting the police had beaten them up meant accepting that the police had beaten false confessions out of them, which meant accepting that the police had framed innocent men for one of the worst IRA atrocities on the British mainland. 'This,' said Denning as he dismissed the case, 'is such an appalling vista that every sensible person in the land would say, "It cannot be right these actions should go any further."'

This amount of trust in the police or security forces now seems quaint and anachronistic. In modern times it is almost accepted that security forces will disregard the law, and abuse it when it suits their purposes.

It is also apparent that the extent of control over the media is important, and it is notable that most governments try to control at least certain aspects of communication, often under the cloak of "security." Control over the opinions of the media is important both prior to and after the terrorist event, as it allows a pre-determined response. An example of this would be, "If a terrorist event occurs we will blame and retaliate against A even if we know A did not instigate or was involved in the terrorist event." The government can use the terrorist event to further its own policy objectives. Terrorism without objectives is anarchy. That is not to argue that the average person classified as a terrorist need necessarily be aware of the objectives, any more than a conventional soldier need know the plan of his generals. It does mean that in order to form a reasoned response the objectives have to be identified. This may change the question from "Who did this?" to "Who benefits from this?"

Another issue that arises in the context of a radiological dispersion device is that it is clearly designed to panic a population. If the population is panicked it becomes more vulnerable to secondary attacks, and dealing with the secondary attacks become more problematic. It also becomes more vulnerable to an authoritarian style government. It would be a challenge for any government not to feel itself pushed towards a more authoritarian response, which may actually be a response that was anticipated and hoped for. Perhaps the only logical response to a terrorist incidence is to stabilize the situation (calm, prevent further casualties) and then assess the situation in depth.

In terms of public policy it might be prudent to legislate in terms of hormetic models rather than linear (no threshold) models. The absolute certainty that any dose of X is harmful is not only scientifically wrong, it is unnecessarily alarmist. Everybody is aware that food is good, unless too much is eaten in which case it becomes bad. By not encouraging the same views for most substances, fear is encouraged because "any dose of radiation is harmful." As we become more regulated as societies, fear becomes a more pervasive weapon for governments to exercise control over the population. It could be argued that over-regulation itself leads to a fearful society, willing to take only sanctioned risks. If the society is fearful, any attacks designed to create fear will be more effective. It is ironic that in trying to control the population more closely the government both emulates and makes terrorism more effective. This quote from Neil Addison, Director of the Thomas More Legal Centre, eloquently supports this conclusion;

> Whether it is fox hunting, smoking, adoption agencies or microchips in rubbish bins, we are a society that is increasingly intolerant, repressive, regulated and untrusting and, in consequence, we have officials who are dictatorial, interfering and untrustworthy.

In this climate terrorism becomes more effective as a weapon.

Reference

1. Radiation Sickness or Death Caused by Surreptitious Administration of Ionizing Radiation to an Individual. Report No. 4 of the Molecular Biology Working Group to The Biomedical Intelligence Subcommittee of The Scientific Intelligence Committee of USIB 27 Aug 1969.

Chapter 8
Bystander Effects and Adaptive Responses Modulate In Vitro and In Vivo Biological Responses to Low Dose Ionizing Radiation

Jie Zhang, Manuela Buonanno, Geraldine Gonon, Min Li, Mariann Galdass, Grace Shim, Sonia M. De Toledo, and Edouard I. Azzam

Abstract We have utilized cellular and molecular approaches to characterize biological effects that are induced in normal mammalian cells and tissues exposed to low doses/low fluences of ionizing radiations that differ in their quality (i.e. linear energy transfer; LET). In human cells exposed to particulate radiations with high, but not low, LET character, the induced stressful effects were not only confined to the cells that have been directly targeted by the radiation, but involved a number of non-targeted and delayed effects. Chromosomal damage and oxidative changes in proteins and lipids were detected in cells exposed to alpha and high charge and high energy (HZE) particles and in their neighboring bystanders. Signaling events mediated via inflammatory cytokines and/or intercellular channels that comprise gap junctions were critical for the expression of the induced non-targeted effects. With relevance to health risks, the stressful changes in bystander cells were propagated to their progeny. In contrast, induced DNA repair and antioxidant defense mechanisms often attenuated the basal level of DNA damage and oxidative stress to below the spontaneous rate in tissues of animals and in cultured rodent and human cells exposed to low dose/low dose-rate γ rays, a low LET radiation. Together, our data suggest that low dose radiation-induced signaling events act to alter the linearity of the dose-response relation that is predicted by biophysical arguments. They show that the nature of the altered responses strongly depend on radiation quality.

Keywords Low dose ionizing radiation • Bystander effect • Adaptive response • Health risks • Linear energy transfer

---

J. Zhang (✉) • M. Buonanno • G. Gonon • M. Li • M. Galdass • G. Shim • S.M. De Toledo • E.I. Azzam (✉)
Department of Radiology, UMDNJ-New Jersey Medical School Cancer Center, 205 South Orange Avenue, Newark, NJ 07103, USA
e-mail: zhangjie78@gmail.com; azzamei@umdnj.edu

C.E. Mothersill et al. (eds.), *Radiobiology and Environmental Security*, NATO Science for Peace and Security Series C: Environmental Security,
DOI 10.1007/978-94-007-1939-2_8, © Springer Science+Business Media B.V. 2012

8.1 Introduction

Although recent advances in biochemical, molecular, and epidemiological techniques have increased our understanding of the biological effects and health risks of low dose ionizing radiation (<100 mSv), great ambiguities in knowledge still remain. The quality (i.e. linear energy transfer; LET), dose and dose-rate of the radiation, the types of irradiated cells, their microenvironment and their metabolic state, as well as the variations in inherent radiation sensitivity are among the factors that can modulate the responses to low dose radiation. These characteristics and others are under intense investigation in many laboratories, including ours; the results are yielding a wealth of novel insights into the mechanisms that underlay cell and tissue responses to low dose/low fluence ionizing radiation [37]. It is hoped that the elucidation of the mechanisms involved may alleviate the uncertainties in estimating low dose radiation effects on human health [13]. Typically, human epidemiological studies would be ideal to assess such effects; however, they currently have limited statistical power due to the small size of the cohorts under study [3].

Understanding the biological effects of low dose radiation is of immense, public, scientific and regulatory interest, as the frequency of human exposure to low dose radiation has been on the increase. In addition to exposures from natural sources (e.g. inhaled radon gas), the human population may be subjected to ionizing radiation during activities related to nuclear technology, mining, air travel and space exploration. Perhaps of greatest significance is the explosive growth in diagnostic radiology use where an increasing number of individuals, including children, are being *repeatedly* exposed to low dose radiation [19].

Currently, for the purposes of radiation protection, the deleterious effects of ionizing radiation are assumed to have a linear dose response with no threshold [13]. Two radiation-induced phenomena that were particularly recognized in the past three decades, namely adaptive and bystander effects, are thought to cause a challenge to these assumptions [28]. The propagation of damaging effects from irradiated to non-irradiated bystander cells would, presumably, result in supra-linear dose-response relationships. In contrast, the expression of adaptive responses that mitigate the initial damaging effects induced by radiation would suggest an infra-linear dose-response relationship or the existence of a threshold dose, below which there would be no risk. Although widely observed, the data confirming the expression of these two phenomena are not universal [17, 33, 74, 75]. Moreover, the exact molecular steps by which adaptive and bystander effects are elicited remain unclear. Elucidation of these steps would clearly increase our understanding of the role of cellular processes that impact the health risks of low dose radiation exposures.

We have been intensely involved in examining the mechanisms underlying radiation-induced bystander and adaptive responses by using model cells and rodent tissues. Here, we summarize some of our findings in the context of a brief review of the latter phenomena. These findings provide strong support for the expression of both bystander and adaptive responses and reveal a critical role of the quality of the radiation in triggering either protective or stressful effects at low doses. Whereas, low doses of low LET radiations (highly energetic X rays, γ rays or protons) triggered processes that mitigated not only stressful effects induced by subsequent challenge doses of radiation, but also stressful effects due to endogenous oxidative metabolism [4–6, 27], high LET radiations (α particles and high charge, high energy (HZE) particles) resulted in the propagation of stressful effects from irradiated to non-irradiated bystander cells [7–9]. The data strongly support a role for intercellular communication and oxidative metabolism in the mediation of these responses. Induction of genomic instability and low dose hypersensitivity are other phenomena that are also thought to impact the health risks of exposure to low dose radiation. These phenomena are being investigated in different laboratories [45, 54, 58, 60].

8.2 Interactions of Ionizing Radiation with Biological Matter

Ionizing radiation is energetic and penetrating. Many of its chemical effects in biological matter are due to the geometry of the initial physical energy deposition events, referred to as the track structure. The transfer of radiation energy to living tissues causes ionization of atoms and molecules and breaks chemical bonds, which initiates a series of biochemical and molecular signaling events that culminate in transient or permanent physiological changes [36].

Ionizing radiation exists in either particulate or electromagnetic types. The ionizations and excitations that it produces tend to be localized, along the tracks of individual charged particles, in a pattern that depends on the type of radiation involved. Whereas the ionization events produced by fast electrons ejected from molecules traversed by high energy X rays or γ rays are well separated in space, those produced by certain charged particles, such as α and HZE particles, occur in dense columns along the particle path [35]. Such differences in ionization patterns mainly arise from differences in charge-to-mass ratio of the impacting particles.

Effects due to the track structure define the quality of the radiation and are commonly called linear energy transfer (LET) effects. In irradiated mammalian cells, which consist mainly of water, single energy deposition events cause bursts of reactive oxygen species (ROS) in and around the radiation track as well as in the intercellular matrix. Depending on the physiological state of the cell, these bursts of reactive species may alter the cellular redox environment, modify signaling cascades and normal biochemical reactions, and generate damage to cellular molecules and organelles [77]. In addition to the damages caused by water radiolysis products (i.e. the indirect effect), cellular damage may also involve reactive nitrogen species (RNS) and other species [80], and can occur also as a result of ionization of atoms on constitutive key molecules (e.g. DNA). The latter is known as the direct effect [36]. The ultimate result, of direct and indirect effects, is the development of biological and physiological alterations that may manifest themselves seconds or decades later. Genetic and epigenetic changes may be involved in the evolution of these alterations [44, 48]. Intercellular communication among the irradiated cells [2], and between irradiated and non-irradiated cells [7, 84], as well as oxidative metabolism and DNA repair mechanisms are major mediators of the *system* responses to ionizing radiation exposure [31].

Because high LET radiation deposits greater amounts of energy per unit length of matter traversed, the possibility of multiple lesions in close proximity and short time frame is high [26]. Consequently, for the same total dose absorbed, high LET radiation is more damaging to cells than low LET radiation [36]. The effects of LET, dose, and dose-rate in the cellular responses to low dose/low fluence ionizing radiation exposures continue to be intensely investigated. Here, we highlight the relevance of the latter characteristics of radiation in the expression of adaptive responses and in the nature of the biological effect propagated from irradiated to bystander cells in the exposed cell populations.

8.3 Ionizing Radiation-Induced Bystander Effects

The ionizing radiation-induced bystander effect is broadly defined as the occurrence of biological effects in unirradiated cells as a result of exposure of other cells in the population to radiation. Bystander effects have been mainly observed in high density cell cultures exposed to low fluences of α particles wherein only a small fraction of cells is irradiated [63]. Emerging data also indicate that bystander effects exist in cell cultures exposed to low doses of HZE particles [39]. They have also been noted in co-cultures of irradiated and unirradiated cells [15, 34]. Stressful effects including up-regulation of stress-responsive proteins, genetic changes, induction of cell cycle checkpoints and cell death occur in both irradiated and non-irradiated cells of human and rodent origin at different stages of growth (reviewed in [10, 37, 60, 62, 67]). More recently, strong evidence for bystander responses *in vivo* has been presented [41, 56]. A few studies have also indicated that radiation-induced protective responses are mediated in a bystander manner in cell cultures exposed to low doses of low LET radiations [47] (and our unpublished data).

By using several biological endpoints to investigate non-targeted effects, including induction of DNA damage and various parameters of oxidative stress, our studies strongly support a role for LET, dose-rate and total absorbed dose in determining the nature and magnitude of the radiation-induced bystander effect and its persistence in progeny cells [18, 27, 66]. Together with results generated by others, the data clearly show that a given cell need not be directly irradiated to experience an ionizing radiation-induced biological response. Depending on cell type and radiation characteristics, distinct molecular interactions lead to propagation of either damaging or protective effects from irradiated to unirradiated cells and between irradiated cells. Gap-junction selectivity, secreted diffusible factors and oxidative metabolism are mediators of these effects [10].

A direct evidence for the role of gap-junction intercellular communication (GJIC) in these processes was shown by our group and by others [7, 8, 84]. The modulation

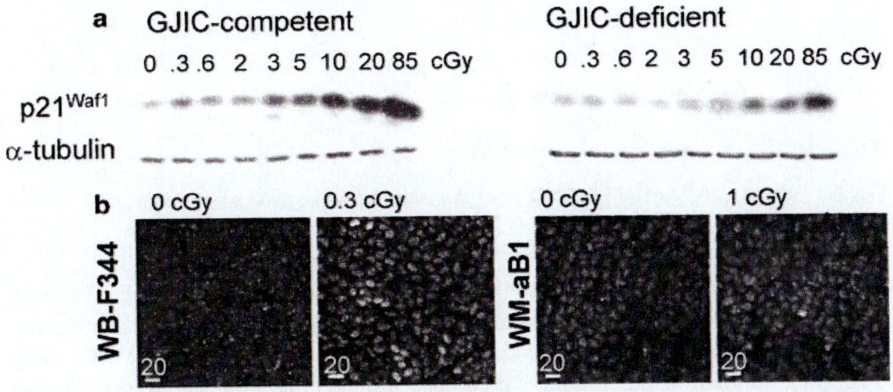

Fig. 8.1 **(a)** Expression of p21Waf1 in protein lysates from gap junction communication-competent WB-F344 or gap junction communication-deficient WM-aB1 confluent cultures following exposure to α particles. Cells were harvested 4 h after the exposure and proteins were examined by western blot analyses. **(b)** *In situ* immunofluorescence detection of p21Waf1 expression in control non-irradiated WB-F344 cultures and in cultures exposed to 0.3 cGy of α particles. Expression of p21Waf1 in control non-irradiated and in 1 cGy-exposed cultures of GJIC-deficient WM-aB1 cells [5]

of proteins involved in the p53/p21^{Waf1} stress-induced signaling pathway and induction of DNA damage in bystander cells were observed only in GJIC-proficient cell cultures. The data in Fig. 8.1 describe p21^{Waf1} expression in sham-exposed and α particle-irradiated cultures from two related rat epithelial cell lines that differ in their ability to communication via gap junctions. The WB-F344 cells are GJIC-competent, a function that is sensitive to inhibition by lindane and other chemicals that block junctional communication [24, 52]. The WM-aB1 cells were derived from WB-F344 cells, but are deficient in GJIC [8, 64]. Similar to WB-F344 cells, WM-aB1 cells express connexin43, a structural protein of gap junctions, however they are deficient in the ability to phosphorylate it, which renders them deficient in functional GJIC [8, 64]. The western blot analyses data in Fig. 8.1a show an increase in p21^{Waf1} levels in confluent WB-F344 cultures exposed to mean doses as low as 0.3 cGy. In WM-aB1 cell cultures, an increase in p21^{Waf1} levels is significant only at mean doses of 5 cGy or higher. Therefore, the magnitude of the response in the GJIC-competent cells and the lack of p21^{Waf1} up-regulation in WM-aB1 GJIC-deficient cultures exposed to low mean doses strongly support the involvement of GJIC in the bystander gene expression response. This is further confirmed by the *in situ* immunofluorescence data in Fig. 8.1b showing induction of p21^{Waf1} in confluent cultures exposed to a mean dose of 0.3 cGy. While small clusters of responding cells were observed in WB-F344 cells, only single isolated and presumably irradiated WM-aB1 cells invariably exhibited up-regulation of p21^{Waf1} after exposure to doses in the range of 0.3–1.0 cGy (Fig. 8.1b). At these mean doses to the monolayer, 1% or less of the WB-F344 or WM-aB1 cells would have their nuclei traversed by an α particle.

The WM-aB1 cells were transformed by mutagenesis of the WB-F344 parental cell line [79]. To exclude effects due to mutagenesis other than loss of GJIC, we

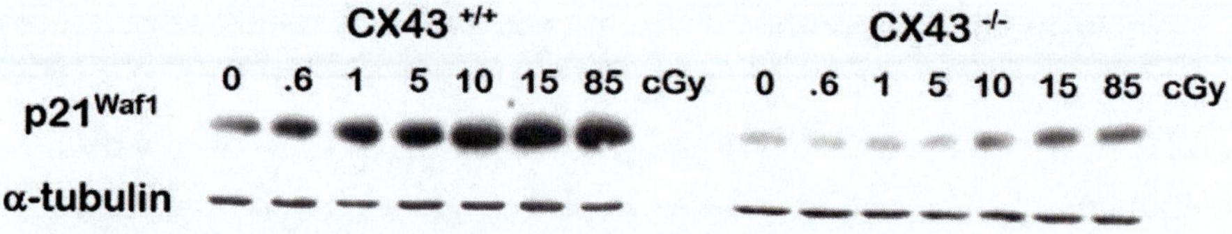

Fig. 8.2 Western analyses of p21Waf1 level in lysates from isogenic wt or connexin43$^{-/-}$ MEF cultures exposed to α particles. Cells were harvested 4 h after the exposure [5]

tested the induction of the p53/p21^{Waf1} signaling pathway after low fluence α particle irradiation of low passage mouse embryo fibroblasts (MEFs) from wt and isogenic knockout embryos for connexin43. Similar to WB-F344 and WM-aB1 cells (Fig. 8.1), the data in Fig. 8.2 indicate a lack of detectable increase in p21^{Waf1} level in connexin43$^{-/-}$ cell cultures exposed to mean doses less than 10 cGy. In contrast, p21^{Waf1} was induced in wt cell cultures exposed to mean doses as low as 0.6 cGy. Collectively, these data support strongly the involvement of GJIC in the bystander gene expression response observed in confluent, density-inhibited cell cultures exposed to low fluences of α particles. They are relevant to estimation of the health risks of exposure to environmental radon. Radon accounts for 55% of the average annual radiation dose to the public in the USA and is considered to be the single largest naturally occurring environmental hazard. In fact, ~10–14% of lung cancer fatalities in the USA may be linked to radon and its α particle-emitting decay [12].

In addition to α particles, and with relevance to space exploration, we have also observed prominent stressful bystander effects in cell cultures exposed to low fluenes of HZE particles [20]. Our published and unpublished data indicate prominent induction of DNA damage, protein oxidation, lipid peroxidation and perturbations in mitochondrial functions, including mitochondrial protein transport and inactivation of the metabolic enzyme aconitase in bystander cells and in their progeny. Gap-junction communication was a major mediator of the propagation of these effects in the tissue culture systems used in our studies.

8.3.1 Gap-Junction Channels and the Cellular Response to Ionizing Radiation

Gap junctions are dynamic structures that are critical for diverse physiological functions [38]. The intercellular channels that comprise gap junctions are formed by *connexin* protein. Each of the ~20 isoforms of connexin forms channels with distinct permeability properties. Though the properties of channels formed by each isoform differ, connexin pores, which vary in diameter, usually allow permeation of molecules up to ~1,000 Da, well above the size of most second messengers. Connexin channels have been shown to be highly selective among molecular permeants [38].

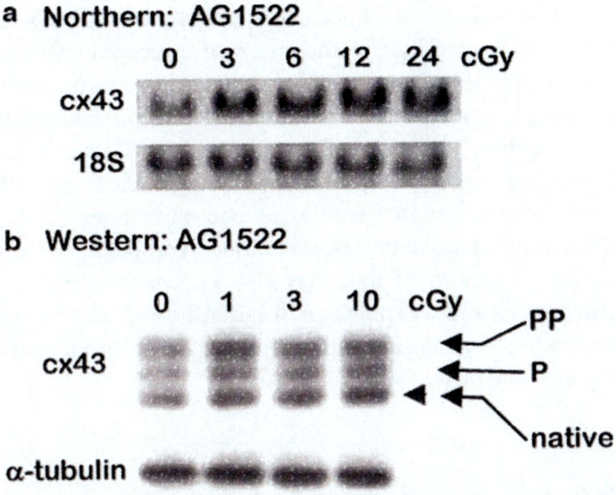

Fig. 8.3 Upregulation of *CONNEXIN43* in α-particle-irradiated cell cultures. (**a**) Northern analyses of *connexin43* expression in AG1522 fibroblast cultures exposed to α-particles doses ranging from 0 to 24 cGy and held at 37°C for 6 h. (**b**) Western analyses of connexin43 in AG1522 confluent cultures at 3 h after exposure to α particle doses ranging from 0 to 10 cGy under normal growth conditions [7]

Evidence for the involvement of GJIC in propagation of bystander effects has been derived from studies with α particle, β particle, γ, and HZE radiations. These studies highlight the relevance of bystander responses to radiotherapy, diagnostic radiology, and risk of environmental and occupational exposures [40]. Manipulation (↓↑) of connexin expression/gap-junction gating by pharmacological agents, forced expression by transfection, and connexin gene knockout studies have provided evidence for the participation of GJIC in radiation-induced bystander effects [10]. This is particularly supported by the stabilization and up-regulation of connexin mRNA and protein by ionizing radiation Fig. 8.3 [11]. Examination by Northern and Western analyses of AG1522 normal human fibroblast cultures exposed to low fluences of α particles indicated that the *CONNEXIN43* gene is indeed activated. Relative to sham-irradiated controls, *CONNEXIN43* mRNA (Fig. 8.3a) and protein levels (Fig. 8.3b) were increased after exposure to mean doses ranging from 1 to 24 cGy. The similar increases at all doses suggest that low fluences of α particles induce molecular pathways that lead to maximal up-regulation of *CONNEXIN43*. The data in Fig. 8.3b indicate increased expression in three protein bands detected by the antibody used. These bands were previously described to represent the native, phosphorylated and hyperphosphorylated isoforms of connexin43 [49].Low fluences of α particles up-regulate both the native and the post-translationally modified isoforms in confluent normal human fibroblasts [11].

Participation of GJIC in stress-induced bystander effects is not unique to ionizing radiation; it has also been described in high density cell populations exposed to chemotherapeutic agents. Toxicity of these compounds was enhanced by functional gap-junction communication in target cells [43]. Thus, many systems show that GJIC enhances the effects of toxic agents on targeted and untargeted cells. Direct intercellular communication may also lead to induction of protective effects that attenuate damage in targeted cells [83]. The determinants and mechanism(s) of these effects, however, remain largely undefined. Our emerging data indicate that permeability properties of gap-junction channels affect the nature of the induced bystander response. Different connexins form channels with *different* selectivities for various molecules including ions and highly similar second messengers [14].

Direct intercellular communication is not unique in propagating radiation-induced non-targeted effects. A wealth of data has also shown the critical importance of secreted diffusible factors in the expression of radiation-induced non-targeted effects [62]. TGF-β, interleukin-8, serotonin and others have been implicated in propagation of bystander effects [50, 61].

8.3.2 Oxidative Metabolism and Bystander Effects

Normal oxidative metabolism is a key endogenous generator of reactive oxygen and nitrogen species [29], and homeostatic control of normal cellular growth pathways is tightly dependent on oxidants [21]. A disruption of the balance between oxidant production and antioxidant defense alters the homeostatic cellular redox environment, resulting in a state of oxidative stress that promotes several pathological conditions including degenerative diseases and cancer [32]. The endogenous targets of oxidants are diverse and include nucleic acids, proteins and lipids.

There is a strong connection between the generation of ROS and RNS and the damage that follows radiation exposure. Whereas ~60 ROS per nanogram of tissue were estimated to be generated from a hit caused by ^{137}Cs γ rays [23, 51], we can estimate that over 2,000 ROS are generated from an α particle traversal, corresponding to a ROS concentration of ~19 nM in the nucleus of a normal human AG1522 cells [2]. Such a ROS concentration can obviously cause extensive oxidative damage and may constitute a signaling event that triggers the spread of stressful effects from irradiated to neighboring bystander cells.

The involvement of ROS in the ionizing radiation induced bystander response was postulated by Nagasawa and Little [63] in their initial report describing the induction of sister-chromatid exchanges (SCE) in bystander Chinese hamster ovary cells present in cultures exposed to fluences of α particles by which less than 1% of the nuclei were directly targeted. Evidence for such involvement was subsequently generated in studies involving various biological endpoints and irradiation modalities [10]. Induction of stress responsive proteins, lethality and genetic changes (SCEs, mutations, chromosomal aberrations) in bystander cells was inhibited by superoxide dismutase (SOD) and other antioxidants [10].

ROS act as second messengers that regulate gene expression by signaling processes that involve activation of redox sensitive transcription factors [1, 16, 51, 53, 70, 71, 76]. Consistent with a role for radiation-induced alterations in redox sensitive transcription factor activation in bystander cells in cultures

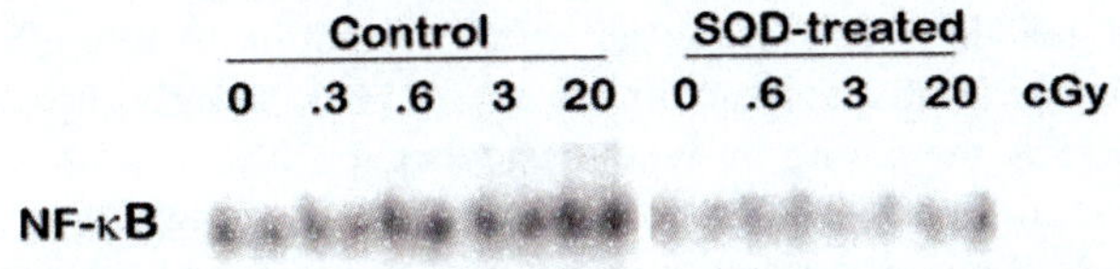

Fig. 8.4 Electrophoretic mobility shift assay in α particle-irradiated AG1522 fibroblast cultures indicates activation of NFκB DNA-binding by low mean doses at 30 min after exposure [10]

exposed to low fluences of α particles, increases in the DNA-binding activity of NFκB were observed in AG1522 normal human cell cultures exposed to doses as low as 0.3 cGy (Fig. 8.4). A similar level of increased DNA-binding activity was observed at 0.6 and 3 cGy although a five-fold greater fraction of cell nuclei are traversed at 3 cGy than at 0.6 cGy. Importantly, when SOD (100 μg/ml, 300 U/ml) was added to the culture 30 min prior to irradiation, the increase in NFκB DNA-binding activity was inhibited (Fig. 8.4), further supporting the role of $O_2^{\bullet -}$ in the bystander response of AG1522 cells.

Extensive data now indicate that the intracellular production of superoxide anions and hydrogen peroxide in both irradiated and bystander cells involves both the plasma bound NADPH-oxidase and mitochondria [9]. Of particular significance, our data strongly indicate that increased ROS levels following cellular exposure to α or HZE particles persist in progeny cells for many generations. This is manifested by increased oxidation of cellular proteins and disruption of mitochondrial physiology. In particular, decreased aconitase activity, which is involved in electron transport and regulation of gene expression, was observed, in bystander cells, 20 population doublings after exposure [20]. Ectopic over-expression at the time of irradiation of the antioxidant enzymes superoxide dismutase or glutathione peroxidase, in bystander or irradiated cells, attenuated DNA damage and induction of stress-responsive proteins in the bystander cells (our data, unpublished). These results show that oxidative metabolism modulates non-targeted effects at the level of the irradiated and bystander cells.

Through *in vivo* experiments consisting of partial body irradiation of male Sprague-dawley rats with low fluences of HZE particles (energetic titanium or oxygen ions), stressful effects involving mitochondrial dysfunction were observed in non-targeted tissues 20 months after the exposure [41]. Decreases in mitochondrial protein import as well as increases in antioxidant defense in non-targeted tissues were associated with perturbations in immune responses and inflammatory cytokines levels (e.g. interleukin-6) (unpublished). These data are consistent with the findings of others who showed that inflammatory-type responses involving oxidative stress occur after exposure to ionizing radiation [23, 55]. In these *in vivo* experiments, activation of macrophages and neutrophil infiltration were not a direct effects of irradiation, but were a consequence of the recognition and clearance of radiation-induced apoptotic cells. The occurrence of such a response has been suggested to provide a mechanism for the interactions between irradiated and non-irradiated haemopoietic cells [23, 55]. Such interaction was also observed in out of field experiments examining the genetic effects of partial organ irradiation. Antioxidants and nitric oxide synthase inhibitors attenuated these effects [46] strongly supporting the role of ROS and RNS in mediating bystander effects [57, 73].

Overall, several studies challenge the traditional paradigm that the important biological effects of ionizing radiation are due to DNA damage induced as a result of direct interaction of the radiation track with the cell nucleus. They indicate that irradiated and non-irradiated cells interact, and oxidative metabolism and intercellular communication have an essential role in signaling events leading to radiation-induced bystander effects. However, clear evidence explaining how these events occur is still lacking. Regardless, the occurrence of bystander effects implies that the modeling of dose response relationships based on the number of irradiated cells may not be a valid approach [54].

8.4 Low LET Radiation-Induced Adaptive Responses

The "adaptive response" is a phenomenon generally induced by low dose/low LET radiation that protects cells and whole organisms against endogenous damage or damage due to a subsequent dose of radiation [81]. Data generated over the last three decades suggest that exposure of mammalian cells, including human cells, to low doses of low LET radiation (e.g. X rays, γ rays, β particles) induces molecular processes that are different from those induced by high dose radiation [30]. Such processes were found to be protective against stress measured by several biological endpoints [28]. Radiation-induced adaptive responses were dependent on the adapting dose, dose rate, expression time, culture conditions and stage of the cell cycle [72]. Adaptive responses seem to be evolutionarily conserved as effects that protect against DNA damage in mammalian cells [65] mirror the evidence of radiation-induced protective mechanisms in prokaryotes and lower eukaryotes [69]. Adaptive responses to ionizing radiation have also been detected *in vivo* [22, 59].

Of particular relevance to risk assessment, it was observed that low-dose/low LET radiation (0.1–10 cGy) decreases the frequency of neoplastic transformation to a level below the spontaneous rate in C3H 10 T½ mouse embryo fibroblasts (MEFs) and in HeLa human hybrid cells [6, 68]. It is noteworthy that these protective effects were seen only in irradiated cells that were allowed to incubate at 37 C before release from contact inhibition (Fig. 8.5), which suggests that time is required for expression of the protective effects.

Chronic exposure of C3H 10 T½ MEFs to cobalt-60 γ-radiation at doses as low as 10 cGy protected the cells not only against damage from endogenous metabolic processes, but also against neoplastic transformation by a subsequent large acute radiation exposure [5]. The induced resistance to neoplastic transformation correlated with increased ability to repair radiation-induced chromosome breaks [4].

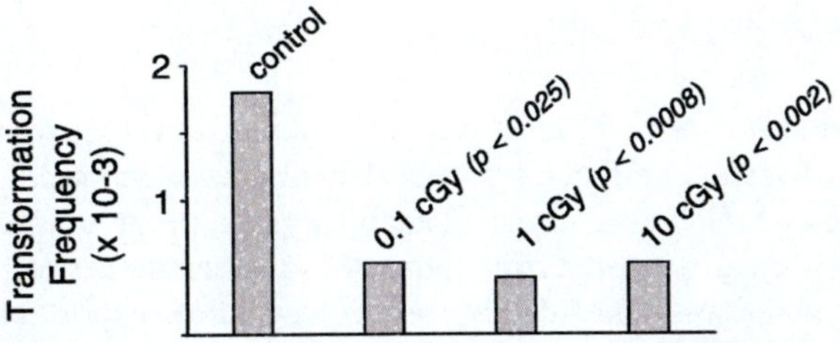

Fig. 8.5 Low dose/low dose rate γ rays decrease the frequency of neoplastic transformation to a level below the spontaneous rate in C3H 10 T½ mouse embryo fibroblasts [9]

Significantly, our recent proteomic studies have identified novel proteins that were distinctly induced by low and not high dose γ rays. We have uncovered a role in DNA repair for the translationally controlled tumor protein, which was induced by almost ten-folds in cell cultures exposed to doses as low as 1 cGy (Zhang et al. manuscript in preparation).

In addition to up-regulation of DNA repair mechanisms [78], other processes may also modulate cellular responses to low dose/low dose-rate irradiation. Irradiation, under such conditions may affect the overall redox-state of the cell and its anti-oxidation potential, and may alter chromatin conformation, hence affecting the accessibility of DNA lesions to components of the DNA repair machinery. It may also induce mechanisms (e.g. apoptosis) that eliminate heavily damaged cells from the irradiated cell population [82]. Our data indicate that direct intercellular communication by gap-junctions [27] is an important modulator of these effects. In addition, the induction of cell cycle checkpoints presumably provides more time for repair of radiation damage. Such effects may involve epigenetic events that could be transmitted to the progeny of low dose irradiated cells (Chaudhry et al. manuscript in preparation).

Similar to its role in modulating bystander effects, oxidative metabolism is also a significant mediator of low dose, low LET radiation effects. Exposure of normal human fibroblasts maintained in 3-dimensional architecture to 10 cGy from γ rays delivered over 48 h reduced the frequency of micronucleus formation (a form of DNA damage) to levels similar or lower than background [27]. The effects correlated with up-regulation of cellular content of the antioxidant glutathione [27]. Extensive data have also shown that whole-body exposure of mice to low dose/low dose-rate γ rays up-regulates superoxide dismutase and alters mitochondrial functions in a manner that attenuates the generation of ROS (Li et al. manuscript in preparation). We predict that such alterations provide a defense mechanism that allows the organism to cope with the radiation-induced oxidative stress. Together, the data suggest that mitochondria, which are active participants in oxidative metabolism, play a crucial role in low dose-induced adaptive responses.

8.5 Conclusions

Some of the mechanisms (e.g. junctional communication, oxidative metabolism) that underlie the bystander effect have been also implicated in the adaptive response to ionizing radiation. However, classical adaptive response protocols involving low LET radiation are clearly distinct from those of bystander studies conducted mainly with high LET radiation. In the adaptive response, cells are exposed to a small dose of low LET radiation. In contrast, cells traversed by an α or a HZE particle receive a substantial dose (10–70 cGy) and undergo a complex type of DNA damage. While similar mediators may modulate the same endpoint in both phenomena, the occurrence of opposite effects, such as pro-survival rather than cytotoxic effect, may reflect changes in concentration of the inducing factor(s). For example, ROS have been shown to be a double-edged sword capable of inducing both proliferative or cell death effects depending on their concentration. Moreover, recent studies emphasized the effect of LET on the yield of water radiolysis products [42]. Prevalence of different radiolysis species at the time of irradiation may induce dissimilar effects. However, the bystander effect and adaptive response could also be mediated by distinct mechanisms/mediating factors.

Due to limitations in the statistical power of current human epidemiological studies in assessing the health risks of low dose radiation exposures, mechanistic studies may be essential to understanding biological effects, and to help evaluating risks at low doses. Coupled with epidemiology, the knowledge of cellular and molecular processes that underlay low dose radiation-induced biological effects should further refine our estimates of radiation risks at low doses. The expression of stressful bystander effects in cell populations exposed to low fluences of high LET particles may contribute to the understanding of lung cancer incidence from environmental radon and degenerative diseases that may occur following deep space travel [25]. Bystander effect studies may also enhance our understanding of biological effects that result from non-uniform distribution of incorporated radioactivity such as α particles emitted from radionuclides used in therapeutic nuclear medicine or released during nuclear accidents or terrorist activities [40]. In particular, they offer avenues to characterize the nature of communicated signaling molecules and formulate strategies to protect normal tissue surrounding irradiated tumor targets. In contrast, the expression of adaptive responses in low dose/low LET exposed cell populations and the propagation of protective effects from irradiated to non-irradiated cells present in these populations may explain reported hormetic effects. They indicate that for some individuals, the risk from very small doses of radiation delivered at low dose-rate may be inexistent.

In conclusion, it is apparent that extensive *in vitro* and *in vivo* experimental evidence suggests that biological responses together with biophysical considerations likely determine the outcome of cellular exposure to ionizing radiation. Collectively, these studies should further contribute to the setting of radiation protection standards that would be effective in different exposure scenarios, applicable to men and women of all ages, and that must protect radiosensitive persons.

Acknowledgements This research was supported by Grants DE-FG02-07ER64344 from the US Department of Energy (Low Dose Radiation Research Program), CA049062 from the NIH and NNJ06HD91G from NASA.

References

1. Adler V, Yin Z, Tew KD, Ronai Z (1999) Role of redox potential and reactive oxygen species in stress signaling. Oncogene 18(45):6104–6111
2. Autsavapromporn N, De Toledo SM, Little JB, Jay-Gerin JP, Harris AL, Azzam EI (2011) The role of gap-junction communication and oxidative stress in the propagation of toxic effects among high dose alpha particle-irradiated human cells. Radiat Res 175(3):347–357
3. Averbeck D (2009) Does scientific evidence support a change from the LNT model for low-dose radiation risk extrapolation? Health Phys 97(5):493–504
4. Azzam EI, de Toledo SM, Raaphorst GP, Mitchel RE (1994) Réponse adaptative au rayonnement ionisant des fibroblastes de peau humaine. Augmentation de la vitesse de réparation de l'ADN et variation de l'expression des gènes. J Chim Phys 91(7/8):931–936
5. Azzam EI, Raaphorst GP, Mitchel RE (1994) Radiation-induced adaptive response for protection against micronucleus formation and neoplastic transformation in C3H 10 T1/2 mouse embryo cells. Radiat Res 138(1 Suppl):S28–S31
6. Azzam EI, de Toledo SM, Raaphorst GP, Mitchel RE (1996) Low-dose ionizing radiation decreases the frequency of neoplastic transformation to a level below the spontaneous rate in C3H 10 T1/2 cells. Radiat Res 146(4):369–373
7. Azzam EI, de Toledo SM, Gooding T, Little JB (1998) Intercellular communication is involved in the bystander regulation of gene expression in human cells exposed to very low fluences of alpha particles. Radiat Res 150:497–504
8. Azzam EI, de Toledo SM, Little JB (2001) Direct evidence for the participation of gap-junction mediated intercellular communication in the transmission of damage signals from alpha-particle irradiated to non-irradiated cells. Proc Natl Acad Sci USA 98(2):473–478
9. Azzam EI, de Toledo SM, Spitz DR, Little JB (2002) Oxidative metabolism modulates signal transduction and micronucleus formation in bystander cells from alpha-particle-irradiated normal human fibroblast cultures. Cancer Res 62(19):5436–5442
10. Azzam EI, de Toledo SM, Little JB (2003) Oxidative metabolism, gap junctions and the ionizing radiation-induced bystander effect. Oncogene 22(45):7050–7057
11. Azzam EI, de Toledo SM, Little JB (2003) Expression of CONNEXIN43 is highly sensitive to ionizing radiation and environmental stresses. Cancer Res 63(21):7128–7135
12. BEIR-VI (1998) Health effects of exposure to radon (BEIR VI). National Academy Press, Washington, DC
13. BEIR-VII (2005) Health risks from exposure to low levels of ionizing radiation. National Research Council of the National Academies, Washington, DC
14. Bevans CG, Kordel M, Rhee SK, Harris AL (1998) Isoform composition of connexin channels determines selectivity among second messengers and uncharged molecules. J Biol Chem 273(5):2808–2816
15. Bishayee A, Hill HZ, Stein D, Rao DV, Howell RW (2001) Free-radical initiated and gap junction-mediated bystander effect due to nonuniform distribution of incorporated radioactivity in a three-dimensional tissue culture model. Radiat Res 155:335–344
16. Blackburn RV, Spitz DR, Liu X, Galoforo SS, Sim JE, Ridnour LA, Chen JC, Davis BH, Corry PM, Lee YJ (1999) Metabolic oxidative stress activates signal transduction and gene expression during glucose deprivation in human tumor cells. Free Radic Biol Med 26(3–4):419–430
17. Blyth BJ, Azzam EI, Howell RW, Ormsby RJ, Staudacher AH, Sykes PJ (2010) An adoptive transfer method to detect low-dose radiation-induced bystander effects in vivo. Radiat Res 173(2):125–137

18. Boyd M, Sorensen A, McCluskey AG, Mairs RJ (2008) Radiation quality-dependent bystander effects elicited by targeted radionuclides. J Pharm Pharmacol 60(8):951–958
19. Brenner DJ, Hall EJ (2007) Computed tomography–an increasing source of radiation exposure. N Engl J Med 357(22):2277–2284
20. Buonanno M, De Toledo SM, Pain D, Azzam EI (2011) Long-term consequences of radiation-induced Bystander effects depend on radiation quality and dose and correlate with oxidative stress. Radiat Res 175(4):405–415
21. Burdon RH (1996) Control of cell proliferation by reactive oxygen species. Biochem Soc Trans 24(4):1028–1032
22. Cai L, Liu SZ (1990) Induction of cytogenetic adaptive response of somatic and germ cells in vivo and in vitro by low-dose X-irradiation. Int J Radiat Biol 58(1):187–194
23. Coates PJ, Robinson JI, Lorimore SA, Wright EG (2008) Ongoing activation of p53 pathway responses is a long-term consequence of radiation exposure in vivo and associates with altered macrophage activities. J Pathol 214(5):610–616
24. Criswell KA, Loch-Caruso R (1995) Lindane-induced elimination of gap junctional communication in rat uterine monocytes is mediated by an arachdonic acid-sensitive cAMP-independent mechanism. Toxicol Appl Pharmacol 135:127–138
25. Cucinotta FA, Chappell LJ (2010) Non-targeted effects and the dose response for heavy ion tumor induction. Mutat Res 687(1–2):49–53
26. Cucinotta FA, Wu H, Shavers MR, George K (2003) Radiation dosimetry and biophysical models of space radiation effects. Gravit Space Biol Bull 16(2):11–18
27. de Toledo SM, Asaad N, Venkatachalam P, Li L, Spitz DR, Azzam EI (2006) Adaptive responses to low-dose/low-dose-rate gamma rays in normal human fibroblasts: the role of growth architecture and oxidative metabolism. Radiat Res 166:849–857
28. de Toledo SM, Azzam EI (2006) Adaptive and bystander responses in human and rodent cell cultures exposed to low level ionizing radiation: the impact of linear energy transfer. Dose-Response 4(4):291–301
29. Droge W (2002) Free radicals in the physiological control of cell function. Physiol Rev 82(1):47–95
30. Feinendegen LE, Pollycove M, Neumann RD (2007) Whole-body responses to low-level radiation exposure: new concepts in mammalian radiobiology. Exp Hematol 35(4 Suppl 1):37–46
31. Feinendegen L, Hahnfeldt P, Schadt EE, Stumpf M, Voit EO (2008) Systems biology and its potential role in radiobiology. Radiat Environ Biophys 47(1):5–23
32. Finkel T, Holbrook NJ (2000) Oxidants, oxidative stress and the biology of ageing. Nature 408(6809):239–247
33. Fournier C, Barberet P, Pouthier T, Ritter S, Fischer B, Voss KO, Funayama T, Hamada N, Kobayashi Y, Taucher-Scholz G (2009) No evidence for DNA and early cytogenetic damage in bystander cells after heavy-ion microirradiation at two facilities. Radiat Res 171(5):530–540
34. Gerashchenko BI, Howell RW (2003) Cell proximity is a prerequisite for the proliferative response of bystander cells co-cultured with cells irradiated with gamma-rays. Cytometry 56A:71–80
35. Goodhead DT (1988) Spatial and temporal distribution of energy. Health Phys 55:231–240
36. Hall EJ (2000) Radiobiology for the radiologist, 5th edn. J. B. Lippincott Co, Philadelphia
37. Hamada N, Maeda M, Otsuka K, Tomita M (2011) Signaling pathways underpinning the manifestations of ionizing radiation-induced bystander effects. Curr Mol Pharmacol 4(2):79–95
38. Harris AL (2001) Emerging issues of connexin channels: biophysics fills the gap. Q Rev Biophys 34(3):325–472
39. Held KD (2009) Effects of low fluences of radiations found in space on cellular systems. Int J Radiat Biol 85(5):379–390
40. Howell RW, Neti PV, Pinto M, Gerashchenko BI, Narra VR, Azzam EI (2006) Challenges and progress in predicting biological responses to incorporated radioactivity. Radiat Prot Dosimetry 122(1–4):521–527

41. Jain MR, Li M, Chen W, Liu T, de Toledo SM, Pandey BN, Li H, Rabin BM, Azzam EI (2011) In vivo space radiation-induced non-targeted responses: late effects on molecular signaling in mitochondria. Curr Mol Pharmacol 4(2):106–14
42. Jay-Gerin JP, Meesungnoen J, Banville P, Mankhetkorn S (2003) Comment on "The radiation-induced lesions which trigger the bystander effect" by J.F. Ward [Mutat. Res. 499 (2002) 151–154]. Mutat Res 525(1–2):125–127
43. Jensen R, Glazer PM (2004) Cell-interdependent cisplatin killing by Ku/DNA-dependent protein kinase signaling transduced through gap junctions. Proc Natl Acad Sci USA 101(16):6134–6139
44. Jirtle RL, Skinner MK (2007) Environmental epigenomics and disease susceptibility. Nat Rev Genet 8(4):253–262
45. Joiner MC, Lambin P, Malaise EP, Robson T, Arrand JE, Skov KA, Marples B (1996) Hypersensitivity to very-low single radiation doses: its relationship to the adaptive response and induced radioresistance. Mutat Res 358(2):171–183
46. Khan MA, Hill RP, Van Dyk J (1998) Partial volume rat lung irradiation: an evaluation of early DNA damage. Int J Radiat Oncol Biol Phys 40(2):467–476
47. Klammer H, Kadhim M, Iliakis G (2010) Evidence of an adaptive response targeting DNA non-homologous end joining and its transmission to bystander cells. Cancer Res 70(21):8498–8506
48. Koturbash I, Kutanzi K, Hendrickson K, Rodriguez-Juarez R, Kogosov D, Kovalchuk O (2008) Radiation-induced bystander effects in vivo are sex specific. Mutat Res 642(1–2):28–36
49. Lampe PD, Lau AF (2000) Regulation of gap junctions by phosphorylation of connexins. Arch Biochem Biophys 384(2):205–215
50. Lehnert BE, Goodwin EH (1997) Extracellular factor(s) following exposure to alpha particles can cause sister chromatid exchanges in normal human cells. Cancer Res 57:2164–2171
51. Li N, Karin M (1999) Is NF-kappaB the sensor of oxidative stress? FASEB J 13(10):1137–1143
52. Li R, Mather JP (1997) Lindane, an inhibitor of gap junction formation, abolishes oocyte directed follicle organizing activity in vitro. Endocrinology 138(10):4477–4480
53. Liou HC, Baltimore D (1993) Regulation of the NF-kappa B/rel transcription factor and I kappa B inhibitor system. Curr Opin Cell Biol 5(3):477–487
54. Little JB (2003) Genomic instability and bystander effects: a historical perspective. Oncogene 22(45):6978–6987
55. Lorimore SA, Coates PJ, Scobie GE, Milne G, Wright EG (2001) Inflammatory-type responses after exposure to ionizing radiation in vivo: a mechanism for radiation-induced bystander effects? Oncogene 20(48):7085–7095
56. Mancuso M, Pasquali E, Leonardi S, Tanori M, Rebessi S, Di Majo V, Pazzaglia S, Toni MP, Pimpinella M, Covelli V, Saran A (2008) Oncogenic bystander radiation effects in patched heterozygous mouse cerebellum. Proc Natl Acad Sci USA 105(34):12445–12450
57. Matsumoto H, Hayashi S, Hatashita M, Ohnishi K, Shioura H, Ohtsubo T, Kitai R, Ohnishi T, Kano E (2001) Induction of radioresistance by a nitric oxide-mediated bystander effect. Radiat Res 155(3):387–396
58. Matsumoto H, Hamada N, Takahashi A, Kobayashi Y, Ohnishi T (2007) Vanguards of paradigm shift in radiation biology: radiation-induced adaptive and bystander responses. J Radiat Res (Tokyo) 48(2):97–106
59. Mitchel RE, Jackson JS, McCann RA, Boreham DR (1999) The adaptive response modifies latency for radiation-induced myeloid leukemia in CBA/H mice. Radiat Res 152(3):273–279
60. Morgan WF (2003) Non-targeted and delayed effects of exposure to ionizing radiation: I. Radiation-induced genomic instability and bystander effects in vitro. Radiat Res 159(5):567–580
61. Mothersill C, Saroya R, Smith RW, Singh H, Seymour CB (57) Serum serotonin levels determine the magnitude and type of bystander effects in medium transfer experiments. Radiat Res 174(1):119–123
62. Mothersill C, Seymour CB (2004) Radiation-induced bystander effects–implications for cancer. Nat Rev Cancer 4(2):158–164
63. Nagasawa H, Little JB (1992) Induction of sister chromatid exchanges by extremely low doses of alpha-particles. Cancer Res 52:6394–6396

64. Oh SY, Dupont E, Madhukar BV, Briand JP, Chang CC, Beyer E, Trosko J (1993) Characterization of gap junctional communication-deficient mutants of a rat liver epithelial cell line. Eur J Cell Biol 60:250–255
65. Olivieri G, Bodycote J, Wolff S (1984) Adaptive response of human lymphocytes to low concentrations of radioactive thymidine. Science 223(4636):594–597
66. Pinto M, Azzam EI, Howell RW (2006) Bystander responses in three-dimensional cultures containing radiolabeled and unlabeled human cells. Radiat Prot Dosimetry 122(1–4):252–255
67. Prise KM, O'Sullivan JM (2009) Radiation-induced bystander signalling in cancer therapy. Nat Rev Cancer 9(5):351–360
68. Redpath JL, Antoniono RJ (1998) Induction of an adaptive response against spontaneous neoplastic transformation in vitro by low-dose gamma radiation. Radiat Res 149(5):517–520
69. Samson L, Cairns J (1977) A new pathway for DNA repair in Escherichia coli. Nature 267(5608):281–283
70. Schreck R, Rieber P, Baeuerle PA (1991) Reactive oxygen intermediates as apparently widely used messengers in the activation of the NF-kappa B transcription factor and HIV-1. EMBO J 10(8):2247–2258
71. Schulze-Osthoff K, Bauer M, Vogt M, Wesselborg S, Baeuerle PA (1997) Reactive oxygen intermediates as primary signals and second messengers in the activation of transcription factors. In: Forman HJ, Cadenas E (eds) Oxidative stress and signal transduction. Chapman & Hall, New York, pp 239–259
72. Shadley JD (1994) Chromosomal adaptive response in human lymphocytes. Radiat Res 138(1 Suppl):S9–S12
73. Shao C, Stewart V, Folkard M, Michael BD, Prise KM (2003) Nitric oxide-mediated signaling in the bystander response of individually targeted glioma cells. Cancer Res 63(23):8437–8442
74. Sowa MB, Goetz W, Baulch JE, Pyles DN, Dziegielewski J, Yovino S, Snyder AR, de Toledo SM, Azzam EI, Morgan WF (2010) Lack of evidence for low-LET radiation induced bystander response in normal human fibroblasts and colon carcinoma cells. Int J Radiat Biol 86(2):102–113
75. Sowa MB, Goetz W, Baulch JE, Lewis AJ, Morgan WF (2011) No evidence for a low linear energy transfer adaptive response in irradiated Rko cells. Radiat Prot Dosimetry 143(2–4):311–314
76. Spitz DR, Sim JE, Ridnour LA, Galoforo SS, Lee YJ (2000) Glucose deprivation-induced oxidative stress in human tumor cells. A fundamental defect in metabolism? Ann N Y Acad Sci 899:349–362
77. Spitz DR, Azzam EI, Li JJ, Gius D (2004) Metabolic oxidation/reduction reactions and cellular responses to ionizing radiation: a unifying concept in stress response biology. Cancer Metastasis Rev 23(3–4):311–322
78. Szumiel I (2005) Adaptive response: stimulated DNA repair or decreased damage fixation? Int J Radiat Biol 81(3):233–241
79. Tsushimoto G, Chang CC, Trosko JE, Matsumura F (1983) Cytotoxicity, mutagenic, and cell-cell communication inhibitory properties of DDT, lindane, and chlordane on hamster cells in vitro. Arch Environ Contam Toxicol 12:721–730
80. Wardman P (2009) The importance of radiation chemistry to radiation and free radical biology (The 2008 Silvanus Thompson Memorial Lecture). Br J Radiol 82(974):89–104
81. Wolff S (1992) Failla Memorial Lecture. Is radiation all bad? The search for adaptation. Radiat Res 131(2):117–123
82. Wolff S (1998) The adaptive response in radiobiology: evolving insights and implications. Environ Health Perspect 106(Suppl 1):277–283
83. Wygoda MR, Wilson MR, Davis MA, Trosko JE, Rehemutulla A, Lawrence TS (1997) Protection of herpes simplex virus thymidine kinase-transduced cells from ganciclovir-mediated cytotoxicity by bystander cells: the good samaritan effect. Cancer Res 57:1699–1703
84. Zhou HN, Suzuki M, Randers-Pehrson R, Chen G, Trosko J, Vannais D, Waldren CA, Hall EJ, Hei TK (2001) Radiation risk at low doses may be greater than we thought. Proc Natl Acad Sci USA 98(25):14410–14415

Chapter 9
Inter-plant Communication of Genome Instability in Radiation Exposed Arabidopsis

Franz Zemp and Igor Kovalchuk

Abstract Bystander effect is the phenomenon of the response of naïve bystanding cells or organisms to the stress experienced by neighboring cells or organisms. It is well documented for animal cells grown in vitro, and there are some reports of similar response on the level of the whole organism and even between stressed and non-stressed organisms. Here we present the evidence of the existence of communication between stressed and non-stressed plants. We show that naïve plants neighboring plants exposed to X-ray or UVC exhibit similar increase in the frequency of homologous recombination as exposed plants. We present the evidence that communication signal is primarily airborne.

9.1 Introduction

'Bystander effects' is a term originally used in cancer therapeutics, and refers to the attempt at affecting a single type of cell within a heterogeneous population resulting in several types of cells being affected by the treatment [11, 25]. Now the term bystander effect has been applied to a number of different phenomena whereby unexposed 'units' exhibit the molecular symptoms of stress exposure when adjacent or nearby 'units' are a subject to a stress. Indeed, a unit seems a peculiar subject; however, the unit here can be referring to neighboring cells, systemic cells, or even entire organisms.

Modern bystander effects are non-targeted effects of ionizing radiation exposure. They refer to naïve cells that were either in direct contact with irradiated cells or

F. Zemp • I. Kovalchuk (✉)
Department of Biological Sciences, University of Lethbridge, Lethbridge, AB T1K 3M4, Canada
e-mail: igor.kovalchuk@uleth.ca

C.E. Mothersill et al. (eds.), *Radiobiology and Environmental Security*, NATO Science for Peace and Security Series C: Environmental Security,
DOI 10.1007/978-94-007-1939-2_9, © Springer Science+Business Media B.V. 2012

received an irradiation 'distress' signals from irradiated cells ([24–27, 29–31, 46, 26]). In these instances, bystander effects can include a wide variety of genetic alterations such as gross genome rearrangements, chromosome aberrations, sister chromatid exchanges, deletions, duplications, gene mutations, and amplifications [13, 14, 16, 41, 42, 46–48]. With further influences such as changes in gene expression, cellular proliferation, senescence, and cell death [20, 21, 36, 37, 39].

Other bystander-like effects are plasma- or blood-borne signals produced upon irradiation that direct chromosomal breakage and abnormalities in unirradiated tissues. These factors have subsequently been termed clastogenic factors [26, 28, 30]. Similar to clastogenic factors are abscopal effects, in which radiation-induced changes occur outside the targeted area of irradiation, often in different organs [25, 26, 28]. A classic example is the bilaterally symmetric radiation-response, where only the left or right organ is exposed, but similar responses occur in both organs [17, 25, 28].

Bystander-like effects have also been shown in separate organisms. This can include the transgenerational transmission of radiation effects to offspring, especially in terms of 'bequeathing' genomic instability to cellular [24, 26, 28] or organismal [3] progeny. Radiation-induced bystander phenomena have been shown to occur between cohabiting animals as well [32, 33].

As such, bystander and bystander-like effects are known to be involved in a number of different processes in different organisms. The commonality lies in non-effected 'units' receiving signals from effected 'units' to produce a response that modifies homeostasis in the non-effected 'units.'

The field of radiation-induced bystander effects in plants, in the sense of cell-cell media transfer experiments, has yet to be explored. However, abscopal bystander-like effects are well known to occur in plants under a variety of biotic or abiotic stresses. The local application of biotic or abiotic stresses to plant tissue can lead to systemic changes in pathogen resistance [12], methylation pattern [7], recombination rate [10], hormone levels [38] and gene expression [44]. UV-triggered intra-plant bystander effect was shown to be in part dependent on the free radicals as application of radical scavenger prior to UVC irradiation resulted in bystander signal of substantially lower intensity [10]. The first published work referring to ionizing radiation-induced abscopal effects in plants occurred in 2007 [45]. This laboratory demonstrated that direct and specific radiation to the shoot apical meristem of embryonic Arabidopsis with 1,000 α-particles resulted in post-embryonic developmental defects in root formation.

Plants have long been known to communicate to one another via diffusible signals, or volatile organic compounds. This phenomenon was first discovered several decades ago when it was observed that non-herbivore attacked (bystander) plants residing next to herbivore-attacked plants were emitting the same defensive signals and activating the same defensive machinery as the attacked plants [2]. The reason for this signalling, or eavesdropping as some would put it, has a number of purposes including the priming of defences before the arrival of the herbivore, the readying of pathogen defences to disease, or to attracting predators or parasitoids of the organisms causing stress [2, 15]. Interestingly, some of the phytohormones implicated in these plant-plant signals are similar to the hormones involved in both the systemic acquired resistance and systemic wound signalling. Again, this plant-plant communication of damage or impending attack could be considered bystander-like phenomenon.

One of the signatures of the influence of bystander signal is the genome instability, including the increase in the homologous recombination frequency (HRF). Homologous recombination in plants plays dual roles as both a double-strand break repair mechanism as well as the mechanism responsible for increased genetic diversity during meiosis [7]. In our laboratory and others, it has been shown that plants treated with ionizing radiation have increased levels of homologous recombination [5, 10, 22, 23, 35]. Further experiments have even shown that the local leaf treatment with UV-C leads to increased recombination rates in both the local and systemic leaves [10]. This systemic recombination signal (SRS) was also found in pathogen- [18] and reactive oxygen stress-treated plants [10].

It is possible that the SRS, as well as some of the other systemic signals, are transmitted not through the plant vasculature, which is the common perception, but through the immediate gaseous environment of the plant. If this was indeed the case, plants exposed to ionizing radiation with increased recombination rates may transmit the signal to neighbouring plants in a closed environment.

In this work, we show that plants neighboring X-ray- and UVC-irradiated plants exhibit similar changes in recombination frequency. We show that the bystander effect has two components: liquid, released through roots and airborne, released by leaves.

9.2 Methods

9.2.1 Plant Lines and Growth Conditions

For the experiments on liquid media, sterilized *Arabidopsis thaliana* (cv. 24) line 11 seeds were spread on sterile Whatman No. 1 filter paper in a divided (Fisher) or undivided (Fisher) Petri plates. A total of 4 mL of liquid MS media was added to the plates (2×2 mL in divided plates). In the solid media experiments, the same sterilized seeds were spread on divided or undivided Petri plates containing 25 mL of solid MS media (liquid media recipe plus 8 g/L agar (Sigma)). The plates were stratified for 48 h at 4°C, and then grown in Enconair (Winnipeg, Canada) growth chambers at 16/8 h light/dark, at 23°C and 18°C, respectively.

Arabidopsis line 11 is a transgenic line that contains a homologous recombination (HR) reporter gene. This reporter line has two, non-functional, overlapping, truncated copies of the *β-glucuronidase* gene (GUS; [6, 43]). If a strand break occurs anywhere in the region of homology in one of the two truncated *GUS* genes, the HR double-strand break repair pathway could repair it using the other strand as a template, potentially restoring the function of the GUS transgene (Fig. 9.1; [6]). In this case, sectors of blue will result after histochemical staining.

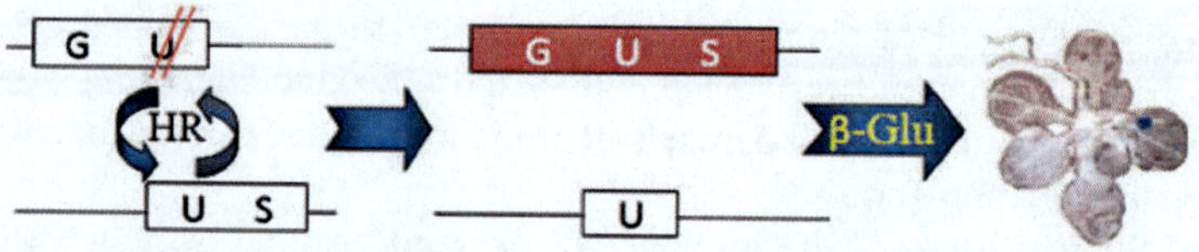

Fig. 9.1 Representation of homologous recombination reporter line #11 in Arabidopsis. *HR* homologous recombination; *β-Glu* GUS staining solution

9.2.2 Experimental Set-Up

Ten-day old Arabidopsis line 11 plants were a subject to either X-Ray (XR) or UV-C (UV) irradiation while growing on either solid or liquid MS media in either undivided or divided Petri dishes. Half of the plates were covered with a shielding that protected the plants underneath, and the plants were irradiated. Seven days post-irradiation the plants were harvested separately as either irradiated (IR; not covered) or bystander (BS; covered) plants. An additional treatment was added, half-bystander (½BS), in which the plates were covered as above, but only media was irradiated. Seventy-five to 150 plants were counted per treatment, and the averages taken to represent the recombination rate for that experiment. The experiment was independently repeated three times, and statistics were performed on those three averages in MS Excel 2003.

9.2.3 Radiation Treatments

Plants were prepared for irradiation at 10 days post-germination. In the instance of UV-C irradiation, the Petri lids were removed and half of the plate was completely covered in tinfoil. The entire plate was than irradiated with 7,000 ergs (~60 ergs/s) of UV-C. A set of control plates were completely covered with tinfoil and irradiated under the same conditions to ensure there was sufficient UV protection. X-ray irradiation was done in a similar manner. Half of the Petri dish was covered with a 2.5 mm thick medical grade lead shield, and whole plates received 20 Gy of X-rays (90 kV, 5 mA, ~18 mGy/s). Again, a set of control plates were irradiated while completely covered with the lead shielding to ensure sufficient protection.

9.2.4 Histochemical Staining

Whole plants were harvested 7 days post-irradiation. These plants were then immediately placed in GUS staining solution (100 mg of 5-bromo-4-chloro-3-indolyl glucuronide substrate (GBT) in 300 mL of 100 mM phosphate buffer (pH 7.0),

0.05% NaN_3, and 1 mL dimethylformamide) and vacuum infiltrated for 10 min. Plants were incubated in the staining solution for 37°C for 48 h, and then destained with 70% ethanol for 24 h. The number of events per plant was than counted under a dissecting microscope, and the number of spots/plant (homologous recombination frequency) was calculated.

9.3 Results

The analysis of HRF in irradiated and non-irradiated bystander plants (Fig. 9.2a) showed the increase in the case of X-ray and UVC exposures, although the effect of X-ray was more pronounced (Fig. 9.2b, c). We hypothesized that the bystander effect on recombination frequency was triggered by the signal released by roots and transmitted through the liquid. To test whether the effect of the signal will be different depending whether plants were grown on liquid or solid media, we performed the experiments using both media (see Sect. 9.2 for details). We found that in general HRF was higher in plants grown on solid media and that the bystander increase in HRF was more pronounced when plants were grown in the liquid media (Fig. 9.2b, c).

To test for the presence of airborne bystander signal, we performed above mentioned experiments using a divider that did not allow liquid exchange but allowed the air exchange (Fig. 9.3a). The analysis showed that there was the increase in HRF in both, irradiated and non-irradiated plants, regardless whether there was a divider present or not (Fig. 9.3d, e). The results were similar for both, X-ray and UVC exposures.

Next, we asked ourselves whether the bystander signal could have originated from liquid media itself as the result of water radiolysis or similar radical producing events. To test this, we placed plants only in one half of the plate that was protected by the shield; no plants was placed in non-shielded area. We used plates that did and did not have a divider (Fig. 9.3b, c). The results showed that in the case of liquid media, there was an increase in HRF in neighboring plants when the divider was not used (Fig. 9.3d, e; data point labelled as "1/2 bystander"). At the same time, no increase in HRF was observed when the divider was used. Similar trend was observed for the solid media, although the data were not significantly different.

9.4 Discussion

Homologous recombination frequency increases can be the result of up-regulation of the homologous recombination double-strand break repair pathway, and has been shown numerous times to respond to direct ionizing radiation [5, 10, 22, 23, 35]. This increase in HFR is a measure of genome instability, which generally refers to the susceptibility of the genome to mutations, rearrangements, and activation of mobile elements [7].

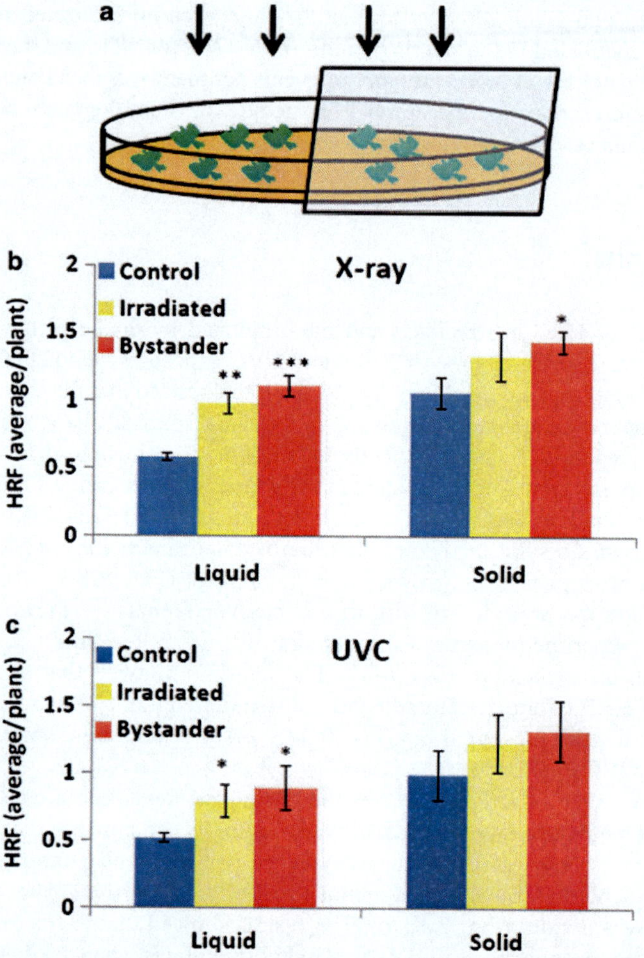

Fig. 9.2 X-ray and UVC induce the HRF increase in both exposed and bystanding plants. Ten-day-old Arabidopsis line 11 plants were a subject to either 20 Gy of X-ray (XR) or 7,000 ergs of UV-C (UV) irradiation while growing on either solid or liquid MS media in either undivided or divided Petri dishes. **(a)** Schematic presentation of the experiment. Half of the plates were covered with a shielding that protected the plants beneath from the radiation (Bystander), while the other half was uncovered (Irradiated). **(b)** X-ray exposure experiment. *Y axis* shows the average recombination frequency in exposed and "bystander" plants grown on liquid or solid medium. *Asterisks* (*) show significant difference from control (* $P<0.05$; ** $P<0.01$; *** $P<0.001$). **(c)** UVC exposure experiment. *Y axis* shows the average recombination frequency in exposed and "bystander" plants grown on liquid or solid medium. *Asterisks* (*) show significant difference from control ($P<0.05$)

Elevated levels of HRF are generally accountable through two mechanisms, the increase in the amount of double strand breaks, and/or the increase in the activity of homologous recombination repair machinery [6]. In the case of ionizing radiation,

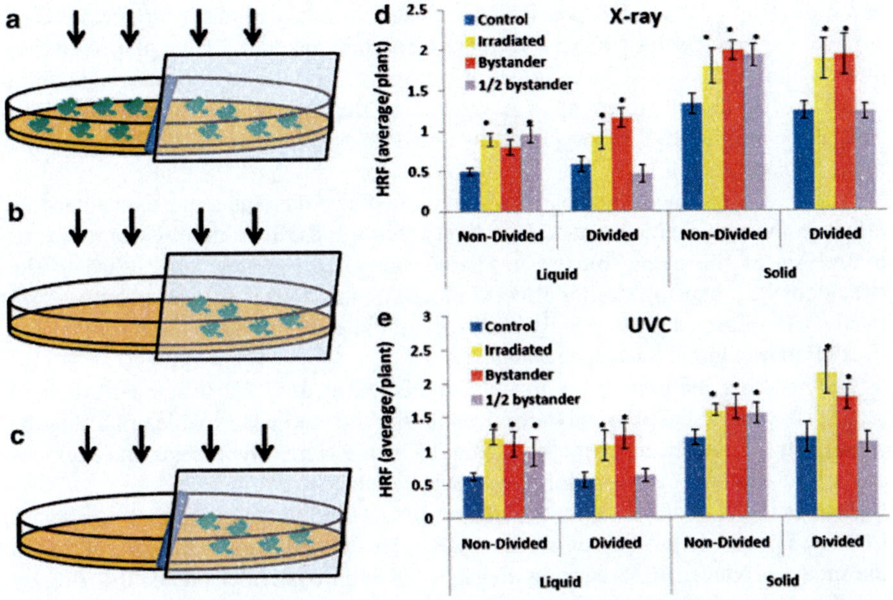

Fig. 9.3 The signal triggering bystander effect is transmitted through liquid and air. Ten-day-old Arabidopsis line 11 plants were a subject to either 20 Gy of X-ray (XR) or 7,000 ergs of UV-C (UV) irradiation while growing on either solid or liquid MS media in either undivided or divided Petri dishes. (**a**) Schematic presentation of the experiment. Half of the plates were covered with a shielding that protected the plants beneath from the radiation (Bystander), while the other half was uncovered (Irradiated). The Petri dishes used contain the dividers separating the medium in which irradiated and non-irradiated plants are grown. (**b**) Schematic presentation of the experiment. Plants were only planted under the shielding and not in the non-protected area. (**c**) Schematic presentation of the experiment. Plants were only planted under the shielding and not in the non-protected area. The Petri dishes used contain the dividers separating the irradiated medium and non-irradiated plants. (**d**) *X-ray* exposure experiment. *Y axis* shows the average recombination frequency in exposed and "bystander" plants grown on liquid or solid medium. *Asterisks* show significant difference from control ($P<0.05$). (**e**) UVC exposure experiment. *Y axis* shows the average recombination frequency in exposed and "bystander" plants grown on liquid or solid medium. *Asterisks* show significant difference from control ($P<0.05$)

it is not surprising that increased HRF occurs upon exposure, as ionizing radiation is well known to induce double-strand breaks in eukaryotic cells, most commonly through the production of genotoxic intercellular reactive oxygen species [9, 10].

However, here our data suggests that direct UV-C and X-ray exposure cause increases in HRF that are more pronounced in liquid than solid media. This is an interesting finding, as it would suggest that plants in liquid media are more susceptible to total plant irradiation. However, if one considers the framework of the experiment, the plants as well as the media are being irradiated. It may be possible that the irradiation of the media is creating free radicals that are affecting the plants global HRF. It is well documented that the radiolysis (X-ray) and photolysis (UV) of water produces substantial quantities of hydroxyl radicals (·OH) and hydrogen peroxide (H_2O_2) [19, 1]. It is possible that irradiation of the media is producing these reactive oxygen species (ROS), which are then being taken in by the plant. Further, as the solid media is a semi-solid gel, it is possible that the ROS produced in solid media are less mobile, resulting in less uptake by the plant and causing less damage. These less-mobile ROS also explains the less dramatic increase in HRF in bystander plants in solid media.

These ROS produced by the irradiation of media could explain the bystander-like effect in the non-divided plates. It is likely that the ROS created by the radio- or hydrolysis of the media on the irradiated side of the plates can diffuse to the bystander side, causing similar effects in these plants. This is supported by the "1/2 bystander" treatments where only media is irradiated resulting in similar increases in HRF in undivided bystander plants.

Further, this increase is completely abolished in the "1/2 bystander" divided plates, suggesting the physical barrier separating the media is capable of halting the signal. This evidence supports the media diffusible signal hypothesis and removes any speculation of media produced gaseous signals.

Interestingly, however, is that bystander plants on divided plates that do contain irradiated plants do produce an elevated HRF. This would suggest that in addition to the media produced ROS causing elevated HRFs in bystander plants, the directly irradiated plants are producing a gaseous signal causing increased genome instability in neighbouring, unexposed plants. This is a truly phenomenal and novel finding. It is curious, however, why there is not an additive result in the increase in HRF in the bystander undivided plates. In this instance, the bystander plants are receiving both the putative aqueous ROS as well as the radiation-derived volatile signal. It may be possible that there is a certain level of saturation in this system, and HRF rates produced in this manner are regulated to a maximal level. Alternatively, as the irradiated plants covered much of the surface of the media, it could be the scenario that the plants are absorbing most of the dose and the media is receiving little radiation, resulting in the generation of only one prominent signal.

The ability of plants to communicate through the release of volatile organic compounds is not novel. In past it was observed that plants not attacked by insects (bystander), residing next to those that were attacked, were able to produce the same volatile signalling molecules and were activating the same defensive pathways as the attacked plants [2]. A number of volatile organics have been studied over the years in an attempt to characterize these plant-plant interactions. Small highly volatile compounds, such as ethylene, methanol, isoprene, acrolein and some monoterpenes, can diffuse in the surrounding environment rapidly, limiting signalling to systemic leaves or very close neighbours [2]. Heavier volatile compounds, such as methyl jasmonate (MeJA), methyl salicylate (MeSA), have been suggested to function over longer distances, as their slow dispersal will allow for the establishment of turbulence-resistant of plumes [2]. In our case, since we are working in a limited air volume of a sealed Petri dish, the signalling of either type of volatile could be possible.

MeSA and MeJA has previously been implicated in intraplant stress responses and systemic signalling, with examples in systemic acquire resistance (SAR; [12, 34]) and systemic response to wounding (SRW; [38]). Interestingly, both the plant-plant signalling of wounding [2] and pathogen infection have been shown before [40].

Similar to these systemic effects is the systemic recombination signal (SRS) discovered in our laboratory several years ago [7, 10, 18]. In this model, the local application of virus, UV-C or rose Bengal (a ROS producing compound) resulted in the local and systemic induction of increased recombination rates. Indeed, this systemic induction of increased HRF may be similar to the plant-plant induction of HRF we show here.

The biological relevance of such a signal between plants is indeed quite perplexing. It has been postulated that the SRS generated in systemic leaves in plants helps drive increasing genetic diversity in plants, as their sessile nature forces them to adapt to survive [4, 8]. However, overwhelming rates of recombination are not 'geno-healthy' either, as homologous recombination can cause the loss of heterozygosity at herterozygous loci. It is possible that this putative plant-plant signal evolved as a systemic signal as opposed to an interplant signal, and it is only in the context of this experiment where close proximately plants in a small gaseous environment are exhibiting this plant-plant-like phenomenon.

References

1. Azrague K, Bonnefille E, Pradines V, Pimienta V, Oliveros E, Maurette MT, Benoit-Marquié F (2005) Hydrogen peroxide evolution during V-UV photolysis of water. Photochem Photobiol Sci 4(5):406–408
2. Baldwin IT, Halitschke R, Paschold A, von Dahl CC, Preston CA (2006) Volatile signaling in plant-plant interactions: "talking trees" in the genomics era. Science 311(5762):812–815
3. Barber RC, Dubrova YE (2006) The offspring of irradiated parents, are they stable? Mutat Res 598(1–2):50–60
4. Boyko A, Kovalchuk I (2008) Epigenetic control of plant stress response. Environ Mol Mutagen 49(1):61–72
5. Boyko A, Greer M, Kovalchuk I (2006) Acute exposure to UVB has a more profound effect on plant genome stability than chronic exposure. Mutat Res 602(1–2):100–109
6. Boyko A, Zemp F, Filkowski J, Kovalchuk I (2006) Double-strand break repair in plants is developmentally regulated. Plant Physiol 141(2):488–497
7. Boyko A, Kathiria P, Zemp FJ, Yao Y, Pogribny I, Kovalchuk I (2007) Transgenerational changes in the genome stability and methylation in pathogen-infected plants. Nucleic Acids Res 35(5):1714–1725
8. Boyko A, Blevins T, Yao Y, Golubov A, Bilichak A, Ilnytskyy Y, Hollander J, Meins F Jr, Kovalchuk I (2010) Transgenerational adaptation of Arabidopsis to stress requires DNA methylation and the function of Dicer-like proteins. PLoS One 5(3):e9514
9. Chatgilialoglu C, O'Neill P (2001) Free radicals associated with DNA damage. Exp Gerontol 36(9):1459–1471
10. Filkowski J, Yeoman A, Kovalchuk O, Kovalchuk I (2004) Systemic plant signal triggers genome instability. Plant J 38(1):1–11
11. Freeman SM, Aboud CN, Whartenby KA, Packman CH, Koeplin DS, Moolten FL, Abraham GN (1993) The "bystander effect": tumor regression when a fraction of the tumor mass is genetically modified. Cancer Res 53:5274–5283
12. Grant M, Lamb C (2006) Systemic immunity. Curr Opin Plant Biol 9(4):414–420

13. Hamada N, Matsumoto H, Hara T, Kobayashi Y (2007) Intercellular and intracellular signaling pathways mediating ionizing radiation-induced bystander effects. J Radiat Res Tokyo 48(2):87–95
14. Huo L, Nagasawa H, Little JB (2001) HPRT mutants induced in bystander cells by very low fluences of alpha particles result primarily from point mutations. Radiat Res 56(5 Pt 1):521–525
15. Jiménez-Martínez ES, Bosque-Pérez NA, Berger PH, Zemetra RS, Ding H, Eigenbrode SD (2004) Volatile cues influence the response of *Rhopalosiphum padi* (Homoptera: Aphididae) to Barley Yellow Dwarf virus–infected transgenic and untransformed wheat. Environ Entomol 33(5):12007–12016
16. Klokov D, Criswell T, Leskov KS, Araki S, Mayo L, Boothman DA (2004) IR-inducible clusterin gene expression: a protein with potential roles in ionizing radiation-induced adaptive responses, genomic instability, and bystander effects. Mutat Res 568(1):97–110
17. Koturbash I, Rugo RE, Hendricks CA, Loree J, Thibault B, Kutanzi K, Pogribny I, Yanch JC, Engelward BP, Kovalchuk O (2006) Irradiation induces DNA damage and modulates epigenetic effectors in distant bystander tissue in vivo. Oncogene 25(31):4267–4275
18. Kovalchuk I, Kovalchuk O, Kalck V, Boyko V, Filkowski J, Heinlein M, Hohn B (2003) Pathogen-induced systemic plant signal triggers DNA rearrangements. Nature 423(6941):760–762
19. LaVerne JA (2000) OH radicals and oxidizing products in the gamma radiolysis of water. Radiat Res153(2):196–200
20. Liu Z, Mothersill CE, McNeill FE, Lyng FM, Byun SH, Seymour CB, Prestwich WV (2006) A dose threshold for a medium transfer bystander effect for a human skin cell line. Radiat Res 166(1 Pt 1):19–23
21. Lyng FM, Maguire P, Kilmurray N, Mothersill C, Shao C, Folkard M, Prise KM (2006) Apoptosis is initiated in human keratinocytes exposed to signalling factors from microbeam irradiated cells. Int J Radiat Biol 82(6):393–399
22. Molinier J, Oakeley EJ, Niederhauser O, Kovalchuk I, Hohn B (2005) Dynamic response of plant genome to ultraviolet radiation and other genotoxic stresses. Mutat Res 571(1–2):235–247
23. Molinier J, Ries G, Zipfel C, Hohn B (2006) Transgeneration memory of stress in plants. Nature 442(7106):1046–1049
24. Morgan WF (2003) Non-targeted and delayed effects of exposure to ionizing radiation: I. Radiation-induced genomic instability and bystander effects in vitro. Radiat Res 159(5):567–580
25. Morgan WF (2003) Non-targeted and delayed effects of exposure to ionizing radiation: II. Radiation-induced genomic instability and bystander effects in vivo, clastogenic factors and transgenerational effects. Radiat Res 159(5):581–596
26. Morgan WF, Sowa MB (2007) Non-targeted bystander effects induced by ionizing radiation. Mutat Res 616(1–2):159–164
27. Morgan WF, Hartmann A, Limoli CL, Nagar S, Ponnaiya B (2002) Bystander effects in radiation-induced genomic instability. Mutat Res 504(1–2):91–100
28. Mothersill C, Seymour C (2001) Radiation-induced bystander effects: past history and future directions. Radiat Res 155:759–767
29. Mothersill C, Seymour C (2003) Radiation-induced bystander effects, carcinogenesis and models. Oncogene 22(45):7028–7033
30. Mothersill C, Seymour C (2004) Radiation-induced bystander effects – implications for cancer. Nat Rev Cancer 4:158–164
31. Mothersill C, Seymour CB (2006) Radiation-induced bystander effects and the DNA paradigm: an "out of field" perspective. Mutat Res 597(1–2):5–10
32. Mothersill C, Bucking C, Smith RW, Agnihotri N, Oneill A, Kilemade M, Seymour CB (2006) Communication of radiation-induced stress or bystander signals between fish in vivo. Environ Sci Technol 40(21):6859–6864
33. Mothersill C, Smith RW, Agnihotri N, Seymour CB (2007) Characterization of a radiation-induced stress response communicated in vivo between zebrafish. Environ Sci Technol 41(9):3382–3387

34. Park SW, Kaimoyo E, Kumar D, Mosher S, Klessig DF (2007) Methyl salicylate is a critical mobile signal for plant systemic acquired resistance. Science 318(5847):113–116
35. Ries G, Heller W, Puchta H, Sandermann H, Seidlitz HK, Hohn B (2000) Elevated UV-B radiation reduces genome stability in plants. Nature 406(6791):98–101
36. Sawant SG, Randers-Pehrson G, Metting NF, Hall EJ (2001) Adaptive response and the bystander effect induced by radiation in C3H 10T(1/2) cells in culture. Radiat Res 156(2):177–180
37. Sawant SG, Zheng W, Hopkins KM, Randers-Pehrson G, Lieberman HB, Hall EJ (2002) The radiation-induced bystander effect for clonogenic survival. Radiat Res 157(4):361–364
38. Schilmiller AL, Howe GA (2005) Systemic signaling in the wound response. Curr Opin Plant Biol 8(4):369–77
39. Sedelnikova OA, Nakamura A, Kovalchuk O, Koturbash I, Mitchell SA, Marino SA, Brenner DJ, Bonner WM (2007) DNA double-strand breaks form in bystander cells after microbeam irradiation of three-dimensional human tissue models. Cancer Res 67(9):4295–4302
40. Shulaev V, Silverman P, Raskin I (1997) Airborne signalling by methyl salicylate in plant pathogen resistance. Nature 385:718–721
41. Smilenov LB, Hall EJ, Bonner WM, Sedelnikova OA (2006) A microbeam study of DNA double-strand breaks in bystander primary human fibroblasts. Radiat Prot Dosimetry 122(1–4):256–259
42. Suzuki M, Zhou H, Hei TK, Tsuruoka C, Fujitaka K (2003) Induction of a bystander chromosomal damage of He-ion microbeams in mammalian cells. Biol Sci Space 17(3):251–252
43. Swoboda P, Gal S, Hohn B, Puchta H (1994) Intrachromosomal homologous recombination in whole plants. EMBO J 13:484–489
44. Truman W, de Zabala MT, Grant M (2006) Type III effectors orchestrate a complex interplay between transcriptional networks to modify basal defense responses during pathogenesis and resistance. Plant J 46:14–33
45. Yang G, Wu L, Chen L, Pei B, Wang Y, Zhan F, Wu Y, Yu Z (2007) Targeted irradiation of shoot apical meristem of Arabidopsis embryos induces long-distance bystander/abscopal effects. Radiat Res 167(3):298–305
46. Zhou H, Randers-Pehrson G, Waldren CA, Vannais D, Hall EJ, Hei TK (2000) Induction of a bystander mutagenic effect of alpha particles in mammalian cells. Proc Natl Acad Sci USA 97(5):2099–2104
47. Zhou H, Suzuki M, Geard CR, Hei TK (2002) Effects of irradiated medium with or without cells on bystander cell responses. Mutat Res 499(2):135–141
48. Zhou H, Randers-Pehrson G, Suzuki M, Waldren CA, Hei TK (2002) Genotoxic damage in non-irradiated cells: contribution from the bystander effect. Radiat Prot Dosimetry 99(1–4):227–232

Chapter 10
Direct and Reverse Bystander Effect Between Irradiated and Unirradiated Organisms: The Modulating Role of Chemosignalling in Ecology

B.P. Surinov, V.G. Isayeva, and A.N. Sharetsky

Abstract It was shown that mice or rats exposed to sublethal dose of ionizing radiation are able to decrease the immune reactivity of intact animals when them kept together in the same cage. Even one individual can cause such a disturbances of immunity in a group of intact animals. The data indicate that at early stage after exposure to sublethal doses of ionizing radiation mice secrete volatile components (VC) in urine which decreased thymus dependent humoral immune response in intact animals (to 60–70% relative control). The stable effect observed as a result of 1-day exposure allowed us to study the time course of secretion of immunosuppressive VC by irradiated animals. These VC possess attractive properties for the intact individuals. The irradiated mice showed an increased attractiveness to intact individuals. The biological significance of a combination of the immunosuppressive and attractive effects of VC is unclear. It is supposed, that mammals possess of the distant immunomodulating chemosignal system, aimed at the immunoreactivity of individuals with immunodeficiency state.

It was established that VC of intact mice restored the humoral immune response and other parameters of immunity in irradiated with a dose of 1 Gy animals. In this case, the irradiated recipients demonstrated an increase of humoral immune response to 140–170%. These VC of intact mice activated at irradiation mice phagocytic activity peritoneal macrophages. Exposure of rats for the third day after irradiation

This work was supported by the Russian Foundation for Basic Research and the Government of Kaluga Region project No 09-04-97505.

B.P. Surinov (✉) • V.G. Isayeva • A.N. Sharetsky
Medical Radiological Research Center of the Russian Academy of Medical Sciences, Kaluga Region, ul. Koroliova, 4, Obninsk 249036, Russia
e-mail: surinov@mrrc.obninsk.ru

C.E. Mothersill et al. (eds.), *Radiobiology and Environmental Security*,
NATO Science for Peace and Security Series C: Environmental Security,
DOI 10.1007/978-94-007-1939-2_10, © Springer Science+Business Media B.V. 2012

(1 Gy) to the VC of intact animals significantly increased the number of red blood cells, lymphocytes and granulocytes in the bloodstream.

Thus, direct and reverse bystander chemosignaling between the irradiated and intact, irradiated and irradiated animals mediate the modification of immunity and behavioral reactions of recipients. Apparently these untargeted effects of radiation spreading from one individual to another can have a significant impact on the viability of the entire population of animals.

Keywords Ionizing radiation • Bystander effect • Chemosignaling • Immunity

10.1 Introduction

As established previously, exposure to ionizing radiation, stress, immunodepressants induced in laboratory animals excretion with urine absent in norm volatile components (VC). Exposure of these VC to intact conspecifics reduced their immunoreactivity and the number of hemocytes [11, 12]. It was found later that such excretions have the properties of chemosignals attracting intact animals.

Combination of attractive and immunosuppressive properties inherent to VC was not explained in terms of biological feasibility a long time. In the published data has been reported only about the role of signaling in natural physiological conditions [1, 3], in particularly [8], in disassortative mating of animals relevant for the viability of offspring or for limitation contact of healthy and infected individuals [4, 5, 9].

In the literature we are not found any information about chemosignalling induced by ionizing irradiation in animals with the exception of works of authors of this paper [10–12].

Meanwhile the advantage of use of such damage as ionizing radiation allowed much deeper to explore the issue of chemocommunication in pathological state and detect a number of previously unknown facts. For example, it was found that the secretion of attractive or aversive chemosignals depended on the dose of ionizing radiation and period after radiation injury [10]. As result is the assumption that the attractive chemosignals induced by sublethal doses of radiation provide with a communication aimed at improving the viability of the animals, whereas the aversive signals appeared in terminal period after exposure to lethal doses of radiation, probably, like an infection, may play a role in limitation of propagation of pathology in the group of animals [14].

This paper presents the results of further study the stimulating effect of urinary volatiles of irradiated and intact animals on the antibody forming capacity of individuals with reduced immune reactivity induced with sublethal dose (1 Gy) of ionizing radiation, or on these parameters in animals immunized with thymus dependent antigen.

10.2 Materials and Methods

10.2.1 Animals and Procedures

Laboratory male mice of the inbred CBA strain and hybrids F1 (CBA × C57Bl/6), 2–3 months old (25–30 g) where kept in standard plastic boxes by groups of 5–6 per box with ad libidum food at room temperature (20–22°C) under a 14 light: 10 h dark cycle. (In certain experiments are used the Wistar rats).

The whole-body irradiation of the animals was carried out with doses of 1 or 4 Gy using a ^{60}C0 source with a dose-rate of 3.2 mGy/s on “Gamma-Cell-220” equipment (Atomic Energy Canada Limited, Canada).

Donors urine volatile components (VC) were intact and irradiated with a dose 4 Gy individuals.

To obtain urine samples at the bottom of the box was placed sheet of filter paper (bedding). Access to this bedding was limited by a performed screen elevated above the bottom at 0.5 cm. Paper bedding containing of absorbed 72 h urine of irradiated or intact mice were transfer for 1 day in box with radiation at a dose of 1 Gy recipients under the mesh screen.

10.2.2 Immunological Investigations

To study the influence VC in the development of the immune response papers with urine samples were placed in boxes to the immunized animals. In 24 h after exposure with bedding animals were injected intraperitoneally with 0.2 ml of suspension of sheep red blood cells (SRBC) in dose 1×10^8 cell/mouse in medium 199. In 4 days mice were decapitated of under ester narcosis and determined the mass and cellularity of the spleen. The number of antibody forming cells (AFCs) in the spleen was determined by Cunningham method. In each of the studied groups was 5–7 individuals.

10.2.3 Phagocytosis

Effect of VC on the phagocytic activity of peritoneal macrophages of mice-recipients was assessed by their ability to absorb latex particles of diameter 1.5 μm (firm DIA*M). Mice were decapitated under ether anesthesia. The suspension of macrophages obtained by washing the abdominal cavity 199 medium supplemented with 10% fetal calf serum. The reaction of phagocytosis set by mixing 1 ml of suspension of macrophages in a concentration 10^6 cells/ml with 1 ml of latex suspension at a concentration of 10 μl/ml. The mixture was incubated for 30 min at 37°C. Smears were fixed 30 min in 96% ethyl alcohol and dyed with Romanovsky-Giemsa. Counted the number of absorbed latex particles per phagocytic macrophage (phagocytic index).

Experiments were reproduced at least three times.
Statistic analysis of results we performed with use of Student's t-test.

10.3 Results and Discussion

10.3.1 Immunological Investigations

The ionizing radiation with 1 Gy causes suppression of immune reactivity in mice. Thus, 3 days after irradiation of animals, significantly reduced spleen weight, cellularity and the number of AFC induced by challenge with SRBC (Table 10.1). As a result of exposure of these mice during the third day after irradiation to VC of intact syngeneic animals led to the fact that the spleen cellularity and the number of AFC significantly exceeded these parameters in the irradiated animals not exposed to VC. Almost the same effects were observed after exposure to VC of individuals irradiated with a dose of 4 Gy.

A similar effect of these VC in the recipient mice on day 7 after irradiation (1 Gy) was also accompanied by a significant increase in the ability to the immune response to antigen (Table 10.1). In this case, unlike the effect observed on day 3 after irradiation, change in spleen mass and cellularity not detected.

Table 10.1 Immunological parameters (M±m) in irradiated 1 Gy male mice F1 (CBA×C57Bl6) after exposition to volatile components (VC) of urine of intact mice in different time post irradiation periods

Groups of animals	Time after irradiation, days	Spleen Mass, mg	 Number of cells, 1×10^6	 AFC, 1×10^3
Control	–	130±6.5 (100±5.0)	189±12.3 (100±6.5)	121±9.0 (100±7.4)
Irradiated 1 Gy	3	94.7±3.8* (72.8±2.9)	137±8.4* (72.5±4.4)	52±3.1* (42.9±2.6)
Irradiated 1 Gy+VC intact mice		82.8±2.5*,** (63.7±1.9)	134±14* (70.9±7.3)	61.7±2.6*,** (50.9±2.2)
Irradiated 1 Gy	7	110±3.8 (84.6±2.9)	164±14 (86.8±7.4)	50±2.2 (41.2±1.8)
Irradiated 1 Gy+VC intact mice		105±5.2 (80.8±4.0)	145±13 (77.2±6.3)	66.2±2.4*,** (54.6±2.0)

Note: In brackets – % to control
*Significant different ($p<0.05$) from control
**Significant differences ($p<0.05$) from 1 Gy

Table 10.2 Dynamics immunological parameters (M ± m) in irradiated 1 Gy male mice CBA after 1-day exposition to volatile components (VC) of urine of intact or irradiated (4 Gy) singenic mice in during the third day post irradiation period

Groups of animals	Time after exposition, days	Mass of spleen, mg	Number of cells at spleen, 1×10^6	AFC in spleen, 1×10^3
Irradiated 1 Gy	1	101 ± 2.3 (100 ± 2.3)	80.0 ± 7.1 (100 ± 8.9)	104 ± 2.9 (100 ± 2.8)
Irradiated 1 Gy + VC intact mice		83.6 ± 4.3* (82.8 ± 4.3)	83.3 ± 3.4 (104 ± 4.3)	126 ± 6.0* (121 ± 5.8)
Irradiated 1 Gy + VC irradiation 4 Gy mice		86.0 ± 3.3* (85.1 ± 3.3)	103 ± 3.4* (129 ± 4.3)	125 ± 15.0 (120 ± 14.4)
Irradiated 1 Gy	3	90.4 ± 1.8 (100 ± 2.0)	74.0 ± 5.1 (100 ± 6.9)	67.4 ± 9.0 (100 ± 13.4)
Irradiated 1 Gy + VC intact mice		83.4 ± 5.2 (92.3 ± 5.8)	84.0 ± 9.8 (114 ± 13.2)	96.7 ± 6.0* (143 ± 8.9)
Irradiated 1 Gy + VC irradiation 4 Gy mice		99.0 ± 4.0 (110 ± 4.4)	82.0 ± 12.4 (111 ± 16.8)	113 ± 6.8* (168 ± 10.0)
Irradiated 1 Gy	7	83.4 ± 2.7 (100 ± 3.3)	90.0 ± 3.2 (100 ± 3.5)	85.0 ± 6.7 (100 ± 7.9)
Irradiated 1 Gy + VC intact mice		102 ± 2.8* (122 ± 3.4)	116 ± 17.5 (129 ± 19.4)	183 ± 4.5* (215 ± 5.3)
Irradiated 1 Gy + VC irradiation 4 Gy mice		115 ± 3.8* (138 ± 4.6)	104 ± 5.1* (115 ± 5.6)	195 ± 10.4* (229 ± 12.2)

Note: In brackets – % to irradiated mice
*Significant differences ($p<0.05$) from irradiated 1 Gy

We evaluated the immune response in mice irradiated with a dose of 1 Gy, at different times after their exposure to VC of intact or irradiated with Gy of individuals (Table 10.2).

As a result of exposure of these mice to VC of intact animals during the third day after exposure to ionizing radiation the spleen cellularity and the number of AFC was significantly higher than these parameters in the group of irradiated mice which is not exposed VC.

Within 3 days after exposure to VC of irradiated with 1 Gy mice, showed a significant, approximately 1.5-fold, increasing of the number of AFC in the spleen.

In more remote periods, within 7 days after exposure, effects of VC on irradiated mice increased – the spleen cellularity and the number of AFC in was more than two times higher than that in animals not exposed to VC.

During this observation period is also increased as the mass and cellularity of the spleen, especially after exposure to VC produced by mice irradiated with 4 Gy.

Therefore, the single exposure of the irradiated with 1 Gy mice to VC of intact or irradiated (4 Gy) was accompanied by stimulation of the immunity. The effect not only on long-preserved, but strengthened over time.

The above data rather differ from previously described ones with respect to the impact of a single exposure with the same VC, but at different times after irradiation of the mice with a dose of 1 Gy.

In this case, stimulation of humoral immunity in the irradiated with the same dose of mice was less pronounced.

This confirms the above data relating to the development of immunopotentiating effect in the course of time after exposure of irradiated animals to VC.

These data are in fact the opposite to results obtained in the process of studying of the immune reactivity of intact mice exposed to VC produced by mice after exposure to 4 Gy. In the latter case, there was inhibition of humoral immunity, which lasted more than 3 days.

In one series of experiments were simultaneously evaluated the impact of VC produced by male mice-CBA after exposure to radiation with 4 Gy (3 days), on intact animals and irradiated with 1 Gy.

It was found that as a result of exposure of VC with intact or subjected to radiation with 1 Gy mice, there was a decrease or increase the number of AFC in the spleen up to 62.0 ± 5.6% ($p<0.05$) and up to 95.6 ± 17.5%, respectively.

In irradiated at a dose of 1 Gy mice not exposed to VC the number of AFC in the spleen was 58.3 ± 1.9% ($p<0.05$) compared with the intact animals (100 ± 2.6%).

Therefore, postradiation VC have different directions in the action depending on the state of recipients. It should be emphasized that the described here immunostimulatory effect was revealed at relatively weakly expressed immunodeficiency, as in this case at deficiency caused by irradiation with a dose of 1 Gy.

Similar experiments, but with mice irradiated with a dose of 4 Gy, resulted in irregular results and in case of exposure to animals with 6 Gy immunostimulatory effect was not found.

Ability of chemosignals produced intact mice to increase the immune reactivity of irradiated animals also demonstrated on a model which has previously been used to confirm the immunosuppressive effect of irradiated mice to intact when the irradiated and intact animals in equal proportions contained in one box.

Assessment of the capacity to antibody genesis of the irradiated with a dose of 2 Gy male mice (10 animals in each group) contained during the 2 weeks together with intact animals showed that the number of AFC in the spleen in 1.4 times ($p<0.05$) higher than that of in only irradiated mice kept separately (Table 10.4).

Immunostimulatory effect observed under the combined housing in one box intact mice with individuals irradiated at a higher dose (4 Gy) was unstable and not reproduced in all series of experiments.

The presence in urine of chemosignals recovering under certain conditions relatively not too deep post-radiation damage was confirmed by experiments with rats in which investigated the number of formed elements in peripheral blood after exposure to the VC (Table 10.5). Exposure of irradiated (1 Gy) of rats during the third day of post-radiation period to VC of intact or irradiated animals caused quantitative changes in the cellular composition of blood.

Thus, 3 days after irradiation (1 Gy) the number of red blood cells in rats was $7.75 \pm 0.24\ 1 \times 10^{12}$/l. Exposure of such animals with the components of irradiated or intact rats increased in recipients compared to 1.2 times.

Influence VC of the intact or irradiated (4 Gy) animals on the intact or the irradiated (1 Gy) rats also resulted in increase of the absolute number of blood lymphocytes and polymorphonuclear leukocytes, respectively, 1.7 and 1.9 times (Table 10.3).

Table 10.3 Cell structure of periferalnblood (M ± m) irradiated in the dose 1 Gy (3 day) rat Wistar after exposition to VC intact or irradiated (4 Gy) animals

Groups of animals	Erythrocytes, 1×10^{12}/l	Leucocytes, 1×10^{9}/l	Lymphocytes, 1×10^{9}/l	Neurophiles, 1×10^{9}/l
Irradiated	7.75±0.24 (100±3.1)	10.7±1.4 (100±13.0)	3.2±0.27 (100±8.4)	2.4±0.45 (100±18.8)
Irradiated + VC intact rat	9.65±0.24* (123±3.1)	14.1±1.5 (132±14.0)	5.5±0.87* (172±27.0)	4.54±0.61* (189±25.4)
Irradiated + VC irradiated rat	9.4±0.48* (121±6.2)	9.8±0.7 (91.6±6.5)	3.4±0.3 (106±9.4)	2.8±0.18 (117±7.5)

Note: *Significant differences (p<0.05) from groups irradiated rat not exposition exposition with VC

Table 10.4 Immunological parameters (M ± m, in brackets – % to control) in irradiated (1 Gy) male mice CBA after co-habitation during the third day with intact male

Groups of animals	Spleen		
	Mass, mg	Number of cells, 1×10^{6}	AFC, 1×10^{3}
Intact	113±5.6 (100±5.0)	106±6.8 (100±6.4)	182±8.3 (100±4.6)
1 Gy	76.6±3.8* (67.8±3.4)	72.0±2.0* (67.9±1.9)	141±19.3 (77.5±10.6)
1 Gy + intact male	86.4±3.8* (76.5±3.4)	108±8.6** (102±8.1)	228±32.4** (125±17.8)

Note: In brackets – % to intact control
*Significant differences (p<0.05) as compared with intact animals
**Significant differences (p<0.05) as compared with irradiating animals

Table 10.5 Immunological parameters (M ± m, in brackets – % to control) in intact male mice CBA after co-contents during the third day with irradiated male

Version experiment	Mass of spleen, mg	Number of cells at spleen, 1×10^{6}	AFC in spleen, 1×10^{3}
Control – content wish intact individual	104±1.7 (100±2.1)	132±9.0 (100+7.1)	90.1±4.9 (100±11.2)
Experiment – content wish irradiated individual	95.3±2.1 (92.1±2.0)	107±6.4 (81.1±5.0)	74.1±4.6 (82.8±6.0)*

Note: In brackets – % to irradiated mice
*Significant differences (p<0.05) from control

Among the findings of particular interest is the fact that the immunostimulatory effect on irradiated mice is intrinsic not only VC that are contained in the urine of mice, but also the intact individuals themselves. It was found that the presence even one intact male-mouse in the group of irradiated mice (1 Gy within 3 days) increased their immune reactivity (Table 10.4), and presence in group of intact animals only one irradiated male depressed their immune reactivity (Table 10.5) Hence, it is probably of VC produced by one animal is sufficient to demonstrate their chemosignal properties aimed at mobilizing the immune system.

In one series of experiments studied the effect of post-radiation VC on animal induced by non-radiation factors. As one of these models has been elected immune response to foreign antigen.

This antigens were sheep red blood cells (SRBC). To some extent this is model of response to infection. Mice recipients were exposed to VC gave off irradiated (4 Gy) individuals up to 1 day, immediately, 3 and 7 days after injection of SRBC.

The number of AFC in the spleen was evaluated 4 days after injection of antigen. As a result of exposure to post-radiation VC before the antigenic stimulus the number of AFC in the spleen decreased to 30.4±6.5% relative to controls.

Such reaction is fully consistent with previously obtained data showing the immunosuppressive properties of post-radiation VC excreted in the urine of irradiated animals [2]. After exposure, VC to the recipient mice immediately after the administration of antigen an immune response is not changed. In that case, when the recipients were exposed to VC through 1, 3 or 7 days after immunization, the number of AFC increased compared with the control to 193±11.0, 122±27.7 and 263±60.7%, respectively.

The presented here results of effects VC of irradiated mice on individuals before and after immunization showed the dual nature of the studied chemosignals – the direction of their influence on syngeneic animals depends on the state of the immune system. If before the immunization was observed suppression then after immunization – stimulation of the humoral immune response (Table 10.6).

10.3.2 *Phagocytic Activity*

Study of the effect of VC on the phagocytic activity of peritoneal macrophages of irradiated mice showed (Table 10.7) that phagocytic index (number of particles absorbed by one of phagocytic macrophages) in mice-recipients after exposure to VC of intact animals increased in 1.4 times in comparison with irradiated unexposed animals.

This information is also, as in the case of antibody genesis, opposite the previously obtained data showing the inhibitory effect of post-radiation VC on phagocytosis.

In the data presented here show that the combination of mutual attractiveness of irradiated and intact with [13] described here immunostimulating effects indicate to the existence of biologically relevant complex of reactions regulated by chemosignals. In this complex participate both animals, producing chemosignals induced by various factors, and individuals whose reaction to the signals depends on their physiological state.

Our data allow to revise the view at some biological traits of the bystander effect at the interorganismal level [6, 7, 10].

If previously considered that such an effect caused by unknown of post-radiation VC were directed only at the suppression of immunity, the above data demonstrate also the opposite effect – the impact VC of intact animals on irradiated individuals resulted in the restoration of immunity. The such phenomenon (immunostimulation) was observed also and after exposure VC to animals after antigenic administration.

Table 10.6 Immunological parameters (M ± m) at male mice F1 (CBA × C57Bl6) after exposition to volatile components (VC) irradiated (4 Gy) individuals in different time relatively immunization

		Spleen		
Groups of animals	Time exposition, day	Mass, mg	Number of cells, 1×10^6	AFC, 1×10^3
Control	1 day before immunization	95.4 ± 3.3 (100 ± 3.5)	122 ± 9.8 (100 ± 8.0)	23.0 ± 0.5 (100 ± 2.2)
Exposition to VC irradiated		102 ± 3.9 (107 ± 4.0)	141 ± 8.0 (116 ± 7.0)	7.0 ± 1.5 * (30.4 ± 6.5)
Control	Immediately after immunization	119 ± 3.7 (100 ± 3.1)	143 ± 5.0 (100 ± 3.5)	27.5 ± 3.1 (100 ± 11.3)
Exposition to VC irradiated		104 ± 6.4 (87.4 ± 5.0)	135 ± 14.0 (94.4 ± 9.8)	30.8 ± 2.9 (112 ± 10.5)
Control	1 day after immunization	103 ± 5.0 (100 ± 4.9)	142 ± 13.0 (100 ± 9.2)	46.6 ± 2.9 (100 ± 6.2)
Exposition to VC irradiated		102 ± 4.8 (105 ± 5.0)	147 ± 10.2 (93.4 ± 6.5)	89.9 ± 5.0 * (193 ± 11.0)
Control	3 day after immunization	77.4 ± 2.6 (100 ± 3.4)	106 ± 6.0 (100 ± 5.7)	18.0 ± 3.8 (100 ± 21.0)
Exposition to VC irradiated		74.6 ± 1.9 (96.4 ± 2.5)	83.3 ± 6.8* (78.6 ± 6.4)	22.0 ± 5.0 (122 ± 27.7)
Control	7 day after immunization	77.2 ± 1.9 (100 ± 2.4)	112 ± 4.9 (100 ± 4.4)	9.5 ± 1.5 (100 ± 16.0)
Exposition to VC irradiated		76 ± 2.1 (98.4 ± 2.7)	88.0 ± 6.7* (78.6 ± 6.0)	25.0 ± 6.0* (263 ± 60.7)

Note: In brackets – % to control
*Significant different ($p < 0.05$) from control

Table 10.7 The phagocytic activity (M ± m) of peritoneal macrophages of irradiated (1 Gy, 3 dais) mice after exposure to the volatile components of urine of intact or irradiated (4 Gy) of individuals

Animals group	Phagocytic index
Intact	9.0 ± 0.4
Irradiated	13.2 ± 1.0*
Irradiated + VC of intact individuals	18.5 ± 0.6*,**
Irradiated + VC of irradiated individuals	12.8 ± 1.2*

Note: *Significant different ($p < 0.05$) from control; **significant different ($p < 0.05$) from irradiated

Hence, in our experiments we found that the orientation of the immunomodulatory effect of chemosignals depended on time of exposure of VC relatively of the effects of ionizing radiation or antigenic stimulus, i.e. the physiological condition of the recipient. This manifests itself most clearly in respect of post-radiation effects chemosignals induced by exposure to radiation with a dose of 4 Gy, which causes immunosuppression in intact animals, but stimulate the immune response in individuals immunized with an antigen, or irradiated with low doses (1 Gy) of radiation.

Under certain conditions, the immunosuppressive chemosignals for intact animals have regenerating and stimulating activity in relation to immunocompromised by the radiative and nonradiative effects individuals.

Therefore, bystander effect in groups of animals can have both suppressive and repaired consequences. Dependence of these effects upon the state of the donor chemosignals, and individuals-recipients can evaluate them as manifestation of biologically advisable of mechanisms that allow animals with chemosignaling to recognize individuals with disturbance of immunity and provide them by stimulating t and reparating effects.

10.4 Conclusion

The presented research findings demonstrate the ability of animals to produce chemosignals with previously unknown properties. Under certain conditions, they show the regenerating or stimulating activity in relation to other individuals who have a disturbance of immunity resulting from radiation and nonradiation effects. Therefore, bystander effect in groups of animals can have both suppress and repaired consequences. The direction of immunomodulatory effect is depends upon the status of immune system of donors and recipients of chemosignals. It is probably argue for existence of biologically advisable mechanisms that allow animals using the chemosignaling to recognize individuals with immunity disorders and provide them a stimulating and repaired effect.

References

1. Beauchamp GK, Yamazaki K (2003) Chemical signaling in mice. Biochem Soc Trans 31:147–151
2. Brennan PA, Kendrick KM (2006) Mammalian social odours: attraction and individual recognition. Philos Trans R Soc Lond B Biol Sci 361(1476):206–278
3. Hurst JL, Beynon RJ (2004) Scent wars: the chemobiology of competitive signaling in mice. Bioessays 26(12):1288–1289
4. Kavaliers M, Colwell DD (1995) Odours of parasitized males induce aversive response in female mice. Anim Behav 50:1161–1169
5. Moshkin MP, Gerlinskaya L, Morosova O, Bakhvalova V, Evsikov V (2002) Behaviour, chemosignals and endocrine functions in male mice infected with tick-borne encephalitis virus. Psychoneuroendocrinology 27:603–608
6. Mothersill C, Seymour CB (2001) Radiation-induced bystander effects: past history and future directions. Radiat Res 155:759–67, Review
7. Mothersill C, Bucking C, Smith RW, Agnihotri N, O'Neill A, Kilemade M, Seymour CB (2007) Communication of radiation-induced stress or bystander signals between fish in vivo. Environ Sci Technol 41:3382–3387
8. Novikov SN (1988) *Feromony i razmnozhenie mlekopitayushchikh* (Pheromones and reproduction in mammals). Nauka, Leningrad

9. Penn DJ, Potts WK (1998) Chemical signals and parasite-mediated sexual selection. Trends Ecol Evol 13:391–396
10. Surinov BP (2007) Mice with radiation or toxic damage or malignant tumors produce aversive chemosignals repelling intact animals. Dokl Biol Sci 414:199–201
11. Surinov BP, Dukhova NN (2004) Postradiation mice's volatile secretions which are attractive for intact individuals. Radiat Biol Radioecologiya 44(6):662–665, In Russian
12. Surinov BP, Isaeva VG, Dukhova NN (2005) Post-radiation immunosuppressing and attractive volatile secretions: "bystander effect" or allelopathy in groups of animals. Dokl Biol Sci 400(5):28–30
13. Surinov BP, Sharetsky AN, Shpagin DV, Isayeva VG, Dukhova NN (2007) Post-radiated and post-stressed volatile secretions: secondary immune and behavioral reactions in groups of animals. In: Mothersill C et al (eds.) Multiple stressors: a challenge for the future. Springer, Dordrecht, pp 127–138
14. Surinov BP, Isaeva VG, Karpova NA (2008) Immunostimulating chemosignaling in animals with secondary immunodeficiency. Dokl Biol Sci 418:41–43

Chapter 11
Molecular-Genetic Analysis of Genetic Predisposition to Myocardial Infarction and Comparison of Risk Factor Population Rates in Different Countries

Alexander Gonchar, Maxim Ameliyanovich, Kristina Zhur, Irma Mosse, and Konstantin Mosse

Abstract Factor V Leiden and Protrombin G20210A mutations as well as some gene polymorphisms, such as Val34Leu of the coagulation factor XIII, 4G/5G of PAI-1 gene and Thr312Ala of fibrinogen alpha chain have been investigated as risk factors of myocardial infarction (MI). The methods of polymerase chain reaction (PCR) with specially designed allele specific primers were used to determine polymorphisms and mutations. DNA extracted from lymphocytes or dried bloodspots was used as matrix for PCR. Blood samples of 175 patients with acute myocardial infarction and 270 people of the control group were investigated. It was obtained that the rate of Thr312Ala heterozygotes among patients with myocardial infarction was 1.3 times as high as in the control group. The frequency of Factor V Leiden and G20210 mutations in patients with myocardial infarction was more than 2 times as high as in the control group. Leu/Leu genotype was shown to be responsible for risk of myocardial infarction development too (OR = 2.02). According to our data 4G/4G genotype of the PAI-1 gene may be predisposing to the myocardial infarction and 5G/5G one – protecting. Population frequencies of MI genetic risk factors in Belarus population have been established. The comparison of MI predisposition genotypes frequencies with the levels of this disease in different countries of Europe has been performed. We revealed a statistically significant positive correlation between the frequencies of MI morbidity and 4G/4G genotype. At the same time, there was no statistically significant correlation

---

A. Gonchar • M. Ameliyanovich • K. Zhur • I. Mosse
Institute of Genetics and Cytology at National Academy of Sciences of Belarus, Akademicheskaya str. 27, Minsk 220072, Belarus
e-mail: i.mosse@igc.bas-net.by

K. Mosse (✉)
National Center of Research and Applied Medicine "Mother and Child", Orlovskaya str. 66, Minsk 220053, Belarus
e-mail: kosmosse@yahoo.com

C.E. Mothersill et al. (eds.), *Radiobiology and Environmental Security*,
NATO Science for Peace and Security Series C: Environmental Security,
DOI 10.1007/978-94-007-1939-2_11, © Springer Science+Business Media B.V. 2012

between frequencies of other risk factors and MI frequency. The obtained data allows conclusion that PAI-1 gene genotype 4G/4G is the most informative for evaluation of genetic predisposition to MI.

Keywords Myocardial infarction • Gene polymorphisms • Factor V Leiden and Protrombin mutations • Genetic predisposition • Population frequencies • European countries

11.1 Introduction

The problem of diagnostics of genetic predisposition to cardiovascular diseases such as infarctions, insults, thrombosis is very important because these diseases are the leading causes of mortality nowadays. The data on genetic risk factors for this kind of diseases are not sufficient enough and are rather inconsistent.

The investigation of genetic predisposition to arterial thrombosis and to myocardial infarction (MI) in particular is worth attention. So, we studied some mutations and gene polymorphisms which influence clot formation at the last stages of coagulation cascade. Scientific publications about genetic predisposition to arterial thrombosis are rather doubtful: some studies declare, for instance, that factor V Leiden is linked with arterial thrombosis [41, 55], while others show that it is not linked [47].

The Factor V Leiden mutation was described in 1994 by Bertina et al. [5]. This mutation results in single nucleotide guanine by adenine substitution in factor V gene in position 1691 and leads to arginine by glutamine substitution in position 506 of polypeptide chain (Arg506Gln). As a result of this substitution, the factor V becomes resistant to cleavage by activated protein C. Later on, this mutation was shown to be a significant risk factor for venous thrombosis and to increase risk of venous thrombosis 7 times for heterozygous individuals and 80 times for homozygous ones [51]. With this end in view, we undertake an investigation of the Factor V Leiden mutation as a risk factor of myocardial infarction.

Prothrombin gene encodes a protein (prothrombin), which is one of the most important factors in the blood coagulation system. Prothrombin is transformed into active thrombin. Thrombin activates factors XI and XIII, influencing formation of fibrin monomers, fibrin-oligomers (or soluble fibrin monomer complex) and fibrin-polymer and subsequent fibrin-polymer stabilization. G20210A mutation is a G to A transition in nucleotide 20210 of the prothrombin gene. The transition takes place at 3′ cleavage site of the mRNA resulting in elevated prothrombin levels in plasma. The level of this protein in plasma of heterozygous individuals can be higher by 30% than in those with wild genotype [36, 44]. This mutation frequency in population is 1–4%. The risk of venous thrombosis and thromboembolism in carriers of G20210A mutation is 2–4 times higher than in those without mutation [52]. This mutation may be responsible for the development of some pathological effects during pregnancy.

The data currently available in the literature do not give a definite answer about the role of G20210A mutation in the development of CVD, although the 2–4 times increase of venous thrombosis and thromboembolism risk in carriers of mutations is not contested [32].

Due to this, we decided to investigate the effect of this mutation on predisposition to myocardial infarction.

XIII blood clotting factor is a heterotetramer, consisting of two catalytic subunits (alpha subunits) with molecular weight of 83.2 kDa and two non-enzymatic subunits (beta – subunits) with molecular weight of 79.7 kDa, which are noncovalently incorporated into the tetrameric molecule with the mass of 325.8 kDa. After the activation by thrombin, the factor XIII catalyzes formation of covalent bonds between the alpha-and gamma-chains of fibrin, stabilizing thereby the fibrin clot. Fibrinogen is a soluble plasma glycoprotein, synthesized in the liver, and its elevated levels in blood plasma may result in the inflammation of the vascular wall, and thereby it can be associated with the risk of cardiovascular disease. Stabilization of the fibrin clot is done through the formation of covalent bridges between lysine and glutamine sites of the fibrin chain. In accordance with the clot-stabilizing role, the lack of the factor XIII turns into serious complications associated with excessive bleeding.

There are five polymorphisms described for alpha-subunits of the factor XIII. The most known one is the Val34Leu polymorphism [4]. It results in valine to leucine replacement in position 34 of the polypeptide chain. The Val34Leu polymorphism influences the structure of fibrinogen – a protein is involved in the process of clot formation at the final stages of the coagulation cascade and, consequently, influences the strength and elasticity of the thrombus [16, 45, 53, 56]. Substitution of valine for leucine is associated with the increased activation of the factor XIII by thrombin and with increased linker activity. The Val34Leu polymorphism does not alter the concentration of the factor XIII, but, however, alter its activity. The 34Leu isoform has increased catalytic efficiency 2.5 times and reduced time of clot formation as compared to the isoform 34Val [39].

Thr312Ala polymorphism of fibrinogen alpha chain – is one of the most important causes of arterial thrombosis because of its contribution to formation of fibrin clots consisting of thinner and compacted fibers with smaller pores between them. The fibrinogen level correlates with traditional risk factors, such as hypertension, smoking, diabetes, age and others [42]. The Thr312Ala polymorphism and the Val34Leu one of the Factor XIII gene are responsible for structural changes in fibrinogen molecule.

There are controversial data in the scientific literature about the Thr312Ala polymorphism relation to genetic predisposition to cardio-vascular diseases. It was shown [10] that this polymorphism influences the poststroke mortality in subjects with atrial fibrilation. In other paper it turned out that poststroke mortality of subjects with sinus rhythm was not dependant on the polymorphism variant, but subjects with atrial fibrillation and the Thr\Thr genotype had survival rate 2.3 times as high as the subjects with the Thr\Ala genotype. However these data were not corroborated in another paper [14].

So, the scientific literature data about the Thr312Ala polymorphism are insufficient for clear conclusion about its relationship with cardiovascular diseases.

4G/5G insertion/deletion in PAI-1 gene was studied too. This gene (the inhibitor of plasminogen activator – an important component of thrombolytic plasminogen-plasmin system) is widely studied in relation to the risk of acute coronary vascular disease, venous thrombosis, eclampsia and other vascular disorders. It is shown that the level of PAI-1 in plasma is associated with polymorphism in the promoter region of the PAI-1 gene, which is a single nucleotide deletion/insertion (4G/5G). Individuals who are homozygous for the 4G allele have a higher level of PAI-1 in plasma. The mechanism underlying allelic differences in the synthesis of PAI-1 was established and the affinity of both alleles with an activator of gene transcription was revealed, whereas the 5G allele has also a binding site for a transcriptional repressor. The absence of the repressor provides a higher level of gene transcription of the PAI-1 allele 4G. An increased PAI-1 concentration in blood is associated with an increased risk of thrombosis [54].

The influence of the 4G/5G polymorphism of PAI-1 on the risk of myocardial infarction was not clear. Some results suggest that the 4G/4G genotype causes an increased risk of myocardial infarction [18], but at the same time there are reports that this genotype is neutral [7] or even protective [17]. In general, there is data on the increased proportion of 4G/4G homozygotes among patients with myocardial infarction as compared with the control samples [25]. The 4G variant leads to increased gene expression and consequently to increased levels of PAI-1 in the blood and to higher risk of blood clots.

That's why these factors were chosen for myocardial infarction (MI) risk factor study.

11.2 Methods

Blood samples of 175 patients with acute myocardial infarction were investigated. These samples were provided on special DNA paper cards by The National Center of Research and Applied Medicine "Cardiology". A control group consisted of 270 people without cardiovascular pathology in anamnesis. Blood samples of the control group were provided by Medical Centre of the National Academy of Sciences of Belarus. The study was complied with the Declaration of Helsinki and was approved by the Local Ethics Committee. All patients and control subjects gave their written informed consent for participation in the study.

Blood samples were used as a matrix for PCR using the method of enhanced direct amplification described by Makowski [38]. The presence of the Factor V Leiden and G20210A mutations was detected by the amplification refractory mutation system (ARMS) method with allele-specific primers which were self-designed. PCR products were visualized on 8% polyacrylamide gel.

To determine the Val34Leu polymorphism of the factor XIII, a technique was developed on the basis of Tetra-primer ARMS. It is based on using two pairs of primers – internal and external in one PCR. Internal primers are allele-specific and the external ones flank polymorphic site in such a way that after the separation of PCR products by electrophoresis bands of different lengths correspond to different alleles.

The Thr312Ala polymorphism was detected using the method described by Carter et al. [10]. with several modifications: the annealing temperature was 62°C, magnesium concentration was 2.5 mmol/L and 1U per sample of RsaI (Fermentas, Vilnius, Lithuania) endonuclease was used for digestion.

To detect the 4G/5G polymorphism of the PAI-1 gene, we developed an original technique using an automated capillary electrophoresis with polychrome laser scanning. This method allows determination of the 4G/5G polymorphism alleles directly by the difference of DNA fragment size equal to one nucleotide.

Statistical analysis was processed using Microsoft Office Excel. The significance in any difference in proportions was tested by the chi-square method.

11.3 Results and Discussion

11.3.1 Molecular-Genetic Study on Myocardial Infarction Risk Factors

11.3.1.1 Factor V Leiden

Three percent of heterozygous carriers of the Factor V Leiden mutation were detected in the control group of Belarus people. This complies well with the population frequencies obtained by other authors in Russia and Belarus [23, 32]. Molecular-genetic analysis revealed the presence of heterozygous (RQ) genotypes in 12 of 175 patients with myocardial infarction (6.9%), none of the patients was homozygous (QQ) and 163 patients had RR genotype. This data was compared to results in the control group: 8 heterozygous genotypes of 270 people (3.0%) and none homozygous (QQ) were found (Table 11.1). So, the rate of heterozygous genotypes in the group with myocardial infarction was 2.4 times as high as in the control group (OR=2.41). This indicates the role of the factor V Leiden as a significant risk factor not only for venous thrombosis but also for arterial one (for myocardial infarction in particular).

The genotype distribution of this mutation did not differ significantly from that expected for Hardy-Weinberg proportions in the myocardial infarction patients as well as in the control group.

11.3.1.2 Factor II (G20210A)

During the investigation of the effect of mutation G20210A on the risk of MI, we found 4 heterozygous genotypes (GA) of the 175 examined samples in the experimental group and 3 heterozygous genotypes in the control group (270 samples).

Table 11.1 Molecular-genetic study on myocardial infarction risk factors

	Factor V (G1691A)	Myocardial infarction	Controls	OR	P value*
Genotype no. (%)	RR	163(91.3)	262(97.0)	0.41 (CI:0.17–1.04)	0.05
	R/Q	12(6.9)	8(3.0)	2.41(CI:0.96–6.02)	
	Q/Q	–	–	–	
Allele no. (%)	R	338(96.6)	532(98.5)	0.42 (CI:0.17–1.05)	0.06
	Q	12(3.4)	8(1.5)	2.36 (CI:0.96–5.84)	
	Factor II (G20210A)				
Genotype no. (%)	G/G	171(97.7)	267(98.9)	0.48 (CI:0.11–2.17)	0.56**
	G/A	4(2.3)	3(1.1)	2.08 (CI:0.46–9.42)	
	A/A	–	–	–	
Allele no. (%)	G	346(98.9)	537(99.4)	0.48 (CI:0.11–2.17)	0.56**
	A	4(1.1)	3(0.6)	2.07 (CI:0.46–9.30)	
	Factor XIII (Val34Leu)				
Genotype no. (%)	Val/Val	79(45.1)	123(45.6)	0.98 (CI:0.67–1.44)	0.19
	Val/Leu	81(46.3)	135(50.0)	0.86 (CI:0.59–1.26)	
	Leu/Leu	15(8.6)	12(4.4)	2.02 (CI:0.92–4.42)	
Allele no. (%)	Val	239(68.3)	381(70.6)	0.90 (CI:0.67–1.20)	0.47
	Leu	111(31.7)	159(29.4)	1.11 (CI:0.83–1.49)	
	Fibrinogen α-chain (Thr312Ala)				
Genotype no. (%)	Thr/Thr	96(54.9)	166(61.5)	0.76 (CI:0.52–1.12)	0.12
	Thr/Ala	76(43.4)	94(34.8)	1.44 (CI:0.97–2.12)	
	Ala/Ala	3(1.7)	10(3.7)	0.45 (CI:0.12–1.67)	
Allele no. (%)	Thr	268(76.6)	426(78.9)	0.87 (CI:0.63–1.21)	0.42
	Ala	82(23.4)	114(21.1)	1.14 (CI:0.83–1.58)	
	PAI-I (4G/5G)				
Genotype no. (%)	4G/4G	65(37.1)	78(28.9)	1.44 (CI:0.96–2.16)	0.19
	4G/5G	77(44.0)	134(49.6)	0.80 (CI:0.54–1.17)	
	5G/5G	33(18.9)	58(21.5)	0.85 (CI:0.53–1.37)	
Allele no. (%)	4G	207(59.1)	290(53.7)	1.25 (CI:0.95–1.64)	0.11
	5G	143(40.9)	250(46.3)	0.80 (CI:0.61–1.05)	

*Calculated for chi-squared test
**Calculated for Yates corrected chi-squared test

We didn't find any samples homozygous for this mutation. The frequency of mutant alleles in the experimental group was 2.3%, which is 2.2 times as high as in the control group (1.1%) (OR = 2.08).

11.3.1.3 Val34Leu

We found no differences in the frequencies of Val34 homozygous and Val34Leu heterozygous genotypes in the experimental and the control groups, but the frequency of individuals homozygous for 34Leu allele in the experimental group was twice as high as in the control group (Table 11.1). It means that the Leu/Leu genotype can be responsible for risk of myocardial infarction development (OR = 2.02).

In fact, the 34Leu isoform of FXIII is known to be activated more rapidly, what leads to shortened clot formation time and generation of clots with an altered structure.

11.3.1.4 Thr312Ala

We observed the following genotype distribution in the group of 175 patients with myocardial infarction: 96 Thr\Thr (54.9%), 76 Thr\Ala (43.4%), 3 Ala\Ala (1.7%). Genotype distribution in the group of 270 control subjects was as follows: 166 Thr\Thr (61.5%), 94 Thr\Ala (34.8%) and 10 Ala\Ala (3.7%) (Table 11.1). Thus, there were a decreased frequency of the Thr\Thr genotypes and an increased frequency of the Ala\Ala ones in patients as compared with the control group. The rates of Thr\Ala heterozygotes in two groups differed more considerable (OR = 1.44). At the same time the frequencies of alleles Thr and Ala in patients with myocardial infarction did not differ significantly from the control rates.

Assessing the results indicating predisposition to thrombosis due to the gene variant, it is necessary to take into account a complex nature of the disease and the possibility of cumulative impact of multiple genes, as well as of environmental factors.

Meanwhile it was obtained in our investigation that the rate of the Thr312Ala heterozygotes among patients with infarction myocardial was 1.3 times as high as in the control group. This fact may be an evidence for the impact of this polymorphism on the genetic predisposition to myocardial infarction.

11.3.1.5 4G/5G Polymorphism

Population-based distribution of PAI-1 alleles 4G/5G was studied. In the control group (270 persons), 28.9% of homozygous 4G/4G genotypes in comparison with 37.1% in patients were observed (OR = 1.44). As for the 5G/5G genotype its rate in the control group was 21.5% and 18.9% of such genotypes were registered in the case (Table 11.1).

These results show that the 4G/4G genotype may be predisposing to the myocardial infarction, and 5G/5G protecting.

11.3.2 Comparison of MI Risk Factor Population Rates with This Disease Levels in Different Countries

We compared population frequencies of genotypes and alleles of Val34Leu, 4G/5G polymorphisms and FVL mutation in different European countries with the incidence of myocardial infarction in those countries.

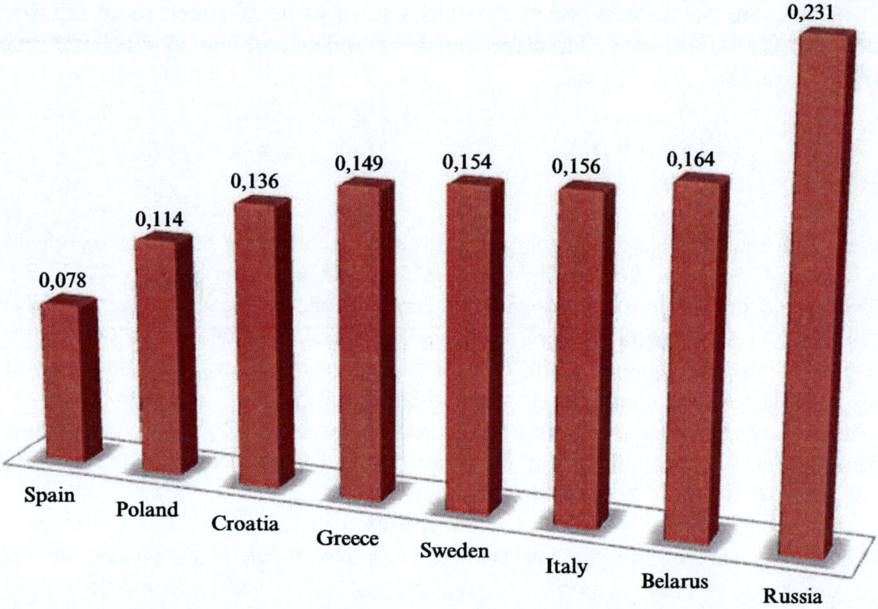

Fig. 11.1 The incidence of myocardial infarction among the inhabitants (percentage) of some European countries according to data of the year 2008 [57]

The frequency of myocardial infarction in the European countries is shown in Fig. 11.1. This diagram is based on the World Health Organization data [57].

The lowest rate of myocardial infarction was found in Spain (0.78%), the highest rate – 2.31% was found in Russia. Greece, Sweden, Italy and Belarus were characterized by the average level of myocardial infarction, while Poland and Croatia had lower one.

Figure 11.2 presents published data which demonstrate the following frequencies for the risk genotype Leu/Leu in different countries: Spain – 3.6% [13], Croatia – 7% [8], Greece – 6% [33], Sweden – 7.1% [15], Italy – 6.5% [28], Belarus – 4.4% [22].

Comparison of these two figures leads to the conclusion that there is no correlation between the incidence of myocardial infarction and the frequencies of the risk genotype Leu/Leu.

The Factor V Leiden mutation is a well-known risk factor of venous thromboembolism. Distribution of this mutation varies from 0% to 27%, with the lowest rate in Asian and African countries [27], and the highest one in Lebanon, Jordan [2, 29] and Israeli Arabs [50]. The incidence of the Factor V Leiden mutation among Caucasians is about 5%. It is reported that the frequency of this mutation is 3% in Belarusian population [32]. Our results corroborate this rate which is rather low in comparison with other western countries. A similar rate of the Factor V Leiden is detected in Russia – 2.6% [23].

Comparative analysis of the Factor V Leiden mutation in different countries and in different continents allows the conclusion about the absence of this mutation in

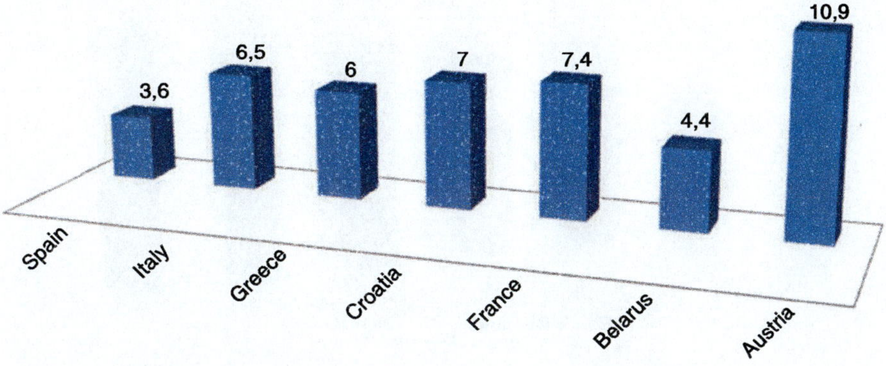

Fig. 11.2 Comparative analysis of the frequencies of the risk genotype Leu/Leu (percentage) in some European countries

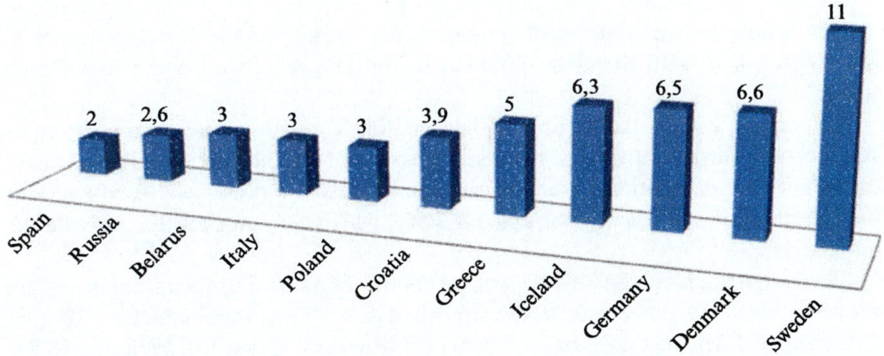

Fig. 11.3 Comparative analysis of the frequencies of the FVL mutation (percentage) in some European countries

African and Eastern Asian countries. Among other peoples a low rate of mutation carriers is revealed in Asian Americans – citizens of the USA (0.45%) [48] and the highest – in Israeli Arabs and Lebanon [29, 50]. No carriers of mutation are described in Africa [27]. Thus, Caucasians are characterized by higher levels of the Factor V Leiden mutation as compared to other races.

Thus, a low mutation frequency is found in Spain [19], Italy [49] and a high frequency in Denmark [34], Germany [24], Iceland [43] and Sweden [26]. It is reported that prevalence of the Leiden mutation is within the average in Croatia [12], Greece [1] and Canada [35].

Data on the frequencies of the Factor V Leiden mutations in populations of different European countries are represented in Fig. 11.3: Spain – 2% [37], Croatia – 3.9% [9], Greece – 5% [20], Sweden – 11% [11], Italy – 3% [46], Belarus – 3% [21], Russia – 2.6% [3]. No correlation between the incidence of myocardial infarction and the frequencies of the FVL mutation was observed.

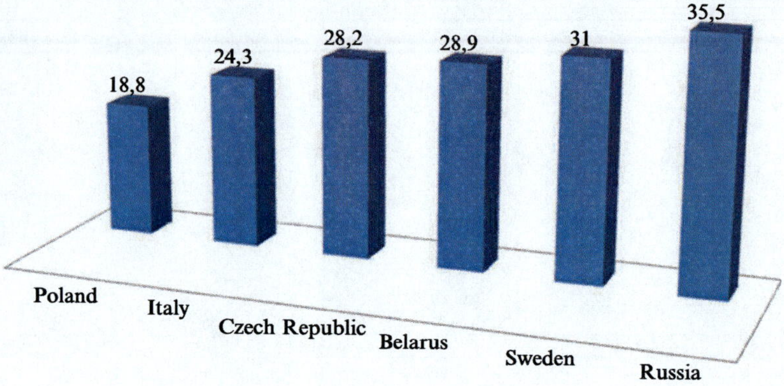

Fig. 11.4 Comparative analysis of frequencies of 4G/4G genotype in some European countries (percentage)

Our results on distribution of polymorphic variants of the PAI-I gene are in good agreement with the data obtained in the U.S.A., Sweden and the Czech Republic [26, 48].

However, a comparison with the data published in scientific literature reveals significant differences from the results in Russia and Poland. The possible cause of this is not only population-based differences, but also the previously used methods for determining allelic polymorphism 4G/5G, which may not always provide the reliable data.

The frequencies of the 4G/4G risk genotype in some European countries are shown in Fig. 11.4. (according to the published data). They are: Sweden – 31% [31, 58], Italy – 24.3% [40], Belarus – 28.9% [22], Russia – 35.5% [3], Poland – 18.8% [6, 30]. This 4G/4G risk genotype is the only one, the rate of which correlates with the level of myocardial infarction in different countries.

The comparison of MI predisposing genotype frequencies with the levels of this disease in different countries of Europe is shown in Fig. 11.5. Statistical analysis revealed a statistically significant positive correlation between the frequency of MI incidence and the frequency of the genotype 4G/4G ($r=0.89$, $P<0.04$).

Thus, the comparison of genetic data showed that a higher 4G/4G genotype level in the population corresponded to a higher incidence of MI, suggesting that this genotype contributed significantly to the development of the disease. The data obtained suggest that the 4G/4G genotype of the PAI-1 gene is the most informative gene variant for determining genetic predisposition to MI.

At the same time a significant correlation of the frequencies of the genotypes Leu34Leu and the Factor V Leiden mutation with the incidence of myocardial infarction has not been revealed.

It is known that development of myocardial infarction is highly dependent on environmental risk factors such as physical inactivity, stress, smoking, diabetes, hypertension and others. Most of these factors are associated with the level of social and economic state of the country, so the frequency of the disease depends largely on

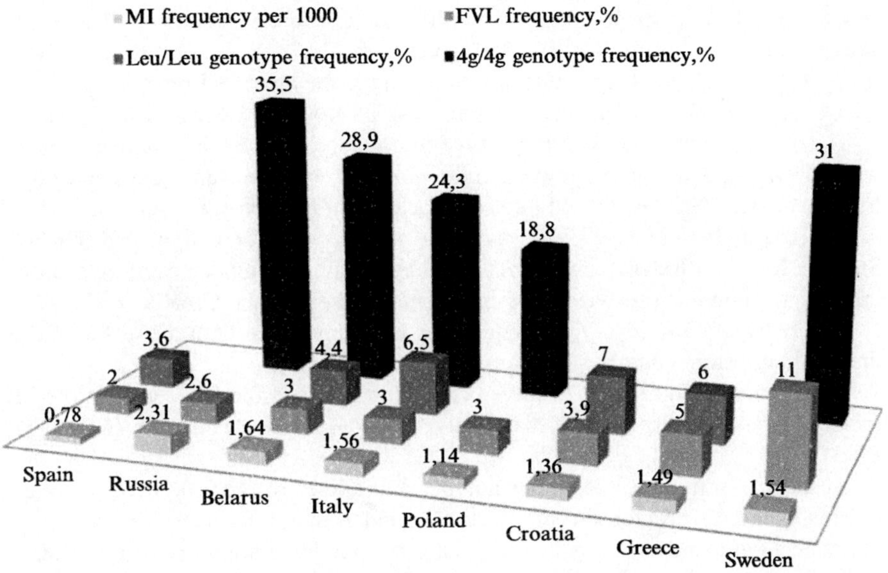

Fig. 11.5 Comparison of the risk genotype frequencies with the levels of MI in different countries

the lifestyle of people in different countries and may not coincide with the frequency of genetic risk factors. It is clear that separate alleles do not substantially increase the risk of myocardial infarction. A combination of the predisposing genotypes, as well as their interaction with environmental factors is the main cause of MI.

11.4 Conclusions

The obtained data show that the Factor V Leiden mutation and Protrombin gene mutation are the factors of an increased risk of myocardial infarction.

It was also shown that the 4G/4G genotype of the PAI-1 gene can predispose to myocardial infarction, and the 5G/5G one can be protective.

Genotype distributions of the Thr\Ala polymorphisms in the groups of patients with myocardial infarction and in the control group were also shown to be different – the frequency of the Thr\Ala genotypes in patients with myocardial infarction was significantly higher than in the control group (43.4% and 34.8% respectively).

The increase of the allele 34Leu homozygote frequency in patients with myocardial infarction twice as much as compared to the control group can correspond to an increased risk of this disease in people with the LL genotype.

Population frequencies of myocardial infarction (MI) genetic risk factors, such as, the 4G/4G genotype of the PAI-1 gene, and the mutation of the Factor V (Leiden) in Belarus population have been compared with the frequencies of identified genetic risk factors in populations of different European countries. The results allowed the conclusion that the frequencies of the risk genotypes Leu34Leu (Fig. 11.2) and 4G/4G (Fig. 11.4) in Belarus are at the average European level and the frequency of the Factor V Leiden mutation is below the average (Fig. 11.3).

The obtained data on the frequencies of risk genotypes in Belarus were compared with the frequencies of myocardial infarction and with relevant data from other European countries. The lowest rate of myocardial infarction was found in Spain – 0.78%, the highest rate – 2.3% was found in the population of Russia. Greece, Sweden, Italy and Belarus are characterized by the average incidence of myocardial infarction, while this rate was below the average in Poland and Croatia.

MI predisposition genotypes frequencies were compared with the levels of this disease in different countries of Europe (Figs. 11.1–11.5).

We revealed a significant positive correlation between the frequency of the MI incidence and the frequency of the genotype 4G/4G of the PAI-1 gene ($r=0.8971$, $P<0.04$).

At the same time the rates of genotypes and alleles of the Val34Leu polymorphisms of the XIII factor and the Factor V Leiden mutation were not revealed to correlate significantly with the myocardial infarction level in different countries.

Thus, it was shown that the higher 4G/4G genotype frequency in a population corresponded to the higher incidence of MI, suggesting that the 4G/4G genotype of the PAI-1 gene was the most informative for determining the genetic predisposition to MI.

It is known that the development of myocardial infarction is highly dependent on environmental risk factors such as physical inactivity, stress, smoking, diabetes, hypertension and others. Most of these factors are associated with the level of social and economic state of a country, so the frequency of the disease depends largely on the lifestyle of people in different countries and may not coincide with the frequency of genetic risk factors.

References

1. Antoniadi T, Hatzis T, Kroupis C, Economou-Petersen E, Petersen MB (1999) Prevalence of factor V Leiden, Prothrombin G20210A and MTHFR C677T mutations in a Greek population of blood donors. Am J Hematol 61:265–267
2. Awidi A, Shannak M, Bseiso A, Kailani MAM, Anshasi B, Sakarneh N, Omar N (1999) High prevalence of factor V Leiden in healthy Jordanian Arabs. Thromb Haemost 81(4):582–584
3. Baranovskaya S, Kudinov S, Fomicheva E, Vasina V, Solovieva D, Khavinson V, Schwartz E (2006) Age as a risk factor for myocardial infarction in Leiden mutation carriers. Mol Genet Metab 84:155–157
4. Bereczky Z, Balogh E, Katona E, Pocsai Z, Czuriga I, Széles G, Kárpáti L, Ádány R, Édes I, Muszbek L (2007) Modulation of the risk of coronary sclerosis/myocardial infarction by the interaction between factor XIII subunit A Val34Leu polymorphism and fibrinogen concentration in the high risk Hungarian population. Thromb Res 120:567–573
5. Bertina RM, Koeleman BPC, Koster T, Rosendal FR, Dirven RJ, de Ronde H (1994) Mutation in blood coagulation factor V associated with resistance to activated protein C. Nature 369:64–67

6. Blasiak J, Smolarz B (2000) Plasminogen activator inhibitor-1 (PAI-1) gene 4G/5G promoter polymorphism is not associated with breast cancer. Acta Biochim Pol 47:191–199
7. Boekholdt SM, Bijsterveld NR, Moons AH, Levi M, Büller HR, Peters RJ (2001) Genetic variation in coagulation and fibrinolytic proteins and their relation with acute myocardial infarction: a systematic review. Circulation 104:3063–3068
8. Broni A, Feren G, Zadro R, Rukavina A (2005) Factor XIII Val34Leu polymorphism in Croatian population. Thromb Haemost 3(1):69–75
9. Camilleri RS (2004) 455G/A [beta]-fibrinogen gene polymorphism, factor V Leiden, prothrombin G20210A mutation and MTHFR C677T, and placental vascular complications. Blood Coagul Fibrinolysis 15:139–147
10. Carter M, Catto AJ, Grant PJ (1999) Association of the α-fibrinogen Thr312Ala polymorphism with poststroke mortality in subjects with atrial fibrillation. Circulation 99:2423–2426
11. Catto A, Carter A, Ireland H, Bayston TA, Philippou H, Barrett J, Lane DA, Grant PJ (2004) Factor V Leiden gene mutation and thrombin generation in relation to the development of acute stroke. Arterioscler Thromb Vasc Biol 17:783–785
12. Coen D, Zadro R, Honovi L, Banfi L, Rukavina AS (2001) Prevalence and association of the factor V Leiden and prothrombin G20210A in healthy subjects and patients with venous thromboembolism. Croat Med J 42(4):488–492
13. Corral J, Iniesta JA (2000) Factor XIII Val34Leu polymorphism in primary intracerebral haemorrhage. Hematol J 1:269–273
14. Curran JM, Evans A, Arveiler D, Luc G, Ruidavets JB, Humphries SE, Green FR (1998) The α-fibrinogen T/A312 polymorphism in the ECTIM study. Thromb Haemost 79:1057–1058
15. de Lange M, Snieder H, Ariëns RAS, Spector TD, Grant PJ (2004) The genetics of haemostasis: a twin study. Lancet 357:101–105
16. de Lange M, Andrew T, Snieder H, Ge D, Futers TS, Standeven K, Spector TD, Grant PJ, Ariëns RAS (2006) Joint linkage and association of six single-nucleotide polymorphisms in the factor XIII-A subunit gene point to Val34Leu as the main functional locus. Arterioscler Thromb Vas Biol 26:1914–1919, 26
17. Doggen CJ, Bertina RM, Cats VM, Reitsma PH, Rosendaal FR (1999) The 4G/5G polymorphism in the plasminogen activator inhibitor-1 gene is not associated with myocardial infarction. Thromb Haemost 82:115–120
18. Eriksson P, Kallin B, van 't Hooft FM, Bavenholm P, Hamsten A (1995) Allele specific increase in basal transcription of the plasminogen activator inhibitor-1 gene is associated with myocardial infarction. Proc Natl Acad Sci USA 92(6):1851–1855
19. Frances F, Portoles O, Gabriel F, Corella D, Sorli JV, Sabater A (2006) Factor V Leiden (G1691A) and prothrombin-G20210A alleles among patients with deep venous thrombosis and in the general population from Spain. Rev Méd Chile 134(1):13–20
20. Gialeraki A, Politou M, Rallidis L, Merkouri E, Markatos C, Kremastinos D, Travlou A (2008) Prevalence of prothrombotic polymorphisms in Greece. Genet Test 12:541–547
21. Gonchar L, Mosse IB, Ivanov AA, Mosse NI, Mosse KA (2009) Association of the PAI-I polymorphism and Factor V Leiden mutation with genetic predisposition to myocardial infarction. Mol Appl Genet Proc 9:114–120
22. Gonchar L, Mosse IB, Ivanov AA, Mosse NI, Mosse KA (2009) Polymorphisms of clot formation genes and their role in myocardial infarction. Arterial Hypertens 15(4):466–469
23. Gusina AA, Gusina NB, Zatsepin IO, Baeshko AA, Naumchik IV (2007) The role of factor II and V mutations in the development of venous thrombosis in Belarusian citizens. Medicinskie novosti 5:85–88, In Russian
24. Herrmann FH, Salazar-Sanchez L, Schröder W, Grimm R, Schuster G, Jimenez-Arce G (2001) Prevalence of molecular risk factors FV Leiden, FV HR2, FII 20210G>A and MTHFR 677C>T in different populations and ethnic groups of Germany, Costa Rica and India. Int J Hum Genet 1:33–39
25. Hoekstra T, Geleijnse JM, Kluft C, Giltay EJ, Kok FJ, Schouten EG (2003) 4G/4G genotype of PAI-1 gene is associated with reduced risk of stroke in elderly. Stroke 34:2822–2828

26. Holm J, Zöller B, Berntorp E, Erhardt L, Dahlbäck B (1996) Prevalence of factor V gene mutation amongst myocardial infarction patients and healthy controls is higher in Sweden than in other countries. J Intern Med 239(3):221–226
27. Hong SH (2003) Genotype distribution of the mutations in the coagulation factor V gene in the Korean population: absence of its association with coronary artery disease. Korean J Biol Sci 7:255–259
28. Iacoviello L, Di Castelnuovo A, De Knijff P, D'Orazio A, Amore C, Arboretti R, Kluft C, Donati MB (2000) Polymorphisms in the coagulation factor VII gene and the risk of myocardial infarction. N Engl J Med 338:79–85
29. Irani-Hakime N, Tamim H, Kreidy R, Almawi WY (2000) The prevalence of factor V R506Q mutation-Leiden among apparently healthy Lebanese. Am J Hematol 65:45–49
30. Izakovicová HL, Bucková D, Fassmann A, Benes P, Znojil V (2002) Plasminogen-activator-inhibitor-1 promoter polymorphism as a risk factor for adult periodontitis in non-smokers. Genes Immun 3:292–294
31. Jood K, Ladenvall P, Tjärnlund-Wolf A, Ladenvall C, Andersson M, Nilsson S, Blomstrand C, Jern C (2005) Fibrinolytic gene polymorphism and ischemic stroke. Stroke 36:2077–2081
32. Kalashnikova EA, Kokarovtseva SN, Kovalenko TF, Sirotkina OV, Sheidina AM, Zorilova IV (2006) The frequencies of mutations in Factor V (FV Leiden), Prothrombin (G20210A) and 5,10-methylentetrahydrofolate reductase (C677T) genes in Russians. Medicinskaya Genetika 7:68–72, In Russian
33. Komitopoulou A, Platokouki H, Kapsimali Z, Moschovi M, Kattamis A, Pergantou H, Aronis S (2006) Mutations and polymorphisms in genes affecting haemostasis components in children with thromboembolic events. Thromb Haemost 35:392–397
34. Larsen TB, Lassen JF, Brandslund I, Byriel L, Petersen JB, Norgaard-Pedersen B (1998) The Arg506Gln mutation (FV Leiden) among a cohort of 4188 unselected Danish newborns. Thromb Res 89(5):211–215
35. Lee DH, Henderson PA, Blajchman MA (1996) Prevalence of factor V Leiden in a Canadian blood donor population. CMAJ 155:285–289
36. Leroyer C, Mercier B, Oger E, Chenu E, Abgrall J, Ferec C, Mottier D (1998) Prevalence of the 20210A allele of the prothrombin gene in venous thromboembolism patients. Thromb Haemost 80:49–51
37. Lucotte G, Mercier G (2001) Population genetics of factor V Leiden in Europe. Blood Cells Mol Dis 27:362–367
38. Makowski GS, Davis EL, Aslanzadeh J, Hopfer SM (1995) Enhanced direct amplification of Guthrie card DNA following selective elution of PCR inhibitors. Nucleic Acids Res 23(18):3788–3789
39. Mansilha A, Araujo F (2009) The association between the Val34Leu mutation in the factor XIII gene and deep venous thrombosis in young people. Dialnet 12:5–8
40. Margaglione M, Cappucci G, Colaizzo D, Giuliani N, Vecchione G, Grandone E, Pennelli O, Di Minno G (1998) The PAI-1 gene locus 4G/5G polymorphism is associated with a family history of coronary artery disease. Arterioscler Thromb Vasc Biol 18:152–156
41. Middendorf K, Göhring P, Huehns TY, Seidel D, Steinbeck G, Nikol S (2004) Prevalence of resistance against activated protein c resulting from factor V Leiden is significantly increased in myocardial infarction: investigation of 507 patients with MI. Am Heart J 147(5):897–904
42. Moller L, Kristensen TS (1991) Plasma fibrinogen and ischemic heart disease risk factors. Arterioscler Thromb 11(2):344–350
43. Olafsson I, Hjaltadottir S, Onundarson PT, Porarinsdottir R, Haraldsdottir V (1998) Prevalence of factor VQ506 and prothrombin 20210 A mutations in an apparently healthy icelandic population and patients suffering from venous thrombosis. Thromb Haemost 79:684–685
44. Poort SR, Rosendaal FR, Reitsma PH, Bertina RM (1996) A common genetic variation in the 3′-untranslated region of the prothrombin gene is associated with elevated plasma prothrombin levels and an increase in venous thrombosis. Blood 88:3698–3703
45. Rallidis S (2008) Factor XIII Val34Leu polymorphism and the risk of myocardial infarction. Thromb Haemost 99:1085–1089

46. Rees DC, Cox M, Clegg JB (2002) World distribution of factor V Leiden. Lancet 346:1133–1134
47. Ridker PM, Hennekens CH, Lindpaintner K, Stampfer MJ, Eisenberg PR, Miletich JP (1995) Mutation in the gene coding for coagulation factor V and the risk of myocardial infarction, stroke, and venous thrombosis in apparently healthy men. N Engl J Med 332(14):912–917
48. Ridker PM, Miletich JP, Hennekens CH, Buring JE (1997) Ethnic distribution of factor V Leiden in 4047 men and women. Implications for venous thromboembolism screening. JAMA 277(16):1305–1307
49. Rodeghiero F, Tosetto A (1999) Activated protein C resistance and factor V Leiden mutation are independent risk factors for venous thromboembolism. Ann Intern Med 130:643–650
50. Rosen E, Renbaum P, Heyd J, Levy-Lahad E (1999) High frequency of factor V Leiden in a population of Israeli Arabs. Thromb Haemost 82:1768
51. Rosendaal R, Koster T, Vandenbroucke JP, Reitsma PH (1995) High risk of thrombosis in patients homozygous for factor V Leiden (activated protein C resistance). Blood 85:1504–1508
52. Rosendaal FR, Doggen CJM, Zivelin A, Arruda VR, Aiach M, Siscovick DS, Hillarp A, Watzke HH, Bernardi F, Cumming AM, Preston FE, Reitsma PH (1998) Geographic distribution of the 20210G to A prothrombin variant. Thromb Haemost 79:706–708
53. Sadasivan C, Yee VC (2000) Interaction of the factor XIII activation peptide with thrombin. J Biol Chem 275:362–368
54. Saksela O, Rifkin DB (1988) Cell-associated plasminogen activation:regulation and physiologic functions. Annu Rev Cell Biol 4:93–126
55. Settin A, Dowaidar M, El-Baz R, Abd-Al-Samad A, El-Sayed I, Nasr M (2008) Frequency of factor V Leiden mutation in Egyptian cases with MI. Hematology 3(3):170–174
56. Shafey M, Anderson JL, Scarvelis D, Doucette SP, Gagnon F, Wells PS (2007) Factor XIII Val34Leu variant and the risk of myocardial infarction: a meta-analysis. Thromb Haemost 97:635–641
57. WHO, Minsk (8 Feb 2010); http://www.euro.who.int
58. Wiklund PG, Nilsson L, Ardnor SN, Eriksson P, Johansson L, Stegmayr B, Hamsten A, Holmberg D, Asplund K (2005) Cohorts replicated findings in two nested case-control studies based on independent plasminogen activator inhibitor-1 4G/5G polymorphism and risk of stroke. Stroke 36:1661–1665

Chapter 12
Genomic Instability in the Offspring of Irradiated Parents

Yuri E. Dubrova

Abstract The review describes the phenomenon of epigenetic transgenerational effects detected in the offspring of irradiated parents. The results of some recent studies showing elevated mutation rates in the non-exposed offspring of irradiated parents are presented and discussed. The possible mechanisms and implications of transgenerational instability are also discussed.

Keywords Instability • Radiation • Germline

12.1 Introduction

The target theory, independently developed in 1947 by Timofeeff-Ressovsky and Zimmer, and Lea [33, 54], still represents the cornerstone of radiation biology and radiation protection. According to the modern interpretation of it, the effects of ionising radiation on the exposed cell are attributed to the initial DNA damage at the sites (targets) either directly hit by photons or affected by free radicals arising as a result of the localised ionisation of water [35]. The main prediction of the target theory is that mutation induction almost exclusively occurs in the directly exposed cells at the non-repaired and mis-repaired damaged sites. In other words, everything, including the recognition of radiation-induced damage, its repair and mutation induction, happens in the exposed cells. There is another important prediction of the target theory postulating that the magnitude of mutation induction is directly related to the amount of radiation-induced DNA damage and the ability of exposed cell to repair it, meaning that that the yield of radiation-induced mutations strongly

Y.E. Dubrova (✉)
Department of Genetics, University of Leicester, Leicester LE1 7RH, UK
e-mail: yed2@le.ac.uk

C.E. Mothersill et al. (eds.), *Radiobiology and Environmental Security*,
NATO Science for Peace and Security Series C: Environmental Security,
DOI 10.1007/978-94-007-1939-2_12, © Springer Science+Business Media B.V. 2012

depends on the dose- and dose-rate of exposure [55]. The target theory also lays the foundation of a linear-no-threshold risk model, whereby the risk of human exposure to ionising radiation is linearly related to the dose of exposure without a threshold [10]. However, a number of recent studies have clearly demonstrated the existence of non-targeted effects of ionising radiation. They include mutation induction occurring at tandem repeat DNA loci, which are not directly targeted by ionising radiation [16, 18, 48], radiation induced genomic instability, detected among the non-exposed progeny of irradiated cells/organisms [13, 44, 45], and bystander effect, whereby an elevated mutation rate is detected in the non-exposed cells, which receive signals from the irradiated neighbours [40, 44, 45]. Taking together, these observations imply that the non-targeted effects of ionising radiation can manifest over a period of time much longer than previously thought and may therefore be regarded as a component of the genetic risk of ionising radiation. Although a recent UNSCEAR report has concluded that the existing experimental data on non-targeted effects 'are insufficient to justify modifications of current risk estimates' [56], further analysis of this phenomenon will be important in improving their accuracy.

In this review the hereditary non-targeted effects of ionizing radiation will be presented and discussed. According to the results of some recent publications, these effects can manifest in the germline of directly exposed parents as well as in their non-exposed offspring. As the phenomenon of non-targeted mutation induction in the germline of irradiated mammals has previously been discussed in detail (see [17]), this review therefore describes the transgenerational non-targeted effects of parental exposure to mutagens, including ionizing radiation. The studies presented here were designed to test the hypothesis that non-targeted effects induced in the germline of directly exposed parents could manifest in the offspring, affecting their mutation rates, cancer predisposition and other characteristics. This paper also presents some recent findings which were not discussed in our previous reviews [3, 13].

12.2 Experimental Data

The ability of ionising radiation to induce mutations in the germline of exposed parents has been known for almost a century. The development of the specific locus test in the late 1950s by William Russell gave rise to a number of comprehensive studies aimed to establish the germline effects of ionising radiation and some chemical mutagens in mice (reviewed by [49]). Although these studies for the first time provided sound estimates of the genetic risk of high-dose exposure to ionising radiation in humans [55], they were designed to evaluate mutation induction in the germline of exposed parents. The detection of transgenerational effects requires a thorough analysis of the offspring of exposed parents. The first evidence for unusually elevated mutation rate in the first-generation (F_1) offspring of exposed parents was obtained in the early 1960th (reviewed by [2]). For example, a substantially elevated incidence of mosaicisms was detected in the F_1 offspring male Drosophila exposed to number of alkykating agents [34, 43]. Another strong evidence for transgenerational effects was obtained by Luning and coworkers, who studied the frequency of dominant lethal mutations in the germline of non-exposed offspring of irradiated male mice [42]. Although the results of these studies provided the first indication for the transgeneration destabilisation of the F_1 genome, for the long time they had remained inconclusive. One of the main reason for this was the lack of a sensitive technique for mutation detection in the germline and somatic tissues.

In the 1990th, we and others proposed that expanded simple tandem repeat (ESTR) loci can provide a sensitive approach for monitoring radiation-induced mutation in the mouse germline [15, 48]. ESTRs consist of relatively short repeats, 5–10 bp long, and belong to the most unstable loci in the mouse genome with mutation rate as high as 0.05 per cell division [11, 29]. In the early studies, the analysis of ESTR mutation induction was restricted to the germline [4, 15, 18, 19, 57], but with the development of single-molecule PCR it was extended to the somatic tissues [61]. Given that according to our results the dose-response of ESTR mutation induction in male mice exposed to ionising radiation and chemical mutagens is remarkably similar to that for protein-coding genes [4, 14, 18, 19, 25, 57], we therefore used them for the analysis of transgenerational instability.

Our first published results show that ESTR mutation rates are substantially elevated in the germline of non-exposed first- and second-generation offspring of irradiated males [4, 20]. This remarkable stimulation of mutation in the F_1/F_2 germline following paternal irradiation (transgenerational instability) is reminiscent of the above mentioned phenomenon of delayed radiation-induced genome instability in somatic cells. Most importantly, our data demonstrate that the phenomenon of radiation-induced genomic instability is not restricted to a certain inbred strain of mice, as roughly equally elevated ESTR mutation rate was detected in the germline of three different inbred strains – CBA/Ca, C57BL/6J and BALB/c [5].

Using single-molecule PCR, we also analysed ESTR mutation frequencies in DNA samples extracted from the germline (sperm) and somatic tissues taken from the F_1 offspring of male mice, exposed to 1 or 2 Gy of acute X-rays [6, 7, 30]. Figure 12.1a summarises the results of these studies, clearly showing that ESTR mutation rates are equally elevated in the germline and somatic tissues of F_1 offspring of irradiated male mice. Transgenerational changes in somatic mutation rates were also observed by studying the frequency of chromosome aberrations [52, 58], micronuclei [22] and *lac*I mutations [41] in the F_1 offspring of irradiated male mice and rats. It should be noted that the most compelling data addressing somatic instability in the F_1 offspring of irradiated males were obtained from the analysis of somatic reversions of the pink-eyed unstable mutation [51] and mutations at the mouse *hprt* locus [6].

The abovementioned results clearly imply that exposure to ionising radiation results in the induction of a transgenerational signal in the germline of exposed parents which can substantially destabilise the genomes of their offspring. As far as potential implications of these data for radiation protection are concerned, the important question is – for how long the exposed males can 'remember' the history of irradiation. In a number of our publication, we have addressed this issue. In these

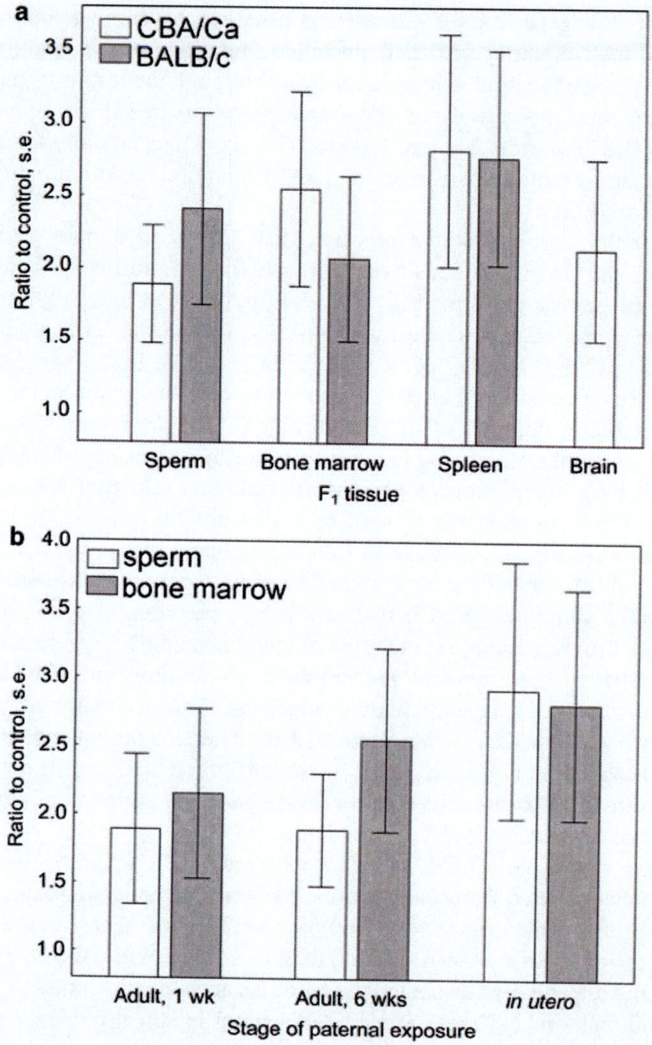

Fig. 12.1 ESTR mutation frequency in the F_1 offspring of irradiated male mice. **(a)** Frequency of ESTR mutation in the germline and somatic tissues in the F_1 offspring irradiated males from two different inbred strains of mice (Data from: [6, 7]). **(b)** Frequency of ESTR mutation in the F_1 offspring BALB/c mice irradiated in utero or during adulthood (Data from [7])

studies, male mice were mated over a considerable period of time following exposure to acute X-rays during adulthood [5, 6, 30]. In our recent study we also analysed the F_1 offspring male mice irradiated *in utero* during the early stage of development [7]. Figure 12.1b summarises the results of these studies, showing that the magnitude of the increases in ESTR mutation frequency in the F_1 offspring of male mice irradiated during adulthood or prenatally does not significantly differ.

These data suggest that some as yet unknown marks of paternal irradiation can survive even massive epigenetic reprogramming during the early stages of development and, being passed to the offspring, destabilise their genomes. Our results therefore imply that the transgenerational effects may be regarded as an important component of the long-term genetic risk of ionising radiation. The analysis of mutation frequencies in the germline and somatic tissues of F_1 offspring of irradiated males has provided important clues on the possible mechanisms of transgenerational instability.

As in many studies the effects of paternal irradiation were studied in the offspring conceived by the exposed males and non-irradiated females, the very design of these studies allowed to establish the F_1 mutation rates at the alleles derived from the both parents [5, 6, 51]. Figure 12.2 presents the results of the two studies analysing the frequency of somatic mutations in the offspring of irradiated male mice. Thus, Shiraishi and co-workers reported equally elevated frequencies of somatic reversions at the mouse pink-eyed unstable locus in both alleles derived from the irradiated fathers and the unexposed mothers (Fig. 12.2a).

In our study, the increased frequency of mutation at the mouse *hprt* locus was found in the male F_1 offspring of male mice (Fig. 12.2b). Given that the *hprt* locus is X-linked, the results of this study are therefore in line with the abovementioned pink-eyed data, as an elevated mutation rate was found at the F_1 locus derived from the non-exposed mothers. These results clearly imply that transgenerational instability is attributed to a genome-wide destabilisation.

12.3 Mechanisms

The data presented in the previous chapter clearly show that mutation rates in the germline and somatic tissues of offspring of irradiated males are substantially elevated. It should be stressed that the abovementioned results are at odds with the main predictions of target theory, as the effects of parental irradiation manifest in the non-exposed offspring, i.e. mutation induction in these animals are clearly off target and therefore the phenomenon of transgenerational instability belongs to the newly discovered non-targeted effects of ionising radiation. The key question is – what are the mechanisms underlying transgenerational instability? In a number of publications is has been suggested that the phenomenon of radiation-induced genomic instability observed either in the clonal progeny of irradiated cells or in the offspring of irradiated parents cannot be ascribed to conventional mechanisms such as a mutator phenotype and is most likely a result of epigenetic events [13, 40, 44, 45]. This conclusion is based on the two sets of experimental data, showing that: (i) *in vitro* and *in vivo* radiation-induced genomic instability persists over a long period of time after the initial exposure; (ii) the number of cells/organisms manifesting radiation-induced genomic instability is too high to be explained by the direct targeting of any group of genes. As far as the transgenerational effects of paternal exposure are concerned, there are two main issues which should be addressed.

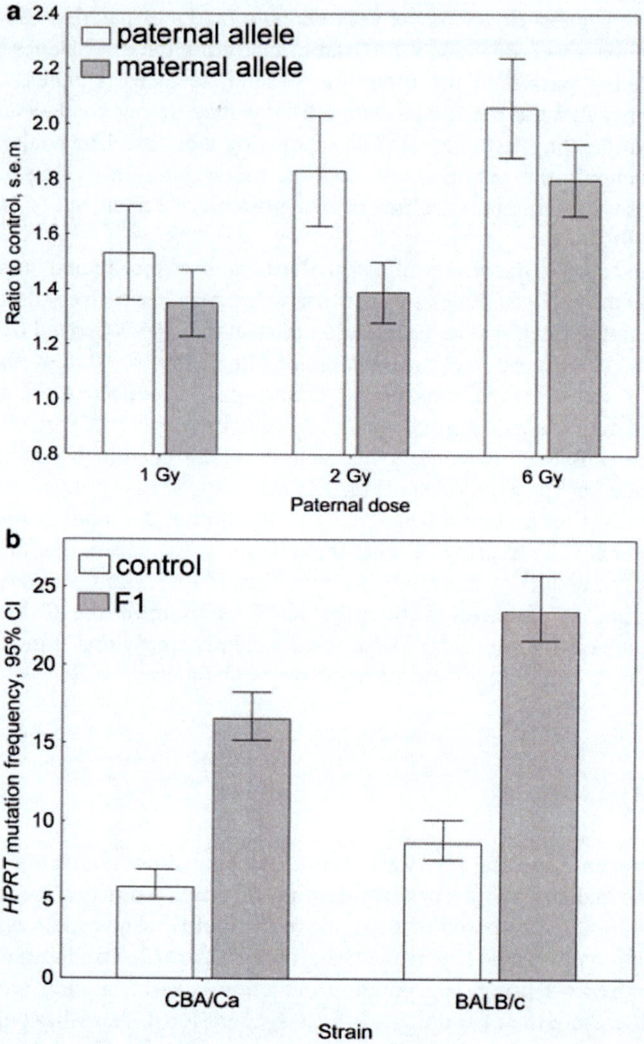

Fig. 12.2 Transgenerational increases in the frequency of somatic reversions of the pink-eyed unstable mutation (**a**) and hprt mutation (**b**) in the F_1 offspring of irradiated male mice (Data from [6, 51])

The first step to elucidate the yet unknown mechanisms is to investigate the initial cellular events triggering an instability signal in the exposed germline. The analysis of the transmission of such a signal to the offspring and its manifestation represents another essential step in the clarification of this issue.

Given that exposure to ionising radiation produces a wide spectrum of DNA lesions [23], the analysis of its transgenerational effects may not provide enough evidence for a specific type of DNA damage triggering an instability signal in the

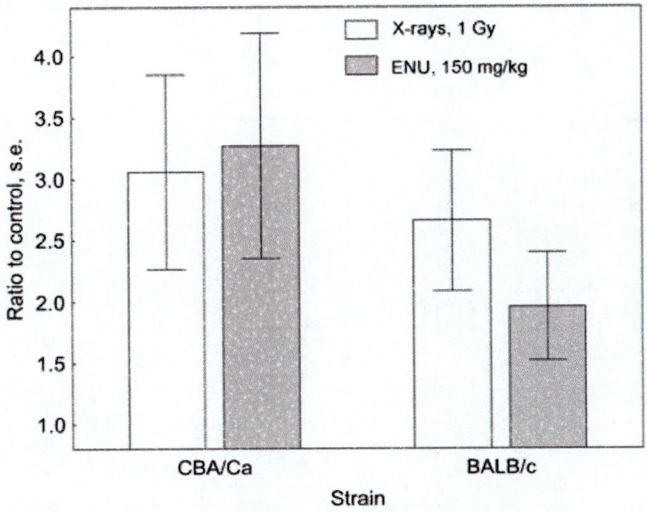

Fig. 12.3 ESTR mutation rates in the F_1 germline of ENU-exposed and irradiated male mice (Data from [5, 21])

directly exposed cells. However, it has been suggested that that radiation-induced complex double-strand DNA breaks (DSBs) may constitute one of the signals that initiate the onset of genomic instability [37]. To verify whether this particular type of DNA damage can trigger an instability signal in the mouse germline, we have studied the effects of paternal exposure to the alkylating agent ethylnitrosourea (ENU) on the manifestation of genomic instability in the offspring of treated male mice. In contrast to irradiation, exposure to ENU mainly causes alkylation of DNA at the N- and O-positions, resulting predominantly in base substitution mutations and seldom leads to any measurable increases in the yield of DSBs [50]. This work revealed a number of striking similarities between the transgenerational effects of paternal exposure to this mutagens and ionising radiation (Fig. 12.3).

Given the profound differences in the spectrum of ENU- and radiation-induced DNA damage, our data suggest that transgenerational instability is not attributed to a specific sub-set of DNA lesions, such as DSBs, but is most probably triggered by a stress-like response to a generalised DNA damage. Our data are in line with the results of previous studies showing that exposure to some chemical carcinogens and mutagens can result in a delayed increase in mutation rate in somatic cells [8, 27, 36, 37] or affect the fitness of the offspring of exposed male rats [28].

Given that the transgenerational changes can affect mutation rates at tandem repeat DNA loci, protein coding genes and chromosome aberrations, it therefore appears that the F_1 genomes could contain a variety of types of DNA damage. To compare the amount of DNA damage in the F_1 offspring and controls, we evaluated the amount of endogenous single- and double-strand DNA breaks, measured by the alkaline Comet and γ-H2AX assays [6]. This analysis revealed an abnormally high

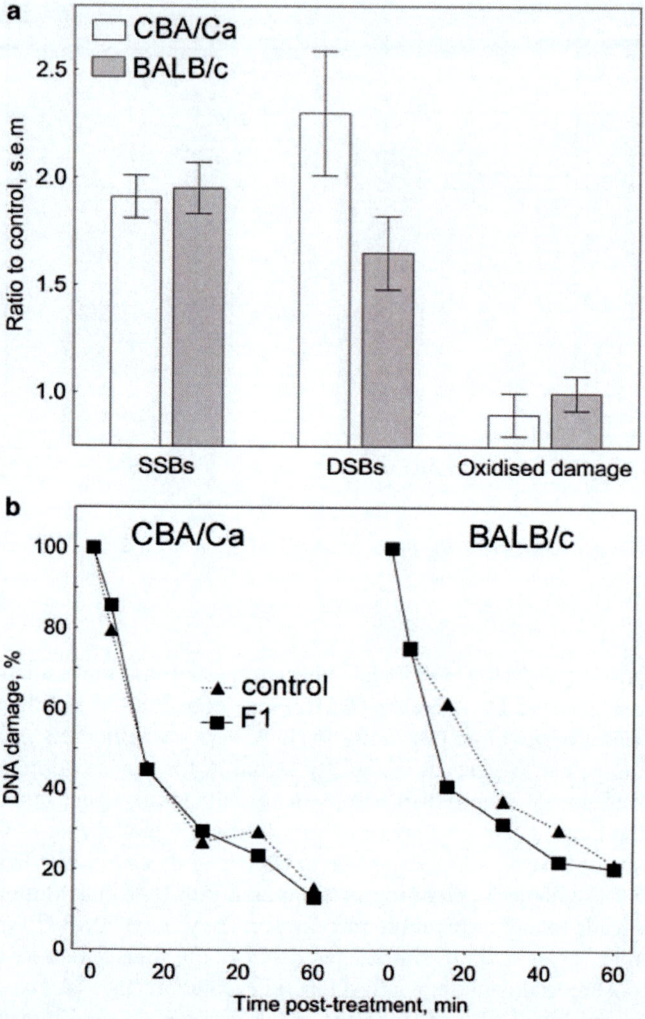

Fig. 12.4 DNA damage and the efficiency of DNA repair in control males and the F_1 offspring of irradiated male mice. (**a**) Endogenous DNA damage (*single-* and *double-strand* breaks) and the amount of oxidative damage in the F_1 offspring of irradiated males. (**b**) Efficiency of DNA repair in the *ex vivo* irradiated bone marrow cells (Data from [4])

level of DNA damage in the F_1 bone marrow and spleen cells (Fig. 12.4a). Given that in tissues with a high mitotic index, such as bone marrow and spleen, the life-span of cells containing deleterious lesions such as single- and double-strand breaks is restricted since these types of DNA damage are not compatible with replication, these data clearly demonstrate that transgenerational instability is an ongoing process occurring in multiple adult tissues.

The presence of a persistent subset of DNA lesions in the offspring of irradiated males may either be attributed to an oxidative stress or reflect their compromised ability to repair of a certain type of endogenous damage. The involvement of oxidative stress/inflammatory-type response in the delayed increases in mutation rates in the progeny of irradiated cells has long been suspected [40, 44]. Reactive oxygen species are the major source of endogenous DNA damage, including single- and double-strand breaks, abasic sites, and a variety of nucleotide modifications [32]. However, according to our data obtained by the FPG Comet assay the F_1 offspring of irradiated males did not show any alteration in the level of oxidatively damaged nucleotides (Fig. 12.4a). Using the alkaline Comet, we also measured the efficiency of repair of single-strand DNA breaks and other alkali labile sites in the F_1 *ex-vivo* irradiated bone marrow samples (Fig. 12.4b). According to our data, the efficiency of DNA repair in the offspring of irradiated parents and controls is similar.

These data are in line with the results of some early studies, comparing the spectra of directly induced and delayed mutations at the *hprt* locus, detected in the progeny of irradiated cells [38]. According to the results of this study, the spectrum of delayed mutations, resulted from the ongoing instability substantially differed from that for directly induced and, in the same time, was very close for the spectrum of spontaneous mutations.

These data imply that radiation-induced genomic instability may result from the enhancement of the process of spontaneous mutation and therefore cannot be attributed to compromised DNA repair, which would affect the structure of delayed mutations. Given this, a genome-wide destabilisation of the F_1 genome could be attributed to replication stress. For example, some recent data suggest that in human precancerous cells the ATR/ATM-regulated checkpoints are activated through deregulated DNA replication, which leads to the multiplicity of DNA alterations [9, 26]. A detailed analysis of the expression profiles in the F_1 tissues should elucidate the still unknown mechanisms underlying the phenomenon of radiation-induced genomic instability.

12.4 Transgenerational Instability and Genetic Risk

The abovementioned results showing that the offspring of exposed parents are genetically unstable may represent a serious challenge to the existing paradigm in radiation protection, according to which mutation induction in directly exposed cells is regarded as the main component of the genetic risk of ionising radiation for humans [55]. Given that in the offspring of irradiated parents an elevated frequency of mutations is detected at a number of fitness-related endpoints, including chromosome aberrations and gene mutations, it would therefore appear that the transgenerational effects could impair the development. As, according to our data [6], the majority of the F_1 offspring of irradiated males manifest transgenerational instability, the number of affected offspring may substantially exceed that predicted by the target theory, which is based on the assumption their elevated mortality and morbidity can only be attributed to mutations induced in the germline of exposed parents. Indeed, the results of some animal studies show that paternal exposure to ionising radiation and some chemical mutagens can affect a number of fitness-related phenotypic traits in the offspring. For example, it has been reported that the rate of dominant lethal mutations is significantly elevated in the germline of F_1 offspring of male mice and rats, exposed either to ionising radiation [42] or cyclophosphamide [28]. The results of other studies showing decreased proliferation of early embryonic cells and increased frequency of malformations in the F_2 offspring of irradiated parents are also consistent with these observations [47, 60].

Given that tumour progression is attributed to accumulation of oncogenic mutations, the long-term destabilisation of the F_1 genome could also predispose them to cancer. In a number of studies, the incidence of cancer in the offspring of irradiated male mice exposed to recognised carcinogens was analysed [31, 39, 46, 59]. Their results show an elevated incidence of cancer among the carcinogen-challenged offspring of irradiated males. The data on significantly elevated incidence of leukaemia in the children of male radiation workers from the Sellafield nuclear reprocessing plant [24] are in line with these results.

It should be stressed that further analysis of the clinical impact of transgenerational instability is currently limited because of lack of the reliable experimental data in humans. So far, just a handful of experimental studies addressing this important issue have been conducted and their results are far from being consistent. For example, a recent publication on the transgenerational effects of post-Chernobyl paternal irradiation showed an elevated frequency of chromosome aberrations among the children of exposed fathers [1], whereas Tawn and coworkers failed to detect significant changes among the children of childhood cancer survivours [53]. On the other hand, the same group of authors reported an elevated G_2 chromosomal radiosensitivity in the children of survivors of childhood and adolescent cancer [12]. Given that the design of these studies, as well as the cohorts of irradiated families analyzed by the authors, dramatically differ, the comparison of their results remains highly problematic. The important issue of the transgenerational effects of parental exposure in humans should therefore be addressed in the carefully planned and executed studies, the results of which will provide experimentally based estimates of the delayed effects of radiation in humans.

References

1. Aghanjanyan A, Suskov I (2009) Transgenerational genomic instability in children of irradiated parents as a result of the Chernobyl Nuclear Accident. Mutat Res 671:52–57
2. Aurebach C, Kilbey BJ (1971) Mutation in eukaryotes. Annu Rev Genet 5:163–218
3. Barber R, Dubrova YE (2006) The offspring of irradiated parents, are they stable? Mutat Res 598:50–60
4. Barber R, Plumb MA, Smith AG, Cesar CE, Boulton E, Jeffreys AJ, Dubrova YE (2000) No correlation between germline mutation at repeat DNA and meiotic crossover in male mice exposed to X-rays or cisplatin. Mutat Res 457:79–91

5. Barber R, Plumb MA, Boulton E, Roux I, Dubrova YE (2002) Elevated mutation rates in the germline of first- and second-generation offspring of irradiated male mice. Proc Natl Acad Sci USA 99:6877–6882
6. Barber RC, Hickenbotham P, Hatch T, Kelly D, Topchiy N, Almeida G, Jones GGD, Johnson GE, Parry JM, Rothkamm K, Dubrova YE (2006) Radiation-induced transgenerational alterations in genome stability and DNA damage. Oncogene 25:7336–7342
7. Barber RC, Hardwick RJ, Shanks ME, Glen CD, Mughal SK, Voutounou M, Dubrova YE (2009) The effects of *in utero* irradiation on mutation induction and transgenerational instability in mice. Mutat Res 664:6–12
8. Bardelli A, Cahill DP, Lederer G, Speicher MR, Kinzler KW, Vogelstein B, Lengauer C (2001) Carcinogen-specific induction of genetic instability. Proc Natl Acad Sci USA 98:5770–5775
9. Bartkova J, Horejsi Z, Koed K, Kramer A, Tort F, Zieger K, Guldberg P, Sehested M, Nesland JM, Lukas C, Orntoft T, Lukas J, Bartek J (2005) DNA damage response as a candidate anticancer barrier in early human tumorogenesis. Nature 434:864–870
10. BEIR (2006) Health risks from exposure to low levels of ionizing radiation: BEIR VII, Phase 2. National Academy Press, Washington, DC
11. Bois P, Williamson J, Brown J, Dubrova YE, Jeffreys AJ (1998) A novel unstable mouse VNTR family expanded from SINE B1 element. Genomics 49:122–128
12. Curwen GB, Winther JF, Tawn EJ, Smart V, Whitehouse CA, Rees GS, Olsen JH, Guldberg P, Rechnitzer C, Schrøder H, Bryant PE, Sheng X, Lee HS, Chakraborty R, Boice JD Jr (2005) G_2 chromosomal radiosensitivity in Danish survivors of childhood and adolescent cancer and their offspring. Br J Cancer 93:1038–1045
13. Dubrova YE (2003) Radiation-induced transgenerational instability. Oncogene 22:7087–7093
14. Dubrova YE (2005) Radiation-induced mutation at tandem repeat DNA loci in the mouse germline: spectra and doubling doses. Radiat Res 163:200–207
15. Dubrova YE, Jeffreys AJ, Malashenko AM (1993) Mouse minisatellite mutations induced by ionizing radiation. Nat Genet 5:92–94
16. Dubrova YE, Nesterov VN, Krouchinsky NG, Ostapenko VA, Neumann R, Neil DL, Jeffreys AJ (1996) Human minisatellite mutation rate after the Chernobyl accident. Nature 380:683–686
17. Dubrova YE, Plumb MA, Brown J, Jeffreys AJ (1998) Radiation-induced germline instability at minisatellite loci. Int J Radiat Biol 74:689–696
18. Dubrova YE, Plumb M, Brown J, Fennelly J, Bois P, Goodhead D, Jeffreys AJ (1998) Stage specificity, dose response, and doubling dose for mouse minisatellite germ-line mutation induced by acute radiation. Proc Natl Acad Sci USA 95:6251–6255
19. Dubrova YE, Plumb M, Brown J, Boulton E, Goodhead D, Jeffreys AJ (2000) Induction of minisatellite mutations in the mouse germline by low-dose chronic exposure to γ-radiation and fission neutrons. Mutat Res 453:17–24
20. Dubrova YE, Plumb M, Gutierrez B, Boulton E, Jeffreys AJ (2000) Transgenerational mutation by radiation. Nature 405:37
21. Dubrova YE, Hickenbotham P, Glen CD, Monger K, Wong H-P, Barber RC (2008) Paternal exposure to ethylnitrosourea results in transgenerational genomic instability in mice. Environ Mol Mutagen 49:308–311
22. Fomenko LA, Vasil'eva GV, Bezlepkin VG (2001) Micronucleus frequency is increased in bone marrow erythrocytes from offspring of male mice exposed to chronic low-dose gamma irradiation. Biol Bull 28:419–423
23. Frankenberg-Schwager M (1990) Induction, repair and biological relevance of radiation-induced DNA lesions in eukaryotic cells. Radiat Environ Biophys 29:273–292
24. Gardner MJ, Snee MP, Hall AJ, Powell CA, Downes S, Terrell JD (1990) Results of case-control study of leukaemia and lymphoma among young people near Sellafield nuclear plant in West Cumbria. Br Med J 300:423–429
25. Glen CD, Smith AG, Dubrova YE (2008) Single-molecule PCR analysis of germ line mutation induction by anticancer drugs in mice. Cancer Res 68:3630–3636
26. Gorgoulis VG, Vassiliou LV, Karakaidos P, Zacharatos P, Kotsinas A, Liloglou T, Venere M, Ditullio RA Jr, Kastrinakis NG, Levy B, Kletsas D, Yoneta A, Herlyn M, Kittas C, Halazonetis

TD (2005) Activation of the DNA damage checkpoint and genomic instability in human precancerous lesions. Nature 434:907–913
27. Gowans ID, Lorimore SA, McIlrath JM, Wright EG (2005) Genotype-dependent induction of transmissible chromosomal instability by gamma-radiation and the benzene metabolite hydroquinone. Cancer Res 65:3527–3530
28. Hales BF, Crosman K, Robaire B (1992) Increased postimplantation loss and malformations among the F_2 progeny of male rats chronically treated with cyclophosphamide. Teratology 45:671–678
29. Hardwick RJ, Tretyakov MV, Dubrova YE (2009) Age-related accumulation of mutations supports a replication-dependent mechanism of spontaneous mutation at tandem repeat DNA loci in mice. Mol Biol Evol 26:2647–2654
30. Hatch T, Derijck AAHA, Black PD, van der Heijden GW, De Boer P, Dubrova YE (2007) Maternal effects of the *scid* mutation on radiation-induced transgenerational instability in mice. Oncogene 26:4720–4724
31. Hoyes KP, Lord BI, McCann C, Hendry JH, Morris ID (2001) Transgenerational effects of preconception paternal contamination with (55)Fe. Radiat Res 156:488–494
32. Jackson AL, Loeb LA (2001) The contribution of endogenous sources of DNA damage to the multiple mutations in cancer. Mutat Res 477:7–21
33. Lea DE (1947) Actions of radiations on living cells. Cambridge University Press, New York
34. Lee WR, Sega GA, Bishop JB (1970) Chemically induced mutations observed as mosaics in Drosophila melanogaster. Mutat Res 9:323–336
35. Lehnert S (2007) Biomolecular action if ionizing radiation. Taylor & Francis, New York
36. Li C-Y, Little JB, Hu K, Zhang L, Dewhirst MW, Huang Q (2001) Persistent genetic instability in cancer cells induced by non-DNA-damaging stress exposures. Cancer Res 61:428–432
37. Limoli CL, Kaplan MI, Phillips JW, Adair GM, Morgan WF (1997) Differential induction of chromosomal instability by DNA strand-breaking agents. Cancer Res 57:4048–4056
38. Little JB, Li C, Nagasawa H, Pfenning T, Vetrovs H (1996) Genomic instability and radiation mutagenesis. J Chim Phys 93:157–164
39. Lord BI, Woolford LB, Wang L, Stones VA, McDonald D, Lorimore SA, Papworth D, Wright EG, Scott D (1998) Tumour induction by methyl-nitroso-urea following preconceptional paternal contamination with plutonium-239. Br J Cancer 78:301–311
40. Lorimore SA, Coates PJ, Wright EG (2003) Radiation-induced genomic instability and bystander effects: inter-related nontargeted effects of exposure to ionizing radiation. Oncogene 22:7058–7069
41. Luke GA, Riches AC, Bryant PE (1997) Genomic instability in haematopoietic cells of F_1 generation mice of irradiated male parents. Mutagenesis 12:147–152
42. Luning KG, Frolen H, Nilsson A (1976) Genetic effects of ^{239}Pu salt injections in male mice. Mutat Res 34:539–542
43. Mathew C (1964) The nature of delayed mutations after treatment with chloroethyl methanesuphonate and other alkylating agents. Mutat Res 1:163–172
44. Morgan WF (2003) Non-targeted and delayed effects of exposure to ionizing radiation: I. Radiation-induced genomic instability and bystander effects in vitro. Radiat Res 159:567–580
45. Morgan WF (2003) Non-targeted and delayed effects of exposure to ionizing radiation: II. Radiation-induced genomic instability and bystander effects in vivo, clastogenic factors and transgenerational effects. Radiat Res 159:581–596
46. Nomura T (1983) X-ray-induced germ-line mutation leading to tumors. Its manifestation in mice given urethane post-natally. Mutat Res 121:59–65
47. Pils S, Muller W-U, Streffer C (1999) Lethal and teratogenic effects in two successive generations of the HLG mouse strain after radiation exposure of zygotes – association with genomic instability? Mutat Res 429:85–92
48. Sadamoto S, Suzuki S, Kamiya K, Kominami R, Dohi K, Niwa O (1994) Radiation induction of germline mutation at a hypervariable mouse minisatellite locus. Int J Radiat Biol 65:549–557
49. Searle AG (1974) Mutation induction in mice. Adv Radiat Biol 4:131–207

50. Shibuya T, Morimoto K (1993) A review of the genotoxicity of 1-ethyl-1nitrosourea. Mutat Res 297:3–38
51. Shiraishi K, Shimura T, Taga M, Uematsu N, Gondo Y, Ohtaki M, Kominami R, Niwa O (2002) Persistent induction of somatic reversions of the pink-eyed unstable mutation in F-1 mice born to fathers irradiated at the spermatozoa stage. Radiat Res 157:661–667
52. Slovinska L, Elbertova A, Misurova E (2004) Transmission of genome damage from irradiated male rats to their progeny. Mutat Res 559:29–37
53. Tawn EJ, Whitehouse CA, Winther JF, Curwen GB, Rees GS, Stovall M, Olsen JH, Guldberg P, Rechnitzer C, Schrøder H, Boice JD Jr (2005) Chromosomal analysis in childhood cancer survivors and their offspring – No evidence for radiotherapy-induced persistent genomic instability. Mutat Res 583:198–206
54. Timofeeff-Ressovsky NW, Zimmer KG (1947) Das Trefferprinzip in der Biologie. S. Herzel Verlag, Leipzig
55. UNSCEAR (2001) Hereditary effects of radiation. United Nations, New York
56. UNSCEAR (2008) Effects of ionizing radiation. Annex C. Non-targeted and delayed effects of exposure to ionizing radiation. United Nations, New York
57. Vilarino-Guell C, Smith AG, Dubrova YE (2003) Germline mutation induction at mouse repeat DNA loci by chemical mutagens. Mutat Res 526:63–73
58. Vorobtsova IE (2000) Irradiation of male rats increases the chromosomal sensitivity of progeny to genotoxic agents. Mutagenesis 15:33–38
59. Vorobtsova IE, Aliyakparova LM, Anisimov VN (1993) Promotion of skin tumors by 12-O-tetradecanoylphorbol-13-acetate in two generations of descendants of male mice exposed to X-ray irradiation. Mutat Res 287:207–216
60. Wiley LM, Baulch JE, Raabe OG, Straume T (1997) Impaired cell proliferation in mice that persists across at least two generations after paternal irradiation. Radiat Res 148:145–151
61. Yauk CL, Dubrova YE, Grant GR, Jeffreys AJ (2002) A novel single molecule analysis of spontaneous and radiation-induced mutation at a mouse tandem repeat locus. Mutat Res 500:147–156

Chapter 13
Low Dose Responses of Bone Marrow to X-Rays In Vivo

Nikolay A. Zyuzikov, Philip J. Coates, Sally A. Lorimore, and Eric G. Wright

Abstract Radiation risk at low doses is determined by linear extrapolation from high dose epidemiological data. In the last decade many non-targeted effects have been reported which may be relevant to low dose risk determination. To investigate cell responses at such low doses we used bone marrow cells of mice. We have not observed non-targeted effects long- or short-term post irradiation. Exposure below 50–100 mGy provides no evidence of a dose response for apoptotic signaling, bystander effects and low responses for p53 and p21 induction with significant individual variability. There is also no evidence for long-term chromosomal instability in the bone marrow at doses below 1 Gy. The data also demonstrate unexpected thresholds above which dose-dependent damage signaling is observed and the chromosomal instability phenotype is induced. The data are consistent with low dose X-irradiation being less damaging than would be expected from the LNT paradigm.

Keywords DNA damage • In vivo • Bone marrow • Individual radiosensitivity • Low doses

N.A. Zyuzikov (✉)
Centre for Oncology and Molecular Medicine, Ninewells Hospital and Medical School, Dundee DD1 9SY, UK

University of Dundee, Dundee DD1 9SY, Scotland, UK
e-mail: n.zyuzikov@dundee.ac.uk

P.J. Coates • S.A. Lorimore • E.G. Wright
University of Dundee, Dundee DD1 9SY, Scotland, UK

C.E. Mothersill et al. (eds.), *Radiobiology and Environmental Security*, NATO Science for Peace and Security Series C: Environmental Security, DOI 10.1007/978-94-007-1939-2_13, © Springer Science+Business Media B.V. 2012

13.1 Introduction

Exposure to high doses of ionizing radiation unequivocally produces adverse health effects including malignancy. At lower doses epidemiology is not powerful enough for the purposes of radiation protection and the carcinogenic risk is estimated by extrapolation linearly proportional to radiation dose down to background levels (LNT model) relying heavily on radiobiological investigations and theoretical biophysical considerations [2]. Importantly, dose is used as a surrogate for risk and averaged across a tissue rather than considered as dose to individual cells. In recent years, many observations of, so called, non-targeted effects indicate that radiation-induced lesions are not necessarily restricted to cells in which energy has been deposited. These observations include radiation induced bystander effects (responses exhibited by non-irradiated cells that have communicated with irradiated cells) and radiation-induced genomic instability (nonclonal effects arising in the descendants of irradiated cells) [16, 29, 32, 40]. The phenomena have been regarded by some as being particularly important at low doses, potentially questioning the validity of a simple linear extrapolation of radiation risks from high to low doses [1, 3, 14, 16, 31, 36]. To study in vivo responses of relevance to low dose effects of low-LET radiation and their potential risk, we investigated short-term (p53 pathway signalling and apoptotic responses) and longer-term (cytogenetic) effects in mouse bone marrow cells after exposure to very low doses of X-rays where, at the lowest doses studied, not all cells are irradiated and after 1.7 mGy approximately 40% of cells would not have received a single electron track [11, 12].

13.2 Material and Methods

13.2.1 *Animals and Irradiation*

C57BL/6 and CBA/Ca mice, susceptible or resistant, respectively, to both radiation-induced myeloid leukaemia [37] and radiation-induced chromosomal instability in the bone marrow [39] and exhibiting differences in the generation of bystander-type signals after exposure to high doses of low-LET radiation [24, 33] were sham- or X-irradiated as described previously [23]. Mice were bred and housed under conventional conditions and experiments conducted using adult mice of 8–12 weeks of age. Mice were sham-irradiated for controls or exposed to different doses of X-rays at a dose rate of 1 mGy/s for low doses (up to 100 mGy) and 7.5 mGy/s for high doses (0.5, 1 and 3 Gy). Experiments were approved by local ethical review and carried out in accordance with Home Office Project License.

13.2.2 Immunohistochemistry

At 3 h post irradiation, femurs were removed, fixed and processed for immunohistochemistry using standard procedures [5]. Femurs were fixed by immersion in 4% buffered formalin for 48 h and later were decalcified with EDTA, pH 7.0 for 10–12 days. All tissues were processed to paraffin wax using standard histological processing schedules. Tissue sections 4 μm thick were stained using a peroxidise avidin–biotin-complex technique (Vector Elite ABC System, Vector Laboratories, UK) detected with an intensified AB/imidazole reaction. Antigen retrieval was performed by boiling in 10 mM citrate buffer, pH 6.0 for 15 min in a microwave oven. The CM5 polyclonal antibody to rodent p53 was provided by D. Lane and used at 1/80,000 (University of Dundee, UK). Purified rabbit antip21/waf1 (Ab5, Oncogene Research Products, Nottingham, UK) was applied at 0.5 μg/ml. Caspase 3 antibody was purchased from Cell Signaling Technology (Cat N. 9664L). Sections were counterstained with haematoxylin. The numbers of positive cells and the total number of cells were counted by light microscopy with the aid of an eye-piece graticule. At least four independent areas were assessed per dose/time point for each individual animal and the decoded data pooled, tested for normality using the Shapiro-Wilk Test and differences between control and single experimental groups analyzed using the Student's t test with multiple comparisons of dose response data analyzed by Scheffé analysis of variance. Differences with $p<0.05$ are stated as significant in the text.

13.2.3 TUNEL Assay

The same tissue sections and cytospined cells as for immunochemistry were treated by Proteinase K (Sigma) at concentration 10 μg/ml for 30 min at 37°C. Then Labeling the DNA ends were done using enzyme TdT (Invitrogen) and biotinated-14-DCTP (Invitrogen) according the supplier instructions. The staining (using a peroxidise avidin–biotin-complex technique), scoring and statistical analysis are performed the same way as for immunochemistry.

13.2.4 In Vitro Assay of Bystander Stress Response

To assay for a radiation-induced p53 bystander effect in vitro, bone marrow cell suspensions obtained from C57BL/6 mice were divided into two aliquots; one was irradiated the other sham-irradiated. The irradiated aliquot was divided to provide two sources of signals for study of bystander effects. Two different assay systems were used: co-culture using transwells with 0.4 μm pores in which irradiated and sham-irradiated cells are separated by a polycarbonate membrane and media transfer in which sham-irradiated cells are suspended in medium obtained from irradiated cells. After incubation for 3 h in 5% CO_2 in air at 37°C, cytocentrifuge preparations were prepared, immunostained, scored and analyzed as for tissue sections.

13.2.5 Assay for Delayed Cytogenetic Aberrations

At 30 and 100 days post-irradiation bone marrow cells were flushed from the femora of the irradiated mice. Metaphases were accumulated for 1 h by adding 0.02 μg/ml colcemid to the marrow cells, suspended in α-MEM supplemented with 20% foetal calf serum at 37°C. These cells were then suspended in 5 ml hypotonic (0.55% w/v) potassium chloride (KCl) for 33 min incubation at 37°C, then 2–3 ml KCl in sodium citrate was added (0.28 g KCl and 0.5 g sodium citrate in 100 ml distilled water) for 8 min. The cells were then fixed in suspension by adding 1–2 ml 3:1 methanol: acetic acid mixture to a final volume of 10 ml at room temperature. After 10–15 min, the cells were resuspended in at least two additional changes of the fixative mixture. Air-dried slides were aged for 10–14 days before Giemsa (15% in distilled water) staining. Chromosomal aberrations in coded metaphase preparations were recorded [39], data obtained from 6 to 12 individual mice pooled after decoding and differences between the proportions of aberrant cells analyzed by the Fisher's exact test as described previously [23, 39].

13.3 Results

13.3.1 Lack of Low Dose-Response and a Threshold for Damage Response Signaling and Apoptosis

Hemopoietic cells express a p53-dependent cell cycle G1 phase arrest mediated by cyclin-dependent kinase inhibitor 1A (p21/Cip1/ CDKN1A) or p53-dependent apoptosis in response to a variety of stress stimuli. Using immunohistochemical techniques for p53 and p21 protein expression, preliminary time-response studies demonstrated that a maximal response could be detected at 3 h after a range of doses in both C57BL/6 and CBA/Ca mouse strains and at 50 mGy irradiation for both endpoints in C57BL/6 mice. Samples for analysis in subsequent investigations were obtained 3 h post-irradiation and there was no significant difference between C57BL/6 and CBA/Ca strains in the proportion of p53- or p21-positive cells in the femoral bone marrow of normal untreated 5 mice. After irradiation with 1.7, 5, 12 or 25 mGy X-rays, there was no significant increase in the proportion of p53-positive cells in C57BL/6 or CBA/Ca bone marrow and significant increases were not detected until exposure of 50 mGy for C57BL/6 bone marrow and 100 mGy for CBA/Ca bone marrow (Fig. 13.1a). After 50 mGy (C57BL/6) and 100 mGy (CBA/Ca) exposure, there were correspondingly significant increases in the proportions of

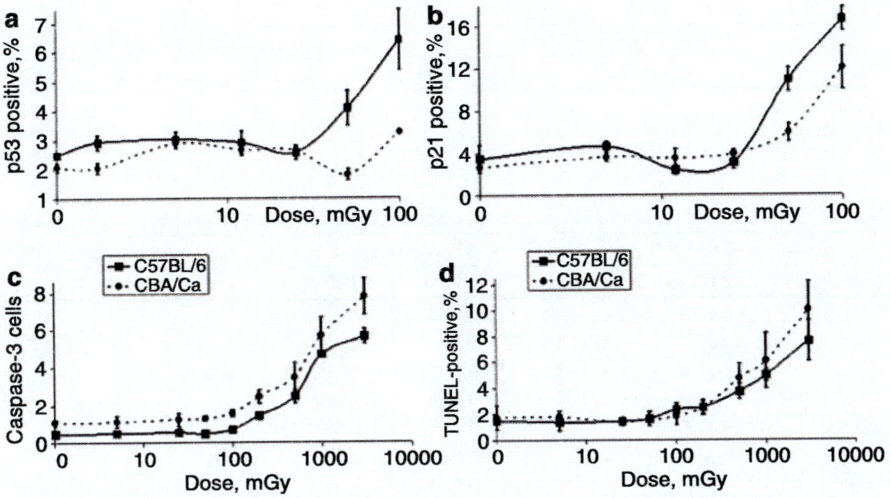

Fig. 13.1 Dose responses in bone marrow after in vivo X-irradiation. C57BL/6 (■) and CBA/Ca (•) strains. Mean ± sem of % positive cells for p53 (**a**), p21 (**b**), Caspase-3 (**c**) or TUNEL (**d**)

p21-positive cells (Fig. 13.1b). Thus, for p53 and p21 expression there was no significant dose response over the range 0–25 mGy.

To determine the proportion of cells responding to radiation-induced damage with an apoptotic response, the proportions of Caspase-3-positive cells and TUNEL-positive cells in the bone marrow of irradiated mice were determined. For both CBA/Ca and C57BL/6 mice, doses below 100 mGy produced no significant increase in Caspase-3-positivity (Fig. 13.1c). After 200 mGy exposures the proportion of Caspase-3-positive cells is greater than controls for both CBA/Ca and C57BL/6 bone marrow and after 500 mGy exposures the proportion of TUNEL-positive cells is significantly increased (Fig. 13.1d). For both strains and for both assays there are significant dose-dependent increases at higher doses. These data indicate that primary bone marrow cells do not show a short-term stress response to low levels of radiation-induced damage.

13.3.2 Lack of an In Vitro Low-Dose Bystander Stress Response

At the lowest X-irradiation doses studied, where not all cells in the bone marrow would have been irradiated, if bystander responses were operating they would be expected to enhance the consequences of irradiation producing a greater than expected response in one or more of the endpoints that we have studied. As there was no evidence of any enhancement at very low doses in vivo and most bystander experiments have used in vitro models an in vitro experiment was performed using a media transfer protocol and also a transwell insert culture dish system such that the non-irradiated cells shared the culture medium but were not in direct contact

Table 13.1 Lack of bystander response in vitro

	p53-Positive cells (%)			Caspase 3 positive cells (%)		
Treatment	Control	1.7 mGy	3 Gy	Control	1.7 mGy	3 Gy
Media transfer	4.1	4.3	3.6	3.8	4.2	4.2
Co-culture	3.0	2.9	2.9	3.1	3.4	3.3

Table 13.2 Dose-response for radiation-induced chromosomal instability 30–100 days post-irradiation

	Cells with aberrations		Aberrations/cell		Chromosome/chromatid
Dose, mGy	30 days	100 days	30 days	100 days	(30 days only)
0	4/908		0.004		3:1
1.7	0/132	0/175	0	0	–
5	1/146	0/193	0.007	0	0:1
12.1	3/346	1/174	0.009	0.006	0:3
25	0/205	–	0	–	–
50	1/446	–	0.002	–	1:0
100	3/611	–	0.006	–	2:1
500	1/384	–	0.003	–	1:0
1,000	6/571	–	0.011	–	1:5
3,000	12/730	6/275	0.016	0.022	2:10

with irradiated cells. Using the media transfer protocol (Table 13.1), the percentage of cells exposed to media obtained from 1.73 mGy irradiated cells expressing p53 (4.3%) was not greater than cells receiving media from sham-irradiated control cells (4.1%).

Similarly, in a co-culture situation there was no evidence for a bystander-mediated increase in p53 expression with 2.9% cells expressing p53 not being greater than for cells in communication with sham-irradiated control cells (3.0%). Increasing the exposure dose to 3 Gy failed to produce a bystander-mediated increase in p53 positive cells. The same results are shown for Caspase-3 staining (Table 13.1). These data indicate that primary bone marrow cells do not show a short-term bystander response to low levels of radiation-induced damage and the question arises as to whether there is a delayed non-targeted effect.

13.3.3 A Threshold for Expression of Radiation-Induced Chromosomal Instability

One of the longer-term radiation-induced non-targeted effects in bone marrow is the induction of a genotype-dependent chromosomal instability phenotype where ongoing chromosome breakage is reflected by significant increases in chromatid-type aberrations [40]. Previously, 3 Gy X-irradiated CBA/Ca mice had been shown to demonstrate such instability [13, 39] and in the present study at 30 and 100 days after whole body irradiation with doses up to 500 mGy (Table 13.2) there was no evidence for the expression of cytogenetic aberrations (overall, 0.0041 aberrations per cell) being greater than the control levels (0.0044 aberrations per cell; p=0.5594). However, consistent with previous data [13, 39], 3 Gy (3,000 mGy) X-irradiation resulted in a significant increase of chromatid-type aberrations (p=0.0020) and the elevation after 1 Gy (p=0.0346) was not significantly different from the elevated levels after 3 Gy (p=0.2887).

13.4 Discussion

The findings of the present study provide no evidence for bystander enhancement of damage signaling and no dose-dependence of short-term (p53, p21 and apoptotic responses) in mouse bone marrow cells after exposure to very low doses of X-rays. Thus, there is no evidence of a linear response at low doses but instead an unexpected threshold where significant p53 pathway responses are initiated (approximately 50 mGy for C57BL/6 and 100 mGy for CBA/Ca bone marrow) which then increase linearly with dose above that threshold.

There may be detection of irradiation at low doses by cells in some mice. Figure 13.1a, b show enhanced levels of p53 and p21 at doses up to the threshold. But this non-significant increase is due to much higher response of some mice (Fig. 13.2). At doses of 50 mGy (the threshold) and 100 mGy all responses are higher than the mean control level indicating that all mice demonstrate response to

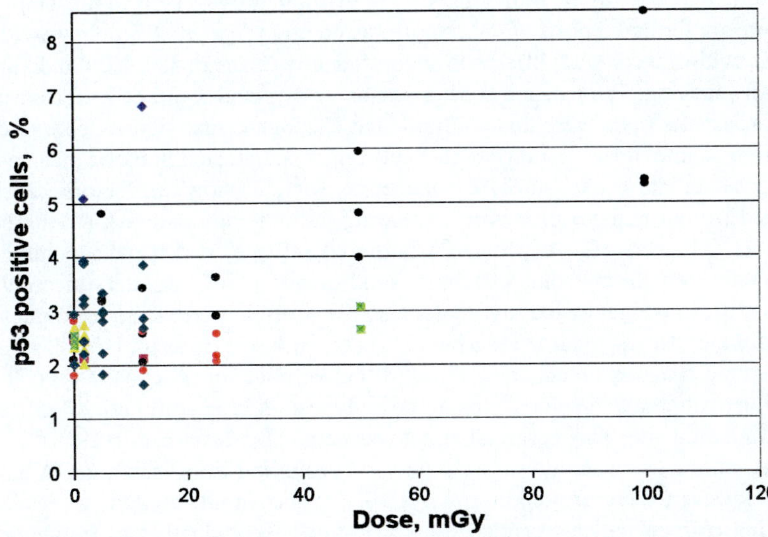

Fig. 13.2 p53 responses in the bone marrow of C57BL/6 mice after exposure to low dose in vivo X-irradiation. Each *dot* is represented by a single mouse

irradiation. At the lower doses most mice respond similarly to controls but in about 15% of mice the proportions of p53 positive cells are higher than in controls and those animals may represent the radiosensitive part of population. So, in the same population of inbred mice variability of responses to low dose radiation is observed.

The chromosomal instability in bone marrow demonstrated in vivo after 3 Gy total body irradiation is consistent with previous studies of high doses of sparsely ionizing radiation [13, 23, 39]. However, at the lowest doses studied in the present study, there was no evidence for such instability indicating that, similar to the damage response signalling, there is a threshold for the detection of this response. The threshold of 1,000 mGy for expression of significant chromosomal instability is approximately an order of magnitude greater than the threshold for p53 pathway signalling. In contrast to these findings for sparsely ionizing low-LET radiation, our previous study of bone marrow cells irradiated with a low dose of densely ionizing high-LET alpha-particles (low fluence with a Poisson distribution and a mean of one particle per traversed cell) resulted in significant expression of chromosomal instability in vitro [21] and in vivo [38]. Like our previous study of bone marrow cells irradiated with a low dose of alpha particles [38], many of the studies of non-targeted effects relate to cells exposed to low fluences of alpha-particles or microbeam-generated charged particles (densely ionizing high LET radiation). In these situations those cells that are traversed by a single particle receive a substantial dose (~0.3 to 0.5 Gy) and sustain damage that is much greater and more complex than is the case for cells irradiated with low doses of sparsely ionizing low-LET X- or γ-rays, where a single radiation track would deliver a dose of the order of 1 to a few mGy to the irradiated cell [11, 12]. It is important to be clear as to what is meant by a low and high dose to the individual cell or to population of cells in this context.

There are limited reports of radiation-induced non-targeted effects in vivo [3, 20, 25] but evidence for such effects in vivo after low doses of low-LET radiation is generally lacking. Prior to the recent interest in bystander effects detected using in vitro systems there were many reports that clastogenic activity, capable of causing chromosome breaks in unirradiated cells, may be detected in blood plasma after high dose radiation exposures. Evidence for such clastogenic factors has been obtained from a number of sources including radiotherapy patients, atomic bomb survivors (31 years after exposure), Chernobyl salvage workers and a variety of individuals with chromosome instability syndromes and inflammatory disorders [7, 22, 28, 30]. Clastogenic activity is associated with lipid peroxidation products [8] and cytokines [9] and regarded as a biomarker of oxidative stress [6, 10]. Clastogenic factors may provide a mechanistic link to 'out of field' or abscopal effects where radiation treatment to one local area of the body results in an antitumor effect distant to the radiation site. This abscopal effect was originally described in 1953 [27] and clinical reports are generally descriptive [17] providing little insight as to mechanism. However, there are reports [4, 18, 19, 35] that implicate damage responses involving oxygen radicals produced as a result of the induction of inflammatory cytokines. It is also of interest the Japanese A-bomb survivors demonstrate evidence of persisting inflammation [15, 34]. Taken together, the various studies implicate inflammatory-type responses as contributing to potential health consequences of radiation exposure.

Considerable progress has been made in understanding the cellular and molecular events that are involved in the acute inflammatory response to infection but responses to tissue injury are much less well understood [26]. The generation of clastogenic factors and induction of abscopal effects are associated with high dose exposures and a case can be made for tissue responses contributing secondary cell damage as a consequence of inflammatory responses to radiation-induced injury rather than any direct interaction between irradiated and non-irradiated cells. It is, therefore, far from clear how readily one may extrapolate simply from the in vitro investigations demonstrating direct interactions between irradiated and non-irradiated cells to in vivo scenarios. In the present study, using relevant exposures and a radiobiologically-relevant tissue, the findings provide no support for arguments that in vivo effects after very low doses of sparsely ionizing, low-LET radiation may be significantly greater than expected by extrapolation from high doses. On the contrary the data demonstrate the opposite as responses are less than expected and show that biological responses at low doses in vivo are not necessarily linearly related to dose and demonstrate, for the first time, a response threshold in primary bone marrow.

Acknowledgements This work was supported by the NOTE IP 036465 (FI6R), Euratom specific programme for research and training on nuclear energy, 6th FP of the EC and by a specialist programme 07003 of Leukaemia and Lymphoma Research.

References

1. Barcellos-Hoff MH, Brooks AL (2001) Extracellular signaling through the microenvironment: a hypothesis relating carcinogenesis, bystander effects, and genomic instability. Radiat Res 156(5 Pt 2):618–627
2. Brenner DJ et al (2003) Cancer risks attributable to low doses of ionizing radiation: assessing what we really know. Proc Natl Acad Sci USA 100(24):13761–13766
3. Brooks AL (2004) Evidence for 'bystander effects' in vivo. (Trans from Eng) Hum Exp Toxicol 23(2):67–70 (In Eng)
4. Camphausen K et al (2003) Radiation abscopal antitumor effect is mediated through p53. Cancer Res 63(8):1990–1993
5. Coates PJ, Robinson JI, Lorimore SA, Wright EG (2008) Ongoing activation of p53 pathway responses is a long-term consequence of radiation exposure in vivo and associates with altered macrophage activities. J Pathol 214(5):610–616
6. Emerit I (1990) Superoxide production by clastogenic factors. In: Crastes de Paulet A (ed.) Free radicals, lipoproteins and membrane lipids. Plenum Press, New York, pp 99–104
7. Emerit I (1994) Reactive oxygen species, chromosome mutation, and cancer: possible role of clastogenic factors in carcinogenesis. Free Radic Biol Med 16(1):99–109
8. Emerit I, Khan SH, Esterbauer H (1991) Hydroxynonenal a component of clastogenic factors? Free Radic Biol Med 10(6):371–377
9. Emerit I et al (1995) Transferable clastogenic activity in plasma from patients with Fanconi anemia. Hum Genet 96(1):14–20

10. Emerit I et al (1997) Clastogenic activity in the plasma of scleroderma patients: a biomarker of oxidative stress. (Trans from Eng) Dermatology 194(2):140–146 (In Eng)
11. Goodhead DT (1989) The initial physical damage produced by ionizing radiations. Int J Radiat Biol 56(5):623–634
12. Goodhead DT (1994) Initial events in the cellular effects of ionizing radiations: clustered damage in DNA. Int J Radiat Biol 65(1):7–17
13. Gowans ID, Lorimore SA, McIlrath JM, Wright EG (2005) Genotype dependent induction of transmissible chromosomal instability by gamma radiation and the benzene metabolite hydroquinone. Cancer Res 65(9):3527–3530
14. Hall EJ (2003) The bystander effect. Health Phys 85(1):31–35
15. Hayashi T et al (2003) Radiation dose-dependent increases in inflammatory response markers in A-bomb survivors. Int J Radiat Biol 79(2):129–136
16. Hei TK et al (2008) Mechanism of radiation-induced bystander effects: a unifying model. J Pharm Pharmacol 60(8):943–950
17. Kaminski JM et al (2005) The controversial abscopal effect. Cancer Treat Rev 31(3):159–172
18. Khan MA, Hill RP, Van Dyk J (1998) Partial volume rat lung irradiation: an evaluation of early DNA damage. Int J Radiat Oncol Biol Phys 40(2):467–476
19. Khan MA, Van Dyk J, Yeung IW, Hill RP (2003) Partial volume rat lung irradiation; assessment of early DNA damage in different lung regions and effect of radical scavengers. Radiother Oncol 66(1):95–102
20. Koturbash I et al (2008) In vivo bystander effect: cranial X-irradiation leads to elevated DNA damage, altered cellular proliferation and apoptosis, and increased p53 levels in shielded spleen. Int J Radiat Oncol Biol Phys 70(2):554–562
21. Lorimore SA et al (1998) Chromosomal instability in the descendants of unirradiated surviving cells after alpha-particle irradiation. Proc Natl Acad Sci USA 95(10):5730–5733
22. Lorimore SA, Coates PJ, Wright EG (2003) Radiation-induced genomic instability and bystander effects: inter-related nontargeted effects of exposure to ionizing radiation. Oncogene 22(45):7058–7069
23. Lorimore SA, McIlrath JM, Coates PJ, Wright EG (2005) Chromosomal instability in unirradiated hemopoietic cells resulting from a delayed in vivo bystander effect of gamma radiation. Cancer Res 65(13):5668–5673
24. Lorimore SA, Chrystal JA, Robinson JI, Coates PJ, Wright EG (2008) Chromosomal instability in unirradiated hemaopoietic cells induced by macrophages exposed in vivo to ionizing radiation. Cancer Res 68(19):8122–8126
25. Mancuso M et al (2008) Oncogenic bystander radiation effects in Patched heterozygous mouse cerebellum. Proc Natl Acad Sci USA 105(34):12445–12450
26. Medzhitov R (2008) Origin and physiological roles of inflammation. Nature 454(7203):428–435
27. Mole RH (1953) Whole body irradiation; radiobiology or medicine? Br J Radiol 26(305):234–241
28. Morgan WF (2003) Non-targeted and delayed effects of exposure to ionizing radiation: II. Radiation-induced genomic instability and bystander effects in vivo, clastogenic factors and transgenerational effects. Radiat Res 159(5):581–596
29. Morgan WF, Sowa MB (2007) Non-targeted bystander effects induced by ionizing radiation. Mutat Res 616(1–2):159–164
30. Mothersill C, Seymour C (2001) Radiation-induced bystander effects: past history and future directions. Radiat Res 155(6):759–767
31. Mothersill C, Seymour C (2003) Radiation-induced bystander effects, carcinogenesis and models. Oncogene 22(45):7028–7033
32. Mothersill C, Seymour CB (2006) Radiation-induced bystander effects and the DNA paradigm: an "out of field" perspective. Mutat Res 597(1–2):5–10
33. Mothersill C et al (2005) Genetic factors influencing bystander signaling in murine bladder epithelium after low-dose irradiation in vivo. Radiat Res 163(4):391–399
34. Neriishi K, Nakashima E, Delongchamp RR (2001) Persistent subclinical inflammation among A-bomb survivors. Int J Radiat Biol 77(4):475–482

35. Ohba K et al (1998) Abscopal regression of hepatocellular carcinoma after radiotherapy for bone metastasis. Gut 43(4):575–577
36. Prise KM, Folkard M, Michael BD (2003) Bystander responses induced by low LET radiation. Oncogene 22(45):7043–7049
37. Rithidech KN, Cronkite EP, Bond VP (1999) Advantages of the CBA mouse in leukemogenesis research. Blood Cells Mol Dis 25(1):38–45
38. Watson GE, Lorimore SA, Wright EG (1996) Long-term in vivo transmission of alpha-particle-induced chromosomal instability in murine haemopoietic cells. Int J Radiat Biol 69(2):175–182
39. Watson GE, Pocock DA, Papworth D, Lorimore SA, Wright EG (2001) In vivo chromosomal instability and transmissible aberrations in the progeny of haemopoietic stem cells induced by high- and low-LET radiations. Int J Radiat Biol 77(4):409–417
40. Wright EG, Coates PJ (2006) Untargeted effects of ionizing radiation: implications for radiation pathology. Mutat Res 597(1–2):119–132

Chapter 14
Pheromones and Adaptive Bystander-Mutagenesis in Mice

Eugene V. Daev, Timofey S. Glinin, and Anna V. Dukelskaya

Abstract The genomic instability has been studied by cytogenetic analysis of chromosome aberrations in somatic and germ cells of CBA mouse strain. It was shown that volatile substances (VS) excreted by unisex groups of animals into the environment induce genomic instability in conspecifics of same sex depending on the state of VS-donors. Different pretreatments of donor animals modified the cytogenetic effect of their VS in recipients. The meaning of such pheromonal "bystander" effects for animal fitness is discussed.

Keywords Mouse • Pheromones • Mitotic disturbances • Meiotic disturbances • Bystander effect

14.1 Introduction

Many social animals share some mechanisms in the formation of their response to stressors with man depending on degree of their genetic homology. Lots of external factors could become stressors if acted too long or too strong, etc. Direct action of some physical or chemical stress-inducing agents on recipient organism of an animal (including human beings) activates its hypothalamic-pituitary-adrenal axis. That is a sign of stress reaction accordingly Selye [27]. In spite of the great importance of studying different negative consequences of the stress state for man the significance of genetic changes during stress is still underestimated. Meanwhile, it is possible to suggest that inside different organs and tissues there are many types of

E.V. Daev (✉) • T.S. Glinin • A.V. Dukelskaya
Department of Genetics & Breeding, Saint-Petersburg State University,
Universitetskaya emb., 7/9, Saint-Petersburg 199034, Russia
e-mail: mouse_gene@mail.ru

C.E. Mothersill et al. (eds.), *Radiobiology and Environmental Security*,
NATO Science for Peace and Security Series C: Environmental Security,
DOI 10.1007/978-94-007-1939-2_14, © Springer Science+Business Media B.V. 2012

sensitive cells which genome serves as target for different stressors. By means of special intraorganismic and intracellular messengers (neuropeptides, hormones, transcriptional factors etc.) these external factors can substantially disturb target cell genomes. That influences, in turn, functional activity of agitated cells, then tissues, organs and finally whole organism viability. Nevertheless we know almost nothing about the precise genetic mechanisms of such stressor action. But what we know that in human beings and in some other animals there are so-called psycho-social stressors and their action can spread rapidly among big groups even without direct contact with these factors. And again neither fine mechanisms nor possible genetic consequences of stress widening are well known till now.

In some animals the social way of their life has led to forming of specific mechanisms of spreading of a stress state among groups and populations. Therefore, impact of genetic disturbances can be amplified drastically also. For studying genetic consequences of stress mouse model of at least one "bystander" mechanisms by means of chemocommunication looks as the ultimate model.

Data obtained show that different chemosignals and especially pheromones serve as stress inducing factors in the house mouse. Some pheromonal treatments in mice activate hypothalamic – hypophyseal axis, induce adrenal hypertrophy and inhibit development of accessory glands, reproductive and immune function. Behavioral changes after these pheromonal exposures also correspond to a stress state in the recipient animals.

It looks like mouse female pheromone 2,5-dimethylpyrazine (DMP) is a good enough example of such stress inducing signal. This puberty-delaying mouse volatile chemosignal, the emission of which is under the control of adrenal glands [14] excreted to the environment by "together caged" (overcrowded) females. The maintenance of laboratory mice under overcrowded condition could be accepted, with the caution, of course, as a modeling of dense natural populations which is considered as a natural stress situation [2, 23].

It is shown in mice that an exposure with female pheromones changes *c-fos* activity in accessory olfactory bulbs of males [18, 26] and in testes [7]. It decreases drastically noradrenalin level in the mucous membrane of the nasal cavity and in the testes tunic of recipient male mice [8]. DMP delays puberty of immature conspecifics [20], decreases significantly the reproductive fitness of male [4, 6] and female mice [19], suppresses immune system [11]. DMP is also aversive chemosignal [9]. There are some other well known pheromonal stress effects in rodents such as activation *c-fos* in hypothalamus and limbic area [16, 21, 31], changes in the level of gonadotropin-releasing factors [13, 15] and pituitary hormones [24, 28, 32], adrenals hypertrophy [1, 30], inhibition of sexual maturation [22] and suppression of immune system [29].

Up until now, little has been known or speculated about pheromonal influences on the genetic machinery and genetic structural integrity of the recipient cells which lies in the very beginning of at least primer-effects of pheromones. The similar is true for a meaning of this "bystander" effect of stress spreading for groups and populations of recipient animals.

Here we wish to report on the effect of chemosignals originating from intact or stressed (by overcrowded caging) mouse donors on genome stability in bone marrow and germ cells of recipient mouse males. Using conventional microscopic techniques, we have observed mitotic disturbances in dividing cells of the bone marrow (anaphase-telophase) and meiotic disturbances in germ cells of the testes of the mouse males.

14.2 Materials and Methods

14.2.1 *Material*

The inbred CBA/LacSto or CBA/LacStoRap mice used in all experiments were initially purchased from "Stolbovaya" or "Rappolovo" animal centers (Moscow region and Saint-Petersburg region, Russia, correspondingly). Isolated animals (donors of chemosignals) or unisex groups of animals (5 per group of tester animals and 7–8 per group of chemosignal donors) were maintained under standard conditions (cage size 22×30×10 cm), food and water provided *ad libitum*. Bedding was changed twice a week. After 2 weeks of adaptation, the animals were used for the experiments.

14.2.2 *Treatment Procedure*

14.2.2.1 Treatment of Chemosignal Donor Animals

We collected pheromonal cues emitted from mouse females or males to the environment. We used mainly the protein fraction of dialyzed urine pooled from 10 isolated or crowded females (VS_{if} or VS_{cf}), correspondingly. Single animals were caged separately at least 2 weeks before urine sample collection. Fresh urine individual samples were immediately pooled and centrifuged (MPW-310, 5,000 rpm, 15 min). Then 2 ml of supernatant were dialyzed against 0.9% NaCl ("Servapore" membrane #44145, t = 1–4°C, 48 h, 2 l×3 changes of salt solution). The volume of samples with main urinary proteins was maintained constant and they were prepared directly before the experiment.

To receive different chemosignals we used diverse pretreatments of donor animals. For some experiments (to escape additional stress of handling of donor animals) we used just soiled bedding from 10 isolated donor females (SB_{if}). Animals were placed separately onto sawdust bedding for 5 days. Then soiled beddings from all donor cages were pooled and mixed. In parallel, we used as stressor soiled bedding from overcrowded females (SB_{cf}). One more stressor was water solution (0.01%) of the mouse female pheromone 2,5-DMP (Aldrich, 98%).

We used also soiled bedding from whole body irradiated CBA males (4 Gy, 0.7 cGy/s, ^{60}Co, "Gammacell 220"). Soiled bedding had been collected for 24 h after irradiation of donor male groups and placed under the wire mesh bottom of standard cage with recipient animals. Direct contact with soiled bedding was prevented. One more donor group was stressed by swimming [9]. Similarly, control animals were exposed to soiled bedding of intact males.

14.2.2.2 Treatment of Chemosignal Recipients

For an exposure of animals to the pheromonal volatile stimuli, cotton balls were soaked in 1.5 ml of a respective VS or DMP solution and placed inside perforated plastic capsules. Each capsule was attached to the external side of the cage lid, at a distance of 2–3 mm from the lid's wiremesh, to prevent a direct contact of the animals with sample. An exposure to distilled water served as a control.

During experiment with soiled bedding half of the recipients own bedding was replaced by the same volume of corresponding sample. The treatment of the recipient animals was continued for 24 h. Clean bedding was used as a control.

All exposures lasted 24 h, after which the animals were sacrificed by cervical dislocation. Their bone marrow was rinsed from femoral bones and fixed in a methanol/glacial acetic acid (3:1) mixture. Testes were fixed the same way after tunic removal. The material was stored at –4°C until its microscopic analysis.

14.2.3 Cytogenetic Analysis

Small pieces of fixed bone marrow tissue (0.2–0.5 mm) were placed in a drop of 2% acetoorcein for 30–45 min. They were subsequently rinsed in 45% acetic acid to remove an extra stain and gently squashed between a cover slip and a microscopic slide. The dividing cells at the anaphase-telophase stages were analyzed microscopically. The bone marrow cells of mice are dividing with a relatively high rate to yield a number of cells sufficient for further statistics. A careful analysis of the anaphase-telophase stage can thus provide an accurate estimation of the factors associated with mutagenic activity, in fixed materials, without any additional pretreatment. In our experiments, the use of the metaphase method was avoided because it demands colchicine or colcemid and hypotonic treatments which all could affect the frequency of chromosomal aberrations.

Anaphase-telophase plates without overlapping or significant shifts were analyzed (more than 200 per each animal). The frequencies of abnormal divisions with a single bridge, a single fragment and a delayed chromosome as well as cells with two and more aberrations were registered. Total frequency of mitotic disturbances (MD) was counted.

Germ cells were removed from seminiferous tubules by shaking and suspended in fixative solution. Then drops of suspension were placed on slides and dried. Slides were stained as it was described earlier.

14.2.4 *Statistics*

Internal homogeneity among each recipient group was checked through the Chi-square contingency test for all types of abnormalities. Differences between groups were estimated by Chi-square criterion [25]. In order to simplify the data for visual clarity, the frequencies (%) and their errors ($m_{\%}$) were subsequently calculated.

14.3 Results and Discussion

Our present results show that chemosignals from crowded females increase frequency of mitotic disturbances in bone marrow cells of recipient mouse males. The results of one experiment are presented in Table 14.1.

DMP induces stronger change in MD frequency than chemosignals of crowded females in their soiled bedding (SB_{cf}). We obtained similar results for the exposure with VS_{cf} (Fig. 14.1b). At the same time pooled chemosignals from isolated females (SB_{if}) act in opposite direction. They decrease the MD level in bone marrow cells of recipient males (Table 14.1, Fig. 14.1b).

Our data provide evidence that the quality of chemosignals excreted by females depend on their state. Stressed by overcrowding donor females emit to the environment chemosignals which affect recipient's organism and destabilize genome of its dividing bone marrow cells. Since genome instability is a sign of stress state [3, 5] we propose that chemosignals from overcrowded females spread the information about unfavorable environmental condition and thus induce stress in their neighbor conspecifics. DMP, which production in mouse females depends on group density and adrenals activity [19, 20], as we suggest, is one of the possible messengers. But except DMP females definitely excrete to the environment many other sex specific substances which "positively" affect recipient males. It could explain more weak influence of SB_{cf} in comparison with pure DMP action. The same explanation is applicable to "positive" effect of SB_{if} (Table 14.1, Fig. 14.1b). Sex-specific "positive" chemosignals stabilize chromosomes structure and mitotic divisions when DMP does not produced by isolated females.

Table 14.1 Frequency of mitotic disturbances in bone marrow cells of CBA mouse males after their exposure with different chemosignals originated from mouse females (%)

Treatment of recipients	N	n	Number of cells with MD	Total frequency of MD ($\% \pm m_{\%}$)	Significant differences
B_{clean}	9	2,994	102	3.4±0.33	
SB_{cf}	10	4,320	225	5.2±0.32	
DMP	10	3,506	306	8.0±0.46	
SB_{if}	10	4,982	87	1.7±0.18	

N number of animals, *n* number of cells, B_{clean} control (clean bedding), *SB* soiled bedding ($_{cf}$, $_{if}$ – after crowded or isolated females, correspondingly), *DMP* 2,5-dimethylpyrazine, ends of vertical lines show on significantly different values (χ^2-criterion, $P<0.01$)

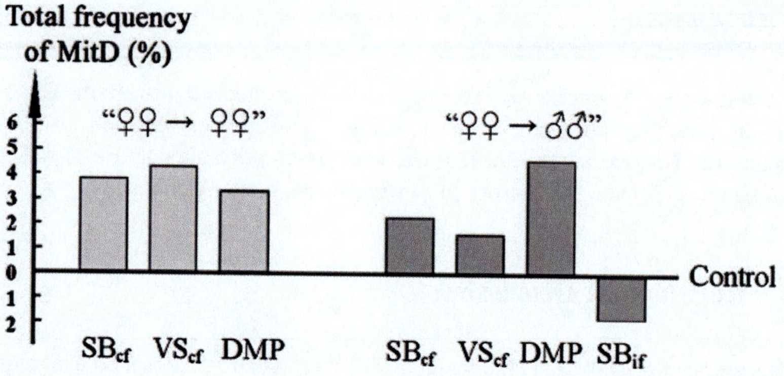

Fig. 14.1 Effect of different chemosignals originating from mouse females in dividing bone marrow cells of mouse females (**a**) and males (**b**) of CBA strain. Differences in total frequencies of mitotic disturbances (MitD) between corresponding control and experimental groups are shown. SB_{cf} – soiled bedding of crowded females; VS_{cf} – volatile substances from urine of crowded females; DMP – 2,5-dimethylpyrazine solution; SB_{if} – pooled soiled bedding of isolated females. All the differences from control are significant; * – difference between effect of SB_{if} and all other effects (Chi-square test, $P<0.01$)

We investigated also action of different female chemosignals on dividing bone marrow cells in mouse female recipients. Our results show that chemosignals in soiled bedding and purified volatiles from urine of crowded females induce mitotic disturbances in bone marrow cells of recipient females. DMP acts in a same manner. The chromosome aberration patterns as well as degree of elevation were not different either in case of soiled bedding or purified volatiles. It was shown earlier that sensitivity of females to DMP depends on their stage of estrous cycle [10].

As we have mentioned previously mouse male chemosignals can serve as stress-factors. Therefore we compare the DMP effect with the influence of other chemosignals originated from stressed or intact males in mitotic and meiotic cells of recipient males. It is shown by ana-telophase analysis that DMP increases total frequency of meiotic disturbances at anaphase-telophase II stage (Fig. 14.2).

Chemosignals of stressed by physical stressor males (after swimming) act in similar manner. Finally, volatiles from irradiated males elevate frequency of meiotic disturbances approximately twice.

Our results are in a good agreement with data that volatile chemosignals from mouse males stressed by swimming or after whole body irradiation induce the same mitotic disturbances in intact recipient males [11, 12]. It was shown also similar genome destabilization in germ cells of stressed mouse males [17].

To our mind cytogenetic disturbances in target cells are a general sign of stress reaction of an organism at an intracellular level. Such influence could be a mechanism by which volatile chemosignals effect development and fitness of recipient animals described earlier. With the aid of that mechanism animals could regulate their population density and structure via "chemosignals – genome integrity of target cells" interactions [3, 5].

Fig. 14.2 Effect of 2,5-dimethylpyrazine (DMP) and volatiles from CBA mouse males stressed by swimming (VS_{sw}) or from irradiated males (VS_{irr}). MeiD – meiotic disturbances. All differences between experimental groups and corresponding controls are significant (Chi-square test, P<0.01)

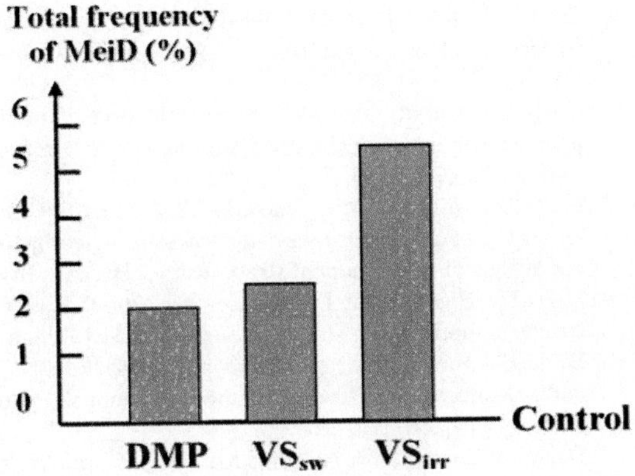

14.4 Conclusion

Our data supports the hypothesis that the integrity of target cell genome is decreased under stress conditions. Cytogenetic disturbances can serve as a universal sign of organismic stress. Pheromonal "bystander-effects" demonstrate that the olfactory mechanism in mice is working effectively to spread a stress state and genetic instability among conspecifics. Definitely there are other different specific and unspecific pathways of indirect widening of environmental influences among communicating animals. But the importance of searching of such pathways is still underestimated as well as mechanisms themselves that are not easily analyzed. Especially it concerns to the genetic consequences of such "bystander-effects". Therefore, we should look for appropriate models to study them at all levels. And research of mouse chemocommunicative mechanisms looks as good possibility to investigate the problem.

Acknowledgements The research is supported by RFBR grant N 09-04-00693.

We are grateful to Prof. B. P. Surinov for his help in conducting our experiments with irradiation and to Dr. Alex Pryor from the Cambridge University (UK) for the manuscript correction.

References

1. Archer JE (1969) Adrenocortical response to olfactory stimuli in male mice. J Mammal 50:836–841
2. Christian JJ (1961) Phenomena associated with population density. PNAS USA 47:428–449
3. Daev EV (1994) Pheromonal regulation of genetic processes: research on the house mouse (*Mus musculus* L.). Russ J Genet 30(8):964–970
4. Daev EV (2003) Induction of dominant lethals in progeny of CBA male mice after pheromonal action. Russ J Genet 39(10):1138–1143
5. Daev EV (2007) Stress, chemocommunication, and the physiological hypothesis of mutation. Russ J Genet 43(10):1082–1092

6. Daev EV, Dukelskaya AV (2004) Reproductive characteristics of mouse males are influenced by the mouse female pheromone 2,5-dimethylpyrazine in the C57BL/6 strain. Ecologicheskaya Genetika 2(1):44–49
7. Daev EV, Dukelskaya AV (2005) Induction of meiotic disturbances in spermatocytes I by pheromones as an inhibiting mechanism of male reproductive function in house mice. Tsitologiia 47(6):505–509
8. Daev EV, Vorob'ev KV, Shustova TI, Zimina SA, Samotokin MB (2000) Genotype-specific changes in functional parameters of immunocompetent cells in laboratory male mice under conditions of pheromoneal stress. Russ J Genet 36(8):872–876
9. Daev EV, Surinov BP, Dukelskaya AV (2007) Chemosignaling in CBA and C57BL/6 mouse strains is modified by stress. Ecologicheskaya Genetika 5(2):37–43
10. Daev EV, Surinov BP, Dukelskaya AV, Marysheva TM (2007) Chromosomal abnormalities and spleenocyte production in laboratory mouse males after exposure to stress chemosignals. Tsitologiya 49(8):696–701
11. Daev EV, Dukelskaya AV, Kazarova VE, Filkina Ya A (2007) Effect of the estrous cycle stage on sensitivity to pheromone 2,5-dimethylpyrazine in the house mouse *Mus musculus* L. Russ J Evol Biochem Physiol 43(6):482–486
12. Daev EV, Filkina Ya A, Surinov BP, Dukelskaya AV (2010) Cytogenetic bystander-effect in mouse after irradiation. In: Modern problems of genetics, radiology, radioecology, and evolution, III international conference, Alushta, Ukraine, 9–14 Oct 2010, p 56
13. Dlusen DE, Ramirez VD, Carter CS, Getz LL (1981) Male vole urine changes luteinizing hormone releasing hormone and norepinephrine in female olfactory bulb. Science 212:573–575
14. Drickamer LC, McIntosh AT (1980) Effect of adrenalectomy on the presence of a maturation-delaying pheromone in the urine of female mice. Horm Behav 14:146–152
15. Dudley CA, Rajendren G, Moss RL (1996) Signal processing in the vomeronasal system: modulation of sexual behavior in the female rat. Crit Rev Neurobiol 10:265–290
16. Fiber JM, Swann JM (1996) Testosterone differentially influences sex-specific pheromone-stimulated Fos expression in limbic regions of Syrian hamsters. Horm Behav 30:455–473
17. Filkina Ya A (2009) Narusheniya meioticheskih delenyi, indutsirovannye letuchimy chemosignalamy stressirovannyh samtsov domovoi myshy, Mus musculus L. Disturbances of meiotic divisions, induced by volatile chemosignals of stressed house mouse (*Mus musculus* L.) males. In: Abstracts of 5th congress of Vavilov's society of geneticists and breeders (In Russian). Part 1, p 355
18. Guo J, Zhou A, Moss RL (1997) Urine and urine-derived compounds induce c-fos mRNA expression in accessory olfactory bulb. Neuroreport 8:1679–1683
19. Jemiolo B, Novotny M (1993) Long-term effect of urinary chemosignal on reproductive fitness in female mice. Biol Reprod 48:926–929
20. Jemiolo B, Novotny M (1994) Inhibition of sexual maturation in juvenile female and male mice by chemosignal of female origin. Physiol Behav 55:519–522
21. Jin BK, Franzen L, Baker H (1996) Regulation of C-Fos mRNA and fos protein expression in olfactory bulbs from unilaterally odor-deprived adult mice. Int J Dev Neurosci 14:971–982
22. Lawton AD, Whitsett JM (1979) Inhibition of sexual maturation by a urinary pheromone in male prairie deer mice. Horm Behav 13:128–138
23. Lecyk M (1967) The influence of crowded population stimuli on the development of reproductive organs in the common vole. Acta Theriol 12:177–179
24. Maruniak J, Coquelin A, Bronson FH (1978) The release of LH in male mice in response to female urinary odors: characteristics of the response in young males. Biol Reprod 18:251–255
25. Motulsky HJ (2003) Prism 4 statistics guide – statistical analyses for laboratory and clinical researchers, GraphPad Software Inc., San Diego, pp 1–148; www.graphpad.com
26. Muroi Y, Ishii T, Komori S, Kitamura N, Nishimura M (2006) Volatile female odors activate the accessory olfactory system of male mice without physical contact. Neuroscience 141:551–558
27. Selye H (1950) The physiology and pathology of exposure to stress. Acta Medical, Montreal, p 203

28. Singer AG, Clancy AN, Macrides F, Agosta WC, Bronson FH (1988) Chemical properties of a female mouse pheromone that stimulates gonadotropin secretion in males. Biol Reprod 38:193–199
29. Surinov BP, Karpova NA, Zhovtun LP (2004) Olfactory stress: immunosuppression dynamics in mice with different genotype. Immunologiya (Russ) 3:183–185
30. Wuensch KL (1979) Adrenal hypertrophy in mice following exposure to crowded males odors. Behav Neural Biol 27(2):222–226
31. Wysocki CJ, Tubbiola M, Lepri LJ (1995) The prairie vole version of puberty: role of the chemical senses. In: 27th annual conference on reproductive behavior (Abstracts), Boston, p 23
32. Xu M, Hill JW, Levine JE (2000) Attenuation of luteinizing hormone surges in neuropeptide Y knockout mice. Neuroendocrinology 72(5):263–271

Chapter 15
Statistical Model of Adaptation Risk Assessment of Chromosomal Instability

V.L. Korogodina, B.V. Florko, and L.P. Osipova

Abstract One of the modern problems of radiobiology and radioecology is risk assessment of chromosomal instability in populations that experienced low radiation effects. For these calculations the probability model of adaptation can be used which we suggested earlier (Korogodina VL, Florko BV: Evolution processes in populations of plantain, growing around the radiation sources: changes in plant genotypes resulting from bystander effects and chromosomal instability. In: Mothersill C, Seymour C, Mosse IB (eds.) A challenge for the future. Springer, Dordrecht, pp 155–170, 2007). Purpose: description of adaptation processes in population and risk assessment of genetic instability. Materials: numbers of chromosome abnormalities in cells and cells with abnormalities in rootlets meristem of seeds collected near the radiation sources; frequency of abnormal blood lymphocytes in samples of persons living in the Tyumen and Irkutsk regions of Russia. Methods: statistical modeling. Results and discussion. An adaptation process can be presented as primary and late damaging processes accompanied by selection which is described by Poisson and geometrical laws. Influence of low radiation intensity on chromosomal instability was analyzed; a continued adaptation process in four generations was shown. Risk of chromosomal instability can be determined by means of the P- and G-distributions parameters. Conclusions. Type and structure of model distributions can indicate adaptation processes in population, or its genetic stability. Accumulation of abnormalities and risk of chromosomal instability in resistant subpopulation can be determined by sample mean of P-distribution. Sample mean and value of G-distribution indicate intensity of damaging and selection in sensitive subpopulation.

V.L. Korogodina (✉) • B.V. Florko
Joint Institute for Nuclear Research (JINR), 141980 Dubna, Moscow Region, Russia
e-mail: korogod@jinr.ru

L.P. Osipova
Institute of Cytology and Genetics SB RAS, Lavrentjeva ave. 10, Novosibirsk, 630090, Russia

C.E. Mothersill et al. (eds.), *Radiobiology and Environmental Security*,
NATO Science for Peace and Security Series C: Environmental Security,
DOI 10.1007/978-94-007-1939-2_15, © Springer Science+Business Media B.V. 2012

Keywords Low dose irradiation • Adaptation process • Chromosomal instability • Plant • Blood lymphocytes • Probability model of adaptation • Continued instability across generation

15.1 Introduction

Chromosomal instability was observed long years ago [26, 30]. In 1970 V.I. Korogodin and his colleagues showed that yeast cells instability was coupled together with accumulation of chromosomal abnormalities [18] induced by low radiation effects or non-optimal environment conditions [4]. A connection of the phenomenon of instability with evolution was mentioned [13, 18].

R.A. Fisher [8], J.H. Gillespie [12], and H.A. Orr [32] investigated fitness to the environment conditions by modeling. Orr showed in evolution models a universal character of distributions with tails and connection of this phenomenon with selection [32]. N.P. Bochkov and A.N. Chebotarev used statistical modeling to analyze an appearance of multiple mutations in blood lymphocytes of persons [5]. These authors suggested a hypothesis based on queues theory that distributions' tails are connected with DNA repair [7]. Analysis of distributions of patients with chromosome instability syndrome on a number of chromosome abnormalities in blood lymphocytes revealed geometrical (G) and Poisson (P) components [3].

Our statistical model of adaptation is based on time-dependent accumulation of abnormalities and selection [9, 17]. To testify the model and investigate features of adaptation processes we used the following data: own studies of lab-irradiated pea seeds, selected line [19] and plantain seeds collected in natural populations growing near the radiation sources [20]; investigations of L.P. Osipova on blood lymphocytes of persons living in the Tyumen and Irkutsk regions [33]; data published by N.A. Chebotarev on blood lymphocytes of radiochemical enterprise' workers [7]. For these objects the modeling of the appearances of abnormal cells, abnormal chromosomes in cells and proliferation activity of cells were performed [18, 21–24].

Our aims were to analyze regularities of adaptation, its processing across generation and to assess the risk of chromosomal instability and selection. To demonstrate these regularities in ecology and epidemiology, we used data on plantain seeds collected near the Balakovo (middle Volga) Nuclear Power Plant (NPP) and on blood lymphocytes of samples of persons living in the Tyumen (Far North) and Irkutsk (Siberia) regions.

15.2 Statistical Methods

We used maximum-likelihood method for approximations. The following statistical criteria were used for the estimation of regression: (i) R_{adj} is the determination coefficient corrected for degrees of freedom [35] (equivalent to the T-criterion known in radiobiology [11]); (ii) the AIC criterion (the Akaike criterion imposes limitation on minimal information distance between the model and experimental distributions) [1]; (iii) the BIC criterion (selection of the most probable models from the ensemble under the condition of a priori equal probability of any of them) [37]. The most sensitive test for our tasks is T-criterion [11], which encourages good efficiency of approximation and fines for use of large numbers of parameters.

The best hypothesis with respect to the majority of criteria was preferred. For close values of criteria, the simpler hypothesis was chosen. For a given number of experimental points, the number of model parameters is, as a rule, larger than optimal; therefore, the stability of distributions was verified. The verification consisted in variation of the length of the partitioning interval upon construction of histograms. The interval length $(D_{max}-D_{min})/M$ was taken as the basis. Here, (D_{min}, D_{max}) is the interval of data variation, $M=[\log_2(N)]+1$ is the number of partition intervals, and N is the number of data points. The verification showed the stability of the distributions.

The program FUMILI [6] was used. The search of optimal values of the model parameters was carried out in two stages. First, the initial approximation for the model parameters was found by the random search method or the simulated annealing method [16]. Then the parameter values were refined using regular procedures BFGS or Newton [39]. Three parameters were optimized.

15.3 Adaptation Model

The adaptive process has three components [9, 17]: primary radiation injury which depends on the intensity of radiation (I_{rad}); late injury which depends on the intensity of intercellular "bystander" (I_{byst}) or intracellular regulatory (I_{reg}) mechanisms; and selection which depends on repair systems and environment conditions. A scheme of adaptation processes in meristem of seeds' rootlets and its cells is presented in Fig. 15.1 [24] because apical meristem of rootlet can be model of bone marrow.

These components can be described by different combinations of P and G laws where P statistics corresponds to statistics of independent events, and G one characterizes it after selection [22]. The independently appeared primary injuries and late damages follow the Poisson or binomial law [9]. The occurring of late damages can be approximated as increasing linearly with time, that is, by the Poisson law with increased sample mean [9].

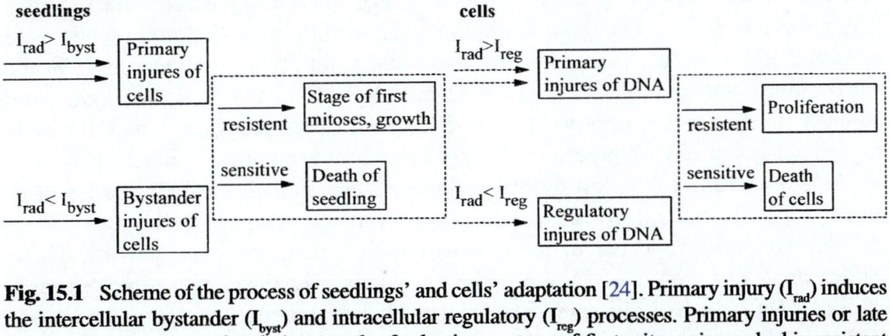

Fig. 15.1 Scheme of the process of seedlings' and cells' adaptation [24]. Primary injury (I_{rad}) induces the intercellular bystander (I_{byst}) and intracellular regulatory (I_{reg}) processes. Primary injuries or late damages can be accumulated. As a result of selection, a stage of first mitoses is reached in resistant seedlings, the sensitive ones die. Intracellular processes (primary or regulatory damaging of DNA) lead to proliferation or death of cells

In practice, the observed P distribution can be a sum of P distributions on the numbers of primary and late damages. The first distribution dominates if radiation intensity exceeds late damaging (for cells, $I_{rad} > I_{reg}$) [17]; therefore its sample mean should be low. For example, a dose rate of 19.1 cGy/h (time between two hits per cell = 5 s) induces hits in cells in the first 2 min with averaged number of hits per cell ≈ 33; this case corresponds to the lowest sample mean of the P distribution of seedlings on the number of damaged cells [17]. Irradiation at dose rates of 0.3 and 1.2 cGy/h (time between two hits = 5 and 1.3 min, respectively) induces bystander effects which should prevail in these cases. Complete analysis of pea data and especially plantain data reveals the correspondence between low P- sample means for seedlings and cells that couple with high damage factor. Increased value of P- sample mean indicates the late processes.

Cell is a more sensitive system and responses to slight signals of environment. We revealed a little constant process of adaptation in intact cells. Environmental stresses affect sensitive cells that form subpopulation of cells G- distributed on the number of DNA-damages. When primary irradiation intensity exceeds intracellular regulatory effects ($I_{rad} > I_{reg}$), Poisson law displaces G-one.

15.4 Irradiated Plant Populations

15.4.1 Plantain Seeds and Ecology Conditions

Seeds of plantain (*Plantago major*) were studied. Seeds of all populations were germinated until seedling roots reached the length that corresponded to the stage of the first mitoses. The chromosome bridges and acentric fragments were registered as chromosome aberrations (CAs) in ana-telophases. Antioxidant activities were studied with a photochemiluminescence method. An amount of seed infusion, which inhibited chemiluminescence by 50% ($C_{1/2}$), was adopted as a measure of antioxidant status (AOS). For each population, 500 seeds were studied. All methods were described in [19, 21].

The tested plantain populations were located in sites within 80 km of the Balakovo NPP and in Chernobyl trace area (Saratov region), and near the accelerator facilities in the Moscow region. The populations were chosen in similar biotopes. Seeds were collected at the end of August in 1998 and 1999 from 20 to 30 plants. In 1999 the temperatures during daylight hours reached 30–32°C in the Moscow region and 38–40°C in the Saratov region (they are extreme for both provinces), and the seeds experienced elevated temperatures during the maturation period in nature [38].

For sites within a 100 km radius of the NPP, the annual γ-radiation dose rates and ^{137}Cs soil concentrations varied little from the ranges ~0.10–0.15 μSv/h and ~5–10 Bq/kg reported in independent radiological surveys [25, 38]. These values do not exceed the average radiation values over the Saratov and Moscow regions [38]. Our examination showed that chemical pollutions do not correlate with results of cytogenetic analysis and cannot imitate NPP radiation factor [23].

Plantain seeds experienced the NPP fallout irradiation in nature (annual fallouts on isotopes: Kr ~2.5 TBq; Xe ~2.5 TBq, and I ~4.4 TBq), the dose rates are controlled by NPP administration [25]. Distribution of the particulate emissions and gases were estimated according to the Smith-Hosker model [14] based on NPP characteristics [25] and winds in summer near the ground in the NPP region [38]. The isotopes fallouts result in γ-irradiation mainly (mean energy ~1.1 MeV/γ-quanta [15]). The relative fallout dose rates (RFRD) values were calculated in the ratio to the dose in site, which was at about 80 km from NPP.

The details of characteristics of sites, determination of radioactivity, and calculation of NPP fallout irradiation of seeds are published in [17].

15.4.2 Adaptation of Plantain Populations to NPP Fallouts in Hot Summer

Analysis of biological values showed the character of radiation effects on seeds' populations collected in years with normal high (1998) and extreme high summertime temperatures (1999) [20, 21]. The statistical modeling was performed on the appearance of CAs in seedling meristem cells and cells with CAs [17, 23]. Adaptive processes induced by combined effects of radiation and high temperatures in plant populations in natural conditions were described in [24]. This synergic effect is verified in lab experiments and modeling [22].

Here, we present data (1999) which demonstrate a strongly synergic process in plantain populations which leads to a high risk of instabilities and selection. The parameters of seed distributions (Fig. 15.2) and cells distributions (Fig. 15.3) are shown as dependents on the calculated fallout dose near the NPP (Figs. 15.2c, d and 15.3c, d) and the seeds' antioxidant status (Figs. 15.2a, b and 15.3a, b) [17]. The G- and P-parameters that characterize sensitive cells' subpopulations are combined into common graph (Fig. 15.3, gr. 2). The parameters that correspond to sensitive seedlings' and cells' subpopulations are correlated ($p<0.001$, $df=5$, $n=7$) [17]. Thus, sensitive subpopulation of seedlings can be considered as a group at risk of instability that leads to selection.

Values of surviving sensitive subpopulations of seedlings (Fig. 15.2d; gr. 2) and cells (Fig. 15.3d; gr. 2) decrease with fallout dose rate ($p<0.1$, $df=3$, $n=5$). At the border (558 and 1,350) or inside (1,500 r.u[1]) the 10-km zone, approximately 70–80% of seeds died, and the value of surviving sensitive seedlings is about 2–3%. Meristem cells of these survived adapted seedlings consist of increased number ($p<0.05$) of late or primary DNA damages [18] (Figs. 15.2c, d and 15.3c, d; gr. 2).

For plantain seeds collected in different populations of neighboring ecosystems (relative NPP fallout dose rate = 82 r.u), values of the P- and G- distributions of

[1]Evaluation of this relative dose rate can be less dependable because this population was located ≈ 100 m from the NPP in the shade of the smokestack.

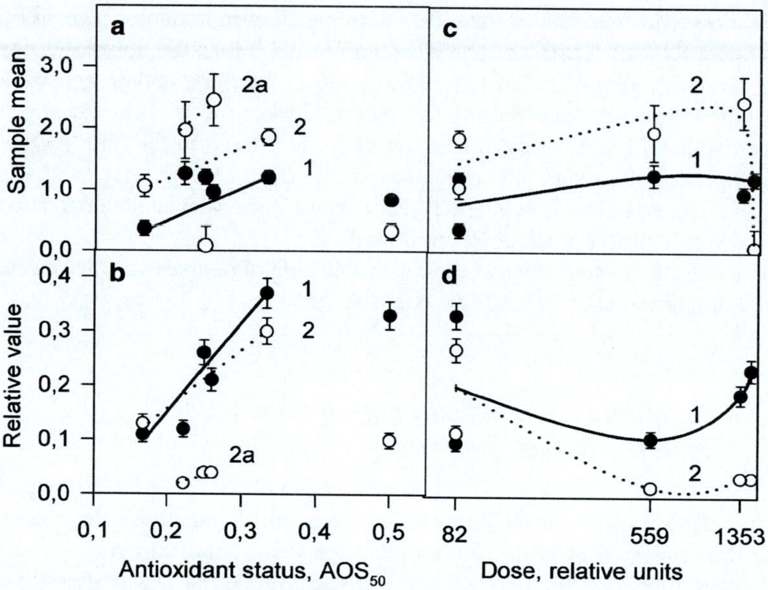

Fig. 15.2 The parameters of distributions of plantain seedlings on the number of cells with CAs in root meristem versus antioxidant status (**a, b**) and calculated relative dose (**c, d**). The sample means (**a, c**) and values (**b, d**) of the P- and G- distributions are shown (curves *1* and *2*, respectively). Points with the same dose rate irradiation are connected (**a, b**). The regressions, a: $y=-0{,}48+5{,}04\cdot x$ ($p<0.05$) (*1*), $y=0.25+4.80\cdot x$ ($p<0.05$) (*2*); $y=-0.92+12.96\cdot x$ ($p<0.05$) (*2a*); b: $y=-0.15+1.43\cdot x$ ($p<0.05$) (*1*), $y=-0.04+1.02\cdot x$ ($p<0.05$) (*2*); $y=-0.10+0.53\cdot x$ ($p<0.05$) (*2a*); c: polynomial fits (*1, 2*); d: polynomial fit (*1*); $y=100/x1.4+0.027$ (*2*)

seedlings (Fig. 15.2b, gr. 1, 2) and G-distributions of cells (Fig. 15.3b, gr. 2) differed ($p<0.05$). We see that antioxidants increase radioprotection; this conclusion agrees with investigations published in [41].

For seedlings, a sample mean of G- distribution increases with AOS value ($p<0.05$) that suggests an enhancement of bystander damaging (Fig. 15.2a, gr. 2) with antioxidant status. This result agrees with investigations which showed the involvement of energy metabolism in the production of bystander effects by radiation [29]. No influence of antioxidant status on the sample means of cells' distributions was revealed in these plantain studies (Fig. 15.3a), perhaps because sample sizes were not sufficient. The hypothesis has been made that the phenomenon of hypersensitivity can be a result of oxidative stress [31]. The authors of this paper investigated effects of low-dose radiation on human blood lymphocytes over a range from 1 to 50–70 cGy, which is higher than around the NPP. In our calculations, each seedling's meristem experiences an influence of even one g-quantum per 3 months in the 20-km NPP zone [17]. We suppose that multiple DNA damages could appear due to affected mass cellular structures. This conclusion agrees with the opinion of C. Mothersill and C. Seymour concerning the role of epigenetic factors in low-dose effects [28].

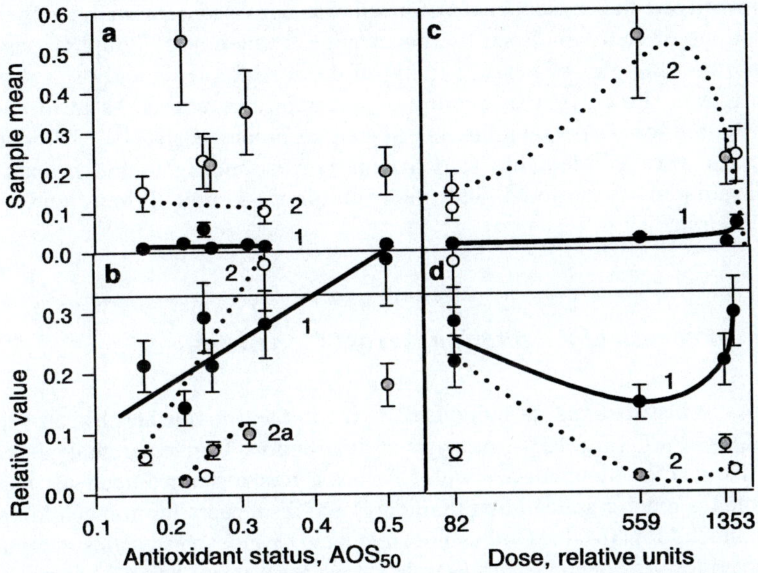

Fig. 15.3 The parameters of distributions of plantain meristem cells on the number of CAs for seeds collected around the NPP versus antioxidant status **(a, b)** and calculated relative dose **(c, d)**. The sample means **(a, c)** and values **(b, d)** of distributions of resistant and sensitive subpopulations are shown (*curves 1 and 2*, respectively). Points with the same dose rate irradiation are connected (a, 1, 2). The regressions, b: $y=0.73 \cdot x+0.02$ ($p<0.05$) (1), $y=0.77 \cdot x - 0.12$ ($p<0.05$) (2); c: polynomial fits (1, 2); d: polynomial fit (1); $y=100/x1.4+0.027$ (2)

15.5 Irradiated Persons and Their Descendants

15.5.1 Samples of Individuals and Their Living Sites

At the beginning of the 1950s, nuclear tests were conducted in the Novaja Zemlja and Semipalatinsk sites, resulting in radioactive fallouts in the Far North and Siberia, including the Tyumen and Irkutsk regions. At present, ^{137}Cs contaminations are 153 and 118 Bq/kg (lichen and venison, respectively) in settlements Samburg of Tyumen region [34], and 55 Bq/m^2 (soil) in Maloe Goloustnoe of Irkutsk region [27]. Random samples of persons living in settlements Samburg and Maloe Goloustnoe were studied. Analyses of chromosomal breakages in blood lymphocytes of persons living in these settlements are published in [33]. Frequency of lymphocytes cells with abnormalities was determined as the ratio of the abnormal lymphocytes number to proliferated ones. As a control, group of individuals from city Novosibirsk, whose blood sample (1 ml) contained not less than 100 activated lymphocytes, was studied. Blood samples of 163 (Samburg), 66 (Maloe Goloustnoe) and 40 (control) individuals were investigated.

To investigate the consequences of irradiation across generations, we analyzed distributions of individuals on the occurrence frequency of lymphocytes with CAs in blood samples of persons living in these settlements and of the control group. It was shown that frequency of lymphocytes with CAs in blood can be considered for statistical modeling of adaptation processes [10]. The samples of persons were divided into four groups corresponding to individuals who were irradiated by fallouts and their children, grandchildren, and great-grandchildren.

15.5.2 *Process of Adaptation Across Generation*

The types of distributions for individuals living in the Tyumen and Irkutsk regions are P, G, and P+G (Fig. 15.4). Analysis of distributions for the parental generation on the frequency of lymphocytes with CAs has a geometrical component, which is negligible for great-grandchildren (Samburg) and disappears for grandchildren and great-grandchildren (Maloe Goloustnoe) regulated by Poisson statistics. A tendency is observed towards decreased P- sample means across generations, but it remains higher (3.1÷4.2, $p<0.05$) than the control value (1.4) (Table 15.1). This tendency may reflect more resistant young individuals than old ones and the deaths of persons with small numbers of lymphocytes (these ones are "G distributed" [2]).

Bystander processes in resistant groups of persons lead to an increasing frequency of lymphocytes with CAs in blood samples, and the risk of these effects can be evaluated by P parameters. The frequency of lymphocytes with CAs is 3.2±0.3 for 50% of persons who are less than 18 years old, whereas the same instability is observed of 5% of tested control individuals only (Table 15.1, Fig. 15.4). Death of sensitive persons is connected with lymphocyte depletion: persons with lymphocyte deficiencies are those with blood diseases, which were accounted in the presented statistics. They form a "group of risk" which can be determined by G value. The P and G values demonstrate that the processes of genome instability and selection prevail in the oldest generation.

A comparison of genome instabilities of persons living in the Tyumen and Irkutsk regions shows that the sample means of P distributions for both regions are approximately equal. Distributions for parents' and children's samples from settlements M. Goloustnoe and for all samples from settlements Samburg reveal a G component. Taking into account the absence of lymphocyte activation in many of the Samburg individuals (Table 15.1), we can surmise that stronger and continued radiation effects can be caused by the traditional food chain of the North nations' "lichen-reindeer-man".

The study of genetic instability across generations of mice irradiated with low-dose-rate radiation indicates a genetic instability in the F1, F2, and F3 generations from irradiated males [40]. Analysis of distributions' structure for samples of persons living in sites Samburg and Maloe Goloustnoe shows chromosomal instability which continues across three generations of irradiated individuals, because the mP values of distributions of youngsters' samples are increased (Table 15.1, Fig. 15.4).

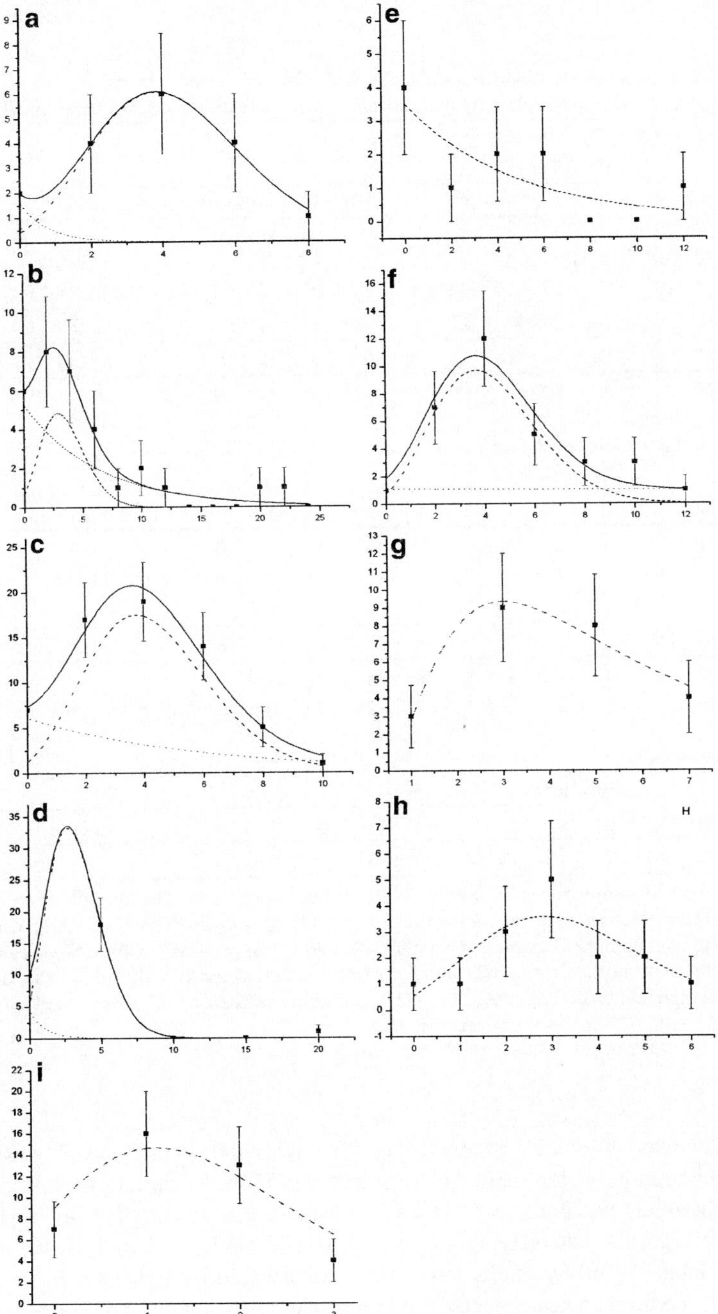

Fig. 15.4 Statistical modeling of distributions of persons of different generations on the frequency of cells with CAs occurrence. X-axis–frequency of cells with CAs, %; Y-axis–frequency of occurrence, %. Sample persons living in settl. Samburg (**a** – **d**); settl. M. Goloustnoe (**e** – **h**); c. Novosibirsk (i). Parents (**a**, **e**); children (**b**, **f**); grandchildren (**c**, **g**); great grandchildren (**d**, **h**). Total curve of approximation (———); Poisson component (— — —); Geometrical one (▪ ▪ ▪). Standard errors of total approximation curve are shown

Table 15.1 Frequency of blood lymphocytes with CAs of persons living in the Tyumen and Irkutsk regions, and Novosibirsk city, parameters of most effective model of approximation

Settlement Samburg, Tyumen region

			Frequency										G, P	
Age	N0	N1	0	2	4	6	8	10	12	14	16	>18	mG/G	mP/P
50–80	24	7	2	4	6	4	1	0	0	0	0	0	0.2/1.7	4.2/15.4
39–49	36	4	6	8	7	4	1	2	1	0	0	2	5.0/19.3	3.5/11.1
18–38	71	7	7	17	19	14	5	1	0	0	0	0	3.2/18.5	4.2/44.7

			Frequency										G, P	
Age	N0	N1	0	5	10	15	20	0	0	0	0	0	mG/G	mP/P
<18	29	–	10	18	0	0	1	0	0	0	0	0	–	3.3/28.1

Settlement Maloe Goloustnoe, Irkutsk region

			Frequency										G, P	
Age	N0	N2	0	2	4	6	8	10	12	14	16	>18	mG/G	mP/P
60–80	10	3	4	1	2	2	0	0	1	0	0	0	3.0/9.9	–
40–60	32	4	1	7	12	5	3	3	1	0	0	0	5.8/7.3	4.2/24.6

			Frequency										G, P	
Age	N0	N2	0	1	2	3	4	5	6	7	8	9	mG/G	mP/P
18–40	23	1	1	1	3	6	4	4	3	1	0	0	–	4.0/23.7
<18	15	–	1	1	3	5	2	2	1	0	0	0	–	3.1/15.3

City Novosibirsk

			Frequency										G, P	
Age	N0	N3	0	1	2	3	4	5	6	7	8	9	mG/G	mP/P
–	40	40	7	16	13	4	0	0	0	0	0	0	–	1.4

Standard error of sample means d(mP) = 10–15%, d(mG) = 30–40%. The control group consisted of persons from Novosibirsk, the number of analyzed lymphocytes exceeding 100 in each blood sample. N0, total number of examined persons; N1, number of persons in whose blood samples the lymphocytes were not activated; N2, number of persons in whose blood samples the number of activated lymphocytes did not exceed 30; N3, the number of persons in whose blood samples the number of analyzed lymphocytes exceeded 100

Model characteristics indicate the significance of resistance in individuals (youngsters' samples) which plays the main role in selection in blood lymphocytes: samples of older individuals are characterized by a G distribution on the number of lymphocytes with CAs (Fig. 15.4) and on lymphocyte depletion (Table 15.1). Finally, it was revealed that the intensity of bystander processes and selection are higher in settlement Samburg, and this can be connected with a stronger radiation influence due to the food chain of "lichen-reindeer-man". All these prevail upon the phenomenon of chromosomal instability and should be considered as components of the adaptive process.

15.6 Conclusions

1. Model of adaptation consists of three components: primary injures; late radiation-induced damages; selection. Statistically it can be described by combination of Poisson and geometrical laws;
2. Type and structure of the distributions indicate a stable state of population or adaptation processes characterized by chromosomal instability and selection;
3. Parameters of G-distribution can be used for assessment of risk of chromosomal instability coupled with selection;
4. Increased value of Poisson sample mean indicates the processes of secondary radiation-induced damaging. Risk of abnormalities' accumulation can be calculated by means of P-parameter;
5. Radiation intensity influence on adaptation processes. Intensities of late damaging and selection increase with dose rate irradiation;
6. Adaptation process continues across generation. Its intensity depends on environmental factor and objects' resistance.

Acknowledgements The hypothesis of adaptation was evolved by V.I. Korogodin and discussed with V.B. Priezzhev. G.A. Ososkov suggested some statistical methods. We are grateful to these researchers, who contributed to these investigations.

References

1. Akaike H (1974) A new look at the statistical model identification. IEEE TransAutomatic Control AC-19:716–723
2. Antonova E, Osipova LP, Florko BV, Korogodina VL (2008) The comparison of distributions of individuals with normal and bad stimulated blood cells' activity on the frequency of aberrant cells' occurrence in blood lymphocytes. Rep Russ Mil-Med Acad. Appl 1(3):73 (Russian)
3. Arutyunyan R, Neubauer S, Martus P, Durk T, Stumm M, Gebhart E (2001) Intercellular distributions of aberrations detected by means of chromosomal painting in cells of patients with cancer prone chromosome instability syndromes. Exp Oncol 23:23–28
4. Bliznik KM, Kapultsevich YuG, Korogodin VI, Petin VG (1974) Formation of radioraces by yeasts. 4. The dependence of the saltant yield on postirradiation cultivation conditions. Radiobiology 14(2):229–236 (Russian)
5. Bochkov NP, Yakovenko KN, Chebotarev AN, Funes Kravioto F, Zhurkov VS (1972) Distribution of the damaged chromosomes on human cells under chemical mutagens effects in vitro and in vivo. Genetika 8:160–167 (Russian)
6. CERN Program Library [http:// www.jinr.ru/programs/cernovr.htm]
7. Chebotarev AN (2000) A mathematical model of origin multi-aberrant cell during spontaneous mutagenesis. Dokl Biol Sci 371:207–209 (Russian)
8. Fisher RA (1930) The Genetical theory of natural selection. Oxford University Press, Oxford
9. Florko BV, Korogodina VL (2007) Analysis of the distribution structure as exemplified by one cytogenetic problem. Phys Part Nuclei Lett 4:331–338

10. Florko BV, Osipova LP, Korogodina VL, (2009) On some features of forming and analysis of distributions of individuals on the number and frequency of aberrant cells among blood lymphocytes. Math Biol Bioinform 4(2):52–65 http://www.matbio.org/journal.php?lang=rus&is_last=1 (Russian)
11. Geras'kin SA, Sarapul'tzev BI (1993) Automatic classification of biological entities by the level of radiation stability. Avtomatika & telemechanika 2:183–189, Russian
12. Gillespie JH (1983) A simple stochastic gene substitution model. Theor Popul Biol 23:202–215
13. Herman RK, Dworkin NB (1971) Effect of gene induction on the rate of mutagenesis by ICR-191 in Escherichia coli. J Bacteriol 106:543–550
14. Hosker RP Jr (1974) Estimates of dry deposition and plume depletion over forests and grasslands. In: Proceedings of physical behavior of radioactive contaminants in the atmosphere, IAEA, Vienna, pp 291–309
15. Ivanov VI, Lyszov VN, Gubin AT (1986) Handbook on microdosimetry. AtomIzdat, Moscow (Russian)
16. Kirkpatrick S, Gelatt CD Jr, Vecchi MP (1983) Optimization by simulated annealing. Science 220:671–680
17. Korogodina VL, Florko BV (2007) Evolution processes in populations of plantain, growing around the radiation sources: changes in plant genotypes resulting from bystander effects and chromosomal instability. In: Mothersill C, Seymour C, Mosse IB (eds) A challenge for the future. Springer, Dordrecht, pp 155–170
18. Korogodin VI, Bliznik KM, Kapultsevich YuG (1977) Regularities of radiorace formation in yeasts. 11 facts and hypotheses. Radiobiology 17(4):492–499 (Russian)
19. Korogodina V, Panteleeva A, Ganicheva I, Lazareva G, Melnikova L, Korogodin V (1998) Influence of the weak gamma-irradiation dose rate on mitosis and adaptive response in meristem cells of pea seedlings. Radiats Biol Radioecol 38:643–649 (Russian)
20. Korogodina V, Bamblevskij V, Grishina I, Gustova M, Zabaluev S, Korogodin V, Kuraeva T, Lozovskaja E, Maslov O (2000) Antioxidant status of plant (Plantago major L.) seeds of the pollution in the Balakovo NPP and chemical enterprisers region. Radiats Biol Radioecol 40:334–338 (Russian)
21. Korogodina VL, Bamblevsky V, Grishina I, Gustova M, Florko B, Javadova V, Korogodin VI, Lozovskaya E, Malikov A, Maslov O, Melnikova L, Shlyakhtin G, Stepanchuk V, Zueva M (2004) Evaluation of the consequences of stress factors on plant seeds growing in a 30-km zone of Balakovo NPP. Radiats Biol Radioecol 44(1):83–90 (Russian)
22. Korogodina VL, Florko BV, Korogodin VI (2005) Variability of seed plant populations under oxidizing radiation and heat stresses in laboratory experiments. IEEE Trans Nucl Sci 52:1076–1083
23. Korogodina VL, Bamblevsky CP, Florko BV, Korogodin VI (2006) Variability and viability of seed plant populations around the nuclear power plant. In: Cigna AA, Durante M (eds) Impact of radiation risk estimates in normal and emergency situations. Springer, Dordrecht, pp 271–82
24. Korogodina VL, Florko BV, Osipova LP (2010) Adaptation and radiation-induced chromosomal instability studied by statistical modeling. The Open Evol J 4:12–22
25. MAPRF (1998) The general information on Balakovo NPP. The Ministry of Atomic Power of the RF, Rosenergoatom concern, Balakovo NPP, pp 3–5 (Russian)
26. McClintock B (1938) The production of homozygous deficient tissues with mutant characteristics by means of the aberrant mitotic behavior of ring-shaped chromosomes. Genetics 23:315–376
27. Medvedev VI, Korshunov LG, Chernyago BP (2005) Radiation effect of Semipalatinsk' nuclear testing area on South Siberia (Long-term investigations on Eastern and Middle Siberia and their comparison with material on Western Siberia). Siberia' Ecol J 12:1055–1071 (Russian)
28. Mothersill C, Seymour C (2003) Radiation-induced bystander effects, carcinogenesis and models. Oncogene 22:7028–7033
29. Mothersill C, Stamato TD, Perez ML, Cummins R, Mooney R, Seymour CB (2000) Involvement of energy metabolism in the production of 'bystander effects' by radiation. Br J Cancer 82:1740–1746

30. Nadson GA (1920) On influence of radium on yeast fungi in connection with the problem of radium effect on living matter. Vestn Rentgenol Radiol 1:45–137 (Russian)
31. Nasonova EA, Shmakova NL, Komova OV, Mel'nikova LA, Fadeeva TA, Krasavin EA (2006) Cytogenetic effects of low dose radiation with different LET in human peripheral blood lymphocytes and possible mechanisms of their realization. Radiats Biol Radioecol 46:457–460 (Russian)
32. Orr HA (2006) The distribution of fitness effects among beneficial mutations in Fisher's geometric model of adaptation. J Theor Biol 238:279–285
33. Osipova LP, Ponomareva AV, Shcherbov BL, Sukhorukov F, Strakhovenko V, Kutsenogij K (2001) Aftermath of radiation effect in population of tundra Nenets inhabitants, Purov region, YaNAD. In: Korogodin VI, Korogodina VL, Dubrovina NI (eds) Proceedings of the first international conference dedicated to the centenary of the birth of N.W.Timofeeff-Ressovsky, 7–9 September 2000. JINR Press, Dubna, pp 200–212 (Russian)
34. Osipova LP, Koutsenogii KP, Shcherbov BL, Strakhovenko VD, Petrov SA (2002) Environmental radioactivity for risk assessment of health status in nature and human population of Northern Siberia. In: Proceedings of the fifth international conference on environmental radioactivity in the Arctic and Antarctic, St. Petersburg, 16–20 June 2002, pp 124–127 (Russian)
35. Pytjev YuP, Shishmarev IA (1983) The course of probability theory and mathematical statistics for physicists. Moscow State University Press, Moscow (Russian)
36. Rakhlin A, Panchenko D, Mukherjee S (2005) Risk bounds for mixture density estimation. ESAIM 9:220
37. Schwarz G (1978) Estimating the dimension of a model. The Ann Stat 6:461–464
38. State Committee on the Environment Protection in the Saratov Region (SCEPSR) (2000) Conditions of the environment in the Saratov region in 1999. Acvarius, Saratov, pp 1–193 (Russian)
39. Turchak LI (1987) The base of numerical methods. Nauka, Moscow (Russian)
40. Zaichkina SI, Rozanova OM, Akhmadieva AKh, Aptikaeva GF, Sorokina SS, Smirnova EN, Romanchenko SP, Vakhrusheva OA (2009) Study of the genetic instability in generations of mice irradiated of a low-dose rate of high-LET radiation. Radiats Biol Radioecol 49:55–59 (Russian)
41. Zhuravskaia AN, Filippov EV, Kershengol'ts BM (2000) The effect of the biochemical adaptations in the Manchu alder (Duschecia fruticosa (Rupr)) to an elevated natural radiation background on the survivability of seedlings and on the radiosensitivity of its seeds. Radiats Biol Radioecol 40:254–260 (Russian)

Chapter 16
Effects of Bystander Factors Induced In Vivo in Blood of People Affected by Chernobyl Accident

Pavel M. Marozik, Irma B. Mosse, Sergey B. Melnov, Mikhail S. Marozik, Carmel E. Mothersill, and Colin B. Seymour

Abstract The purpose of this work was the analysis of the effects of bystander factors from blood sera of people affected by the Chernobyl accident on human keratinocyte cell culture (HPV-G cells). A new method was developed for evaluation of bystander factor presence *in vivo* in blood of people irradiated by the Chernobyl accident. Affected population groups included liquidators of the Chernobyl accident and people living and working in areas of Gomel region contaminated by radionuclides. The analysis has shown that bystander factors persist in Chernobyl liquidator blood samples for more than 20 years since irradiation. The data suggest that blood sera contain bystander factors, which are able to induce micronuclei and decrease metabolic activity of HPV-G cells.

Keywords Bystander effect • Chernobyl • Liquidators • Keratinocytes • Micronuclei • Alamar blue • Melanin • Melatonin

16.1 Introduction

For a long time it was considered that genetic effects of irradiation are related only to direct DNA damage. But a lot of evidence suggests the existence of a phenomenon of information transfer from irradiated cells to non-irradiated (so-called radiation

P.M. Marozik (✉) • I.B. Mosse
Laboratory of Human Genetics, Institute of Genetics and Cytology, NAS of Belarus, 27, Akademicheskaya Str., Minsk 200072, Belarus
e-mail: P.Marozik@igc.bas-net.by

S.B. Melnov • M.S. Marozik
International Sakharov Environmental University, 23, Dolgobrodskaya Str., Minsk 220009, Belarus

C.E. Mothersill • C.B. Seymour
McMaster University, 1280, Main Street West, Hamilton, ON L8S4K1, Canada

C.E. Mothersill et al. (eds.), *Radiobiology and Environmental Security*, NATO Science for Peace and Security Series C: Environmental Security,
DOI 10.1007/978-94-007-1939-2_16, © Springer Science+Business Media B.V. 2012

induced bystander effect, RIBE) [7, 8]. Such bystander factors could be induced not only by ionizing radiation, but also observed under different pathological conditions [5]. The specific nature of these factors and mechanism of RIBE are still unknown [1].

In recent years some evidence of RIBE induced *in vitro* was obtained, while the effects *in vivo* are not enough studied.

Effects of bystander factors induced in vivo in blood of people affected by the Chernobyl accident are of particular interest.

For the first time such studies were performed by Dr. I. Emerit [2], when it was shown that blood sera from liquidators of the Chernobyl accident or residents of areas contaminated by radionuclides contained an increased level of bystander factors. These factors increased the frequency of chromosome aberrations and micronuclei in own lymphocytes of serum donors even decades since irradiation.

In Emerit studies own lymphocytes of serum donors were used as a test-system. But these cells were already exposed to ionizing radiation, as a result, the level of chromosome aberrations and micronuclei was increased there. At the same time, human peripheral blood lymphocytes under *in vitro* conditions could be subcultured only for 2–4 days, therefore it is impossible to evaluate serum effects for a longer period.

In the present study a new method was developed which allows evaluation of the effects of bystander factors from sera of different groups of populations on immortalized culture of human keratinocytes. This method has serious advantages as compared to that which was used earlier: cells could be cultivated for a longer period of time (re-cultivation once a week), they are more sensitive as compared to human peripheral blood lymphocytes, experiments could be easily repeated, allowing simple comparison and interpretation of the obtained results.

Using this method, we evaluated the level of damaging bystander factors in blood of areas irradiated by the Chernobyl accident.

The study of the nature and possibility of modification of bystander factors, circulating in blood serum of populations, affected by the Chernobyl accident, will facilitate better understanding of radiation damage mechanisms. Knowledge of these mechanisms is increasingly important for cancer radiation therapy – for using special correcting coefficients taking into consideration effects of bystander factors. The anti-tumor therapy could be more effective and safe. Revealing the nature of the damaging bystander factors will promote more exact dose evaluation and radiation risk for people, exposed to ionizing radiation, as the risk of radiation exposure at low doses, calculated using direct and bystander effects, may be higher as compared to that calculated only from direct effects. The ability to neutralize and modify these factors may help to decrease significantly the effects of whole body radiation exposure.

16.2 Materials and Methods

16.2.1 Cell Culture

In the present study HPV-G cells (human keratinocytes, immortalized by human papilloma virus transfection), deficient in p53 were used as a test system. Cells were cultured in Dulbecco's MEM: F12 (1:1) medium supplemented with 10% Foetal bovine serum, 1% penicillin-streptomycin (1 g per 100 ml), 1% L-glutamine and 1 μg/ml hydrocortisone. The cells were maintained in an incubator at 37° centigrade, with 95% humidity and 5% carbon dioxide and routinely subcultured every 8–10 days.

16.2.2 Affected Populations

The objects of investigation were blood sera from people affected by the Chernobyl accident: Chernobyl liquidators of 1986–1987 (22 persons), Polessky State Radiation Environmental Reserve workers[1] (PSRER, 21 persons) and people, living in areas of Gomel region (GR) contaminated by radionuclides(15 persons). The analysed groups also included four persons with acute virus infection (flu) living in areas contaminated by radionuclides. The control group included clinically healthy people from non-contaminated areas, corresponding to the main group in age and sex (36 persons).

16.2.3 Serum Extraction

The blood samples were taken and placed in Vacutainers for serum extraction (Becton Dickinson, USA), centrifuged at 2,000 g for 10 min, and the serum was frozen and stored at –20°C before use. Before freezing, the sera were filtered through 0.22 μm filters (Nalgene, USA) in order to remove all residual cell components of the blood.

16.2.4 Radioprotective Substances

Antioxidant substances melanin and melatonin were used as radioprotectors. Melanin (Belarusian pharmaceutical association, Minsk) was used at 10 mg/l concentration, melatonin (Sigma, Germany) – at 10 mg/ml.

16.2.5 Micronucleus Assay

Micronuclei assay. After plating, cells were left at 37°C in the CO_2 incubator to be attached for 12 h. The blood serum from affected populations was placed into 25 cm^2 flasks (NUNC, USA) (6,000 cells per flask) 1–2 days after plating, and cells were

[1] Polessky State Radiation and Environment Reserve (founded in 1988) is a territory of Gomel region, where humans cannot live because of the very high levels of radiation contamination. The territory of this reservation is 215.5 ha.

placed again in the incubator for 1–2 h. Then cytochalasin B was added (7 μg/ml concentration) and the cells were incubated for 24 h. The cell culture medium was removed, the cells were washed with PBS and fixed with chilled Karnua solution (1 part of glacial acetic acid and 3 parts of methanol, 10–15 ml three times for 10–20 min). Later flasks were dried and stained with 10% Giemsa solution. The micronuclei count was carried out under inverted microscope. The data are presented as the micronuclei (cells with micronuclei) frequency recorded per 1,000 binucleated cells ± standard error.

16.2.6 Alamar Blue Assay

Alamar blue (AB) assay. HPV-G cells were plated on 96-well microplates (NUNC, USA) at the concentration of 2×10^4 cells/well. After plating, cells were incubated for 24 h to allow attachment to the bottom of the well. Then the medium was removed, cells were rinsed with phosphate buffered saline (PBS) and the blood serum from the Chernobyl accident populations was added to the cells together with melanin and melatonin as appropriate. Microplates were placed into the incubators. Twenty-four hours later, serum was removed, cells were rinsed with PBS and 100 μl of a 5% solution of Alamar Blue prepared in phenol red free DMEM media were added. Microplates were placed again into the incubators. Three hours later, fluorescence was quantified using a microplate reader (TECAN GENios, Grödig, Austria) at the respective excitation and emission wavelength of 540 and 595 nm, respectively Wells containing medium and Alamar Blue without cells were used as blanks. The mean fluorescent units for the three replicate cultures for each exposure treatment were calculated and the mean blank value was subtracted from these results.

16.2.7 Statistical Analysis

All experiments were repeated at least three times, and within each experiment cultures were set up in triplicate. Results are expressed as the means +/– standard errors.

16.2.7.1 t-Test

When the distribution was normal, significance was determined using the *t* test. The level of significance was chosen as 95%; at $t \geq 2.67$, the difference is highly significant at $p<0.01$, at $t \geq 1.96$, the difference is significant at $p<0.05$. At $p>0.05$, the difference is not significant.

16.2.7.2 Mann-Whitney U-Test

The Mann-Whitney U test is a nonparametric alternative to the *t*-test for independent samples and was calculated using STATISTICA 8.0 (Statsoft, USA). The U statistics is accompanied by a *z* value (normal distribution variate value), and the respective *p*-value.

16.3 Results

16.3.1 *Study of Effects of Blood Sera from People, Affected by the Chernobyl Accident on Micronuclei Frequency in HPV-G Cells*

Table 16.1 presents data on the effects of blood serum samples from different population groups on the total micronuclei frequency in HPV-G cells (average data presented).

The micronuclei frequency in the controls indicates the level of spontaneous mutagenesis (it is comparatively low). The data from Table 16.1 show that the number of the cells with two and especially three micronuclei is very low as compared with the number of cells with one micronucleus.

As seen from Table 16.1, people exposed to chronic radiation (PSRER workers) have an increased level of bystander factors, expressed as a considerable increase in micronuclei frequency (almost three times as high as the control level – 248.03‰ ± 20.77 as compared with 80.30‰ ± 13.14, $p<0.01$).

At the same time an increase in the number of the cells with more than one micronucleus was observed. Thus, the data clearly indicate that intensive chronic irradiation of PSRER workers significantly promoted accumulation of bystander factors in blood.

Table 16.1 The effects of blood serum samples from different groups of population on HPV-G micronuclei (MN) frequency (average data are presented)

	Frequency of cells with MN, ‰			Total number of cells with MN, ‰	Total MN frequency, ‰
	1 MN	2 MN	3 MN		
Control	69.5±11.1	5.0±1.1	0.3±0.1	74.8±12.4	80.3±13.1
Liquidators	200.6±12.2	30.6±2.6	4.6±1.1	235.6±14.0*, †	273.7±22.4*, †
PSRER workers	196.5±10.9	18.8±3.8	4.4±1.1	219.7±18.3*, †	248.0±20.8*, †
Residents of GR	130.5±7.7	11.8±1.8	0.8±0.3	143.1±9.3*	156.5±11.2*
Residents of GR with acute virus infection	190.2±6.6	87.7±4.8	23.3±2.3	301.2±7.8*, †	435.6±8.4*, †

* $p<0.01$ (compared to controls); † $p<0.01$ (compared to GR residents)

Similar results were observed after comparative analysis of the micronuclei frequency between the control group and the liquidators group (people exposed to acute radiation). The total micronuclei frequency of 273.7‰±22.4 and the frequency of cells with micronuclei of 235.6‰±14.0 in cells treated with serum from liquidators were significantly higher than those in the control group (in both cases $p>0.01$). Also, an increase in the number of cells with more than one micronucleus was observed.

The level of micronucleus frequency induced by serum samples from the residents of contaminated areas of Gomel region is statistically significantly different from the control (156.47‰±11.22 vs. 80.30‰±13.14 $p<0.01$), but much lower than in people, exposed to additional radiation influence (as compared to liquidators and PSRER workers in all cases $p<0.01$).

At the same time, the level of the micronucleus frequency induced by serum samples from the residents with acute virus infection at the active stage is higher than in all previous cases – the micronucleus frequency induced by the serum from these patients is 435.6‰±8.4, and the number of cells with micronuclei is 301.2‰±7.8. These figures are much higher than those for liquidators and PSRER workers (in both cases $p<0.01$).

16.3.2 Study of the Effects of Blood Sera from Population Groups, Affected by the Chernobyl Accident on Metabolic Activity of HPV-G Cells

The study was run to understand if there were any bystander factors observed in serum samples of these population groups which could affect cell metabolic activity. The blood serum samples for Alamar Blue analysis were taken from Chernobyl liquidators and residents of contaminated areas of Gomel region.

Figure 16.1 presents the average data for all these groups (metabolic activity of intact cells is taken as 100%). The viability of the cells treated with serum samples from non-irradiated individuals is very close to intact levels ($t=0.33$, $p<0.01$ – the difference is not significant). It means that the metabolic activity of these cells is not damaged by serum samples from healthy people, not increasing or decreasing significantly the viability of HPV-G cells.

Treatment of the cells with serum samples from Chernobyl liquidators clearly reduces the viability of HPV-G cells more than 1.5 times – from $24.89\pm0.25\times10^3$ FU (intact cells) and $24.67\pm0.62\times10^3$ FU (non-irradiated individuals) to $15.65\pm0.82\times10^3$ FU (liquidators); $p<0.01$ in both cases ($t=10.79$ and 8.77, respectively). Treatment of HPV-G cells with serum samples from residents of contaminated areas of Gomel region also reduces the viability of cells ($19.16\pm0.71\times10^3$ FU, $t=7.62$ compared to intact cells, $p<0,01$), but not as significantly as sera from liquidators.

Table 16.2 presents the results of the comparison between all groups using the Mann-Whitney U-test.

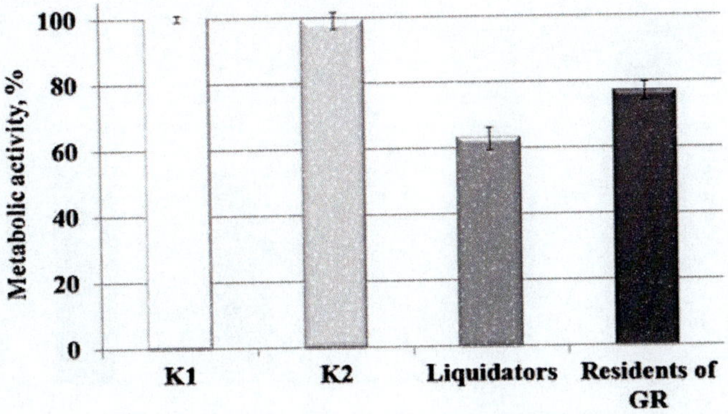

Fig. 16.1 Cytotoxic effects of serum samples from the control groups, Chernobyl liquidators and residents of GR on metabolic activity of HPV-G cells (as a percentage of intact cells, average data for all groups of populations are presented)

Table 16.2 Comparative analysis of four independent groups (intact cells K_1, non-irradiated individuals K_2, liquidators and residents of GR) using Mann-Whitney U-test

Groups of comparison	Z	p
K1 vs. K2	0.39	>0.6
K1 vs. Liquidators	3.92*	<0.00001
K1 vs. GR residents	3.40*	<0.001
K2 vs. Liquidators	3.54*	<0.0005
K2 vs. GR residents	3.07*	<0.005
Liquidators vs. GR residents	3.03*	<0.005

* Statistically significant at $p<0.01$

The analysis of differences between all groups showed that there was statistically significant difference (with at least $p<0.005$) between all groups excluding intact cells vs. non-irradiated population group ($p>0.05$). The highest difference was observed between liquidators vs. intact cells.

16.3.3 *Study of the Effects of Antioxidant Substances on Bystander Factors*

In previous studies [6] we have shown that bystander effect when induced *in vitro* by transfer of culture medium from irradiated cells to non-irradiated could be decreased using radioprotectors with antioxidant activity (melanin, melatonin). In the present investigation we have studied the possibility of these substances to neutralize bystander factors, induced *in vivo.*

Figure 16.2 presents the results of an attempt to modify the effects of blood serum samples from Chernobyl liquidators on metabolic activity of HPV-G cells using melanin and melatonin (average data).

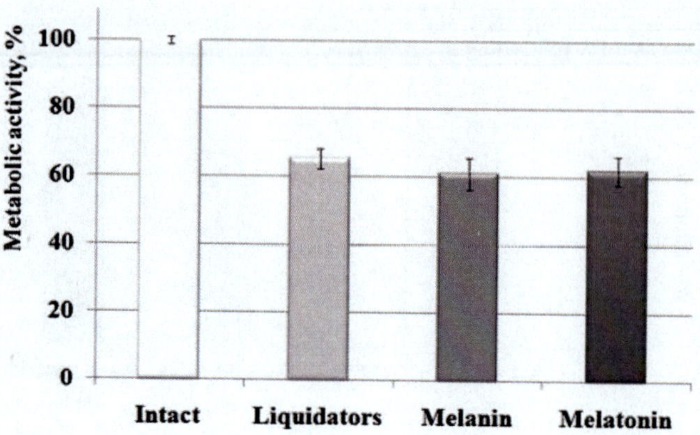

Fig. 16.2 Effects of melanin and melatonin on HPV-G cells treated with blood serum samples from Chernobyl liquidators (control is taken as 100%)

As it can be seen from Fig. 16.2, addition of melanin and melatonin to the medium together with serum samples from Chernobyl liquidators doesn't have any protective effect. The viability of cells treated with only serum samples ($16.24 \pm 0.73 \times 10^3$ FU) is almost the same as viability of cells treated with serum samples and melanin ($15.21 \pm 1.13 \times 10^3$ FU) or melatonin ($15.45 \pm 1.04 \times 10^3$ FU) – the difference is not significant ($t=0.77$ and $t=0.62$, respectively; $p>0.05$ in both cases).

16.4 Discussion

In similar studies on liquidators from Armenia [3, 4] it was shown that only 42% of liquidators had an increased level of clastogenic factors (chromosome aberrations) as compared to the level of spontaneous mutations. In our experiments 100% of liquidators, 95% of PSRER workers and 82% of residents from contaminated areas had an increased level of factors as compared to the control levels. A possible explanation of such difference could be the different protocols used in analysis. As mentioned above, HPV-G cells are much more sensitive to mutagenic factors as compared to peripheral blood lymphocytes. Another explanation could be that the cohort from Dr. Emerit studies [3] included liquidators of 1986–1988. After 1987 the dose limits for people engaged in clean-up works significantly decreased. Therefore liquidators from 1986 expressed higher levels of damaging factors as compared to liquidators of 1987–1988. In the present study liquidators of the consequences of the Chernobyl accident in 1986–1987 were analyzed, which were the most affected group and received the highest doses of radiation (higher than 250 mGy).

According to the results of the present study, bystander factors persist in the blood of Chernobyl liquidators for more than 20 years after the accident. In previous studies it was shown that irradiation of blood *in vitro* at a radiation dose of

500 mGy resulted in significant clastogenic activity and irradiated cells incubated in fresh culture medium continued to produce factors in culture as it was shown in studies [3].

Thereby, performed studies have shown that blood sera from population groups affected by the Chernobyl accident contain bystander factors, which are able to induce micronuclei and decrease metabolic activity of recipient cells. Pathological processes, such as acute virus infection, could significantly influence the level of bystander factors, increasing their damaging effect. The developed new method of evaluating of effects of damaging factors from blood sera has obvious advantages as compared to the assays which were used in previous studies [3] and may be applied in long-term monitoring of biological consequences of irradiation.

The results of the study of the mechanisms and nature of bystander factors circulating in blood stream of irradiated individuals make a major contribution in understanding non-direct effects of radiation *in vivo* and allows obtaining of new data on intercellular communications.

References

1. Baverstock K, Belyakov OV (2010) Some important questions connected with non-targeted effects. Mutat Res 687:84–88
2. Emerit I (1990) Superoxide production by clastogenic factors. In: Crastes de Paulet A, Douste-Blazy L, Paoletti R (eds) Free radicals, lipoproteins and membrane lipids. Plenum Press, New York, pp 99–104
3. Emerit I, Cernjavski L, Arutyunyan R, Oganesian N, Pogosian H, Mejlumian H, Sarkisian T, Gulkandanian M, Quastel M, Goldsmith J, Riklis E, Kordysh R, Poliak S, Merklin L (1994) Transferable clastogenic activity in plasma from persons exposed as salvage personnel of the Chernobyl reactor. J Cancer Res Clin Oncol 120:558–561
4. Emerit I, Oganesian N, Sarkisian T, Arutyunyan R, Pogosian A, Asrian K, Levy A, Cernjavski L (1995) Clastogenic factors in the plasma of Chernobyl accident recovery workers. Anticlastogenic effect of *Ginkgo biloba* extract. Radiat Res 144:198–205
5. Emerit I, Oganesian N, Arutyunyan R, Pogossian A, Sarkisian T, Cernjavski L, Levy A, Feingold J (1997) Oxidative stress-related clastogenic factors in plasma from Chernobyl liquidators: protective effects of antioxidant plant phenols, vitamins and ligoelements. Mutat Res 377:239–246
6. Mosse I, Marozik P, Seymour C, Mothersill C (2006) The effect of melanin on the bystander effect in human keratinocytes. Mutat Res 597:133–137
7. Mothersill C, Seymour C (1997) Medium from irradiated human epithelial cells but not human fibroblasts reduces the clonogenic survival of unirradiated cells. Int J Radiat Biol 71:421–427
8. Nagasawa H, Little JB (1992) Induction of sister chromatid exchanges by extremely low doses of alpha-particles. Cancer Res 52:6394–6396

Chapter 17
Association Study of the Frequencies of Spontaneous and Induced Chromosome Aberrations in Human Lymphocytes

Luybov Sal'nikova, Anastasia Chumachenko, and Alexander V. Rubanovich

Abstract The data on the frequencies of chromosome aberrations in lymphocytes of peripheral blood of 99 volunteers depending on genotypes by genes of xenobiotics detoxification before and after γ-irradiation with the dose of 1 Gy *in vitro* are presented. The frequencies of aberrations were estimated by analyzing not less than 500–1,000 metaphases per person. The data of the cytogenetic analysis were compared with the results of PCR genotyping by the genes of xenobiotics detoxication (*CYP1A1, CYP2D6, GSTM1, GSTT1, GSTP1, COMT, NAT2*), genes of DNA repair (*XRCC1, XPD, ERCC1, APEX1, RAD23B, OGG1, ATM*) as well as by the genes of oxidative response and cell cycle regulation (*SOD2, CAT, GCLC, MTHFR, Tp53*). The frequency of spontaneous aberrations of chromosome type was reduced for homozygotes by the deletion of *GSTM1* locus, especially for double homozygotes by the deletions of *GSTM-GSTT1* genes. The mentioned frequency increased additively with the number of copies of the minor allele variants *XPD**2251 G and *XPD**862A of gene of excision repair *XPD* ($p=0.025$). The frequency of γ-induced chromosome aberrations proved to be elevated for the carriers of a minor allele *OGG1**977 G ($p=0.011$). The significantly elevated number of γ-induced chromosome aberrations was also observed for the carriers of major alleles *XRCC1**G1996 and *XRCC1**C580 ($p=0.002$). The frequency of γ-induced chromosome aberrations proved to be reduced for G/G homozygotes by a minor allele of poorly studied site *CYPA1*T606G: 0.094 ± 0.006 against 0.112 ± 0.002 for the carriers of allele T ($p=0.004$). The results of meta-analysis of the data of 14 works on the effect of genotype by *GSTM1* locus on the frequency of cytogenetic anomalies in the control and exposed populations are discussed.

Keywords Genetic polymorphism • DNA repair genes • Associative studies • Chromosome radiosensitivity • Gamma radiation in vitro • Chromosome aberrations

L. Sal'nikova • A. Chumachenko • A.V. Rubanovich (✉)
Vavilov Institute of General Genetics, Russian Academy of Sciences,
Gubkina, 3, Moscow 119991, Russia
e-mail: rubanovich@vigg.ru

C.E. Mothersill et al. (eds.), *Radiobiology and Environmental Security*,
NATO Science for Peace and Security Series C: Environmental Security,
DOI 10.1007/978-94-007-1939-2_17, © Springer Science+Business Media B.V. 2012

17.1 Introduction

In this work, the results of our investigations [11, 12] aimed at the systematic search for genotypic associations of the radiosensitivity of human chromosomes are reviewed. These works are carried out using the bank of DNA of young healthy volunteers on the base of the previous thorough cytogenetic analysis of spontaneous and γ-induced (1 Gy *in vitro*) chromosome aberrations of blood lymphocytes. The existence of the DNA bank allows the recurrent control of new groups of polymorphic candidate genes for the presence of the association with the frequency and spectrum of chromosome aberrations.

It is well known that the revealed dependence of the radiosensitivity on the polymorphism of candidate genes often is not confirmed in other works. In our opinion, this is to a great extent related to a low statistical stipulation of the corresponding association studies (the analysis of no more than 100 metaphases per person for sampling of 20–60 donors). A distinctive feature of this work is the estimation of the frequencies of aberrations of chromosome type via scanning of many metaphase cells as well as a significant (for such studies) number of donors.

By now we have performed the search of genetic markers of the radiation risk among the genes of xenobiotics detoxication (*CYP1A1, CYP2D6, GSTM1, GSTT1, GSTP1, COMT, NAT2*), genes of DNA repair (*XRCC1, XPD, ERCC1, APEX1, RAD23B, OGG1, ATM*) as well as genes of oxidative response and cell cycle regulation (*SOD2, CAT, GCLC, MTHFR, Tp53*). All loci studied were characteristic of the functional polymorphism associated with the change in activity and/or in the amount of the corresponding enzyme as well as of associations with different biological effects and diseases.

17.2 Materials and Methods

The correlations of the frequencies of spontaneous and induced chromosome aberrations were studied with the samples of peripheral blood taken from 115 young (20–25 years old) healthy males (students of the Military Technical University, Balashikha). The investigation was sanctioned by the Ethic Commissions of the Institute of General Genetics, RAS and the Military Technical University.

The cytogenetic analysis was successfully carried out for 99 donors. Each sample of pure blood for the cytogenetic analysis was poured into two tubes: the first one was exposed to γ-irradiation (Co^{60}, 1 Gy, dose rate of 1.37 Gy/min), and the second tube was used for analyzing the spontaneous chromosome aberrations.

The preparations with metaphase cells for estimation of chromosome aberrations were obtained by the standard method [11, 12]. 500 (for γ-induced aberrations) or 1,000 (for spontaneous aberrations) of metaphases of the first mitosis per person were analyzed. The differential assessment of aberrations of the chromosome type (dicentric and ring chromosomes, acentrics, atypical monocentrics) and of chromatid type (single and isochromatid fragments and exchanges) was carried out.

The genotyping was accomplished using the allele-specific tetraprimer PCR. The method allows the amplification of DNA fragments corresponding to the alternative alleles in one tube. The amplification products are separated via electrophoresis on an agarose gel without fluorescence labels. The studied polymorphic sites are listed in Table 17.1. In the case of deletion-insertion polymorphism (*GSTM1, GSTT1*) two genotypes were revealed: a "zero" one – homozygote by deletion (D/D) and a "positive" one carrying a functional allele in the homo- or heterozygote state (I/*). From here on * is related to an arbitrary allele.

Table 17.1 Average frequencies of spontaneous and induced aberrations per 100 cells (±SE) for carriers of different genotypes

Loci and genotypes		Spontaneous aberrations			γ- induced aberrations (1 Gy *in vitro*)		
		#	Chromosome type aberrations	All aberrations	#	Chromosome type aberrations	All aberrations
CYP1A1	T/T	40	0.26±0.04	0.94±0.09	42	11.0±0.3	12.1±0.4
T606G	T/G	41	0.25±0.05	0.87±0.12	43	11.4±0.3	12.9±0.4
rs2606345	G/G	12	0.15±0.06	0.68±0.10	12	9.4±0.6	10.4±0.8
CYP1A1	T/T	81	0.24±0.03	0.90±0.08	84	11.1±0.2	12.4±0.3
T3801C	T/C	11	0.23±0.07	0.74±0.15	11	10.7±0.5	12.0±0.6
rs4646903	C/C	1	0.35	0.47	2	8.9±2.9	9.8±3.4
CYP1A1	A/A	88	0.24±0.03	0.89±0.07	91	11.0±0.2	12.3±0.3
A4889G							
rs1048943	A/G	5	0.29±0.10	0.62±0.10	6	10.3±0.9	11.7±1,1
CYP2D6	G/A	28	0.27±0.05	0.98±0.11	28	11.3±0.4	12.4±0.5
A1934G	G/G	65	0.23±0.04	0.83±0.08	65	10.8±0.3	12.4±0.3
rs3892097							
GSTM1	D/D	35	0.20±0.05	0.80±0.09	39	11.2±0.4	12.3±0.4
	I/*	58	0.27±0.03	0.92±0.10	58	10.9±0.3	12.3±0.4
GSTT1	D/D	30	0.21±0.05	0.97±0.16	30	10.9±0.4	12.2±0.5
	I/*	62	0.26±0.04	0.84±0.07	67	11.0±0.3	12.3±0.3
GSTP1	A/A	47	0.24±0.04	0.83±0.08	48	10.8±0.4	12.0±0.4
A313G	A/G	37	0.26±0.05	0.87±0.08	40	11.4±0.3	12.7±0.4
rs1695	G/G	9	0.17±0.11	1.13±0.49	9	10.5±0.5	12.3±1.2
COMT	A/A	23	0.28±0.07	0.95±0.12	26	11.5±0.5	12.9±0.5
G1947A	G/A	49	0.24±0.04	0.80±0.07	50	10.9±0.3	12.1±0.3
rs4680	G/G	21	0.21±0.06	0.97±0.22	21	10.6±0.5	12.1±0.7
NAT2	A/A	5	0.37±0.16	0.92±0.22	5	11.6±1.1	12.6±1.2
G590A	G/A	36	0.26±0.05	0.94±0.13	37	11.4±0.4	12.9±0.5
rs1799930	G/G	52	0.22±0.04	0.83±0.08	55	10.7±0.3	11.8±0.3
SOD2	C/C	27	0.22±0.05	0.82±0.10	28	10.7±0.4	11.7±0.4
C47T	C/T	43	0.24±0.05	0.93±0.12	45	11.3±0.4	12.7±0.4
rs4880	T/T	23	0.26±0.05	0.84±0.10	24	10.9±0.5	12.1±0.5
CAT	A/A	12	0.15±0.08	0.88±0.14	12	10.9±0.5	12.2±0.6
T21A	T/A	44	0.26±0.09	0.84±0.09	46	10.6±0.3	11.8±0.4
rs7943316	T/T	37	0.25±0.05	0.92±0.13	39	11.5±0.4	12.9±0.5

(continued)

Table 17.1 (continued)

Loci and genotypes		Spontaneous aberrations #	Chromosome type aberrations	All aberrations	γ- induced aberrations (1 Gy *in vitro*) #	Chromosome type aberrations	All aberrations
GCLC	C/C	78	0.25±0.03	0.90±0.08	81	11.1±0.2	12.5±0.3
C129T	C/T	15	0.21±0.06	0.74±0.10	16	10.3±0.6	11.4±0.7
rs17883901							
MTHFR	C/C	43	0.22±0.04	0.80±0.06	46	10.8±0.3	11.9±0.4
C677T	C/T	41	0.29±0.05	1.02±0.13	42	11.4±0.3	12.8±0.4
rs1801133	T/T	9	0.15±0.07	0.60±0.25	9	10.4±0.9	11.8±1.0
XRCC1	C/C	84	0.21±0.03	0.82±0.07	87	11.10±0.23	12.41±0.28
C590T	C/T	11	0.25±0.09	0.62±0.11	11	10.51±0.79	11.38±0.88
rs1799782	T/T	1	0.00	0.68	1	7.80	9.00
XRCC1	G/G	44	0.23±0.04	0.88±0.12	45	11.53±0.31	13.02±0.41
G1996A	G/A	44	0.20±0.04	0.72±0.06	46	10.64±0.34	11.73±0.38
rs 25487	A/A	8	0.14±0.04	0.75±0.17	8	10.07±0.56	11.05±0.71
XPD	T/T	35	0.14±0.03	0.72±0.08	38	10.99±0.33	12.29±0.38
T2251G	T/G	44	0.23±0.04	0.79±0.11	44	11.04±0.34	12.28±0.44
rs13181	G/G	17	0.29±0.07	0.96±0.13	17	10.92±0.61	12.17±0.73
XPD	G/G	37	0.15±0.03	0.86±0.13	39	11.02±0.35	12.35±0.46
G862A	G/A	43	0.22±0.04	0.72±0.07	44	11.04±0.33	12.27±0.38
rs1799793	A/A	16	0.31±0.07	0.85±0.11	16	10.86±0.62	12.02±0.70
ERCC1	G/G	77	0.23±0.03	0.83±0.07	80	11.12±0.25	12.39±0.31
G262T	G/T	17	0.14±0.03	0.67±0.10	17	10.50±0.53	11.67±0.60
rs2298881	T/T	2	0.20±0.20	0.54±0.06	2	10.50±1.30	12.10±1.30
ERCC1	T/T	46	0.22±0.04	0.85±0.10	48	10.74±0.35	12.05±0.44
T354C	T/C	30	0.20±0.04	0.79±0.10	31	11.60±0.29	12.80±0.38
rs11615	C/C	20	0.20±0.05	0.67±0.09	20	10.70±0.51	11.94±0.59
APEX1	T/T	35	0.24±0.05	0.77±0.06	35	10.90±0.40	12.09±0.45
T444G	T/G	38	0.17±0.03	0.72±0.08	41	11.07±0.30	12.33±0.34
rs1130409	G/G	23	0.23±0.05	0.94±0.20	23	11.03±0.53	12.42±0.73
RAD23B	C/C	75	0.20±0.03	0.82±0.07	78	11.09±0.25	12.40±0.30
C 746 T	C/T	21	0.23±0.05	0.69±0.11	21	10.68±0.50	11.76±0.59
rs1805329							
OGG1	C/C	63	0.19±0.03	0.77±0.09	65	10.75±0.25	12.03±0.33
C977G	C/G	27	0.24±0.05	0.81±0.08	28	11.16±0.48	12.33±0.53
rs1052133	G/G	6	0.23±0.10	0.95±0.12	6	13.01±0.58	14.43±0.74
ATM	G/G	49	0.22±0.04	0.78±0.05	51	11.00±0.31	12.11±0.34
G5557A	G/A	46	0.20±0.04	0.79±0.11	47	11.02±0.33	12.43±0.43
rs1801516	A/A	1	0.10	1.60	1	10.40	1.80
Tp53	G/G	57	0.21±0.04	0.85±0.09	59	11.17±0.29	12.55±0.35
G215C	G/C	33	0.22±0.04	0.73±0.07	34	10.66±0.38	11.75±0.45
rs1042522	C/C	6	0.15±0.07	0.64±0.17	6	11.25±0.97	12.29±1.19

The cases of significant genotype differences are distinguished

The statistical analysis was carried out by the standard methods with a WinSTAT 2003.1 package integrated into Excel. All assessments of the group frequencies of aberrations were obtained as a result of the averaging of the individual frequencies for persons with the specific genotype. The corresponding errors reflected the intragroup variability of the frequencies of aberrations. For intergroup comparison, a nonparametric Mann–Whitney test was used.

17.3 Results

The frequencies of spontaneous and γ-induced chromosome aberrations for the carriers of different genotypes are given in Table 17.1. The distributions of allele frequencies for all loci studied were conform to the Hardy–Weinberg equilibrium and did not differ from the data for the groups of the residents of the Central region of Russia examined previously.

17.3.1 Association Studies of Spontaneous Chromosome Aberrations

Genotype by *GSTM1* locus had the most significant effect on the frequency of spontaneous chromosome aberrations. The frequencies of aberrations of the chromosome type for "positive" (I/*) and "zero" (D/D) genotype had very different distributions (Fig. 17.1c). In particular, in 57% persons, which are homozygote by *GSTM1* deletion, the cells with the aberrations of chromosome type were not found. For "positive" genotypes I/*, the corresponding index was equal to 27% (OR=3.1; $p=0.0095$ by the precise Fischer criterion). However, the differences between the average frequencies of the aberrations of chromosome type proved to be at the limit of significance: 0.0020±0.0005 for homozygotes D/D against 0.0027±0.0003 for I/* ($p=0.049$ by the Mann–Whitney test).

An essentially reduced frequency of spontaneous aberrations of the chromosome type was observed in donors, which are double homozygotes by the deletions of *GSTM1-GSTT1* loci (11 persons). In this group, only three donors showed one aberration and eight donors had no aberrations of the chromosome type. In double homozygotes by deletions, the differences in the average values comprise 0.0006±0.0003 against 0.0027±0.0003 for the rest of genotypes ($p=0.018$ by the Mann–Whitney test).

The additive trend to an elevated frequency of aberrations of the chromosome type was observed in the carriers of minor alleles of gene *XPD* (excision repair of nucleotides – NER) in sites that were in linkage disequilibrium T2251G (Lys751Gln) and G862A (Asp312Asn). In this connection the regression analysis of the dependence of the frequency of chromosome aberrations on the total number of copies of minor allele variants *XPD**2251 G and *XPD**862A was performed. Figure 17.2 shows the results of this analysis: the correlation coefficient for 96 observations $r=0.228$ at

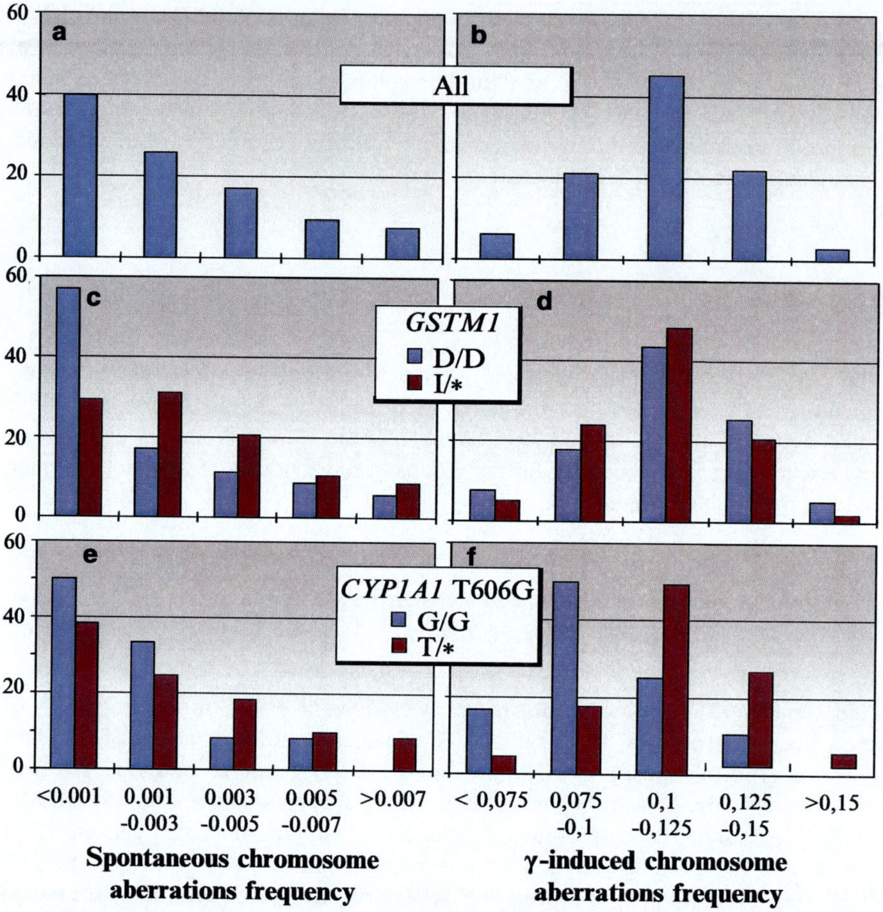

Fig. 17.1 Distributions of frequencies of spontaneous and γ-induced aberrations of the chromosome type for all sampling (**a**, **b**) and for different genotypes by loci GSTM1 (**c**, **d**) and CYP1A1 T606G (**e**, **f**). In histograms (**a**, **c**, **e**), the extreme left group is composed uniquely of donors, in which spontaneous aberrations of the chromosome type were not found

$p=0.025$. For group 4 (both sites are homozygote by minor alleles), the frequency of induced chromosome aberrations is significantly higher compared to group 0 (both sites are homozygote by major alleles): 0.0031±0.0017 against 0.0014±0.0003 at $p=0.026$ by the Mann–Whitney Test. The combinations with both major and minor homozygotes (i.e., 2251 G/G-862 G/G or 2251 T/T-862A/A) were not found that results from a strong linkage disequilibrium of sites ($D'=0.772$; $p=2\cdot10^{-16}$).

Thus, the regression analysis allowed one to reveal the significant additive effects of minor alleles in two sites of *XPD* gene as to the frequency of spontaneous aberrations of the chromosome type, with the maximum effect being achieved in the carriers of double minor homozygotes. The residual loci showed no significant associations with the frequency of chromosome aberrations at the given level of sampling.

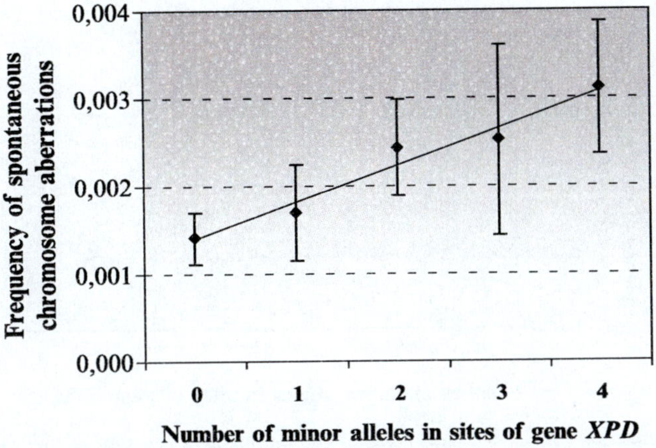

Fig. 17.2 Frequency of spontaneous chromosome aberrations as a function of the total number of minor alleles in sites T2251G and G862A of gene XPD. The average aberrations frequencies (±SE) for the carriers of 0, 1, 2, 3, and 4 minor alleles are given. Class 4, for example, is composed of double homozygotes by minor alleles

17.3.2 Association Studies of Induced Chromosome Aberrations

The frequency of aberrations induced by γ-irradiation dose of 1 Gy *in vitro* was independent of genotype by *GSTM1* locus (Fig. 17.1d), but proved to be reduced for homozygotes G/G by a minor allele of gene *CYP1A1* T606G: 0.094±0.006 against 0.112±0.005 for the carriers of a major allele T ($p=0.004$ by the Mann–Whitney test). The differences in the distributions of the frequencies of chromosome aberrations for these two genotypes are impressive (Fig. 17.1f). For the frequencies of spontaneous chromosome aberrations, the analogous tendency proved to be insignificant (Fig. 17.1e).

The significantly elevated level of induced aberrations of the chromosome type was revealed for homozygotes T/T by major allele of locus *CAT* T21A: 0.115±0.004 against 0.106±0.003 for the carriers of a minor allele A ($p=0.0269$ by the Mann–Whitney test).

The increase in the frequency of induced aberrations of the chromosome type was associated with major variants of two sites of excision repair gene of bases (BER) *XRCC1*: G1996A (Arg399Gln) and C580T (Arg194Trp). In spite of the strong linkage disequilibrium (D'=0.998; $p=0.006$), the statistically significant results were obtained for site G1996A only. In this case a major variant of site *XRCC1**G1996 was associated with the elevated frequencies of chromosome aberrations: 0.115±0.003 against 0.106±0.003 for the carriers of minor allele *XRCC1**1996A ($p=0.007$). The analogous results were obtained for all types of aberrations: 0.0130±0.004 against 0.116±0.003 ($p=0.006$). For other site of the same gene C580T (Arg194Trp), the effects were of the same type, but the differences were not significant.

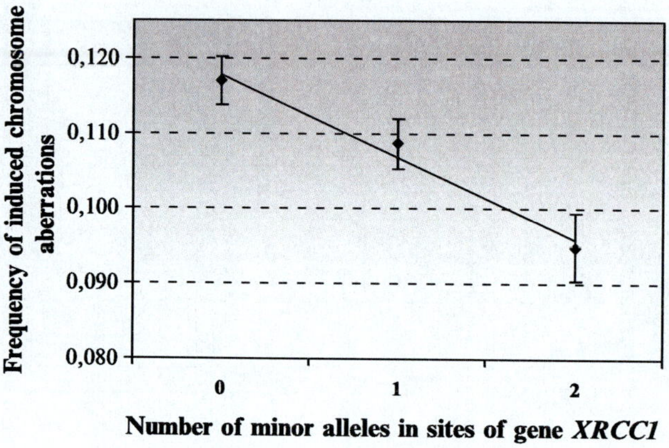

Fig. 17.3 Frequency of induced chromosome aberrations as a function of the total number of minor alleles in sites G1996A and C580T of gene XRCC1. The average aberrations frequencies (±SE) for the carriers of 0, 1, and 2 minor alleles are given. No carriers of 3 and 4 minor alleles were found

Taking into account the obvious additivity of the observed effects, the regression analysis of the dependence of the aberration frequency on the total number of copies of minor allele variants in sites G1996A and C580T was carried out (Fig. 17.2). The regression line is characteristic of correlation coefficient $r=0.303$ at $p=0.002$. The extreme groups (Fig. 17.3) differ significantly by the frequencies of induced aberrations of the chromosome type: 0.117±0.003 against 0.095±0.005 ($p=0.0008$).

Locus *OGG1* (BER) in site C977G (Ser326Cys) showed the association of a specific aberrations (dicentrics and centric rings): 0.074±0.005 for minor allele with increasing frequency of induced aberrations. The frequency of aberrations of the chromosome type for homozygotes by minor allele *OGG1**977 G comprised 0.130±0.006 against 0.109±0.002 for the carriers of a major allele in homo- or heterozygote state ($p=0.011$). For all aberrations, the corresponding differences were equal to 0.144±0.007 against 0.121±0.003 ($p=0.024$). Most significant differences were observed for radiation-homozygotes by minor allele *OGG1**977 G against 0.058±0.001 for the rest of genotypes ($p=0.009$ by the Mann–Whitney test). For the frequencies of spontaneous aberrations, the analogous tendencies were not significant.

17.4 Discussion

By now 24 polymorphic sites functionally associated with the chromosome damage are used in our study of radiosensitivity. The found effects are listed schematically in Table 17.2.

While on the subject of general tendencies, the following observations should be noted.

Table 17.2 Associations of allele variants of polymorphic genes with the frequency of aberrations of the chromosome type in blood lymphocytes by the results of this work and previous publications

Locus	Spontaneous chromosome type aberrations	γ- induced chromosome type aberrations (1 Gy *in vitro*)
XRCC1 G1996A rs 25487	–	G/G↑ G/* [brace with C/* below, ↑]
XRCC1 C580T rs1799782	–	C/* [brace with G/* above, ↑] (p=0.002)
XPD T2251G rs13181	*/G↑ [brace with */A↑ below, ↑]	–
XPD G862A rs1799793	*/A↑ [brace with */G↑ above, ↑] (p=0.025)	–
OGG1 C977G rs1052133	–	G/G↑ (p=0.011)
CYP1A1 T606G rs2606345	–	G/G↓ (p=0.005)
GSTM1 Ins-Del	I/*↑ (p=0.044)	–
CAT T21A rs7943316	–	T/T↑ (p=0.017)

The upward arrow indicates an elevated level of aberrations for the carriers of the specific genotype. The braces indicate the combined action of allele related to different sites

1. The association of the DNA polymorphism with the frequencies of cytogenetic damages is manifested to the greatest extent for aberrations of the chromosome type. In this case the significant associations of chromatide aberrations with the genetic polymorphism can be absent.
2. The frequencies of induced *in vitro* and spontaneous chromosome aberrations are associated with different groups of polymorphic genes.
3. The relative differences in the average levels of aberrations for the carriers of different genotypes are rather small: 10–20% for γ-induced aberrations ($p=0.002 \div 0.01$) and 30–50% for spontaneous aberrations of the chromosome type ($p=0.02 \div 0.04$). The shifts and changes in the type of distributions of the aberrations frequency are seen to a much greater extent as the different genotypes are compared (Fig. 17.1).

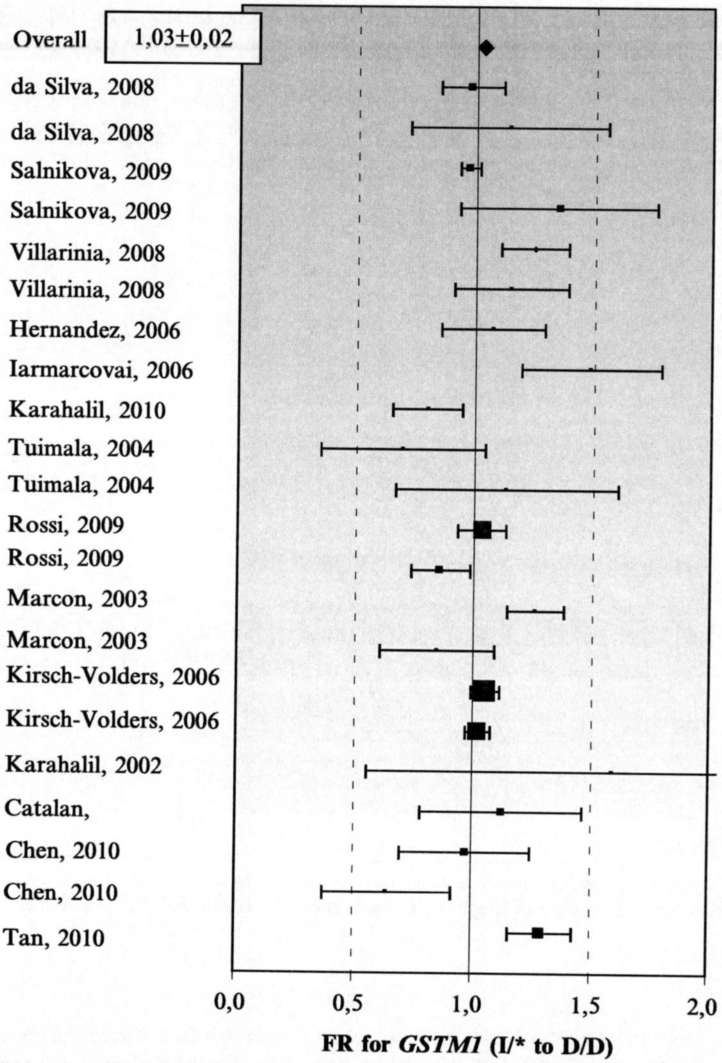

Fig. 17.4 Meta-analysis of the data of 14 works (22 samplings) on the relationship of cytogenetic disorders in the carriers of different genotypes by gene GSTM1. The average values (±SE) of the ratio of FR of cytogenetic effects for genotype I/* to the analogous value for D/D are given. The point size is proportional to the sampling volume

4. The differences between genotypes are often manifested while considering the combined effects of several sites in linkage disequilibrium related to the same gene, for example, as correlations of the aberrations number with the total number of minor alleles in these sites.

5. The elevated level of spontaneous chromosome aberrations is associated with the presence of minor allele variants (*XPD, GSTM1*). In this case the elevated radiosensitivity under irradiation *in vitro* is often associated with the presence of more common (major) allele variants (*XRCC1, CAT, CYP1A1*). *OGG1*, site C977G (Ser326Cys) was the unique gene, which showed the association of a minor rather than a major allele with the increasing frequency of induced aberrations.
6. Many polymorphic sites, for which an essential association with the radiosensitivity was suggested, did not show significant tendencies in the association with the frequency of aberrations (for example, *APEX, ATM, RAD23B, ERCC1, Tp53*).

In conclusion, the results of meta-analysis of the dependence of chromosome aberrations on genotype by locus *GSTM1* will be presented. The repeated attempts to find the association between the radiosensitivity and DNA polymorphism are related to this xenobiotics detoxication gene. It was suggested that the homozygotes by deletion (D/D – "zero" genotypes) are more radiosensitive than the carriers of a functional allele (I/* – "positive" genotypes). However, even in the first works on this subject [6, 9], the contradictory results were obtained.

For meta-analysis, the results of 11 studies of the control and/or exposed populations as well as the data of three works on the irradiation *in vitro* were chosen. Figure 17.4 shows the forest plot for the frequency ratio ($FR = p_+/p_0$), where p_+ and p_0 are the frequencies of cytogenetic disorders (aberrations of chromosomes or micronuclei) in carriers of "positive" and "zero" genotypes, respectively. The above data demonstrate no significant heterogeneity. The proportion of variation attributable to the heterogeneity (Higgins–Thompson I^2): $I^2 = 10.6\%$ (95%CI = 0.0 ÷ 45.1%). According to the fixed-effect model the average value *FR* (±*SE*) by overall data is equal to 1.03 ± 0.02 (95%CI = 0.98 ÷ 1.07). Thus, by overall data, significant differences of the average *FR* from unity are not observed (two-taled $p = 0.11$ for *Z* test).

References

1. Catalan J, Heilimo I, Falck G, Jarventaus H, Rosenstrom P, Nykyri E, Kallas -Tarpila T, Pitkamaki L, Hirvonen A, Norppa H (2009) Chromosomal aberrations in railroad transit workers: effect of genetic polymorphisms. Environ Mol Mutagen 50:304–316
2. Chen L, Guo H, Yuan J, He M, Chen D, Shi J, Yang J, Bai Y, Ju Y, Liu A, Yu Z, Li L, Sheng G, Fu J, Wu T, Chen X (2010) Polymorphisms of GSTT1 and GSTM1 and increased micronucleus frequencies in peripheral blood lymphocytes residents at an e-waste dismantling site in China. J Environ Sci Health A Tox Hazard Subst Environ Eng 45(4):490–497
3. da Silva J, Moraes CR, Heuser VD, Andrade VM, Silva FR, Kvitko K, Emmel V, Rohr P, Bordin DL, Andreazza AC, Salvador M, Henriques JA, Erdtmann B (2008) Evaluation of genetic damage in a Brazilian population occupationally exposed to pesticides and its correlation with polymorphisms in metabolizing genes. Mutagenesis 23(5):415–422
4. Hernández A, Xamena N, Gutiérrez S, Velázquez A, Creus A, Surrallés J, Galofré P, Marcos R (2006) Basal and induced micronucleus frequencies in human lymphocytes with different GST and NAT2 genetic backgrounds. Mutat Res 606(1–2):12–20

5. Iarmarcovai G, Sari-Minodier I, Orsière T, De Méo M, Gallice P, Bideau C, Iniesta D, Pompili J, Bergé-Lefranc JL, Botta A (2006) A combined analysis of XRCC1, XRCC3, GSTM1 and GSTT1 polymorphisms and centromere content of micronuclei in welders. Mutagenesis 21(2):159–165
6. Karahalil B, Sardaş S, Kocabaş NA, Alhayiroğlu E, Karakaya AE, Civelek E (2002) Chromosomal aberrations under basal conditions and after treatment with X-ray in human lymphocytes as related to the GSTM1 genotype. Mutat Res 515:135–140
7. Karahalil B, Emerce E, Kocabaş NA, Akkaş E (2010) Associations between GSTM1 and OGG1 Ser326Cys polymorphisms and smoking on chromosomal damage and birth growth in mothers. Mol Biol Rep 38:2911–2918. doi:10.1007/s11033-010-9953-0
8. Kirsch-Volders M, Mateuca RA, Roelants M, Tremp A, Zeiger E, Bonassi S, Holland N, Chang WP, Aka PV, Deboeck M, Godderis L, Haufroid V, Ishikawa H, Laffon B, Marcos R, Migliore L, Norppa H, Teixeira JP, Zijno A, Fenech M (2006) The effects of GSTM1 and GSTT1 polymorphisms on micronucleus. Cancer Epidemiol Biomarkers Prev 15(5):1038–1042
9. Marcon F, Andreoli C, Rossi S, Verdina A, Galati R, Crebelli R (2003) Assessment of individual sensitivity to ionizing radiation and DNA repair efficiency in a healthy population. Mutat Res 541(1–2):1–8
10. Rossi AM, Hansteen IL, Skjelbred CF, Ballardin M, Maggini V, Murgia E, Tomei A, Viarengo P, Knudsen LE, Barale R, Norppa H, Bonassi S (2009) Association between frequency of chromosomal aberrations and cancer risk is not influenced by genetic polymorphisms in GSTM1 and GSTT1. Environ Health Perspect 117(2):203–208
11. Sal'nikova LE, Akaeyva EA, Elisova TV, Kuznetsova GI, Kuzmina NS, Vesnina IN, Lapteva N, Chumachenko AG, Romanchuk VA, Rubanovich AV (2009) The effect of polymorphism of genes of xenobiotics detoxication on the frequencies of spontaneous and induced chromosome aberrations in human lymphocytes. Radiat Biol Radioecol 49(5):543–551 (in Russian)
12. Sal'nikova LE, Chumachenko AG, Vesnina IN, Lapteva NSh, Kuznetsova GI, Abilev SK, Rubanovich AV (2009) The relationship between genotypes and frequencies of chromosome aberrations in human lymphocytes under irradiation in vivo and in vitro. Radiat Biol Radioecol 50(3):1–5 (in Russian)
13. Tan H, Wang Q, Wang A, Ye Y, Feng N, Feng X, Lu L, Au W, Zheng Y, Xia Z (2010) Influence of GSTs, CYP2E1 and mEH polymorphisms on 1, 3-butadiene-induced micronucleus frequency in Chinese workers. Toxicol Appl Pharmacol 247(3):198–203
14. Tuimala J, Szekely G, Wikman H, Järventaus H, Hirvonen A, Gundy S, Norppa H (2004) Genetic polymorphisms of DNA repair and xenobiotic-metabolizing enzymes: effects on levels of sister chromatid exchanges and chromosomal aberrations. Mutat Res 554(1–2):319–333
15. Villarini M, Moretti M, Fatigoni C, Agea E, Dominici L, Mattioli A, Volpi R, Pasquini R (2008) Evaluation of primary DNA damage, cytogenetic biomarkers and genetic polymorphisms for CYP1A1 and GSTM1 in road tunnel construction workers. J Toxicol Environ Health A 71(21):1430–1439

Chapter 18
Radioecology of the Future: The Problem of Final Geological Disposal of Radioactive Wastes

Arrigo A. Cigna

Abstract The ethical principle according which the concept of "sustainable development" implies "*satisfying the needs of the present, without compromising the ability of future generations to meet their own needs*" is considered to be and absolute principle. Nevertheless also absolute principles have a their own range of application. Outside such a range they may lead to unwanted situations. With reference to radioactive wastes they should be managed in a way that secures an acceptable level of protection for human health and the environment, and affords to future generations at least the level of safety, which is acceptable today, But by referring to the rate of evolution of the human society with an extremely great increase in the last centuries, it is not possible to consider the far future generations as equivalent to the current one. In particular, the trend of evolution of the human society forecasts a strong acceleration never observed till now. Consequently the habits of the far future generations will be totally different from the today. Therefore also the criteria to be adopted for the safe containment of a geological repository must be take into account this fact. In fact it must be stressed that an excessive degree of protection implies, very often, only a waste of resources without any advantage and implying an absolute negative balance.

Keywords Sustainable development • Radioactive waste management • Protection • Human health and environment • Evolution

A.A. Cigna (✉)
Past President of the Union Internationale de Radioécologie, Str. Bottino 2 – Fraz. Tuffo, I-14023 Cocconato (Asti), Italy
e-mail: arrigocigna@tiscali.it

C.E. Mothersill et al. (eds.), *Radiobiology and Environmental Security*, NATO Science for Peace and Security Series C: Environmental Security,
DOI 10.1007/978-94-007-1939-2_18, © Springer Science+Business Media B.V. 2012

18.1 Introduction

The problem of the radioactive wastes is generally considered as probably the most relevant one for the development of nuclear energy. Frequently there is confusion between the perception of public and the achievements of science and technology. The result of such confusion is the statement that problem of the radioactive wastes has not yet found a solution.

Another source of confusion is the inclusion in the big basket of the radioactive wastes any kind of them without any difference among them: low, medium and high levels, alpha emitters and long living radionuclides.

For the purpose of this paper the High-Level Wastes (HLW) generated by the reprocessing of spent fuel from nuclear power reactors to recover uranium and plutonium, and the Alpha-Bearing Wastes (ABW) that are contaminated with long lived, alpha emitting nuclides are considered. Spent fuel that is not reprocessed is also considered a High-Level waste [11].

Such High-Level and Long-Lived wastes could be disposed of in mined deep underground repositories, while the medium and low level wastes do not require such a long and stringent isolation. But sometimes, under the pressure of the public opinion, which is frightened qualitatively by radioactivity without any consideration for the quantities involved, the authorities concerned with the problem of the radioactive wastes choose to avoid any distinction and any type of radioactive waste is disposed of in the same repository. Other radioactive wastes are normally kept in surface deposits and do not imply relevant problems.

It is never stressed enough that the problem of radioactive wastes is purely political because the solutions from a scientifical and technological point of view have already been found.

18.2 Ethical and Environmental Considerations

The timescales over which the intrinsic toxicity of long-lived radioactive wastes can extend go well beyond the life spans of current or forthcoming generations, for many thousands of years into the future. Therefore, for fairness and equity considerations, the current generations are responsible for the risks and burdens to future generations [11].

Such an ethical principle has been illustrated by the World Commission on Environment and Development using the concept of "sustainable development" put forward by the so called "Bruntland Commission" [43] and defined as: "*satisfying the needs of the present, without compromising the ability of future generations to meet their own needs*".

This principle was taken into account as the basis for assessing the acceptability of strategies for the long-term management of radioactive wastes [38]. The practical implementation of this principle, when applied to very long time intervals, implies a number of factors and requires a special attention. The Radioactive Waste Management

Committee of the OECD Nuclear Energy Agency reported that in the management of wastes having a long term potential for harm, interest focuses on two classes of ethical concern.

The first is the achievement of "**intergenerational equity**" to minimize the resource and risk burdens passed to future generations by the current generations which produce the wastes. The second concern is the achievement of "**intragenerational equity**" and in particular an ethical approach to the handling, within current generations, of questions of resource allocation and of public involvement in the decision-making process [38].

While the latter class does not rise particular objections provided the public involvement is assured within the frame of persons with at least a minimum of competence on the matter, the former must be based not only on a theoretical principle but also the effect of the evolution of the human society must be taken into account, as it will be discussed later with more details.

18.3 Systematic Comparison of Emplacement Options

After an interim storage to obtain the elimination of the short-lived fission products, the radioactive wastes have to be disposed of in a suitable repository, which can be placed in different environments. According the long-term isolation strategy, the high-level wastes can be emplaced in deep geological formations; three land-based options can be considered: clay, hard rock and salt. In the past, as a possible alternative to land disposal, the sub-seabed emplacement was considered. But in 1972 the contracting parties of the London Convention decided the total ban of immersion of radioactive wastes into the sea, and the sub-seabed alternative was abandoned. The same parties voted a total ban on the disposal at sea of radioactive wastes and other radioactive matter in November 1993 [39].

18.3.1 The Pagis Project

In 1982, within the European Community, it was started a project for Performance Assessment of Geological Isolation Systems (PAGIS). This exercise was conceived as an intermediate step between the generic site independent safety assessment and those made in view of the licensing of particular disposal sites. A large body of scientists in EC countries have worked together for 6 years producing both the data and the methods required for the analysis. In particular sensitivity analysis of model parameters has shown the areas where additional research will be more effective for decreasing the uncertainty. In this chapter the results and conclusions obtained by PAGIS are reported [8].

The reference waste arising refer to the high level waste from Light Water Reactors spent fuel reprocessed by the Purex process, vitrified as borosilicate glass and contained in stainless steel canisters; in PAGIS Project the interim storage before disposal was assumed to be from 30 to 100 years. For continental repositories, the reference designs are based essentially on the shaft-gallery-borehole concept. For the sub-seabed option, the information has been obtained from the Seabed Working Group of OECD-NEA [34]; the free fall penetrator is the reference emplacement technique, the drilled hole being the variant.

The multibarrier concept for describing the pathway to man of the radionuclides has been adopted. For each option, "normal evolution scenarios" representing the gradual changes, which may be expected on the basis of the geological trends and the information available on the repository effects, were firstly considered. Then, "altered evolution scenarios", representing perturbation by events largely of probabilistic nature, which can modify the parameters determining the normal evolution, or generate new evolution scenarios, were also taken into account.

18.3.1.1 Clay Formations

The low permeability of clay results in negligible pore water movements and its high retention capacity for many radionuclides further reduces radionuclide migration. In the PAGIS study, the Mol site has been chosen as reference site where a sub-horizontal Boom clay formation 110 m thick is covered by at least 160 m sediments.

In a normal evolution scenario the majority of the radioisotopes, which would be released from the waste packages, do not travel beyond a few metres from the repository and disappear through decay. Only a few long-lived nuclides poorly absorbed in the clay matrix are able to induce a contaminated plume in the aquifer in at least half a million years before any significant radioactivity would be noticed at the surface.

Altered evolution scenarios have been analyzed. A fault crossing the repository would not change the situation found for the normal evolution scenario because only a very limited portion of the repository would be affected. If heavy climatic changes would occur, the radionuclides dilution would be reduced by a factor of four compared with that computed for the normal evolution.

As a conclusion, a clay formation can ensure long-term protection for a high-level waste repository, provided that the clay layer has a sufficient thickness and care is taken in characterising the overlying aquifers.

18.3.1.2 Granite Formations

The very low permeability of granite, if not exceedingly fractured, and the large and stable formations commonly found, allows a very slow groundwater movement with long pathways to the biosphere. Auriat, in France, was the reference site for PAGIS since it has been the subject of geological and hydrogeological investigations.

No appreciable exposure is expected before 250,000 years for a normal evolution scenario. The case of a human intrusion was considered as an altered evolution scenario. By assuming that a cavity would be excavated in the vicinity of the repository, the underground water system would be perturbed and workers and surface dwellers exposed. The probability of such an event is expected to be low and the consequences of an intrusion at 100,000 years would result in an exposure of the same order of magnitude of that due to miners by the natural background from the rock itself. Vertical faults would also have negligible effects unless connected by horizontal pathways through the repository.

Uncertainties are relatively large as compared with the other continental options but the high-level waste would be adequately confined in granite if an appropriate repository design is made and good rock properties in a thickness of 10 m are available.

18.3.1.3 Salt Formations

The almost complete absence of water in the salt formation and the capacity of rock salt to quickly fill any open cavity prevent radionuclides migration. The reference site for PAGIS was the salt dome at Gorleben in Germany where a large amount of information has been collected.

Under normal evolution scenario, no contamination can attain the biosphere unless dissolution of the salt dome itself would allow the aquifers to reach the repository level. It was assumed a dissolution rate of 0.033 mm/a and therefore the time span involved in the release would be of the order of million of years.

An altered evolution scenario involving brine intrusion via an anhydrite vein which may intersect the access shaft would not result in radionuclides release unless the presence of an undetected brine pocket near to a corner borehole is postulated together with an early intrusion time within 200 years after the closure of the repository. Also under all these very pessimistic assumptions, the release would occur 20,000 years after disposal, and the exposure of local people would be similar to that occurring in the normal evolution scenario. These results show that the long-term protection can be ensured by rock salt, since extreme situations are needed in order to allow the radioactivity to reach the biosphere.

18.3.1.4 Sub-seabed

In the seabed there are very thick sedimentary layers with long-term stability where radioactive waste, conditioned in suitable containers, could be emplaced. Two alternative methods were considered: the freefall penetrators containing the canisters of vitrified waste or their emplacement in boreholes drilled in the lithified sediment below the unconsolidated one; the reference area was the Great Meteor East on the Madeira Abyssal Plain in NE Atlantic.

The normal evolution scenario has been defined as the slow degradation of the waste packages, diffusion of radionuclides through the sediments, dispersion in the ocean and exposure pathway to man mainly via marine foodstuff. The maximum exposure would occur 100,000 years after disposal with values less than one millionth lower than the natural background.

Altered evolution scenarios covering a range of natural events, human actions and accidents including damaged penetrators on the seafloor have been considered. In the worst case the maximum exposure would always be one thousandth lower than the natural background.

Therefore the post emplacements doses and risks from the disposal of high level waste in the sub-seabed are negligible. As it was reported above, this solution was abandoned after the decision of the London Convention.

It must be emphasised that the option of the sea dumping of both large and heavy blocks of concrete and iron pipes (which in principle could be contaminated essentially by ^{60}Co by neutron activation) would have been the cheaper solution with a negligible detriment of the marine environment.

18.3.2 *The Radiological Effects of Radioactive Wastes*

The International Nuclear Fuel Cycle Evaluation (INFCE) was organized at a conference held in Washington, DC, on 19–21 October 1977. The participants defined a programme based upon the principles that nuclear energy for peaceful purposes should be made widely available to meet the world's energy requirements, the danger of proliferation of nuclear weapons should be minimized without jeopardizing the development of nuclear energy and the specific need and conditions in developing countries.

Since then on, INFCE was supported by IAEA and a series of meetings took place from 1978 to 1980. Environmental, health and safety issues were addressed by the INFCE Working Groups within their terms of reference, to determine whether specific fuel cycle activities could be carried out in conformity with accepted standards and study of these problems showed that these standards could be respected [16].

The radiological effects have been extensively examined by INFCE with reference to the different fuel cycles taken into account [17] and by PAGIS Project with reference to different disposal options [8].

In addition, it must be stressed that the assessment presented by INFCE are not site-specific and they are attempts to obtain representative values of the collective dose commitments resulting from disposal of radioactive wastes for comparison between different fuel cycle alternatives. On the other hand the PAGIS Project aimed to obtain data from reference sites, which are, therefore, site-specific. Therefore any comparison between these two sets of data must take into account their intrinsic differences.

The calculations to evaluate the radiological impact have been made by INFCE on the basis of ICRP Publication 26 [18]. It was not deemed necessary to update the results for the purpose of this work by taking into account the changes (e.g., in the weighting factors) introduced by ICRP in its more recent publications. The collective dose commitment from different sources and fuel cycle are now reported.

18.3.2.1 Mill Tailings

The radiological contribution of thorium mill tailings can be neglected in comparison to that of uranium mill tailings [3]. Therefore only the latter will be taken into account. In addition to the assumptions listed before in the descriptions of the different nuclear fuel cycles, calculations were carried out assuming that the tailings are covered with 2 m of soil and that the population distribution around mills is of 3 person/km^2 at short distances (less than 80 km) and 25 person/km^2 at greater distance [42].

In the present assessment it is also assumed that the mean residence time of ^{226}Ra and ^{230}Th in mill tailings is 1,000 years, these elements being gradually removed from the tailings area to circulating waters.

18.3.2.2 Geological Repositories

The behaviour of the radionuclides in the wastes depends on the characteristics of the containers [3]. Notwithstanding that, in principle, the life of such containers could be much longer and for the purpose of the collective dose assessment it is assumed that the mean life of containers is over 1,000 years. At the same time it is assumed a leach rate of about $4 \cdot 10^{-4}$ of the borosilicate glass. Therefore the complete dissolution of the glass would take over $3 \cdot 10^4$ years. The conservative estimate of the delay before release would be, then, in the range 10^4–10^5 years. Successively the radionuclides would start their pathway according to the characteristics of the repository.

In the previous chapters fours options have been described (clay; granite, salt and sub-seabed). As no realistic values of the delay provided by geological isolation for each waste radionuclide are available, the assessment presented by INFCE cannot be carried out by using sophisticated compartment analysis of the time functions of the resulting contamination of the biosphere. This type of analysis is more applicable to site-specific assessments.

Therefore, the collective dose commitment has been roughly estimated to range between two extreme values, corresponding to the entry into circulating waters of an undisturbed mixture of waste radionuclides of an age taken to be equal, respectively, either to the shortest or to the longest delay found for the nuclides of the mixture.

Taking into account the engineered isolation factors and the time of water migration, it was assumed that 10^5 years is representative of the delay without sorption of radionuclides and a delay of the order of 10^6 years was taken to represent the case where sorption is the main retardation factor.

An assessment carried out for INFCE by a geosphere model [6] for an hypothetical repository in a reference salt formation involving a geological event as a release assumption, found estimated arrival times for fission products and actinides ranging from $6.7 \cdot 10^5$ to $8.1 \cdot 10^7$ years. Such values agree acceptably well with the order of magnitude of 106 years reported before.

It is interesting to compare the assumptions made by INFCE for the delay of the release of radionuclides with the evaluation found by PAGIS for the reference sites and reported in a previous chapter: once again the agreement is quite acceptable. This means that the results obtained by INFCE are still valid after 10 years and a large amount of rather detailed studies have not modified the main INFCE conclusions.

18.3.2.3 Assessment of Dose Commitments from Waste Repositories: INFCE

For the assessment of the dose commitments from waste repositories, the transport of the long-lived radionuclides by groundwater was considered to be the normal mechanism by which they return to the biosphere notwithstanding that disposal repositories are located in formations with a very scarce amount of migrating groundwater [3].

It was found that the drinking water and fish consumption pathways give the most important contribute to the dose to man. Also irrigation of crops, when practiced, can contribute significantly while other fresh water pathways are in general negligible for the collective dose.

The ocean contribution may result either from the fresh waters flowing into the sea or when the exposure pathway starts directly from the sea as, e.g., in the case of the sub-seabed disposal. Several pathways are considered in the assessment of the human exposure from radionuclides dispersed into the marine environment: fish and seafood consumption and sediment resuspension inhalation. It was found that external exposure to sediments in general contributes negligibly to the collective dose commitment compared to the other pathways.

When unconventional pathways are considered, because they may be added in the future, also desalted water contribution appears to be negligible because desalination processes will separate most radionuclides from the water with varying decontamination factors, probably in the order of 100. On the contrary small zooplankton might have conceptually a significant contribution to the collective dose commitment, as the concentration factors for many radionuclides in plankton are substantially higher than in fish.

18.3.2.4 Collective Dose Commitments

In Table 18.1 a summary of collective dose commitments (man•Sv) from waste arisings of reference fuel cycles per $GW_e \cdot a$ is reported.

Table 18.1 Summary of collective dose commitments (man • Sv) from waste arisings of reference fuel cycles per GW_e • a [17]

	LWR		FBR	HWR			HTR
Operation	Once through	U-Pu cycle	U-Pu cycle	Once through	U-Pu cycle	U-Th cycle	U-Th cycle
Ore processing	370	220	2	330	130	13	100
Refining, conversion	270	160	–	–	–	9.9	83
and enrichment [a]	90	59	–	–	–	3.6	30
Fuel element fabrication [a]	<1	<1	<1	<1	<1	<1	<1
	1	1	3	<1	1	5	<1
Reactor operation	<1	<1	<1	<1	<1	<1	<1
Unreprocessed	89	–	–	310	–	–	–
spent fuel [a]	270	–	–	450	–	–	–
Reprocessing [a]	–	31	18	–	137	26	25
	–	50	52	–	117	36	12
Sum total [a]	730	410	22	640	270	61	210
	730	330	58	780	250	280	250

[a]Upper and lower figures for migration times of 10^6 and 10^5 years, respectively

Table 18.2 Summary of the incomplete (500 years) collective dose commitments (man•Sv) from waste arisings of reference fuel cycles per GW_e•a [3]

	LWR		FBR	HWR			HTR
	Once through	U-Pu cycle	U-Pu cycle	Once through	U-Pu cycle	U-Th cycle	U-Th cycle
From:	3.7	0.2	0.4	11	0.1	1.2	0.6
To:	7.9	0.8	1.3	16	0.9	5.7	3.5

As it was stated in advance, the estimation of the collective dose commitment has been carried out by integration to infinity. This procedure implies knowledge of the future, which is totally unrealistic because the evolution of the human society is continuously accelerated and therefore it is absolutely impossible to argue what will be the size, the habits and the needs of human kind in, e.g., 10^5 or 10^6 years.

Therefore, when the exposures are delivered over very long time after the practice has ceased, the estimation of the incomplete collective dose commitments, i.e. the collective dose commitments integrated over only a few hundred years, provide a useful tool for the evaluation of impact which can be expected in the near future. In Table 18.2 are reported the ranges of the incomplete collective dose commitment (man•Sv) from waste arisings of reference fuel cycles per GW_e•a integrated over 500 years.

18.3.2.5 Individual Doses

The doses to the most exposed individual could obviously be higher than the population average doses. The actual values would be site-specific and it is not possible to have a generic estimate. Within the INFCE programme two assessments were carried out to evaluate at least the order of magnitude for somewhat idealized sites.

The first one refers to a repository in a salt formation at 250 m below the surface located in a large sedimentary basin. The release assumption involves a geological event creating a fracture, which results in brine flowing through the repository. The maximum annual effective dose equivalent ranges between the lowest $2 \cdot 10^{-7}$ Sv/a for fuel cycle strategy "Light Water Reactor, U-Pu with reprocessing" and $3 \cdot 10^{-5}$ Sv/a for fuel cycle strategy "Light Water Reactor, once through". The time of occurrence of all doses is of the order of 10^6 years [3, 6].

In another study, the individual doses from a repository in granite were assessed for possible sites in Sweden. The release was postulated to occur by groundwater flow and different pathways were considered including also ingestion, inhalation and external irradiation. The maximum value was estimated to be of the order of 10^{-4} Sv/a after periods exceeding 10^5 years [23].

It can be assumed, therefore, that the maximum doses to the critical group in the vicinity of repositories will be very low and occur in a very far distant future, well beyond the time interval, which can be reasonably taken into account.

18.3.2.6 Assessment of Dose Commitments from Waste Repositories: PAGIS

The models adopted by PAGIS are more sophisticated than those utilized by INFCE because in the decade elapsed between the two projects the knowledge of the behaviour of radionuclides in the global environment largely increased [8].

The concept of near-field and far-field compartments was adopted. The former extends to the limit where the host rock is unaffected by the disposal techniques or alterated by the presence of waste; beyond this, there is the far-field region which include the remaining host rock and the surrounding geosphere.

Two sets of models were used in general. The first set represents at the best the current knowledge of the various physical-chemical phenomena involved in the release and migration of radionuclides. Such models are used for deterministic evaluations of doses. But, when repeated calculations are needed in order to statistically simulate the behaviour of the repository, another set of models, more appropriate for a stochastic approach, was elaborated. Multidimensional geosphere models have mostly been used in the deterministic calculations, while simplified 1-D models have been employed for the stochastic evaluations. A crucial work of calibration linked the two sets of models to ensure an agreement between them. An advantage of using such advanced models was the possibility to evaluate, in addition to doses, also the local sensitivity (response of the dose rate to perturbations of single parameters) and to carry out the uncertainty (distribution of dose-time functions) and the global sensitivity analysis.

All relevant pathways from the underground aquifers to man were considered. Since there are no justifiable procedures to forecast human habits and diets in the long term, the biosphere parameters were assumed to be constant with time, neglecting spatial variability as well. The "hypothetical critical group" was characterized by very high food consumption rates that are unlikely to be reached in future generations' diet. As it was anticipated in a previous chapter, both a normal evolution dose rate and altered evolutions dose rates were evaluated.

18.3.2.7 Collective Dose Commitments

The collective dose commitment was calculated only for the case of the sub-seabed reference repository of the Great Meteor East area in the Madeira Abyssal Plain (Atlantic Ocean). It was found to be $1.2 \cdot 10^4$ man Sv, most of which being delivered by 10^6 years, with the collective dose rate reaching a peak at 10^5 years [28].

18.3.2.8 Individual Doses

These values are available for each repository and some results are summarized in Table 18.3.

Table 18.3 Maximum individual dose rate for the reference repositories (data from [8])

Repository	Evolution	Pathway	Max individual dose rate (Sv/year)	Time years	Radionuclides
Clay [26]					
Mol	Normal	Well	$1.2 \cdot 10^{-8}$	10^{6}	^{99}Tc
			10^{-7}	10^{7}	^{237}Np: 70%; ^{135}Cs:30%
	Normal	River	$1.1 \cdot 10^{-11}$	10^{6}	^{99}Tc
			$0.8 \cdot 10^{-11}$	10^{7}	^{237}Np; ^{229}Th
	Climatic change	Well	$2.9 \cdot 10^{-7}$	10^{7}	^{237}Np: 55%; ^{135}Cs:45%
	Faulting	Well	10^{-8}	$3 \cdot 10^{5}$	^{99}Tc
			$5.0 \cdot 10^{-8}$	$2 \cdot 10^{6}$	^{237}Np
Granite (Van Kote et al. 1988)					
Auriat	Normal	Water	$5.7 \cdot 10^{-7}$	$3 \cdot 10^{6}$	^{237}Np: 80%; ^{229}Th:20%
	Intrusion	(Miner)	$1.6 \cdot 10^{-2}$	10^{5}	^{237}Np; ^{229}Th
Salt (Storck et al. 1988)					
Gorleben	Normal	Water	10^{-6}	$1.5 \cdot 10^{-7}$	^{237}Np
	Early intrusion	Water	$9 \cdot 10^{-6}$	$2 \cdot 10^{4}$	^{135}Cs
	Late intrusion	Water	$3 \cdot 10^{-5}$	$7 \cdot 10^{5}$	^{237}Np
Bedded salt (France)	Sealing failure	Water	$5 \cdot 10^{-6}$	$3 \cdot 10^{4}$	^{135}Cs
Sub-seabed (Mobbs et al. 1988)					
GME	Normal	Seafood	$1.8 \cdot 10^{-10}$	10^{5}	^{79}Se
			$5.1 \cdot 10^{-10}$	$2 \cdot 10^{5}$	^{99}Tc
	Mining	Seafood	$1.4 \cdot 10^{-7}$	10^{6}	^{99}Tc; ^{135}Cs; ^{229}Th
	Damaged Penetrator on the seabed	Seafood	$1.3 \cdot 10^{-10}$	$5 \cdot 10^{3}$	$^{239+240}Pu$; ^{243}Am; ^{126}Sn

Note. Dose rates from altered evolution scenarios (i.e. those not marked "Normal") are conditional on the occurrence of the event and are associated with a probability factor. They cannot be compared, therefore, with those from normal evolution

It must be emphasized the great detail achieved in the case of the sub-seabed repository in comparison with the other options. This fact is due mainly to two reasons. Firstly, on account of the models which have been developed for the marine environment in the framework of CRESP [29, 30, 32, 33, 35, 36, 39]. In addition, a relevant research on ocean-floor burial of high-level waste have been carried out since 1977 in the framework of the Seabed Working Group established under the Radioactive Waste Management Committee of the OECD Nuclear Energy Agency.

The work of the Seabed Working Group provided scientific and technical information to assess the safety and engineering feasibility of seabed disposal. The results were published in a series of eight volumes [7, 12, 14, 24, 25, 34, 40, 41]. They are consistent with those obtained within the PAGIS framework.

18.4 Environmental and Ethical Aspects

The basic principles of radiation protection underwent a certain evolution to take into account all the possible implications, which became evident with time. In particular, when radioactive wastes containing long-lived radionuclides are involved, there is the problem of far future detriments. It is evident that such detriments could be treated differently from the present ones. This would imply the assumption of a weighting factor with respect to the present or the immediately foreseeable future.

Because of social time preferences less weight could be assigned to far future detriments. But, for ethical considerations, the opposite assumption could be made because people who neither have an advantage from the practice would incur such far future detriments nor had any influence on the decision leading to their exposure [19, 31].

Calculations of doses resulting several thousands of years from now are based on current living habits and any estimate of far future living conditions would be largely speculative. Therefore these calculations are an illustration of what the doses would be if the release occurred today, rather than as a prediction of the actual dose to some human living in the far future [38].

In ICRP Publication 60 [20] it is reported that when the current practices give rise to doses, that will be received in the future (and sometimes in the far future), these future doses should be taken into account in the protection of both populations and individuals, although not necessarily on the same basis as is used for current doses. Also the probability of incurring such future exposures is important, in addition to the magnitude of the exposures.

On the practical side, there are difficulties in predicting future conditions, whether demographic, biological, physical or other aspects, with sufficient confidence to make useful estimates of long term radiation impacts [19]. Another aspect of the problem of exposures in the far future is related to the evaluation of collective doses over long timescales of 10^4–10^9 years. This aspect has been studied by Barraclough et al. [4] and their conclusion was that in decision-making, less significance should be attached to collective dose estimates relating to periods beyond 500 years into the future than those relating to shorter time periods.

If the extrapolation of the present society is taken into account, also a time interval of 500 years is probably too long. The habits and the problems of mankind 500 years ago, i.e. at the time of the discovery of the New World, were rather different from today and any extrapolation from that time into our present time would have resulted in a substantially wrong description of the present situation. Since the evolution of the society is faster now than in the past, another unpredictable factor adds its contribution in the general uncertainty.

When much longer time intervals are involved, any extrapolation becomes totally unreliable. In fact, if one of our ancestors (e.g. a Neanderthal man) had forecasted the needs of our society based on his available knowledge, he would have identified the flint as the limiting factor for society's development and the degree of civilisation attainable [10]. Of course the results of such an analysis would have been completely wrong, notwithstanding a set of correct starting data, because other factors have substantially changed the situation in the meantime.

There is no reason to consider ourselves in a better position to assume that ionizing radiations will be a threat for populations in the far-future: for example cancer will probably not be so important in the future as it is today [9, 22, 27]. A comparison with the past, as given in Fig. 18.1, clarifies the concept of the impossibility to obtain reliable forecast over very long periods.

The scenarios and the models adopted by the different projects lead to dose assessments, which are rather low. Nevertheless it must be emphasized that there is a good chance that such assessments are wrong because they are too much conservative, i.e. pessimistic. In fact there are some examples resulting from special geological conditions, which give results more optimistic.

The Oklo phenomenon has been studied extensively [5, 13, 15] and the natural reactor zones contain essentially all of the radioactive material present at the end of the reaction period. Isotopic analysis of the uranium grains in a microscopic scale showed that fission products and ^{239}Pu had remained immobile until they became harmless by radioactive decay.

Another case is given by the ore deposit of Koongara at Alligator Rivers (Australia) where the radionuclides in the superficial layer subject to the leaching by the meteoric water moved downstream some ten of meters in millions of years but they did not move at all in the deeper protected layers [37].

Also an open pit uranium mine at Poços de Caldas, Minas Gerais, Brazil, was extensively studied with similar results. Therefore, when the natural analogues are taken as a kind of validation of the models [2], it should be concluded that the evaluations obtained by such models are to be considered as a pessimistic upper level of what can be really expected.

18.5 The International Recommendations

The recommended limit of the effective dose for the public is presently established in 1 mSv in a year (the average over 5 years must not exceed 1 mSv per year) [20]. This dose limit applies to the sum of the relevant doses from external

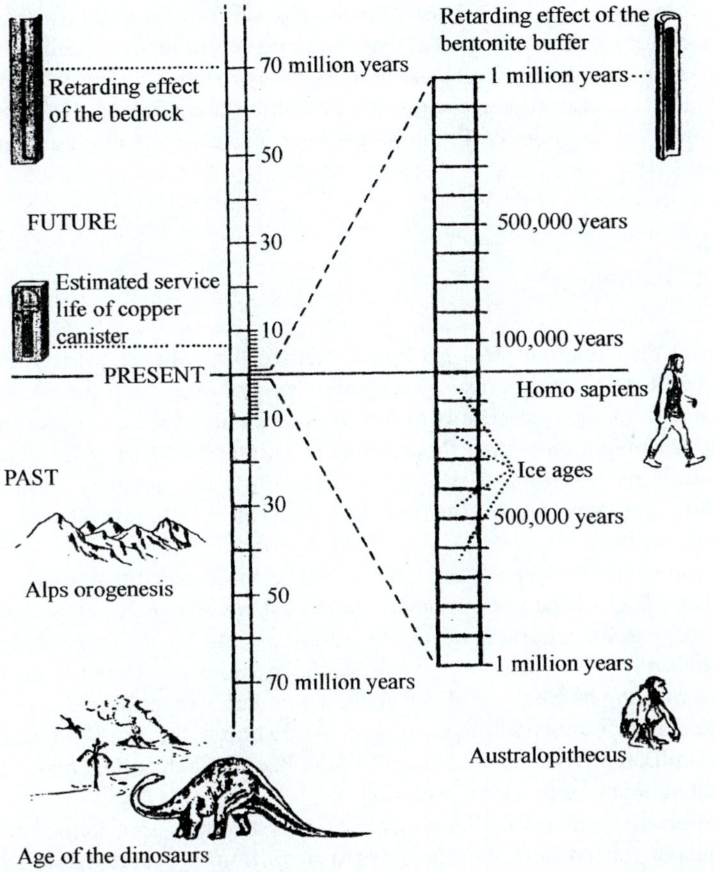

Fig. 18.1 A comparison between the geological repository life and the geological time scale (From [1], modified)

exposure during 1 year and the 50-year committed dose from that year's intake of radionuclides. When both individual and collective exposures are small and not in excess of the dose limits, it is adequate to use the product of the expected dose and its probability of occurrence as if this were a dose that was certain to occur [20].

To avoid that limited resources are wasted on trivial problems at the cost of neglecting major problems it is necessary have exemption rules. An annual probability of death of the order of 10^{-6} per year is largely accepted, because it is not taken into account by individuals in their decision as to actions that could influence their risks. This level of risk corresponds to an annual dose of the order of 10^{-5} Sv. In order to ensure that the total annual dose to a single individual will not exceed this value, it is almost certain that this result is achieved if the annual individual dose exemption criterion is reduced by a factor of ten [19].

Therefore, most regulatory systems include provisions for granting exemptions: the grounds for exemptions are that the source gives rise to small individual doses (of the order of 10^{-6} Sv per year) and regulatory provisions will produce little or no improvement in dose reduction since the protection is optimised. If the collective dose is small (of the order of 1 man Sv per year) protection is often assumed to be optimised [21].

18.6 Conclusion

The ethical principle adopted can be summarized as follows: wastes should be managed in a way that secures an acceptable level of protection for human health and the environment, and affords to future generations at least the level of safety which is acceptable today, but there seems to be no ethical basis for discounting future health and environmental damage risks [38]. This strategy should aim at bequeathing a passively safe situation, which places no reliance on active institutional controls [11].

As reported above, such assumptions do not take into account anything else than the characteristics of the present environment and are designated to assure a strict compliance with the requirements of the present society.

Now let's move back into the past in a prehistoric time, e.g. 10,000 years ago, and imagine a meeting at the Grotta Romanelli, a well known prehistoric cave in South Italy where a Mesolithic culture developed. At that time, the world population was about ten million people and the "Gotha" of the Romanellian culture had their G1 to discuss the problem of mankind evolution.

A number of person inhabited caves or rock shelters and, if the number of individuals increased too much (let's say to six billions…) there would have been a shortage of caves suitable for living in. But the solution of rock shelters suggested a good way out, i.e. by building artificial equivalents to rock shelters as huts or pile-dwellings according the local environmental characteristics.

But another important issue, concerning technology, had to be considered. By taking into account the trend of the number of individuals, which was assured to have a steady increase, the participants to their G1, who were extremely careful about the life style of the future generations, focused on the identification of the limiting factor after a very long time interval. On the basis of their living habits the answer was easily obtained. At that time they were already compelled to look for flint even in distant locations, therefore if the mankind evolution would have resulted in a very large number of individuals, as for instance the six billions quoted above, the availability of flint-stones was unanimously identified as the limiting factor: after, for example. 10,000 years, would there have been enough flint-stones for the needs of everyone? Certainly not, at least on the basis of their knowledge and experience.

Now, contrary to the wise, but wrong, forecast of our ancestors, flint-stones cannot be considered a limiting factor for us, because life style and habits evolved in a way totally different from the view of our ancestors. This fact confirms that an extrapolation into an age far removed from our own leads to completely wrong conclusions.

The evolution of mankind in the past 10,000 years was absolutely not uniform. In a mathematical slang the second derivative of the evolution is positive, i.e. its speed increases with time. Life habits changed very slowly from Mesolithic to Neolithic. Later on, in historical time, the life styles of every day during the Egyptian, the Greek, the Roman, the Middle Age cultures kept rather similar.

But after the industrial revolution the changes were every day more important, up to the present time when we perceive a great change within a single generation. In fact, most of the world's industrial production has taken place in the last few centuries.

Many problems facing the present society are already totally different from those of few tens years ago. Therefore the life habits of human kind evolve presently faster and faster and in a rather close future they may become quite different from the previous ones. Already beyond 100 years from now it is very hard to foresee the development of ecology, medicine, technology, economy and social structure.

Therefore, it is absolutely not justified to choose a management of radioactive wastes based only on radioactive decay without any consideration for the biosphere and living habits of the humans that are expected to change drastically in less than a few hundred years.

Someone could object that the biological characteristics of human individuals with respect to radiation could (or would) be the same of today, but as pneumonia is no longer a mortal illness as it was only some tens of years ago, it is unlikely that cancer will keep the same degree of gravity of today after some hundreds of years.

The uncertainties associated with assessment results increase with time and at present it is not possible to make any reliable forecast of the life style of far future generations. Nevertheless, it is possible to identify two likely main trends: a continuation of the fast evolution observed in the last centuries or a disruption of this mechanism followed by a return to rather primitive conditions.

In the first case, it is obvious to imagine that mankind should have such a competence and technological capacity to face any possible contact with the radioactive waste both as a consequence of an intrusion and a release for a natural event.

In the second case, with a mankind decayed to habits and life style similar to those of prehistoric times (it could be defined a "post historical" mankind) the chance of an intrusion would not exist for lack of a suitable technology while the risk from a release due to a natural event would be unimportant on account of the small probability to have a population in the area interested by the release. But, what is much more relevant, the danger of this event would be negligible in comparison to the dangers in the environment inhabited by our poor descendants.

The validity of the ethical principle of offering to society a satisfactory level of safety for both current and future generations is not questioned but it must be emphasized that the same current needs cannot be considered valid for any future generation. Previously it was pointed out the large difference between the Romanellian population and the present generation, which would have obviously implied at present an absurd statement, about the limiting factor for future development.

Since the speed of the evolution of human kind is every day faster, the ethical principle reported above must be applied by taking into account the potential evolving requirements of the future generations which will be affected by the decision of the current generation. A servile application of the principle by attributing the same current requirements to far away future generations would imply unjustified burdens, which would be, not only useless to the future generations but could also result in a waste of resources.

If these considerations are taken into account, the ethical principle reported above implies an important change of some current positions. In particular 95% of the potential energy is still contained in the used fuel after the first cycle. Therefore, since uranium is a limited resource (even if more abundant than oil), the recycle of used fuel should be adopted to achieve a better efficiency. The advantage would be double because, by so doing, a very long-lived radioisotope ($T_{1/2}$=24,000 years for ^{239}Pu) would be no longer discarded as a waste but used in the new fuel. At the same time this radionuclide, relevant from the point of view of proliferation, would be no longer present among the waste.

The criterion of having a number of safe requirements implemented for 10,000 or 100,000 years after disposal as currently assumed by some regulatory authorities, is absolutely not justified since in such a long time interval the human society (if still existing!) is expected to be totally different from the present generation.

Therefore the safe requirements to be applied to a geological repository should be assured for a much more limited time interval, e.g. a few generations. Even by adding a coefficient of safety to take care of unforeseen events, a safe containment for few centuries would assure to the future generations the same treatment accepted for the present one. Obviously such time lengths belong to the human scale and not to the geological one, and the burden charged to the current generation is, therefore, perfectly acceptable.

In addition it must be emphasised that the problem of the radioactive waste disposal is purely political since the technological solutions now available comply perfectly with the radiation protection requirements and the ethical principles.

The point of view described here can be adequately summarized by a passage of an Italian writer Giuseppe Tomasi di Lampedusa in his book "The leopard" (Chap. 18.1): *"For us a palliative promising to last for one hundred years is equivalent to eternity. We may even be worried about our sons, perhaps about our grandchildren; but beyond those we may hope to caress with there hands, we ha no obligation to fulfil".*

References

1. Ahlstroem PE, Loefverberg S, Nilsson LB, Papp T (1980) Safe handling and storage of high level radioactive waste. Radioactive Waste Manage (USA) 1:57–103
2. Andersson K, Nicholson T, Grundfelt B, Larsson A (1990) INTRAVAL as a n integrated international effort for geosphere model validation–A status report. In: Proceedings of a symposium on the safety assessment of radioactive waste repositories, Paris, 9–13 Oct 1989, IAEA, Vienna, pp 725–736

3. Argentina (CNEA) (1979) Radiological health and safety impact. INFCE/DEP/WG.7/15, IAEA, Vienna
4. Barraclough IN, Robb JD, Robinson CA, Smith VR, Cooper JR (1996) The use of estimates of collective dose to the public. J Radiol Prot 16(2):73–80
5. Blanc PL, Von Maravic H (1996) OKLO Working Group. In: Proceedings of the fourth joint EC-CEA progress and final meeting held in Saclay, France, on 22 and 23 Jun 1995. European Commission, D.G. XI, Report EUR 16704
6. Bond FW, Cole CR, Reisenauer AE, Washburn JF (1979) Release consequence analysis for a hypothetical geologic radioactive waste reporitory. INFCE document CC/WG.7/39 Rev. 2
7. Brush LH (1988) Review of laboratory investigations of radionuclide migration through deep-sea sediments. In: Feasibility of disposal of high-level radioactive waste into the seabed. OECD, Paris, 7
8. Cadelli N, Cottone G, Orlowski S, Bertozzi G, Girardi F, Saltelli A (1988) PAGIS, Performance assessment of geological isolation systems for radioactive waste, summary. Commission of the European Communities, DG. XII, Report EUR 11775 EN
9. Chapman NA, Côme B (1990) Long-term predictions: making proper use of geological evidence. In: Proceedings of symposium on the safety assessment of radioactive waste repositories, Paris, 9–13 Oct 1989, IAEA, Vienna, pp 318–329
10. Cigna AA (1996) Origin and aim of radioecology. In: Luykx FF, Frissel MJ (eds) Radioecology and the restoration of radioactive-contaminated sites. NATO ASI Series, 2. Environment, 13. Kluwer Acad. Publ, Dordrecht/Boston, pp 1–15
11. Cigna AA (2010) The problem of long lived radioactive waste management. In: Lavino JG, Neumann RB (eds) Psychology of risk perception. Nova Science Publishers, Inc., Hauppauge, pp 179–187
12. de Marsily G, Berhendt V, Ensminger D, Flebus C, Hutchinson B, Kane P, Karpf A, Klett R, Mobbs S, Poulin M, Stanner D (1988) Radiological assessment. In: Feasibility of disposal of high-level radioactive waste into the seabed. OECD, Paris, 2
13. Hagemann R, Nief G, Roth E (1974) Etudes chimiques et isotopiques du réacteur naturel d'Oklo. CEA Bull Inf Sci Technol 193:65–76
14. Hickerson J, Freeman TJ, Boisson JY, Murray CN, Gera F, Nakamura H, Nieuwenhuis JD, Schaller KH (1988) Engineering. In: Feasibility of disposal of high-level radioactive waste into the seabed. OECD, Paris, 4
15. IAEA (1975) The Oklo phenomenon. In: Proceedings of a Symposium Libreville, Gabon, 23–27 June 1975, IAEA, Vienna
16. IAEA (1980a) INFCE summary volume. IAEA, Vienna
17. IAEA (1980b) Waste management and disposal. Report of INFCE Working Group 7, IAEA, Vienna
18. ICRP (1977) Recommendations of the international commission on radiological protection. ICRP Publication 26, Pergamon Press, New York
19. ICRP (1985) Radiation protection principles for the disposal of solid radioactive waste. ICRP Publication 46, Annals of the ICRP 15(4). Pergamon Press, Oxford
20. ICRP (1991) 1990 Recommendations of the international commission on radiological protection. ICRP Publication 60, Annals of the ICRP 21(1–3)
21. ICRP (1993) Protection from potential exposure: a conceptual framework. ICRP Publication 64, Annals of the ICRP 23(1)
22. Johansson G, Haegg C (1990) The disposal of high level radioactive waste and the need for assessing the future radiological impact. In: Proceedings of a symposium on the safety assessment of radioactive waste repositories, Paris, 9–13 Oct 1989, IAEA, Vienna, pp 309–317
23. KBS (1977) Handling of spent nuclear; fuel and final storage of vitrified high level reprocessing wastes. Solna, Stockholm
24. Lanza F (1988) Review of processes near a buried waste canister. In: Feasibility of disposal of high-level radioactive waste into the seabed. OECD, Paris, 8
25. Marietta MG, Simmons WF (1988) Dispersal of radionuclides in the oceans: models, data sets and regional descriptions. In: Feasibility of disposal of high-level radioactive waste into the seabed. OECD, Paris, 5

26. Marivoet J, Bonne A (1988) PAGIS: Disposal in clay formation. Commission European Communities, DG XII, Report EUR 11776 EN
27. Merz E, Schifferstein K (1990) Time period of concern for longterm safety assessment of a geological radioactive waste repository. In: Proceedings of a symposium safety assessment of radioactive waste repositories, Paris, 9–13 Oct 1989, IAEA, Vienna pp 299–307
28. Mobbs SF, Charles D, Delow CE, McColl NP (1988) PAGIS: disposal in the sub-seabed. Commission European Communities, DG. XII, Report EUR 11779 EN
29. NEA (1980) Review of the continued suitability of the dumping site for radioactive waste in the North-East Atlantic. OECD, Paris
30. NEA (1983) Gurbutt PA, Dickson RR (eds). Interim oceanographic description of the North-East Atlantic site for the disposal of low-level radioactive waste. OECD, Paris, 1
31. NEA (1984) Long-term radiation protection objectives for radioactive waste disposal. NEA-OECD, Paris
32. NEA (1985) Review of the continued suitability of the dumping site for radioactive waste in the North-East Atlantic. NEA-OECD, Paris
33. NEA (1986) Dickson RR, Gurbutt PA, Kershaw PJ (eds). Interim oceanographic description of the North-East Atlantic site for the disposal of low-level radioactive waste. OECD, Paris, 2
34. NEA (1988) Overview of research and conclusions. In: Feasibility of disposal of high-level radioactive waste into the seabed. OECD, Paris, 1
35. NEA (1989) Nyffeler F, Simmons W (eds). Interim oceanographic description of the North-East Atlantic site for the disposal of low-level radioactive waste. OECD, Paris, 3
36. NEA (1990) CRESP activity report 1986–1990. NEA-OECD, Paris
37. NEA (1994) ARAP: The International Alligator Rivers Analogue Project – Background and results. NEA-OECD, Paris
38. NEA (1995) The environmental and ethical basis of geological disposal of long lived radioactive wastes. NEA/OECD, Paris
39. NEA (1996) CRESP final report, 1981–1995. OECD, Paris
40. Pentreath RJ, Hargrave BT, Roe HSJ, Sibuet M (1988) Deep-sea biology, biological processes and radiobiology. In: Feasibility of disposal of high-level radioactive waste into the seabed. OECD, Paris, 6
41. Shephard LE, Auffret GA, Buckley DE, Schüttenhelm RTE, Searle RC (1988) Geoscience characterization studies. In: Feasibility of disposal of high-level radioactive waste into the seabed. OECD, Paris, 3
42. UNSCEAR (1977) Sources and effects of ionizing radiations. Report to the General Assembly Annexes, New York
43. WCED (1987) Our common future. Oxford University Press, New York

Chapter 19
Reproductive Effects from Chronic, Multigenerational, Low Dose Rate Exposures to Radiation

Thomas G. Hinton, Daniel Coughlin, Yi Yi, Travis Glenn, and John Zimbrick

Abstract Relatively few experiments have been conducted on the effects to organisms following long-term exposures to low levels of radiation. Even fewer studies have examined the effects of radiological exposures to multiple generations of organisms. Speculations that damage will accumulate and be greater with each passing generation are plausible. Alternative, opposing views that adaptive response and repair mechanisms will counter the effects, such that damage does not increase with each generation, are equally plausible. Few data exist to support one hypothesis over the other, particularly for chronic, low-level exposures to vertebrate organisms. Our research explored exposures of low-dose irradiation to multiple generations of a model vertebrate organism, Japanese medaka (*Oryzias latipes*), one of the most widely used fish in comparative mutagenesis and carcinogenesis studies. A unique outdoor irradiation facility allowed us to examine effects to five generations of medaka that were continuously irradiated to different dose rates. The dose rates bracketed the IAEA guideline for acceptable chronic exposures to aquatic wildlife (10 mGy day^{-1}), and thus were a test to see if the guidelines were applicable for multigenerational exposures. The effects on reproductive

T.G. Hinton (✉)
Department of Radioecology, Environmental Modeling and Ecotoxicology, Institute of Radiation Protection and Nuclear Safety, Cadarache, Bât 159-BP3, Saint-Paul-Lez-Durance 13115, France
e-mail: thomas.hinton@irsn.fr

D. Coughlin • Y. Yi
Savannah River Ecology Laboratory, University of Georgia, Aiken, SC, USA

T. Glenn
Department of Environmental Health Science, University of Georgia, Athens, GA, USA

J. Zimbrick
Environmental and Radiological Health Sciences, Colorado State University, Fort Collins, CO, USA

C.E. Mothersill et al. (eds.), *Radiobiology and Environmental Security*, NATO Science for Peace and Security Series C: Environmental Security, DOI 10.1007/978-94-007-1939-2_19, © Springer Science+Business Media B.V. 2012

endpoints that might influence population dynamics were examined. Our intention was to test the hypothesis that multigenerational exposures to low dose rate irradiation are no more damaging, as measured by reproductive characteristics that could potentially impact a population, than damage incurred from exposure to a single generation. The data suggest that total accumulated dose may be more meaningful than dose rate when trying to predict effects to populations of chronically exposed organisms. Such knowledge is needed because chronic exposure to low levels of radiation is a more likely scenario for nuclear workers, and to wildlife exposed to routine releases from nuclear facilities.

Keywords Medaka • Radioecology • Dose-effect relationship • Chronic exposure • Low dose-rate • Reproduction

19.1 Introduction

Exposure regimes influence the biological effects caused from ionizing radiation. A dose received over a few seconds will have a different biological effect compared to the same dose received over an organism's life time. Current knowledge of how responses vary from different radiological exposure regimes is inadequate to allow confident predictions about the effects that might result from low-level, chronic exposures. Most effects data have been derived from acute, high dose rate experiments. Relatively few experiments have been conducted on the effects to vertebrates following long-term exposures to low levels of radiation. Even fewer studies have examined the effects of radiological exposures to multiple generations of organisms. Speculations that damage will accumulate and be greater with each passing generation are plausible. Alternative, opposing views that adaptive responses, repair and compensating mechanisms will counter the effects, such that damage does not increase with each generation, are equally plausible. Few data sets exist to support one hypothesis over the other, particularly for chronic, low level exposures to vertebrate organisms. Such knowledge is needed because chronic exposure to low levels of radiation is a more likely scenario for nuclear workers, and to wildlife exposed to routine releases from nuclear facilities. Regulation of the latter is becoming a necessity with recent changes in regulations [4] and as international committees place an emphasis on developing protection criteria specific to non-human biota [9].

Our research explored multigenerational exposures of low dose rate irradiation to a model vertebrate organism. A unique outdoor irradiation facility allowed us to examine effects to five generations of vertebrate animals that were continuous irradiated to different dose rates. The dose rates bracketed the IAEA guideline for acceptable chronic exposures to aquatic wildlife, and thus were a test to see if the guidelines were applicable for multigenerational exposures. The effects on reproductive endpoints that might influence population dynamics were examined. Our intention was to test the hypothesis that multigenerational exposures to low dose rate irradiation are no more damaging, as measured by reproductive characteristics that could potentially impact a population, than damage incurred from exposure to a single generation. Additionally, the vertebrate model that we chose, Japanese medaka (*Oryzias latipes*), are one of the most widely used fish in comparative mutagenesis and carcinogenesis studies. Shimada and Shima [17] showed that the mutational response of the medaka male germ cell was close to that of the mouse, and therefore medaka are a viable non-mammalian model system for risk assessments from environmental mutagens.

19.2 Methods

19.2.1 Low Dose Rate Irradiation Facility (LoDIF)

LoDIF is an outdoor, gamma-irradiation array consisting of 40 fiberglass, open-air tanks designed to house aquatic organisms such that all stages of their life cycle (from eggs to breeding adults) can be continuously irradiated [6]. This aspect of the LoDIF, as well as the ability to replicate the experimental design and irradiate large numbers of organisms, makes it a unique facility. Each tank, or mesocosm, is parabolic in shape, 2.4 m in diameter and holds approximately 965 L of water with a maximum depth of 41 cm (Fig. 19.1). The mesocosms are of a flow-through design and receive water from a nearby lake at rates that were adjusted to about 1 L$\,min^{-1}$.

Fig. 19.1 Photo of a single mesocosm, showing the dimensions of the tank and a ^{137}Cs irradiator with collimated shielding positioned above it. The LoDIF has 40 such mesocosms installed over a 0.4 ha area. A single container is shown directly underneath the irradiator. Such containers house 10–20 fish. Several containers can be placed in a circle around the central beam of the irradiator such that each is exposed to the same dose rate. Precise geometry measurements and mapping of the irradiation field with TLDs allowed accurate estimates of the dose received by the fish

Fig. 19.2 Photo of a portion of the LoDIF facility, showing a pad with five mesocosms, three of which have sealed Cs-137 sources above them (cylinders mounted on frames that cross the mesocosm). Eight replicate pads exist within the LoDIF, resulting in 40 mesocosms

The mesocosms allow manipulative experiments to be conducted on whole organisms in conditions more natural than laboratory settings, similar to what has been successfully used in ecology and population biology [13, 19]. Mesocosms allow the researcher to apply specific treatments in a more controlled environment than large-scale field tests. The LoDIF facility is arranged into eight 6.5 × 10.5 m blocks, with each block containing five mesocosms, suitable for randomized block experimental designs (Fig. 19.2).

Irradiation treatments can be applied via specially designed irradiators mounted on steel frames and placed over three of the five individual mesocosms within each of the eight blocks (Fig. 19.2). The two unirradiated mesocosms within each block were intended to serve as controls. (The controls tanks received a dose rate of 0.1 mGy day^{-1} due to scatter of radiation from the nearby irradiated mesocosms; discussed below). Each irradiator contains a 0.74, 7.4 or 74.0 MBq sealed ^{137}Cs source within a lead container, collimated to deliver an exposure to animals residing in the tank below. The 7.4 MBq source strength and associated geometry was designed to deliver a mean dose rate of approximately 10 mGy day^{-1}. The other two source strengths give mean dose rates that are factors of 10 less than, and greater than 10 mGy day^{-1}. Thus, the facility was designed specifically to determine if populations of aquatic organisms are truly protected by what is currently thought to be appropriate dose rates by the International Atomic Energy Agency (10 mGy day^{-1}).

Our experiments used a fish species as a vertebrate model. All life stages of the fish, from egg to reproductive adult, were continuously irradiated within the LoDIF under controlled conditions. Living conditions within the LoDIF were favorable for the species to grow and reproduce. Fish were maintained in mixed gender groups of ~20 individuals per container (×8 blocks ~160 adult fish per treatment), and placed within the gamma field to obtain specific dose rates.

19.2.2 Model Organism and Life Cycle Exposures

We selected a species of small fish, the Japanese medaka (*Oryzias latipes*), as our model vertebrate organism. The genetics, developmental biology, embryology, and specific developmental stages of medaka have been extensively characterized [23]. Medaka is the leading fish species used to identify and predict human health effects following toxicant exposure. Medaka are recognized as important comparative animal models for addressing questions related to a variety of processes and diseases, including cancer, DNA repair, mutations, infectious diseases, aging, developmental biology, endocrine disruption, and genetics. The use of medaka in biomedical and environmental research, especially as a carcinogenesis model related to identifying and predicting human health effects from toxicant exposure, has received considerable attention [16, 20, 21].

The species has numerous advantages for experimental laboratory and field studies including: small size (about 2.5 cm in length); a short maturation time of 8–10 weeks; and a prolific capacity to reproduce (6–30 eggs day^{-1}). The life history characteristics of medaka are such that several generations of fish can be exposed within a relatively short time period. We used a Kyoto-Cab Medaka wild-type strain, courtesy of Hiroshi Mitani (Tokyo University), with a low background of embryonic abnormalities [5].

Female medaka spawn daily and retain their egg clutch externally on their abdomen for several hours. Males fertilize the eggs externally while the clutch adheres to the female. Fertilization occurs at dawn. To obtain the embryos, clusters of eggs were removed from females that were chronically exposed to the various dose rate treatments within the LoDIF. The eggs were immediately brought into the laboratory for cleaning and evaluation. Egg clutches were gently rubbed to remove attached chorion filaments and to isolate individual eggs. Eggs were examined under a stereoscopic microscope to determine if fertilized. The number of viable eggs, and number of dead eggs per gravid female were recorded. Cleaned, viable eggs were placed in petri dishes with 1 ppm methylene blue water, returned to the LoDIF within hours after the formation of the clutch, and exposed to the same irradiation dose rate treatment as the parents (Fig. 19.3). The health status of eggs was monitored daily until hatching. Newly hatched fry remained in the LoDIF, also under the same radiation treatment as their parents. Hatching success per radiation treatment was recorded. Fry were contained in small buckets with fine screen mesh to allow in- and out-flow of water (Fig. 19.4). Subadults were moved to larger buckets. Densities of fish were kept similar among the irradiation treatments to normalize potential confounding variables of crowding, food resources, lighting, competition, etc. Subadults matured, reached breeding conditions, and were monitored accordingly, while continuously being irradiated.

Fig. 19.3 Special containers housed fish eggs and were exposed under known geometries with doses measured by TLDs. The photo depicts one of several egg exposure arrangements that were tested

Fig. 19.4 Several medaka fish are observable within a hatchling container used for irradiating small fry. The *grey circle at the top* of the inner bucket is a screened port for water to flow. Similar, *rectangular ports* are seen as *dark patches* on the *right* and *left* of the larger container. The *circular object at the bottom* of the inner container is a water temperature recording device

The process described above was repeated such that five generations of medaka were exposed to the radiation treatments; from eggs through breeding adults. Several containers of fish were placed within each mesocosm, each under a known and specific exposure geometry in which the dose rate was known from measured TLD readings. The numbers of fish within each treatment were sufficiently large that cohorts of fish could be removed for specialized analyses, while leaving sufficient stock for producing the next generation of fish. Samples of fish were routinely removed for genetic and molecular analyses (results to be reported elsewhere). The irradiators were shut down for a cumulative time of about 2 days per month for maintenance or special care of the fish. Feeding was done remotely and did not require turning off irradiators.

19.3 Experimental Design

These experiments started with the introduction of the founding generation, as subadult medaka, into the LoDIF on 17 February 2006. These fish were termed G0; subsequent generations were designated as G1 through G5. Fish were continuously exposed to four dose rate treatments (0.1; 2.4; 21; and 221 mGy/day). Treatments were replicated 8 times within the eight pads (Fig. 19.2), and ~20 fish of mixed gender were housed within each holding container (Fig. 19.1) for each replicate. Thus a total of ~640 fish were used for the G0 group (160 fish per dose rate treatment, with 8 blocks of 20 fish each, ×4 treatments). G0 fish were exposed for 94 days at which time their eggs were collected and the G1 generation exposure began. The experiment was completed on 17 September 2008, with the collection of G5 eggs from the fifth generation of chronically exposed adults (G4).

19.3.1 Dosimetry

Thermoluminescent dosimeters (TLDs) were placed within the exposure field to obtain mean dose rate estimates. The precision with which dose rates can be estimated within the facility reduces the uncertainties often associated with dose–response relationships. Dose rates varied according to the life history stage in which the fish were exposed (egg, fry, adult) due to the type of enclosures required to house the fish and their geometries relative to the ^{137}Cs sources.

The total dose received by each generation of fish was dependent on the dose rate of the treatment and the length of exposure period. The latter varied, particularly for fish that over-wintered within the LoDIF. Their exposure period was longer. Over-wintering was necessary to maintain the fish lines for the following spring. Natural temperature and light conditions were not conducive for the medaka to breed in the winter, thus we were required to wait until spring for the next generation of fish to be produced. The accumulated doses for each generation [G(n)] at the time eggs were collected for start of the next generation [G(n+1)] are given in Table 19.1.

Table 19.1 Mean dose received by each parental group [G(n)] at the time the next generation of eggs [G(n + 1)] were produced as a function of the four dose rate treatments (0.1; 2.4; 21 and 221 mGy/day). Total dose from one generation to the next were not the same due to different exposure durations (required over-wintering of fish before the next generation could be produced)

G(n) producing G(n + 1)	Date G(n) placed in LoDIF	Date G(n + 1) collected from LoDIF	Exposure days of G(n) in LoDIF	Dose rate treatments (mean mGy/day) Control 0.1	Low 2.4	Medium 21	High 221
				Resulting cumulative dose (mGy)			
Go–G1	17-Feb-06	25-May-06	94	**9**	**220**	**1,970**	**20,800**
G1–G2	25-May-06	8-Sep-06	102	**10**	**240**	**2,150**	**None**[a]
G2–G3	8-Sep-06	29-May-07	254	**25**	**590**	**5,340**	**None**[a]
G3–G4	29-May-07	30-Aug-07	90	**9**	**210**	**1,890**	**None**[a]
G4–G5	30-Aug-07	17-Sep-08	371	**37**	**850**	**7,800**	**None**[a]

[a]Effects at the highest dose rates were sufficient that subsequent generations could not be maintained

19.3.2 Reproductive Endpoints

Reproduction is regarded by the IAEA to be a more sensitive endpoint than mortality following exposure to radiation [8]. Changes in reproductive output are also thought to impact populations of biota. With the exception of endangered species, or humans, populations are the level of biological organization managed for health from contaminant exposures. Thus, we chose to examine several reproductive-related endpoints in the medaka exposed to chronic irradiation within the LoDIF.

A suite of data was collected in the LoDIF facility at the time eggs were harvested for propagation of the next generation. Procedures for processing the eggs were described under Sect. 19.2.2. Data collected included: (1) number of gravid females; (2) total number of eggs produced; (3) total number of live eggs produced; (4) total number of eggs that hatched; and (5) total number of hatchlings that survived for 15 days. Each of these reproductive parameters is related to the successful recruitment of young into the next generation of animals. The sensitivity of each parameter varies in response to contaminants and other stressors. Compensating mechanisms and adaptation can result in non-linear responses, and in what may seem like opposing responses to stress. A good example is the work of Blaylock [3], where fish exposed to radiation suffered a reduction in the number of fertile eggs; but an increase in the number of eggs produced, such that the net effect in total fertile eggs produced did not differ from controls. Ultimately, it is the integration of all these reproductive parameters that determines the size of the next population cohort.

Thus, herein, we elected to combine the above parameters into an *Index of Reproductive Success* (IRS), by multiplying the number of eggs produced per female, times the percent fertility of the eggs, times the percent of fertile eggs that were still alive as fry 15-days post hatching. 15 days is a critical threshold for medaka. Fry that survey that period have a reasonable probability of reaching adulthood. We used the following parameter to compare treatments and to test our hypothesis:

$$\text{IRS} = \text{Eggs} \times \text{Fertility} \times \text{Survival}$$

Where: IRS = Index of Reproductive Success; Eggs = mean # of eggs per female; Fertility = mean percent fertility; and Survival = mean percent survival of fry (15 days post-hatching). Uncertainties, expressed as standard deviations (Stdv), were propagated as:

$$Stdv\ IRS = \sqrt{\left(\frac{Stdv\ Eggs}{Eggs}\right)^2 + \left(\frac{Stdv\ Fertility}{Fertility}\right)^2 + \left(\frac{Stdv\ Survival}{Survival}\right)^2} \times mean\ IRS$$

Effects to reproductive success, as measured by the reproductive index described above, were compared among the various LoDIF dose rate treatments (Table 19.1); and also compared among the various generations of exposed fish.

19.4 Results and Discussion

After 94 days of exposure to the various dose rate treatments, the reproductive abilities of the founding G0 fish were examined. The mean numbers of eggs produced were not affected by the irradiation treatments. However, significant reductions occurred in the number of live eggs and in the number of hatchlings produced from fish exposed to the highest dose rate. These reductions were so strong that we were unable to populate the high dose rate group for an additional generation. Their mean Index of Reproductive Success was only 0.05 ± 0.1. Thus, the remaining multigenerational experiment within the LoDIF involved fish from the three lower dose rate treatments (0.1; 2.4 and 21 mGy day^{-1}).

19.4.1 Low Dose-Rate Field Data

Figure 19.5 shows the mean IRS for five generations of fish continuously exposed to three dose rate treatments within the LoDIF. No differences appear to exist in IRS among the three dose rate treatments for generations G0, G1 or G3. A reduced IRS appears to have occurred at the highest dose rate of 21 mGy/day for generations G2 and G4.

The overall exposure times in days are also shown for each generation in Fig. 19.5. Generations G0, G1 and G3 were exposed for similar lengths of times (~90–100 days); whereas G2 and G4 were exposed considerable longer (G2=254 days and G4=371 days). The latter two are the generations in which the IRS decreased for the fish exposed to the highest dose rates. The longer exposure periods resulted in fish in G2 and G4 acquiring greater total doses (Table 19.1 presents total dose per generation). The fact that the 21 mGy day^{-1} treatment appears to have been impacted during the generation (G2) were total doses were twice as great; "recovered" the following generation (G3) when total doses were halved; and then suffered damage again in G4 when total dose were three time greater suggests that reproductive effects, as measured by the IRS, are likely driven by total dose, and less by a multigenerational mechanism or dose rate.

Note that even at the lowest dose rate of 0.1 mGy day^{-1} there is considerable differences in IRS between generations (Fig. 19.6). Differences in temperatures, food resources, and seasons are factors known to influence egg production [10]. Such factors definitely occurred within the outdoor conditions of the LoDIF, and contributed, to some unknown extent, in fish from the same dose rate treatments having differences in IRS from one generation to the next.

The lowest dose rate of 0.1 mGy day^{-1} occurred in the "controls" due to scatter radiation from the nearby irradiated mesocosms. 0.1 mGy day^{-1} is two orders of magnitude lower than the IAEA guideline considered to be safe for populations of aquatic organisms. It is also less than the 0.2 mGy day^{-1} benchmark value proposed as protective of the environment, based on species sensitivity distributions [1]. If we

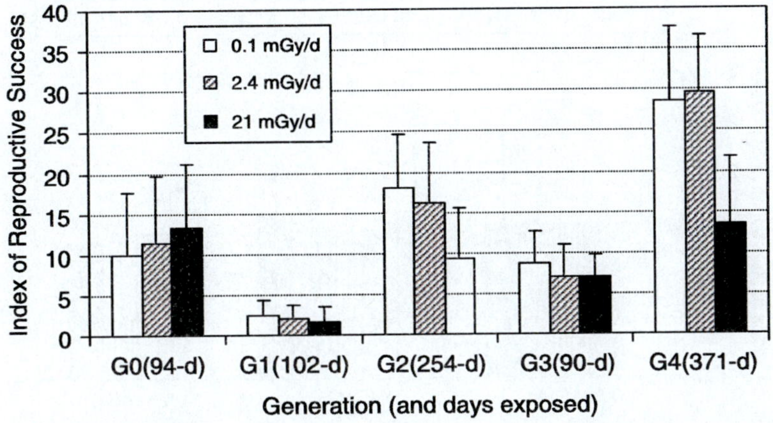

Fig. 19.5 LoDIF field data. Five generations of medaka, continuously exposed to three dose rate treatments. Data show an Index of reproductive success (IRS) for each generation. The exposure days for each generation are also shown

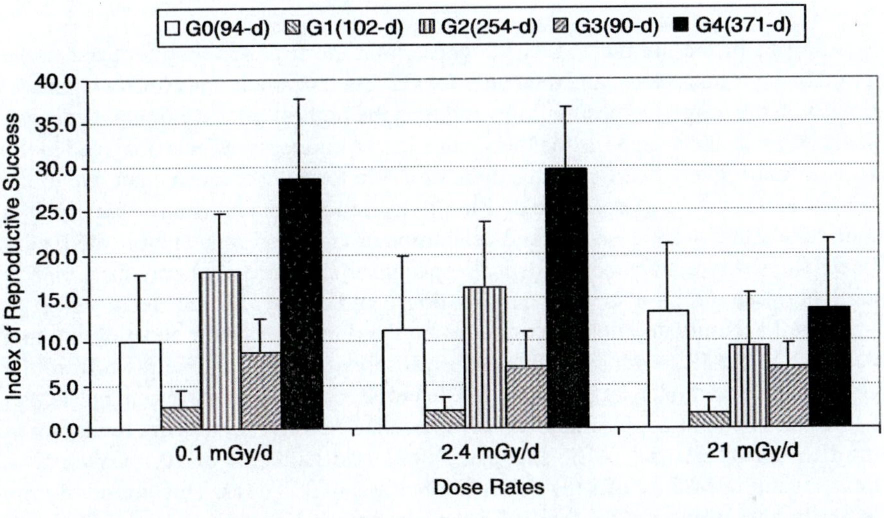

Fig. 19.6 LoDIF field data. Index of reproductive success for five generations of medaka exposed to three dose rate treatments

consider the 0.1 mGy day^{-1} to represent control conditions, then the data can be normalized to the same IRS across generations for the 0.1 mGy day^{-1} treatments, and we can thus attempt to remove the variation caused from temperature, light regimes, and food resources. The direction and amount of normalization required for each generation within the 0.1 mGy day^{-1} treatment was then applied to the other two dose rate treatments, resulting in the normalized data graphed in Fig. 19.7.

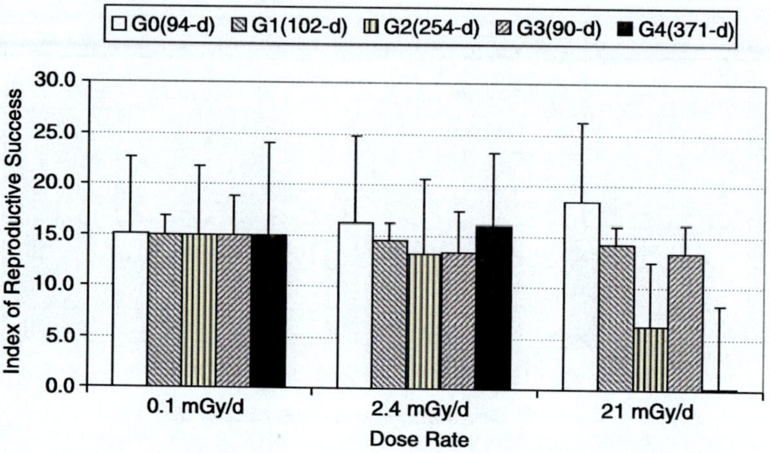

Fig. 19.7 LoDIF field data normalized to remove seasonal outdoor temperature, lighting and food resource variation. All data were normalized such that the IRS for the lowest dose rate treatment for each generation were equal. The normalized mean IRS value for G4 at 21 mGy d^{-1} was 0.0 ± 8.1

Analysis of the normalized data suggests that the IRS was only reduced in the 21 mGy day^{-1} treatments, and then only for G2 and G4 populations (the two generations with the longest exposure times and thus the highest total accumulated doses). Thus normalization supports the theory that the reproductive effects observed in the IRS are more likely due to the accumulation of a total dose, rather than due to the confounding factors of dose rate and multigenerational phenomena. Actually, it is more likely that dose, dose-rate and generation are interacting in complicated, non-linear ways. More advanced statistical procedures are currently being used on this data set to see if their respective contributions can be better illuminated.

If total accumulated dose turns out to be the dominant factor contributing to a decrease in the IRS endpoint, then a dose response curve can be drawn from the entire normalized data set (Fig. 19.8). The curve, on log-linear axes, suggests that a threshold dose of approximately 3 Gy is needed before a significant reduction in the IRS occurs. At the IAEA suggested dose rate guidance of 10 mGy day^{-1}, a total accumulated dose of 3 Gy could be obtained in 300 days. This period of time is easily achievable within the 2–5 year life time typical of medaka living outdoors [15].

Overall, these data represent a valuable contribution to the science of radiation-effects studies. They suggest that total accumulated dose may be more meaningful than dose rate when trying to predict effects to populations of chronically exposed organisms. The data are rare in that few other studies have examined multiple generations of a vertebrate animal under different chronic dose rate treatments (Table 19.2). The data from several other LoDIF experiments (including some acute exposures to medaka at the same doses) are being analyzed. We are actively conducting more rigorous statistical analyses of the data presented herein.

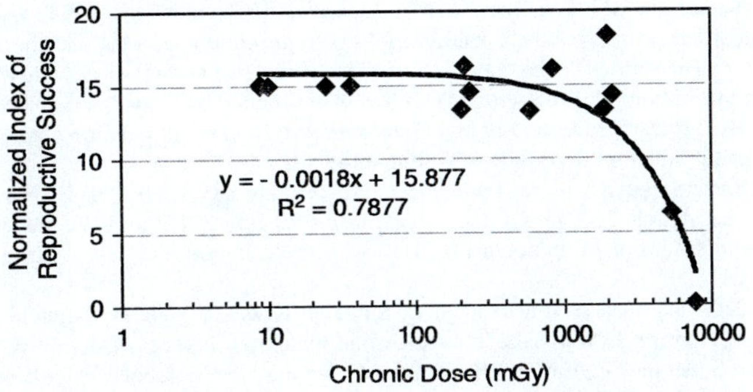

Fig. 19.8 Dose response curve for the normalized LoDIF data when all treatments and generations are combined

Table 19.2 Other reproductive effects documented in fish following chronic exposures to radiation

Fish species	mGy/day	Reproductive effects	Reference
Gambusia affinis	4	Increased frequency of dead embryos, offset by larger brood sizes	[3]
Rutilus rutilus	>5	Reduced fecundity	[11]
Hypophthalmichthys molitrix	10	6% sterility when living in Chernobyl cooling reservoir	[2]; as cited by [14]
Tilapia mossambica	30–40	Total sterility from chronic ^{90}Sr exposure	[18]; as cited by [14]
Poecilia reticulata	40	57% decrease in fecundity, chronically exposed	[22]
Oryzias latipes	65	Sterility and unfertilized eggs increased	[7]
Ameca splendens	185	Complete sterility when exposed for 190 days (~35 Gy)	[12]

Acknowledgments Funding was provided by the Environmental Remediation Sciences Division of the Office of Biological and Environmental Research, U.S. Department of Energy, through Financial Assistance Award DE-FC09-96SR18546 to the University of Georgia Research Foundation. All experimental protocols were approved by the University of Georgia's Animal Use and Ethics Committee (# A2005-10169). The animal husbandry assistance of Laura Marsh was most appreciated.

References

1. Andersson P, Garnier-Laplace J, Beresford NA, Brown JE, Copplestone D, Hingston JL, Howard BJ, Howe P, Oughton D, Whitehouse P (2009) Protection of the environment from ionising radiation in a regulatory context (PROTECT): proposed numerical benchmark values. J Environ Radioact 100:1100–1108

2. Belova NV, Verigin BV, Yemeljanova NG, Makeeva AP, Ryabov IN (1993) Radiobiological analysis of silver carp Hypohthalmichthys molitrix in the cooling pond of the Chernobyl NPP in the post-accident period. I. The condition of the reproductive system of dish survived the accident. Boprosy ithtiologii (Probl Icthyol) 33:814–838 (as cited by Sazykina and Kryshev, 2003)
3. Blaylock BG (1969) The fecundity of a *Gambusia affinis affinis* population exposed to chronic environmental radiation. Radiat Res 37:108–114
4. Euratom Basic Safety Standards Directive, issued 24 February 2010. Title X, Protection of the Environment, Article 102, pp 48. (accessed on-line 20 July 2011) http://ec.europa.eu/energy/nuclcar/radiation_protection/doc/art31/2010_02_24_draft_euratom_basic_safety_standards_directive.pdf
5. Furutani-Seiki M, Sasado T, Morinaga C, Suwa H, Niwa K, Yoda H, Deguchi T, Hirose Y, Yasuoka A, Henrich T, Watanabe T, Iwanami N, Kitagawa D, Saito K, Asaka S, Osakada M, Kunimatsu S, Momoi A, Elmasri H, Winkler C, Ramialison M, Loosli F, Quiring R, Carl M, Grabher C, Winkler S, Del Bene F, Shinomiya A, Kota Y, Yamanaka T, Okamoto Y, Takahashi K, Todo T, Abe K, Takahama Y, Tanaka M, Mitani H, Katada T, Nishina H, Nakajima N, Wittbrodt J, Kondoh H (2004) A systematic genome-wide screen for mutations affecting organogenesis in Medaka, *Oryzias latipes*. Mech Dev 121:647–658
6. Hinton TG, Coughlin DP, Yi Y, Marsh LC (2004) Low dose rate irradiation facility: initial study on chronic exposures to medaka. J Environ Radioact 74:43–55
7. Hyodo-Taguchi Y (1980) Effects of chronic gamma-irradiatin on spermatogenesis in the fish, (Oryzias latipes) with special reference to regeneration of testicular stem cells. In: Egami N (ed) Radiation effects on aquatic organisms. University Park Press, Baltimore, pp 91–104
8. IAEA (1992) Effect of ionizing radiation on plants and animals at levels implied by current radiation protection standards. IAEA Tech. Report Series #332, Vienna, Austria
9. ICRP (2009) Environmental protection: the concept and use of reference animals and plants. Publication 108. Elsevier. ISBN-13: 978-0-444-52934-3
10. Koger CS, The SJ, Hinton DE (1999) Variations of light and temperature regimes and resulting effects on reproduction. Biol Reprod 61:1287–1293
11. Peshkov SP, Shekhanova IA, Romanov GN, Prister BS, Sheyin GP (1978) Radioecological problems of atomic power plant cooling reservoirs. Urals Scientific Center, USSR Academy of Sciences, Sverdlovsk (as cited in IAEA, 1992)
12. Rackham BD, Woodhead DS (1984) Effects of chronic gamma-radiation on the gonads of adult Ameca splendens (Osteichthyes: Teleostei). Int J Radiat Biol 45:645–656
13. Rowe CL, Dunson WA (1994) The value of simulated pond communites in mesocosms for studies of amphibian ecology and ecotoxicology. J Herpetol 28:346–356
14. Sazykina TG, Kryshev AI (2003) EPIC database on the effects of chronic radiation in fish: Russian data. J Environ Radioact 68:65–87
15. Shima A, Mitani H (2004) Medaka as a researach organism: past, present and future. Mech Dev 121:599–604
16. Shima A, Shimada A (2001) The medaka as a model for studying germ-cell mutagenesis and genomic instability. Mar Biotechnol 3:S162–S167
17. Shimada A, Shima A (1998) Combination of genomic DNA fingerprinting into the medaka specific-locus test system for studying environmental germ-line mutagenesis. Mutat Res 399:149–165
18. Voronina EA (1973) Growth and fertility of Tilapia fish under chronic exposure to radiostrontium. Abstract of the dissertation for an academic degree of candidate of biological sciences. VNIRO, Moscow, pp 1–27 (as cited by Sayzkina and Kryshev, 2003)
19. Wilbur HM (1987) Regulation and structure in complex systems: experimental temporary pond communities. Ecology 68:1437–1452
20. Winn RN, Norris MB, Brayer KJ, Torres C, Muller SL (2000) Detection of mutations in transgenic fish carrying a bacteriophage lambda cII transgene target. Proc Natl Acad Sci USA 97:12655–12660
21. Wittbrodt J, Shima A, Schartl M (2002) Medaka—a model organism from the far East. Nat Rev Genet 3:53–64
22. Woodhead DS (1977) The effects of chronic irradiation on the breeding performance of the guppy, (*Poecilia reticulata*) (Osteichthyes: Teleostei). Int J Radiat Biol 32:1–22
23. Yamamoto TO (1975) Medaka (killifish): biology and strains. Keigaku Publishing Company, Tokyo

Chapter 20
Aquatic Biota within the Chernobyl Accident Exclusion Zone: Consequences of the Long-Term Radiation Exposure

Dmitri Gudkov, Elena Dzyubenko, Natalia Shevtsova, Natalia Pomortseva, Sergey Kireev, and Alexander Nazarov

Abstract The effects of chronic radiation exposure on aquatic biota within the Chernobyl accident exclusion zone during 1998–2009 were studied. The absorbed dose rate for hydrobionts registered in range from 1.3 mGy/year to 3.4 Gy/year. The heightened chromosomal aberration rate in the embryo tissue of snails (up to 27%) and in the root meristems of higher aquatic plants (up to 18%) was determined. In hemolymph of snails from contaminated lakes the quantity of death cells averages 36–44%; the part of phagocytic cells averages 41–45%, as well as decrease of the young amoebocytes quantity to 10–20%. In peripheral blood of fish the high rate of abnormal red cells (up to 29‰) as different type of invaginations, ramifications, micronuclei etc. was marked. The high level of parasitic fungi and gall-producing arthropods of the common reed in the most contaminated lakes within the exclusion zone was registered. Above mentioned phenomenon may testify upon the decreasing of the parasitical stability of plant under impact of long-term radiation exposure.

Keywords Chernobyl NPP accident • Radioactive contamination • Aquatic ecosystems • Hydrobionts • Absorbed dose rate • Chromosomal aberrations • Haematological indices • Parasitical instability

---

D. Gudkov (✉) • E. Dzyubenko • N. Shevtsova • N. Pomortseva
Freshwater Radioecology Department, Institute of Hydrobiology,
Geroyev Stalingrada Ave. 12, Kiev 04210, Ukraine
e-mail: digudkov@svitonline.com

S. Kireev • A. Nazarov
Chernobyl Radioecological Centre, Shkolnaya Str. 6, Chernobyl 07720, Ukraine

C.E. Mothersill et al. (eds.), *Radiobiology and Environmental Security*,
NATO Science for Peace and Security Series C: Environmental Security,
DOI 10.1007/978-94-007-1939-2_20, © Springer Science+Business Media B.V. 2012

20.1 Introduction

In spite of 25 years, past after the Chernobyl NPP accident in 1986, self-purification of closed water bodies in the Chernobyl exclusion zone, defined as a roughly circular area of 30-km radius around the destroyed unit, is extremely slow process. Therefore, ecosystems of the majority of lakes, dead channels and crawls possess high level of radionuclide contamination of all the components.

The basic problems of radiation safety of the Chernobyl exclusion zone concern radionuclide wash-off with surface drainage water to river network, their export outside the exclusion zone and affection of the water quality in the Dnieper River. Undoubtedly one of the most important and still insufficient studied problems of aquatic ecosystems within the exclusion zone is research of chronic impact of ionizing radiation on non-human biota.

Thus the investigation of biological effects of long-term exposure on hydrobionts, inhabiting the radioactive contaminated water bodies, has a theoretical and applied importance for ecological hazard understanding, associated with changing of environmental radioactivity state due to human activity.

Our main tasks were: (1) dose rate estimation due to external and internal sources of irradiation for different groups and species of hydrobionts; (2) evaluation of cytogenetical, hematological and parasitological effects dynamics due to long-term radioactive impact on hydrobionts in conditions of water bodies within the Chernobyl exclusion zone.

20.2 Materials and Methods

Our researches were carried out during 1998–2009. The water bodies of research were the flood plain reservoirs of the Pripyat River: Azbuchin Lake, Yanovsky Crawl, Dalekoye-1 Lake, Glubokoye Lake as well as Uzh River and Pripyat River within the Chernobyl accident exclusion zone. The results of the cytogenetical and haematological analyses compared to the data received for hydrobionts from the control lakes, located in the neighbourhood of the Kiev City: Vyrlitsa, Opechen', Pidbirna, Goloseevo as well as Kiev Storage Reservoir.

External gamma irradiation dose rate was measured by DKS01 dosimeter and by Na-I field radiometer SRP-68-03. The estimation of the internal dose rate was carried with use of dose conversion coefficients [6]. The ^{137}Cs content was measured by γ-spectrometry complex: PGT IGC-25 detector (France), "Nokia LP 4900 B" analyser ("Nokia", Finland), low-volt feeding source – crate NIM BIN, amplifier NU-8210 ("Elektronicus Merokeszulekek Gyara", Hungary) and 100 mm thickness leaden protection. The ^{90}Sr content was measured on low-background NRR-610 β-radiometer ("Tesla", Czech). Minimal detectable activity was 0.04 Bq under 1,000 s sample exposition. ^{238}Pu and $^{239+240}$Pu content in electrolytic samples was determined by α-spectrometric tract by NUC-8192 impulse analyser ("Elektronicus Merokeszulekek

Gyara", Hungary). The ^{241}Am content was measured by x-ray-spectrometric line including x-ray detector EG&G Ortec LOAX-51370/20 CFG-SU-GMX ("EG&G Ortec", USA) and analyser "Nokia LP 4900 B". The results were measured in Bq/kg at natural humidity and the mistake of estimated radionuclide concentration fell within 15–20%.

The chromosomal aberration rate was measured in embryo cells of gastropod pond snails (*Lymnaea stagnalis*) by the standard anaphase method [10] and in the apical root meristems of the four species of higher aquatic plants: common reed (*Phragmites australis*), arrowhead (*Sagittaria sagittifolia*), flowering rush (*Butomus umbellatus*), fresh-water soldier (*Stratiotes aloides*), narrow-leaved cat's-tail (*Typha angustifolia*), broad-leaved cat's-tail (*Typha latifolia*), branched bur-reed (*Sparganium erectum*), rush flower (*Butomus umbellatus*) and manna (*Glyceria maxima*) by the modified method [12]. Haematological studies were carried out in mantle liquid of gastropod snails (*Lymnaea stagnalis*) by the analysis of dead cells, young amoebocytes and phagocytic cells quantity [2, 9]. The rate of abnormal red cells as different type of invaginations, ramifications, micronuclei, amitosis etc. was analysed in blood of pike (*Esox lucius*), perch (*Perca fluviatilis*), roach (*Rutilus rutilus*), redeye (*Scardinius erythropthalmus*) and crucian carp (*Carassius carassius*) from the most contaminated lakes of the exclusion zone.

20.3 Results and Discussion

20.3.1 Chromosomal Aberration Rates

The freshwater snails have great importance for the processes of radionuclide biogenic migration in aquatic ecosystems. Due to ability to accumulate practically all of radionuclides which registries in water these invertebrates can be considered on the one hand as bio-indicators of radioactive contamination of aquatic ecosystems and from another hand - as organisms which can take active part in the processes of radionuclide redistribution in hydrobiocenosa. In this part of studies was evaluated the cytogenetical effects level in embryo tissue of gastropod snail as chromosomal aberration rate, considering it as reaction of snails on radiotoxicological condition of environment.

The absorbed dose rate for snails from lakes within the Chernobyl exclusion zone was registered in the range from 0.8 to 3.4 Gy year^{-1}. The highest rate was found in snails from lakes Dalekoye-1 and Glubokoye located within the dammed territory on the left-bank flood lands of the Pripyat River, the lowest - from the Pripyat River and Uzh River. Molluscs from the control lakes were characterised by absorbed dose rate about 0.3 mGy year^{-1}. The rate of chromosomal aberration was found in snails from lakes Dalekoye-1 and Glubokoye – 21–22%. In the embryo tissue of snails from Yanovsky Crawl the chromosomal aberration rate was about 18% and in Azbuchin Lake – about 23%. About 3.3% of aberrant cells were registered in snail's embryo

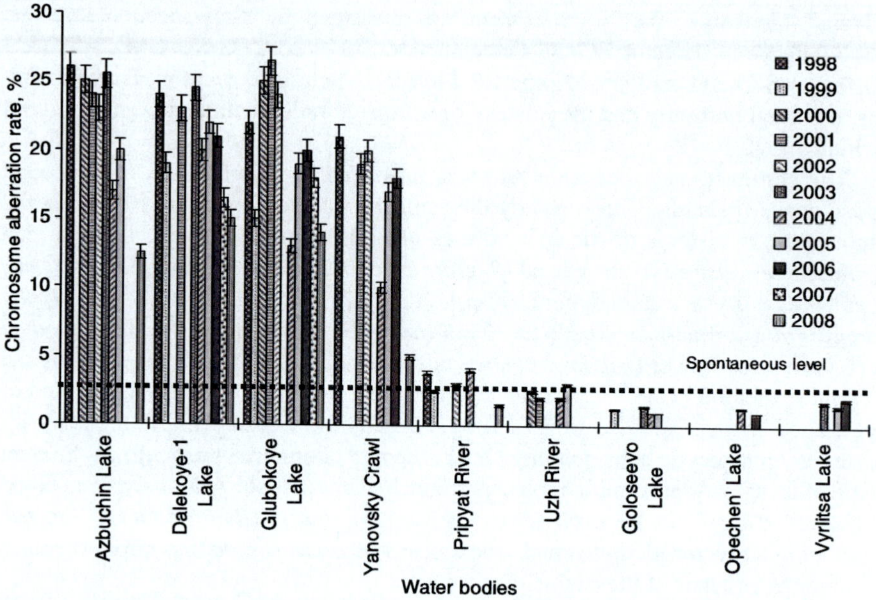

Fig. 20.1 Chromosomal aberration rate in pond snail's embryos in water bodies within the Chernobyl exclusion zone and control lakes within the Kiev City area

from the Pripyat River and in snails of Uzh River the aberration rate was about 2.3%. The rate of chromosomal aberration for snails from the control lakes was about 1.1–2.0% (Fig. 20.1).

During 1998–2008 a tendency to decrease of chromosomal aberration level in molluscs from all lakes of the exclusion zone was registered. The probabilistic prediction of the chromosomal aberration rate for gastropod snails in lakes of the Chernobyl exclusion zone have shown that spontaneous mutagenesis level (2.0–2.5%) [13] can be reach in Azbuchin Lake and Yanovsky Crawl in 2020s–2030s and in Dalekoye-1 Lake and Glubokoye Lake – in 2060s–2070s.

The studies of the different species of plants within the exclusion zone have revealed a numerous morphological anomalies as repeated organs, gigantism or dwarf, underdevelopment or sterility of reproductive organs, excessive branching, growth inhibition of the secondary points of growth etc. All of this variety of plant anomalies of development is testify that the vegetation within the exclusion zone has undergone to strong damage of the genotype, which consequence is the long genetic instability and thus increased variability of many species. The spontaneous rate of aberrant cells for higher aquatic plants from the control lakes does not increase 2.0–2.5%. We found 7.5% aberrant cells in common reed of Yanovsky Crawl. The highest rate of chromosomal aberration was registered in plants from Azbuchin, Glubokoye and Dalekoye-1 lakes - 9.0%, 10.8% and 17.8% respectively. In comparison, the data received for the common reed from Goloseevo Lake amount

Table 20.1 Average rate of aberrant cells in root meristem of higher aquatic plants of water bodies within the Chernobyl exclusion zone and Kiev Storage Reservoir in 2008

Water body	Species	Quantity of analyzed roots	Quantity of analyzed ana-, telophases	Rate of aberrant cells, %
Glubokoye Lake	*Stratiotes aloides*	8	1,098	11.8±1.4
	Sagittaria sagittifolia	15	2,684	11.8±1.1
	Phragmites australis	26	8,048	7.6±0.6
	Glyceria maxima	7	897	7.6±1.6
	Sparganium erectum	6	1,244	7.5±1.2
	Typha angustifolia	9	1,026	6.0±1.1
Dalekoye Lake	*Sparganium erectum*	11	1,654	7.1±1.4
	Glyceria maxima	11	1,209	7.0±1.0
	Butomus umbellatus	10	1,288	6.5±0.9
	Phragmites australis	30	9,260	5.7±0.6
	Typha angustifolia	8	1,152	5.2±0.9
Azbuchin Lake	*Stratiotes aloides*	9	1,065	10.3±1.6
	Glyceria maxima	10	2,218	6.1±0.9
	Butomus umbellatus	11	1,128	5.9±0.4
	Phragmites australis	29	7,620	5.7±0.6
	Typha angustifolia	12	1,522	4.4±0.7
Yanovsky Crawl	*Stratiotes aloides*	6	698	6.7±0.9
	Sagittaria sagittifolia	14	1,365	6.1±0.6
	Glyceria maxima	9	976	5.2±0.2
	Phragmites australis	24	8,080	4.0±0.3
	Butomus umbellatus	11	954	3.2±0.4
	Typha angustifolia	6	868	3.6±0.9
Krasnensky Dead Channel	*Glyceria maxima*	5	1,452	6.1±1.5
	Phragmites australis	9	2,644	4.2±0.5
	Typha latifolia	8	964	4.2±1.0
Kiev Storage Reservoir (vlg Strakholesye)	*Glyceria maxima*	10	2,384	2.1±0.6
	Phragmites australis	10	2,314	1.6±0.1
	Typha angustifolia	9	1,008	1.6±1.0

to 1.0%. The rate of chromosomal aberration in the common reed from closed water bodies within the left-banked flood plain of the Pripyat River was in 5–8 times higher than spontaneous mutagenesis level.

During 2008 a sufficiently comprehensive cytogenetical data concerning rate of aberrant cells in root meristems of six species of higher aquatic plans from different water bodies within the exclusion zone, as well as from the part of Kiev Storage reservoir (near vlg Strakholesye) adjacent to the exclusion zone as a control (Table 20.1).

The positive correlation between absorbed dose and rate of aberrant cells of apical root meristems of the common reed and embryos of the gastropod pond snail was found. According to our data this dependence more corresponds to a power-mode function (Fig. 20.2).

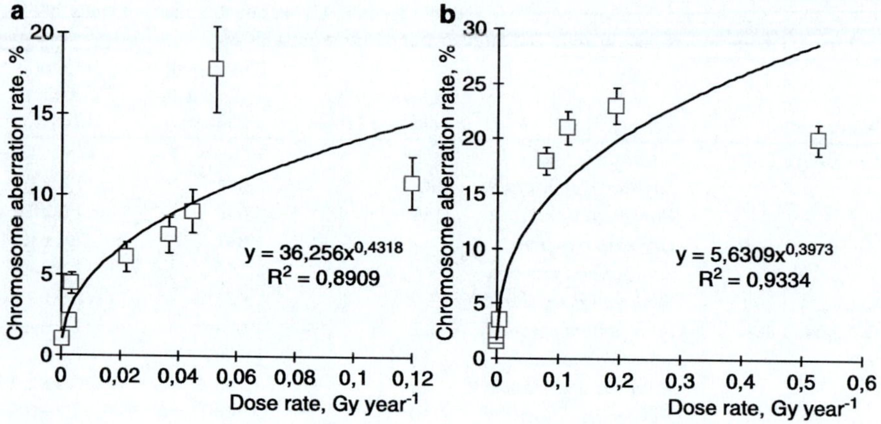

Fig. 20.2 The dependence of chromosomal aberration rate from absorbed dose rate in cells of the common reed (**a**) and embryos of the gastopod pond snails (**b**) from water bodies within the Chernobyl exclusion zone

20.3.2 *Haemotological Effects*

Hematopoietic and immune system of animals are the most sensitive ones to the impact of ionizing radiation. Molluscs is one of the first group of animal that have got cell immunity during the evolution. The protective function has the cells of mantle liquid (hemolymph), which peculiar to high proliferation, heterogeneity and sensitivity to radiation impact. The most functional and active elements of hemolymph of the snails are the granular amoebocytes, which have morphological and quantitative reaction on the physiological changes of organism. The decrease of adsorbed dose rate coses the decrease of young amoebocytes quantity in hemoliph. The decrease of total agranulocyte quantity in cell population occurs due to decrease of amoebocytes quantity and due to increase of granulocyte quantity, which involves in phagocytic reaction.

In hemolymph of snails from Dalekoye-1 Lake, Azbuchin Lake and Glubokoye Lake the quantity of dead cells averages 36.2%, 39.2% and 43.8% respectively, the part of phagocytic cells averages 44.3%, 41.2% and 45.0%, as well as decrease of the young amoebocytes quantity to 13.2%, 20.1% and 9.5% respectively (Fig. 20.3).

The insignificant quantity of abnormal cells and micronuclei has been observed here as well. In conditionally "clean" water bodies the part of dead cells averages from 2.2% to 5.3% and the quantity of phagocytic was at level 3.0–4.2%. The quantity of young amoebocytes has increased here to 79.7–89.6%.

Fishes is object of special interest at the study of chronic radiation impact on aquatic species as occupying in aquatic ecosystems top trophic levels and characterized by comparative high radio-sensitivity. However, radiobiological researches of fishes in the reservoirs of the Chernobyl exclusion zone were limited mainly by the analysis of morphometric indexes, including fluctuating asymmetry of pair organs,

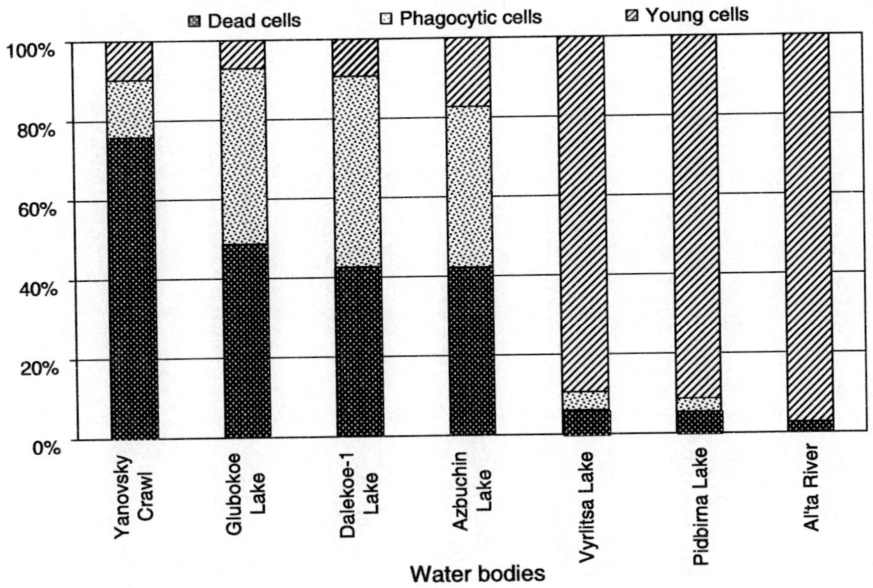

Fig. 20.3 Ratio of different types of amoebocytes in the mantle liquid of snails in water bodies within the Chernobyl exclusion zone and control lakes within the Kiev City area

and also estimation of the state of the reproductive system of representatives of ichthyofauna mainly in the Chernobyl NPP cooling pond. Unfortunately, hematopoietic and immune system remained out of attention of researchers both for fishes of the Chernobyl NPP cooling pond and other reservoirs of the exclusion zone, characterized by the extremely low rates of natural purification and by increased chronic radiation dose rates on hydrobionts.

The comparative leukogram of peripheral blood analysis of fishes from the lakes of the Chernobyl exclusion zone shows the presence of increased part of lymphocytes, presence of pseudobasophiles and foamy cells, as well as low maintenance of mature neutrophils, due to young granulocytes (mainly myelocytes and promyelocytes), that can testify to probable adaptation of white blood to the long-term radiation exposure in low doses. Correlation of red blood cells to leucocytes and thrombocytes of fishes from the exclusion zone is higher, than in the control Kiev Storage Reservoir.

Drawing attention the high rate of red cells aberrations and abnormalities in peripheral blood of fishes from the lakes within the exclusion zone, where dose rate on the fish organisms due to incorporated radionuclides on three orders higher in comparison with background water bodies – 20 and 0.01 $\mu Gy\ h^{-1}$ respectively. It can testify to certain mutagenous of environment and possible display of radiation-induced genetic instability of fishes in the conditions of chronic radiation impact. Micronuclei and double-nucleus red cells in the blood of control fishes of the Kiev Storage Reservoir are not discovered. Making a point of the amount of red cells with

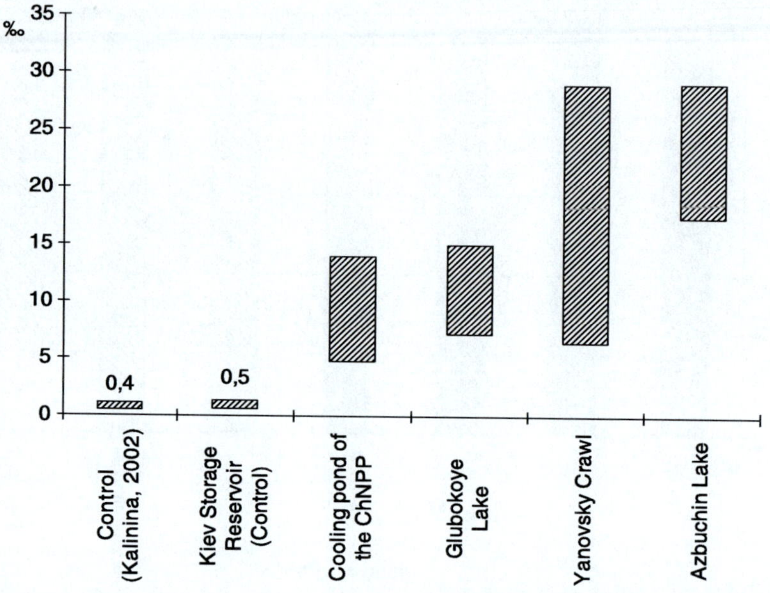

Fig. 20.4 Part of red cell abnormalities in peripheral blood of fishes from the Chernobyl exclusion zone

atypical shape of nucleus as different type of invaginations, ramifications etc. as well as cells without nucleus in the blood of fishes from the lakes of the exclusion zone - from 10 to 60 times exceeding a similar index for fishes from the control reservoir and reference (Fig. 20.4).

The atypical shape of red cell nucleus in blood of healthy fishes, from data of row of authors, meets with the frequency 0.4‰. The increase of frequency of red cells with the deformed nucleus (invagination of nuclear envelope) is estimated by different authors as degenerative changes of red cells, arising up as a result of negative action of environmental factors on the organism of fishes [7, 8].

The analysis of the conducted researches allows to assume that the qualitative indexes of red cells of peripheral blood of fishes are more sensible to chronic radiation influence in comparison with the elements of white blood and can be used for conducting of the haematological monitoring of radioactive contaminated water bodies.

20.3.3 Arthropod-Borne and Parasitical Fungi Diseases

In reservoirs with the increased level of radioactive contamination (lakes of the dammed territory of the left-bank flood plain of the Pripyat River) the high level of the common reed damage by gall-producing arthropods, in particular by mites *Steneotarsonemus phragmitidis* since 2000 is observed. This phenomenon for the first time was registered at the territory of Ukraine within the Chernobyl exclusion zone [4]. We suppose the scales and speed of distribution of this phenomenon in reservoirs of the exclusion zone deserve the special attention. As the common reed is almost cosmopolitan it is quite logical to predict wide moving of mites in other reservoirs, which the Polesye region (woodlands) is so rich. Thus if in 2000 and 2001 the damaged individual registered only in one of seven reservoirs of sampling in the exclusion zone – Dalekoye-1 Lake, during 2002–2009 the damaged individuals of the common reed began to meet in all other researched water objects. At that if in some reservoirs we fond a single affected individuals only, in Yanovsky Crawl and Azbuchin Lake the described phenomenon quickly has received distribution and in 2005 the share of damaged plants was accordingly 74% and 32%.

The highest percent of affected plants observed in Dalekoye-1 Lake (practically 100%), located on the dammed territory of Krasnensky flood-lands. This territory is characterized by the maximal density of radioactive contamination in the exclusion zone. The specific activity of radionuclides in the reed's tissue (at natural humidity) during the researched period reached to 10,000 Bq/kg for ^{137}Cs and 2,000 Bq/kg for ^{90}Sr [5]. At that the absorbed dose, caused by external gamma-radiation and radionuclides incorporated in meristem tissue of plants, was reached more than 4.0 Gy/year during last 10–15 years after Chernobyl accident. As have shown cytogenetical studies the rate of chromosomal aberration in cells of meristem tissue of the common reed in Dalekoye-1 Lake reaches to 18% and is maximal among the investigated reservoirs of the exclusion zone [11]. The damage events of common reed by larvae of gall fly of family *Chloropidae*, genus *Lipara* has been registered as well.

The influence of parasitic fungi *Claviceps purpurea* (ergot) lesions on the seed production of the common reed was determined. The positive correlation between registered abnormalities and parasitic fungi lesions of the common reed with levels of radiation exposure on plants, sampled from lakes within the Chernobyl exclusion zone was registered. Registered high level of parasitic fungi lesions ratio for the common reed plants sampled from the lakes of the left-bank dammed floodplain of the Pripyat River, which are the most radioactive contaminated water bodies of the exclusion zone. Such significant damage of the common reed's panicles by the parasitical fungi in lakes of the Krasnensky flood land was observed on the background of the most low parameters of seed production in water bodies of the exclusion zone.

The increase of the parasitic fungi lesions ratio simultaneous the enhancement of absorbed dose rate of ionizing radiation, which may conclude of chronic low-dose radiation influence upon the populations of the common reed within the Chernobyl exclusion zone, was determined. At that the Pearson correlation coefficient for the parasitic fungi lesions ratio and absorbed dose rate of ionizing radiation was 0.913 and error coefficient was 0.030.

Relation of potential seed production and effective seed output is one of the significant parameter characterizing rate of reproductive performance suppression of plant. Thus for the cultural cereal explorers observed complete or partial sterility [3] but for the common reed we found that panicle sterility ratio does not exceed 72%. In normal case 1 panicle produces about 5,000 viable seeds [1] but in our case even potential seed production was much lesser. Above mentioned phenomenons may testify upon the decreasing of the parasitical stability of the common reed under impact of long-term irradiation.

20.4 Conclusion

Different radiation effects of ionizing radiation on hydrobionts within the Chernobyl exclusion zone have been registered in post-accident period. Some of these effects appear shortly, while an increasing importance is expected by the remote consequences - genetic damages induced by a long-term irradiation. These remote consequences are a long-drawn realization of changes in molecules of heredity, where the initial molecular damages have a latent period without any display and can be transferred through the many generations of cells to be a reason of genome instability in future.

The absorbed dose rate for hydrobionts of litoral of the researched water bodies within the Chernobyl exclusion zone during 1998–2009 was registered in a range 1.3 mGy year^{-1} to 3.4 Gy year^{-1}. The highest levels were registered in lakes of the dammed territory of the left-bank floodplain of the Pripyat River - Glubokoye Lake and Dalekoye-1 Lake, the lowest one - for running water objects. The ratio of doses, caused by an external and internal irradiation of hydrobionts in different reservoirs, essentially varies and depends on the contents of gamma-emitting radionuclides in bottom sediments of littoral zone and soils of shoreline. At their high level of radionuclide contamination up to 95% of radiation dose can form due to external sources and only 5% due to radionuclides, incorporated in the tissues. The rate of external dose on hydrobionts of the same water body can change in range of three exponents and depend from radioactive contamination level of ecological niche. At present time the main dose-formed radionuclide in the majority water bodies within the exclusion zone is ^{90}Sr, which part of total internal radiation dose amount 90–95%.

The long-term impact of low dose irradiation in aquatic ecosystems, especially in lakes within the inner (10-km) Chernobyl exclusion zone exhibits in increased level of chromosomal aberrations and, connected with it, in reproductive death of cells. Cytogenetical research of hydrobionts in the exclusion zone testify to a high level of chromosomal aberration in embryo tissues of gastropod snails and in root meristems of higher aquatic plants in the most contaminated lakes. High chromosomal aberration rate in tissue of hydrobionts from the closed water bodies of the exclusion zone repeatedly exceeds a spontaneous mutagenesis level for aquatic biota and can be display of radiation-induced genetic instability.

Haematological research of hydrobionts in the exclusion zone shows essential changes of hemolymph structure of molluscs and high amount of red cells with atypical shape of nucleus as different type of invaginations, ramifications etc. as well as cells without nucleus in the blood of fishes from the water bodies with high levels of radioactive contamination that allows to assume that the qualitative indexes of red cells of peripheral blood of fishes are more sensible to chronic radiation influence in comparison with the elements of white blood.

For the common reed in lakes of the inner exclusion zone, the high level of affection by parasitic fungi *Claviceps purpurea* and by gall-producing arthropods, in particular by mites *Steneotarsonemus phragmitidis* (sometimes up to 100% of a vegetative population of lake), was discovered. During 2002–2009 affection by mites quickly propagated in other closed reservoirs of the exclusion zone, essentially reducing rates of growth, seed efficiency and bioweight of plants. Now there are no bases finally to approve, that the damage of a reed causes by the impact of ionizing radiation, however we are anxious about the fact, that this mite species for the first time is registered in Ukraine within the Chernobyl exclusion zone, in the territory which is the most contaminated by radionuclides. In this connection it is supposed, that one of the possible reasons of a total plants disease can be a loss of them a parasitical stability in conditions of chronic radiation influence.

Acknowledgements This study was financially supported by the Ukraine Ministry for Emergency Situations and by the National Academy of Sciences of Ukraine (Projects 0101U004987, 0102U004665 and 0102U003541). The authors also wish to thank the personnel of Radioanalytical Laboratory of the State Specialised Scientific and Production Enterprise "Chernobyl Radioecological Centre" for the radionuclides measuring procedure.

References

1. Dubina DV, Stoyko SM, Sytnik KM (1993) Macrophytes - indicators of environmental changes. Naukova Dumka, Kiev (Rus)
2. Dzyubo SM, Romanova LG (1992) Amoebocyte morphology in hemoliph of Japanese scallop. Cytol 10:52–58, (Rus)
3. Grodzinsky DM, Kolomiets KD, Kutlakhmedov YuA, Bulakh AA, Dmitriev AP, Khomlyak MN, Bubryak II, Zezina NV, Mikheev AN, Kravets AP (1991) Anthropogenic radioactive anomaly and plants. Lybid', Kiev (Rus)
4. Gudkov DI, Kuzmenko MI, Derevets VV, Nazarov AB (2005) Aquatic ecosystems within the Chernobyl NPP exclusion zone: The latest data on radionuclide contamination and absorbed dose for hydrobionts. In: Brechignac F, Desmet G (eds) Equidosimetry–ecological standardization and equidosimetry for radioecology and environmental ecology, vol 2, Series C: environmental security. Springer, Dordrecht, pp 333–342
5. Gudkov DI, Uzhevskaya CF, Nazarov AB, Kolodochka LA, Dyachenko TN, Shevtsova NL (2006) Lesion in common reed by gall-producing arthropods in water bodies of the Chernobyl NPP exclusion zone. Hydrobiol J 42(1):82–88
6. Handbook for assessment of the exposure of biota to ionising radiation from radionuclides in the environment (2003) Brown J, Strand P, Hosseini A, Børretzen P (eds). Project within the EC 5th Framework Programme, Contract FIGE-CT-2000-00102. Framework for Assessment of Environmental Impact, Stockholm
7. Kalinina MV (2002) Feature of young chum salmon blood as indicator water bodies contamination by heavy metals. In: Proceedings international conference "New technologies for the protection of biodiversity in aquatic ecosystems", Moscow, 27–29 May 2002. p 123 (Rus)
8. Lugas'kova NV (2003) Species specific of cytogenetical stability of fish in condition of eutrophic water body. Ecology 3:235–240, (Rus)
9. Majone F, Brunetti R, Gola I, Levis AG (1987) Persistence of micronuclei in the marine mussel, Mytilus galloprovincialis, after treatment with mitomycin. Mutat Res 191(3–4):157–161
10. Pausheva ZP (1974) Practical work on cytology of plants. Kolos, Moscow (Rus)

11. Shevtsova NL, Gudkov DI (2009) Cytogenetic effects of long-term radiation on higher aquatic plants within the Chernobyl accident exclusion zone. Radioprotection 44(5):937–940
12. Shevtsova NL, Gudkov DI, Stoyko YuA, Syvak EV (2005) To the method of determination of chromosome damages of higher aquatic plants at the example of common reed and arrowhead. Sc Acta of the Ternopil State Teacher's Training University 26(3):478–479 (Rus)
13. Tsytsugina VG (1998) An indicator of radiation effects in natural populations of aquatic organisms. Radiat Protect Dosim 75(1–4):171–173

Chapter 21
Endless Pursuit of DNA Double-Strand Break Ends

Michael A. Resnick

Abstract In the pursuit of radiation-induced cellular damage that manifests in cell killing and genome instability, but is reversible by the phenomenon of liquid holding, DNA double-strand breaks (DSBs) have been shown to be the major culprit. It is now widely accepted that DSBs provide both beneficial and deleterious effects and that elaborate systems were evolved to address their detection, processing and repair. Understanding the mechanisms of DSB induction, processing and eventual biological consequences as well as just the detection of DSBs remain a challenge – especially for randomly induced breaks. Given the large number of extensive reviews available and the limited space here, I have chosen to focus on summarizing recent novel findings from our group using the budding yeast *Saccharomyces cerevisiae* that are relevant to many molecular and genetic aspects of DSB repair and impact on genome stability. Included are the systems we have developed that address (1) induction of random primary and secondary DSBs; (2) processing of DSB ends; (3) genetic control of DSB induction and repair; (4) genome instabilities associated with DSBs including rearrangements and hypermutability associated with resected ends; and (5) physical factors that determine the transition from DSB to chromosome break or recombination. Using a circular chromosome, we find that resection of random, dirty-end DSBs induced by ionizing radiation or derived from MMS single-strand damage is rapid and is primarily due to the MRX complex. Interestingly, the transition from DSB to chromosome break at a unique DSB in yeast is largely prevented by the nuclease function of exonuclease 1 as determined from separation of fluorescent markers that flank a DSB. Based on a tetraploid gene-dosage model, the role of the chromosome structural complex cohesin is not only to enhance DSB repair between sister chromatids, but it also directs recombinational repair events to

M.A. Resnick (✉)
Chromosome Stability Group (CSG), Laboratory of Molecular Genetics,
National Institute of Environmental Health Sciences, NIH, Box 12233,
Research Triangle Park, NC 27709, USA
e-mail: resnick@niehs.nih.gov

C.E. Mothersill et al. (eds.), *Radiobiology and Environmental Security*,
NATO Science for Peace and Security Series C: Environmental Security,
DOI 10.1007/978-94-007-1939-2_21, © Springer Science+Business Media B.V. 2012

sisters thereby preventing loss of heterozygosity. Notably, DSBs greatly sensitizes cells to localized mutability as well as gross rearrangements. Overall, these findings demonstrate the genomic vulnerability to DSBs and the genetic investment in their orderly processing.

Keywords DNA repair • Double-strand breaks • Resection • MRX • RAD52 • Mutations • Genome stability • Ionizing radiation • UV • Chromosome breaks • Tetraploidy • Recombination

21.1 Overview

All organisms possess intricate networks for genome duplication and maintenance that assure stability in the face of internal and external environmental stresses or during normal growth and development. The stability is provided through gene expression networks and protein interactions that assure coordination of normal replicative processes and responses to potential genome threats. In addition, all organisms are capable of detecting double-strand breaks (DSBs) and correcting them in an efficient and timely manner. In fact, several years prior to the molecular characterization of DSBs, repair and genetic controls it was clear that yeast cells had a tremendous potential to withstand ionizing radiation insults either through liquid holding [7] or change in cell cycle [2].

It is important to have a full understanding of how DSBs and other types of lesions arise and contribute to genome instability. While there is an upside to DSBs including antibody diversity and chromosome segregation in meiosis, the downside is that DSBs can lead to a variety of genome destabilizing events including incorrect restitution of broken molecules, chromosome breaks, mutation, growth inhibition, disease and death. In the multistep process leading from normal to cancer cells, human chromosomal DNA frequently exhibits genetic instability resulting in mutations, rearrangements and/or aneuploidy. Chromosome breakage, which might arise from a DSB, has been hypothesized to be an initiating event in many of these processes. DSBs at various stages of the cells cycle checked and can generate checkpoint responses. Abrogation of checkpoint functions, for example loss of p53 in human cells, may contribute to genetic instability and eventual neoplastic progression.

Three pathways of DSB repair have been characterized in yeast and human cells: recombination, single-strand annealing and end-joining. Efficient repair of DSBs in yeast generally involves *RAD52*-dependent recombination provided that there is homologous or related (diverged) DNA available. This repair mechanism was originally proposed to require a resected, invasive 3' DNA end supplied by the DSB and a homologous chromosome or sister chromatid as template [14]. Several features of this model have proven correct, as for the case of synthesis dependent, single strand annealing, as well as resection after radiation-induced DSBs as described below. (For a discussion of models and implications see [5].) While it was envisioned as

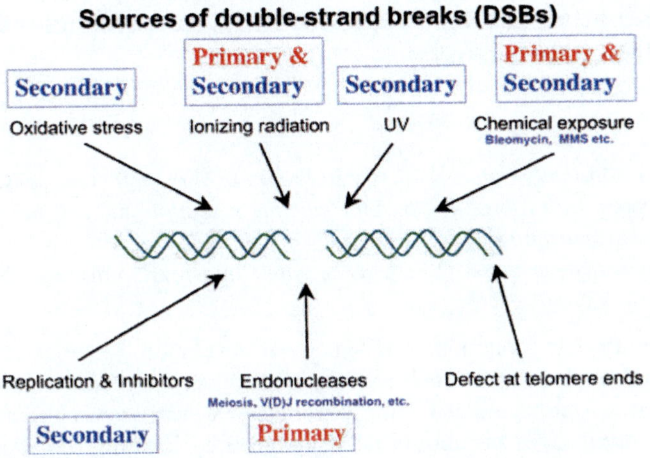

Fig. 21.1 Sources of double strand breaks (DSBs)

being largely error free, subsequent experiments have clearly demonstrated that the repair may be at-risk for errors. Deletions and rearrangements can also occur from the *misrepair* of DSBs. For example, deletions can occur between homologous, directly repeated sequences flanking a DSB via single-strand annealing in the presence or absence of *RAD52*. Misrepair of DSBs can also result from illegitimate recombination between related DNA sequences thereby reshaping the genome (as described below). Another pathway involving direct nonhomologous end-joining (NHEJ) was described first in mammalian systems and subsequently characterized in yeast. This DSB repair pathway is mediated by the yeast homologues of human Ku70 and Ku80.

The choices in DSB repair pathways may depend on the nature and number of breaks as we have shown for the case blunt-end [8] vs staggered-end DSBs [20] produced by endonucleases, opportunities for homologous interactions and chromosome organization. Figure 21.1 describes many of the sources of DSBs either as primary events, secondary during processing of lesions, or even during replication resulting from collapsed forks. Our Chromosome Stability Group has developed multi-pronged approaches to understanding the appearance of a variety of DSBs created by different agents and situations, including replication and ARMs (at-risk motifs) associated DSBs [4, 6, 9] along with mechanisms of repair and genomic changes resulting from DSBs in budding yeast. There are several reviews over the past few years that address many of the issues related to DSB induction, repair and consequences. In light of this, I have chosen to briefly review our recent approaches with the budding yeast *Saccharomyces cerevisiae* which have led to novel systems and findings that address many molecular and genetic aspects of DSB repair and impact on genome stability. [This summary is based on a recent unpublished report to the Scientific Board of NIEHS that also addresses studies on the human p53 damage response network [12].]

21.2 Radiation-Induced Random DSBs and Subsequent Molecular Events

Highlights

- *Repair of randomly induced, dirty-end breaks is highly efficient in G2 cells*
- *Development of a robust system for detecting resection and recombinants using circular chromosomes*
- *Resection at γ− induced DSB ends is rapid and predominantly controlled by RAD50 (MRX)*

Armed with new tools, many of which we developed, we revisited our early studies that provided the first evidence for DSB repair via recombination [14, 15] with a view towards understanding chromosomal variation in repair and the genomic consequences of random DSBs induced by ionizing radiation (IR) and other agents. Radiation, which is commonly used in cancer therapy, has proven a model agent for instantaneous induction of random DSBs (unlike *in vivo* enzymatic induction of DSBs). For most chemical and IR inducers of DSBs, there is the issue of repair at "dirty" ends and the formidable challenge of studying molecular events associated with random DSBS.

The 16 chromosomes of haploid yeast are between ~200 and 3,000 kb. Since the individual chromosomes can be displayed with neutral pulse-field gel electrophoresis (PFGE), we have chosen to address repair in individual chromosomes (*i.e.,* detected as complete chromosome restitution) and genetic controls. DSB induction by IR and subsequent repair has been extensively examined in haploid and diploid, logarithmically growing cells arrested in G2, since they have the highest potential for recombinational repair (either sister chromatids or homologous chromosomes; also discussed below in terms of the role for cohesin). A unique addition to defining events at random DSBs is our use of circular chromosomes (Fig. 21.2). A single DSB leads to a unique, unit-size linear molecule detectable by PFGE unlike the circular chromosomes that remains in the starting well. This innovation has provided accurate measurements of DSBs as well as closely-spaced single-strand breaks (SSBs) [11] and the opportunity to study resection.

We established that there are 0.07 DSBs/mb-krad [19]. Repair can be detected at about 30–60 min after IR, with about half the DSBs repaired in 1 h at a dose inducing ~140 DSB in the haploid G2 genome. The time scale for complete restitution to full-size chromosomes is related to size, consistent with a time-dependent component in the random repair of DSBs. Thus, the smallest chromosomes are repaired the fastest, which may indicate differences in ability of radiation to induce chromosome changes. The ribosomal DNA region located on Chr XII contains nearly 200 large rDNA repeats is subject to different repair rules for IR induced DSBs than the rest of the genome, consistent with spontaneous alterations in rDNA. We investigated further the roles of *RAD52, RAD51* and both *RAD50* and *MRE11* (MRX) in repair and restitution.

We wanted to address events shortly after IR, a task that we anticipated would be particularly difficult given that DSBs are random. Surprisingly, the circular chromosomes that received a single break (*i.e.,* a unique linear band in PFGE)

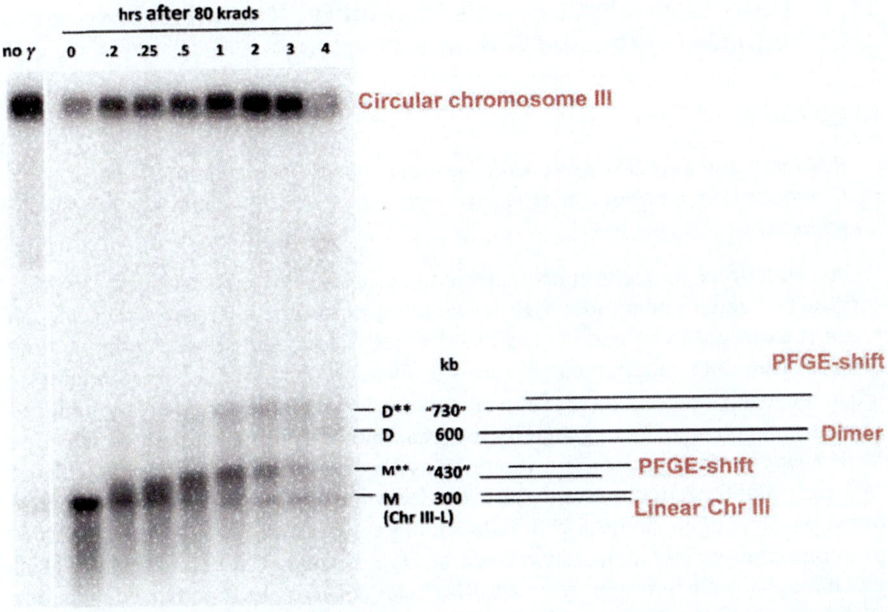

Fig. 21.2 Single DSBs in circular chromosomes lead to linear molecules which exhibit PFGE-shift with time of incubation

developed a unique property—slower mobility – within minutes after irradiation as diagrammed in Fig. 21.2. We established that the slower mobility, referred to as PFGE-shift, was due to single strand tails generated by resection.

This has provided a novel, robust assay and led to the first direct measurement of resection at random breaks, opening opportunities to address the efficiency, consequences and genetics of resection [19]. Within 10 min after IR of WT cells arrested at G2/M, there is a near synchronous PFGE-shift of the linearized circular molecules, corresponding to resection of a few hundred bases. Resection continues so that by the time of significant repair of DSBs at 1 h there is about ~1–2 kb resection per DSB end. The PFGE-shift is comparable in WT and recombination-defective *rad52* and *rad51* strains. However, in *rad50* and *mre11* null mutants the initiation and generation of resected ends at radiation-induced DSB ends is greatly reduced. Thus, the Rad50/Mre11/Xrs2 complex is responsible for rapid processing of most damaged ends into substrates that subsequently undergo recombinational repair. Among the few molecules exhibiting shift in the *rad50* mutant, the residual resection is consistent with resection at only one of the DSB ends (also recent unpublished results).

In addition to these findings, we were able to directly assess sister chromatid recombination associated with the DSBs. Within 1 h after irradiation, double-length linear molecules are detected in the WT and *rad50,* but not *rad52,* strains providing the first real time description of recombination induced by random DSBs. These recombinants were largely resection and *RAD50* independent and provide a unique opportunity to address DSB-induced recombination.

21.3 DSB-Associated Genome Instability: Radiation-Induced Random DSBs and Genome Plasticity

Highlights

- *Randomly induced DSBs frequently generate chromosome aberrations*
- *IR-induced aberrations are primarily associated with breakpoints in repeat (Ty) sequences*

Investigations into chromosomal changes in unselected survivors (20–40% survival) revealed remarkably high levels of chromosome alterations even though repair is expected to be maximal in 2n−4 C cells [1]. Because we employ diploid cells and the only requirement is survival, our approach enables us to address a broad range of changes. Studies from other labs of spontaneous and damage-induced chromosome changes have generally used haploid cells in which there is selection for changes in small regions and where many chromosome aberrations are lethal.

Nearly half the radiation survivors had at least one apparent chromosomal change based on karyotypic analysis with pulse-field gels. This led us to investigate chromosomal changes by Comparative Genome Hybridization (CGH) microarray analysis through a collaborative effort with Tom Petes at Duke. We found that duplication or deletions were common (~100 among 37 colonies). Surprisingly, the incidence of chromosome aberrations was not related to chromosome size and, therefore, frequency of DSBs. Nearly 90% of the breakpoints occurred in the vicinity of multiply repeated (although diverged) sequences, with some repeats being much hotter than others and subsequent sequencing demonstrated that the breakpoints occurred within the repeats. Thus, while repair between sister chromatids is highly efficient, repeat regions—particularly TY elements – in the genome can compete in the repair process.

Based on the appearance of tripartite events and our measured frequency of DSBs, we concluded that the two ends of a radiation-induced DSB in a repeat sequence can independently interact with other repeats in the genome. This has obvious implications for damage in the human genome. Thus, randomly-induced DSBs can have profound genomic effects and lead to nonrandom genomic changes opening the diploid genome to dramatic changes.

21.4 DSB Associated Genome Instability: DSBs, ssDNA and Localized Hypermutability

Highlights

- *Single-strand DNA formed at DSBs and uncapped telomeres is at risk for hypermutation and multiple mutations*
- *Strong potentiation of weak mutagens in single strand DNA*

A common mechanism of repair of DNA lesions involves excision and resynthesis utilizing the intact template strand. This implies that damaged single-strand DNA (ssDNA) might be at high risk for loss or genome instability. Based on the systems we had developed to study DSB repair by oligonucleotides [16, 17] including RNA-containing oligonucleotides [18], we were uniquely poised to address the vulnerability of ssDNA to spontaneous and damage-induced changes including mutagenesis. Unlike the lesions induced in dsDNA, most lesions in ssDNA are not repairable. In these studies which were initiated and developed by Dmitry Gordenin [see his report in this volume], we found that multiple lesions caused by ultraviolet light-C (UV) and MMS can be tolerated in ssDNA and that they lead to extremely high frequencies of single and multiple mutations.

Using our uniquely designed yeast systems [21], up to 20 kb of persistent ssDNA can be generated at the ends of a chromosomal DSB or at uncapped telomeres in *cdc13-1* temperature-sensitive mutants. Cells are then exposed to a DNA damaging agent. The mutagenic consequences of the lesions in the ssDNA can be assessed with mutation reporters that are included in the regions that experienced ssDNA.

Since nearly all the UV-induced mutations correlated with pyrimidines in the nonresected strand [21] and the majority of MMS-induced mutations could be related to damaged cytosines in the nonresected strand which provided a unique mutation signature [22], ssDNA was concluded to be highly vulnerable to mutagenesis. Consistent with this finding, there was a striking multiplicity of strand-biased mutations, with up to 6 mutations separated by hundreds of nucleotides after a non-lethal dose of damage. These results established that long stretches of transient ssDNA can be restored to the dsDNA state even when they contain multiple lesions. The ssDNA was greater than 1,000-fold more mutable than the dsDNA in the rest of the genome. The high levels of induced, as well as spontaneous, mutagenesis were largely attributable to bypass by the DNA polymerase zeta (controlled by *REV3* as well as *REV1*).

These findings are contrary to commonly held views of mutation independence across the genome. They reveal novel mechanisms of mutation and biological consequences. Environmental toxicants and drugs that can lead to and/or generate lesions in single strand DNA are of special health concern since single-strand regions in the genome may be highly prone to mutation and subsequent disease. This is particularly relevant since transient stretches of ssDNA are formed during 5'-end-resection of DSBs, as discussed above, as well as during DNA replication. Abnormalities in DNA metabolism, such as replication fork uncoupling, delay in recombination repair of a DSB or in telomere recapping after replication could be sources of long ssDNA stretches (thousands of bases). Our findings establish long ssDNA as extremely risky to genome stability as well as a potential source of adaptive evolution and cancer [see report by Gordenin in this volume] without severe mutation load in the rest of the double-strand genome, where repair of lesions is efficient.

21.5 Base Damage-Associated DSBs and Implications for SSB Repair

Highlights

- *System developed to monitor induction and repair of DSBs caused by DNA base damage*
- *Chromosomal DSBs can result from interrupted or misrouted base-excision repair*
- *Coordinated polymerase δ and Rad27/Fen prevent transition of closely-opposed lesions to DSBs*

Endogenous metabolism or environmental factors such as oxidizing and alkylating agents produce a wide variety of lesions in cellular DNA. The genomes of mammalian cells experience from 10,000 to 200,000 modifications per day. Most lesions are repaired by a network of proteins that are part of an elaborate, multi-step base excision repair (BER) system that generate single-strand break (SSB) intermediates. Importantly, defects in BER can lead to malignancies and are associated with age-associated disease, including neurodegeneration. While BER enzymes have been characterized biochemically, BER functions within cells are much less understood, in part because replication bypass and DSB repair can impact resistance to base damage. To investigate *in vivo* BER, we examined repair of methyl methanesulfonate (MMS) induced DNA damage in haploid G1 cells in the absence of replication bypass and recombinational DSB repair [11].

Following methylation of DNA bases, abasic sites (AP sites) are generated by glycosylases. The AP sites are heat-labile and give rise to SSBs during sample preparation for PFGE at a high temperature (55°C) while few SSBs are formed at lower temperatures (30°C). If closely spaced on a chromosome, SSBs will result in a secondary chromosomal DSB that is detectable by PFGE. Assuming that pairs of single strand lesions separated by less than 10 nt are detectable as DSBs, the number of secondary DSBs appearing after MMS treatment corresponded to 2,000–10,000 base lesions per haploid genome. These DSB studies were facilitated by our approach of simultaneous probing of a circular and a linear chromosome using a common sequence; the system can detect 2–40 DSBs per yeast genome. (Note that in this system DSB measurements are independent of the amount of DNA entering the gel.)

The heat-labile sites associated with DSBs were efficiently repaired after incubating G1 stationary cells for 24 h in buffer. Because the cells were also haploid, the repair could not occur through recombination. Consistent with a proposed role for BER in repair of heat-labile sites, the repair required the glycosylase *MAG1*. Simultaneous deletion of AP-endonuclease genes *APN1* and *APN2* led to a large number of DSBs even for samples processed at 30°C, indicating that heat-labile sites can be converted into SSBs within the cell by other BER enzymes (possibly Ntg1 or Ntg2 lyases) and are not efficiently repaired thereafter. Consistent with this, DSBs formed in *apn1 apn2*-deletion strains were not repaired and even increased in number during post-treatment incubation in buffer. These DSBs were suppressed by a *mag1* deletion suggesting the DSBs resulted from downstream processing of AP-sites *in vivo*. Thus our approach provided direct physical evidence that Apn1 and Apn2 not only repair cellular base damage but also prevent break accumulation that can result from AP sites being channeled into other BER pathway(s).

Our findings provide the first direct evidence that interrupted or misrouted base-excision repair of clustered lesions can lead to the *in vivo* accumulation of chromosomal DSBs. In subsequent experiments [11], we described a pathway for the prevention of DSBs from clustered lesions through the coordinated repair of closely-opposed lesions. Using single and multiple mutants that impede interaction of DNA polymerase δ and 5'-flap endonuclease Rad27/Fen1 with the PCNA sliding clamp, we found that lack of coordination between these components during long-patch base excision repair of alkylation damage can result in the generation of many DSBs within nondividing haploid cells. This contrasts with efficient repair of nonclustered MMS-induced lesions, as measured by quantitative PCR and S1 nuclease cleavage of SSB sites.

We conclude that closely-opposed single-strand lesions are a unique threat to the genome and that repair of closely-opposed strand damage requires greater spatial and temporal coordination between the participating proteins than widely-spaced damage in order to prevent generation of DSBs. In our recent unpublished studies (Ma et al. in preparation) we have gone on to describe how MMS can lead to DSBs in *apn1 apn2* cells and that the DSBs are subject to resection and recombinational repair, similar to findings with radiation-induced DSBs.

21.6 Chromosome Structure and DSBs: Single Molecule Analysis of DSB to Chromosome Aberration Transitions

Highlights

- *Chromosomal continuity is not disrupted by a DSB*
- *The transition of nearly all DSBs to a chromosome break is controlled by MRX & EXO1*

Since DSBs are powerful sources of chromosome instability and rearrangements, we set out to understand the relationship between a DSB and a cytologically detectable chromosome break (CRB) and what factors might influence the transition from DSB to CRB. In collaboration with Kerry Bloom (University of North Carolina, Chapel Hill) we developed a yeast-based system that provides for chromosome analysis in real time following the induction of a single DSB by an inducible I-SceI endonuclease [10]. The system (Fig. 21.3) utilizes the fluorescent labeled proteins tetR-CFP and LacI-GFP marking each side of a DSB and Spc29-RFP fusion to identify the spindle poles. The tetR-CFP and LacI-GFP proteins bind multiple repeats of their operator target sequences ~5 kb from the DSB site. This arrangement enabled

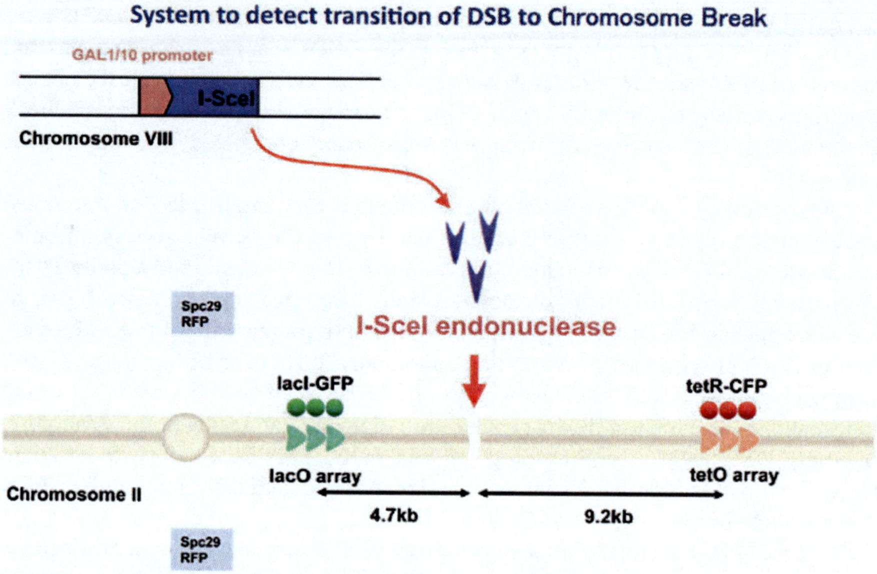

Fig. 21.3 System to detect transition of DSB to chromosome break

us to investigate the development of a CRB following DNA breakage and the relation to spindle pole separation and sister chromatid separation in wild type and various end-joining or end-processing RMX mutants.

Previously, we showed that nearly all wild type cells were arrested at the G2/M phase in response to an unrepaired DSB signaling event [10]. The centromere-containing and acentric broken DNA fragments remained in close proximity, which led us to conclude that a break in DNA does not develop into a CRB. (Similar results were reported using an identical flourescent markers nearly 50 kp from an induced DSB.) This physical association may be important in DSB repair. We went on to identify some factors that help prevent a DSB to CRB transition. There is a requirement for the complex, since 10–20% of chromosomes exhibited a CRB when any of these genes were mutated. However, it is not the end-joining or end-processing MRX function that prevents chromosome breakage. Instead, it is the tethering feature of the complex through the Rad50 component. We proposed that this function, in addition to repair, is likely to play an important role in preventing chromosome instability. These studies were recently extended to other mutants. Remarkably, deletion of the exonuclease 1 (*EXO1*) gene increases the frequency of cells with a CRB to ~40% [13]. Importantly, in the *exo1 rad50* double mutant nearly all DSBs transition to CRBs. Since the exo1 nuclease function specifically affects the DSB to CRB transition, the findings have led us to conclude that some aspect of resection plays a major role in preventing a DSB to CRB transition.

21.7 Chromosome Structure and DSBs: Limiting Cohesin Leads to Genome Instability

Highlights

- *Development of genetic system to identify limiting components controlling genome stability*
- *Cohesin is limiting for DSB repair.*
- *Cohesin defines the sister chromatid as a donor for DSB repair, preventing recombination between homologous chromosomes and loss of heterozygosity (LOH)*

The repair of DSBs via homologous recombination (HR) is an evolutionarily conserved process that is relatively error free. While HR may have a deleterious effect on genome stability through nonallelic recombination or LOH, these risks might be reduced by restricting HR to sister chromatids. Although the sister chromatin cohesion complex (cohesin) facilitates DSB repair between chromatids, its role in HR and repair when homologous chromosomes are present has not been addressed. We examined the consequences of changes in cohesin level on HR and DSB repair as well as the role that cohesin might play in directing events between sister chromatids vs homologous chromosomes [3]. Using tetraploid yeast, gene dosage could be varied four-fold for the essential sister chromatid cohesin sub-unit *MCD1* as well as other key players in HR. Survival and global DSB repair are compromised in *MCD1* simplex (single-copy) strains that are γ-irradiated at G2/M when sister chromatids are present. Importantly, γ-induced HR between homologous chromosomes was actually increased ~10 times over that in WT strains (4 copies), while there was no difference in spontaneous HR rates. The survival and HR frequency were comparable to WT tetraploid yeast irradiated in G1, when only homologous chromosomes are present. The results for UV and IR were similar although to a lesser extent for the DNA replication inhibitor hydroxyurea. This system has provided the first direct evidence that the cohesion complex is both limiting in DSB repair and that it has a dual role in HR, namely, promotion of recombinational repair between sister chromatids and suppression of recombination between homologues (Fig. 21.4). The suppression would prevent loss of heterozygosity (LOH) through recombination and segregation of sister chromatids.

Thus, reduction in cohesin may have important consequences to genome change and evolution. Cells with moderately reduced cohesin levels (*MCD1* simplex) have a reasonable fitness but exhibit increases in DNA damage-induced recombination between homologous chromosomes, even when survival is high. The promiscuous recombination which can cause elevation in LOH may also lead to more nonallelic recombination, reshaping the genome through structural variations [3]. While the present results indicate a general role of cohesin in control of HR, our approach can provide useful insights into genome dynamics as well as genetic processes associated with tetraploidy and consequences of cell-to-cell variation in limiting proteins. In addition, the experimental design can be used to search for subtle changes in factors

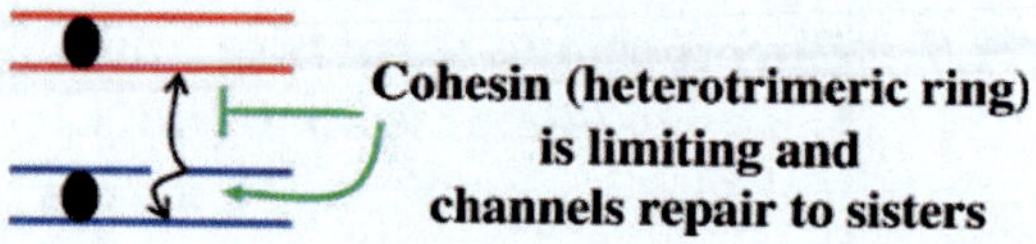

Fig. 21.4 Cohesin (Heterotrimeric ring) is limiting and channels repair to sisters

that are limiting for genome stability, especially when combined with modest (*i.e.*, high survival) levels of genotoxic environmental stress. Our system utilizes normal, wild type proteins and native gene expression regulation, eliminating the uncertainty associated with mutations and controlled gene expression.

21.8 Summary

Overall, our approaches have established the efficiency of repair of random DSBs in yeast and the genomic vulnerability associate with these breaks. Based on results with a unique cut site, broken DNA ends are held in close contact which could facilitate recombinational repair. In addition the physical interaction between sister chromatids appears to enhance opportunities for repair. Importantly, the ends have a strong potential for genome instability either by interacting with homologous chromosomes and/or sequences across the genome or through resection of ends, making them highly vulnerable to mutagenesis. These approaches have helped elucidate a portion of the considerable genetic investment that all cells have in the detection and orderly processing of DSBs.

Acknowledgements I would like to thank all the past and present members of the Chromosome Stability Group who have contributed to the many studies and papers summarized here, many of whom are indicated in the references. Included among these exceptional researchers are Wenjian Ma, Yong Yang, Shay Covo, Jim Westmoreland, Steven Roberts and especially Dmitry Gordenin for development of several ideas and systems that have been employed in this work. Special thanks to Kerry Bloom at the University of North Carolina, Chapel Hill, and Tom Petes at Duke University as well Lucas Argueso. This work was supported by intramural research funds from NIEHS to M.A. Resnick under projects Z01-ES021016 and Z-01-ES065073.

References

1. Argueso JL (2008) Double-strand breaks associated with repetitive DNA can reshape the genome. Proc Natl Acad Sci U S A 105(33):11845–11850 (in eng)
2. Beam CA, Mortimer RK, Wolfe RG, Tobias CA (1954) The relation of radioresistance to budding in Saccharomyces cerevisiae. Arch Biochem Biophys 49(1):110–122 (in eng)
3. Covo S, Westmoreland JW, Gordenin DA, Resnick MA (2010) Cohesin Is limiting for the suppression of DNA damage-induced recombination between homologous chromosomes. PLoS Genet 6:e1001006
4. Gordenin DA, Resnick MA (1998) Yeast ARMs (DNA at-risk motifs) can reveal sources of genome instability. Mutation Res 400:45–58

5. Haber J (2008) Evolution of models of homologous recombination. In: Egel R, Lankenau DH (eds) Recombination and meiosis: models, means and evolution. Springer, Berlin/Heidelberg, pp 1–64
6. Jin YH et al (2001) The 3'–>5' exonuclease of DNA polymerase delta can substitute for the 5' flap endonuclease Rad27/Fen1 in processing Okazaki fragments and preventing genome instability. Proc Natl Acad Sci U S A 98(9):5122–5127
7. Korogodin VI (1958) Some regularities of the post-irradiation changes in resting yeast cells. Biofizika 3(6):703–710
8. Lewis K et al (2005) Use of a restriction endonuclease cytotoxicity assay to identify inducible GAL1 promoter variants with reduced basal activity. Gene 363:183–192
9. Lobachev K, Gordenin G, Resnick MA (2002) The Mre11 complex is required for repair of hairpin-capped double-strand breaks and prevention of chromosome rearrangements. Cell 108:183–193
10. Lobachev K, Vitriol ES, Stemple J, Resnick M, Bloom K (2004) Chromosome fragmentation after induction of a double-strand break is an active process prevented by the RMX repair complex. Curr Biol 14:2107–2112
11. Ma W, Resnick MA, Gordenin DA (2008) Apn1 and Apn2 endonucleases prevent accumulation of repair-associated DNA breaks in budding yeast as revealed by direct chromosomal analysis. Nucleic Acids Res 36(6):1836–1846
12. Menendez D, Inga A, Resnick MA (2010) Estrogen receptor acting in cis enhances WT and mutant p53 transactivation at canonical and noncanonical p53 target sequences. Proc Natl Acad Sci U S A 107(4):1500–1505
13. Nakai W, Westmoreland JW, Yeh E, Bloorm K, Resnick MA (2011) Chromosome integrity at a double-strand break requires exonuclease 1 and MRX. DNA Repair 10(1):102–110
14. Resnick MA (1976) The repair of double-strand breaks in DNA: a model involving recombination. J Theor Biol 59:97–106
15. Resnick MA, Martin P (1976) The repair of double-strand breaks in the nuclear DNA of Saccharomyces cerevisiae and its genetic control. Mol Gen Genet 143(2):119–129
16. Storici F, Durham CL, Gordenin DA, Resnick MA (2003) Chromosomal site-specific double-strand breaks are efficiently targeted for repair by oligonucleotides in yeast. Proc Natl Acad Sci U S A 100(25):14994–14999
17. Storici F, Snipe JR, Chan GK, Gordenin DA, Resnick MA (2006) Conservative repair of a chromosomal double-strand break by single-strand DNA through two steps of annealing. Mol Cell Biol 26(20):7645–7657 (in eng)
18. Storici F, Bebenek K, Kunkel TA, Gordenin DA, Resnick MA (2007) RNA-templated DNA repair. Nature 447(7142):338–341
19. Westmoreland J et al (2009) RAD50 is required for efficient initiation of resection and recombinational repair at random, gamma-induced double-strand break ends. PLoS Genet 5(9):e1000656
20. Westmoreland JW, Summers JA, Holland CL, Resnick MA, Lewis LK (2010) Blunt-ended DNA double-strand breaks induced by endonucleases PvuII and EcoRV are poor substrates for repair in Saccharomyces cerevisiae. DNA Repair 9(6):617–626
21. Yang Y, Sterling J, Storici F, Resnick MA, Gordenin DA (2008) Hypermutability of damaged single-strand DNA formed at double-strand breaks and uncapped telomeres in yeast *Saccharomyces cerevisiae*. PLoS Genet 4(11):e1000264
22. Yang Y, Gordenin DA, Resnick MA (2010) A single-strand specific lesion drives MMS-induced hyper-mutability at a double-strand break in yeast. DNA Repair 9(8):914–921

Chapter 22
Facultativeness Principle and Generalized Approach to the Genome and Hereditary Variability

Michael Golubovsky

Abstract Facultativeness in the genome structure and function reflects the general principle of the life organization and evolution: *the unity of the whole and the freedom of the parts*. The genome or cell hereditary system codes, maintains and transfers the hereditary information in both structurally and dynamically. Generalized genome concept presents a genome as an ensemble of both obligate and facultative elements. Template processes (Replication, Transcription and Translation) and Basic genetic processes (Repair, Recombination and Segregation) are capable of functional facultativeness and dynamic (epigenetic) alterations. It is reasonable to discriminate between three kinds of heritable changes – mutations, variations and epigenetic alterations. Structural genome facultativeness is expressed as a subdivision of cell DNA and RNA elements on two subsystems: Obligate genetic elements (OGE) and Facultative genetic elements (FGE). FGE include various kinds of repeated sequences, mobile elements, amplicons, inserted viral and foreign DNA, B-chromosomes, plasmids and cytobionts. The number and intracell topography of FGE varies from cell to cell, in different tissues and individuals. Changes in structure or order of OGE are mostly correspond to classical mutations. For diverse changes in number or cell/tissue topography of FGE (as it is well known for mobile elements) the term *variations* seems appropriate. Mutations and variations significantly differ on the character and mode of their occurrence. Variations may occur simultaneously in many cells/individuals and are induced by non-mutagenic factors. Spontaneous hereditary changes in nature usually occur in the system ENVIRONMENT – FGE – OGE via two-step mechanism. FGE are the first to react to both internal and external environmental challenges. Then their activation induces

---

M. Golubovsky (✉)
Department Molecular Cell Biology, University of California,
Berkeley, CA 94720, USA
e-mail: mdgolub@berkeley.edu

C.E. Mothersill et al. (eds.), *Radiobiology and Environmental Security*,
NATO Science for Peace and Security Series C: Environmental Security,
DOI 10.1007/978-94-007-1939-2_22, © Springer Science+Business Media B.V. 2012

gene/chromosomal mutations. Both variations and epigenetic alterations (epimutations) may have non-mendelian features, occur simultaneously in many individuals, sometimes reminding phenomenon of inheritance of acquired characteristics.

Keywords Genome • Facultative elements • Mutation • Variation • Epigene • Heritable changes

22.1 Introduction

The genome, which in the broad sense is the cell's hereditary system, codes, stores and transfers information both structural and dynamic ways. Thus we need discriminate between different forms of structural heritable changes of genome and dynamic ones (or epigenetic). Amusing absence of the strict correlation of DNA content in the haploid nuclear DNA content with taxonomic status of species and 2–10 fold differences in genome size in closely related species was called in 1970s the "C-value paradox" [13, 21]. It appears that in most eukaryotes protein-coding sequences constitute only small part of nuclear DNA. For example, in humans they occupy less than 5% of the haploid nuclear DNA. Other parts comprise regulatory DNA regions (about 15% of genomic DNA) and diverse kinds of repeated sequences including mobile elements [16]. The recent study of 270 human individuals from four populations in Europe, Africa and Asia revealed a total of 1,447 copy number variable regions covering 360 megabase or about 12% of the genome. These variable regions contained 100 genes, functional elements, segmental duplications and repeats [25].

To describe such unusual eukaryote genome organization and its variability we suggest to discriminate between two genome subsystems: OGE or Obligate Genetic Elements (genes and their families) and FGE or Facultative Genetic Elements. FGE are predominant fraction in the genome content of most eukaryotes. FGE include various kinds of repeats, mobile elements, amplicons, viral and foreign DNA, B-chromosomes, plasmids an nuclear/cytoplasmic cytobionts. The number and intracell topography of FGE varies from cell to cell, in different tissues and individuals. Mutations in their classical Morgan's sense (point and chromosomal) are mainly changes in the OFG subsystem. Various changes in the number or cell topography of facultative elements were coined as *variations* [9, 11, 12].

Mutations and variations tremendously differ in the rate of their occurrence, level of genotype/environmental dependence, stability and pattern of genome topography. FGE are the first to react to genetic background or environmental challenges. They constitute an operational genome memory. Activation of OFG leads directly or undirectly to both gene mutations and chromosome rearrangements. The process of occurrence of heritable changes in nature has two-step pattern being mainly mediated by mobile elements and other members of FGE. Recurrent global and local bursts of definite mutations in wild Drosophila populations observed during long-term observations [1] appeared to be result of activation and insertion of diverse mobile elements [7, 10]. In the case of P-M hybrid dysgenesis in Drosophila multibreak rearrangements occur mostly in the sites of insertions of P-mobile elements [22, 3]. Similarly, in the species *D. virilis*, that normally has very stable karyotype, activation and amplification of *Penelope* mobile element in the situation of hybrid dysgenesis is accompanied by multisite rearrangements. Important that about 50% of them discriminate various species of this Drosophila subgenera [6]. Such facts indicate on involvement of mobile elements in the speciation process.

Subdivsion of the hereditary system on Obligate and Facultative subsystems reflects the universal principle in organization and evolution of life organisms: the unity of the whole and the freedom of the parts. Or similarity in design but differences in details. FFG elements may be mobile regulators of gene expression in many chromosomal sites. Thus chromosome/cell topography of FGE may predetermine the direction of genomic evolutionary changes. In this paper we discuss how diverse interactions in the OFG and FGE subsystems lead to non-canonical (non-mendelian) heritable changes.

Facultativeness characterizes also the function of two triades of genetic processes universal for all living organisms. (1) Template processes – Replication, Transcription and Translation and (2) Basic genetic processes – Recombination, Repair and Segregation. These two triades determine coding, maintenance and transfer of genetic information during individual development and during inter-generation transmission.

Dynamic aspects of coding, storage and transfer of genetic information are called epigenetic. The spectrum of epigenetic changes is wide: from the transformation of serotypes in paramecium to chromosomal and gene imprinting. The terminology in this field is not yet stabilized. We argue a heuristic value of an epigene concept developed both in theoretical ad experimental aspects [29].

22.2 Facultativeness of Structure and Function of the Genome

22.2.1 Genome and Cell Information System

The term "genome" was coined by cytologist Winkler in 1920 to designate the haploid set of chromosome for a species. The term was used for an analysis of allopolyploid species or for such mutations as an amplification in chromosome numbers. Then its meaning has widened to include the entire hereditary constitution of the cell: both structural and dynamic modes of coding, storage and transfer of species-specific hereditary information.

Necessity of including in the genome concept of stochastic dynamics comes from the discovery of regulatory genes and specific chromosome sequences (like enhancers) that control the level and stability. The genome contains not only blueprints, but a coordinate program of protein synthesis and cell function. The holistic aspects of species-specific hereditary systems might be viewed metaphorically as the structural design of a temple that cannot understood by studying separate breaks – genes – at a fixed point in time. The discoverer of the operon the principles of gene regulation entitled their classical paper as "Teleonomic mechanisms in cellular metabolism, growth and differentiation" [23]. To preserve intracellular homeostasis and the adaptive response of he genome to environmental challenges, they emphasized the biological purposefulness or teleonomy of cell's regulatory systems. Recent molecular discoveries of signal transduction pathways and chromosome organization has shifted focus from genes as units of inheritance and function to the genome as a complex dynamic system.

The cell's ability to analyze external and internal conditions (and to control growth, movement and differentiation) can be compared to an information computing network and check-points. By means of signal transduction pathways a cell receives external signals and transmits, amplifies and directs them internally. The 17 discovered pathways are highly conserved in nematodes, flies and vertebrates [24].

Each pathways includes signal accepting receptor, membrane or cytosolic proteins including kinases and phosphatases to convey the signal and key transcription factors capable of switching its state, activating or suppressing transcription of definite genes.

DNA repair systems remove damages. Multiple proofreading mechanisms recognize and remove errors that occur during DNA replication or due to mutagens. Repair systems allow cells not be passive victims of random physical and chemical forces. They control the level of mutability by modulating repair system activity [5].

Mobile genetic elements found now in all eukaryotic genomes, can move from one chromosomal position to another and induce chromosome rearrangements (review: Kazazian [18]). ME contain genetic punctuation signs (promoters, enhancers, stop signals of transcription etc.) that regulate expression of genes and promote an appearance of new constructs. The term "Natural Genetic Engineering" coined by Shapiro [26, 27] emphasizes that biotechnology uses the same enzymes (nucleases, ligases, reverse transcriptases and polymerases) that living cells use to reshuffle the genome and its function. Though ME are repetitive and dispersed on different chromosomes, they can be activated simultaneously by cell signal (e.g., HSP90, the genetic "capacitor") resulting in outbreaks of non-random genetic variability.

22.2.2 *Genome Structure: Obligate and Facultative Elements*

The eukaryotic genome can be naturally subdivided on two subsystems: Obligate and Faculative genetic elements. FGE include the hierarchy of intra and extra chromosomal elements in nucleus and cytoplasm. Nuclear FGE comprise a highly repeated and nucleated DNA sequences, pseudogenes and retrotranscripts, transposons, amplicons and even an additional or B-chromosome. In cytoplasm FGE include plasmids, amplified rod and circular segments, and endosymbionts (Fig. 22.1).

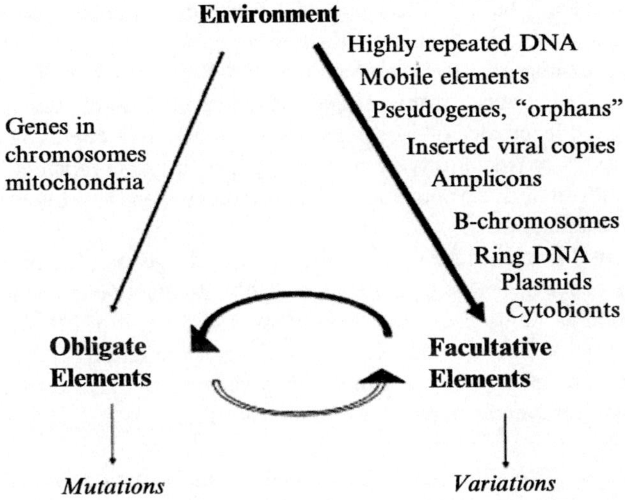

Fig. 22.1 Obligate an Facultative elements of the genome and two types of hereditary changes: mutations and variations. *Arrow* indicates of the link, while their *width* corresponds to the intensity of their force

OGE and FGE exhibit different patterns of heritable changes. Mutations in their classical sense are changes in structure, position and number of genes. These events are connected with OGE. Diverse changes with FGE are titled as *variations*. Let' see human genome [16]. Coding sequences comprise less than 5% of all DNA; 15–20% is connected with gene/chromosome activity regulations. FGE occupy about 50% of the genome and include highly repetitive sequences, duplication of chromosome segments and distinct ME of four types: three kinds of retroelements (LINE, SINE and Alu) and one class of transposons. Segmental duplications of 1–200 kb blocks are a remarkable feature of the human genome and comprise about 3,3% of all DNA. Other repetitive elements are simple sequence repeats (SSRs) : a short repeat units or microsatellite (1–11 b.p) and longer SSR as minisatellites (14–500 b.p). SSR comprising ~3% of the genome are important in human genetic studies because they show a higher degree of length polymorphism in populations and promote to molecular localization. The genome includes also several families of human endogenous retroviruses dispersed on chromosomes [18]. The LINE-1 retrotransposon, L1, has in the human genome 3,000–5,000 full length copies and 500,000 truncated copies, ~15–17% of the genome. About million copies of Alu retrotransposons comprise 10–12% of the genome. One of every 100–200 human births has a de novo Alu insertion. A similar rate is found for L1. More 30 L1–mediated insertion mutations leading to diseases were observed (hemophilia A and B, thalassemia, DMD).

Changes in number and chromosome topography of intra-cell population of FGE are drastically different from gene mutations. Jacob and Wollman [17] were the first who studied similar hereditary changes in the system phage-bacteria and called them as variations. It became clear now that variations in eukaryotes arc frequent phenomena and embrace diverse changes in an intra-cell populations of diverse FGE. Typical example is the phenomenon of hybrid dysgenesis established in Drosophila. In F1 hybrid from crosses of paternal P-stock containing active P-transposons with females of M-stocks devoid of P-active copies of cytoplasmic repressor mass P-transpositions occur in the germ line accompanied by multiple insertion mutations and rearrangements. Their incidence in F1 progeny of dysgenic crosses may reach about 10%.

Chromosomal breaks in the cases of P-M hybrid dysgenesis are ordered and site specific: they occur near P-site localization. Multisite inversions occur in dysgenic hybrids as often as single ones [2, 22]. Activation of diverse of ME in nature result in mutation burst of definite genes in distant natural populations [10].

Variations or hereditary changes in the FGE subsystem may be induced by non-mutagenic environmental factors such as interline crosses, food/temperature fluctuations.

There is a two-step mechanism of spontaneous mutation process in nature (Fig. 22.1). First, an activation of mobile elements (significant part of FGE) in a reply on diverse environment challenges. Second, insertional mutations and chromosome rearrangements.

Typical example of variation are changes in the ratio of OGE and FGE. These changes accompany phenomenon of amplification of definite chromosomal segments during development or an adaptation of somatic cells to drugs that block cell division.

Amplified segments or amplicons might be as tandem duplications or be transformed into plasmids or even mini-chromosomes capable to autonomous replication in cytoplasm. Both the number and topography of amplicons varies over cell lines. Amplification of chromosomal segments containing *myc* oncogene occurs in 25% human neuroblastomas. The exact number of amplified segments cannot be determined even in daughter cloned cells [19].

22.2.3 Template and Basic Genetic Processes and Facultativeness

An occurrence and fixation of new hereditary information are realized via two triades of genetic events acting both in prokaryotes and eukaryotes: template and basic genetic processes. Template processes include Replication, Transcription and Translation. Basic genetic processes include Recombination, Repair and Segregation [9]. To be hereditary, all DNA changes need go through template and basic genetic processes. Among the 4,228 genes of bacteria *E.coli* the number of genes involved in these processes are following. Replication, recombination and DNA repair – 115 (2.7%). Transcription, synthesis and modification of RNA – 55 (1,3%). Translation and post-translation protein modification – 182 (4,2%). Ribosomal r-RNA synthesis- 21 and tRNA – 86 genes [4].

There are a lot of examples of facultativeness in function of template and basic genetic processes. First, facultative hyper=replication or under-replication of chromosome segments enriched by DNA repeats (heterochromatin areas). Amplification is another example of facultative local DNA hyper-replication during development or environmental challenges.

Near 60% of genes in humans are capable to alternative transcription and alternative splicing, depending on specific tissue or cell/tissue physiology. This ability is based on existence of two or more promoters and exon/intron structure of eukaryotic genes. Thus, in humans due to an alternative gene splicing there are average of 2,6 transcripts per gene.

Facultative translation is well established in yeast. Thus, in the presence of protein Sup35 which control a subunit of the translation termination complex and exhibit prion traits, ribosomes begin to read through stop codons an appreciable fraction in time. This releases a hidden genetic variation and creates a variety of new phenotypes especially in the stress conditions [30].

DNA repair is the main guardian of diverse errors and damages of DNA structure.

In addition to normal mechanisms of the repair process there are facultative ones: photoreactivation, excision and postreplicative repair. Facultative recombination includes such its variants as site-specific recombination and replicative transposition of LTR-containing mobile elements. Segregation process as the necessary final of both mitotic and meiotic divisions might be also facultative as in the case of some genetic factors like Segregation Distortion or some chromosomal rearrangements [20].

22.3 Dynamic Inheritance and the Logic of an Epigene

Dynamic aspects of coding, storage and transfer of genetic formation are called as an epigenetic. The spectrum of epigenetic inheritance is wide: from the transformation of serotypes in the paramecium to gene and chromosome imprinting. The term "epigenotype" was introduced in 1960s. Holliday [14, 15] first used the term "epimutation" and indicated on variations in DNA methylation profile as the main its source in diverse organisms.

Since the middle of 1970s the concept of an *epigene* as a unit of epigenetic inheritance is developing. This fruitful idea was experimentally validated by artificial epigene synthesis [28, 29]. The epigene is an auto-regulatory hereditary unit, a genetic system with cyclic links or feedback, having two or more functional states and capable of maintain each other over cell generations. A one–component epigene is shown on Fig. 22.2. Feedback might be positive as in the system bacteria-phage with autoregulatory *cl* gene function. The state of *cl* gene determines the option between lambda phage lysogenic or lythic regimes. Similarly, transposons *P* in Drosophila and *Ac* and *Spm* in corn are organized as epigenes with positive auto-regulation.

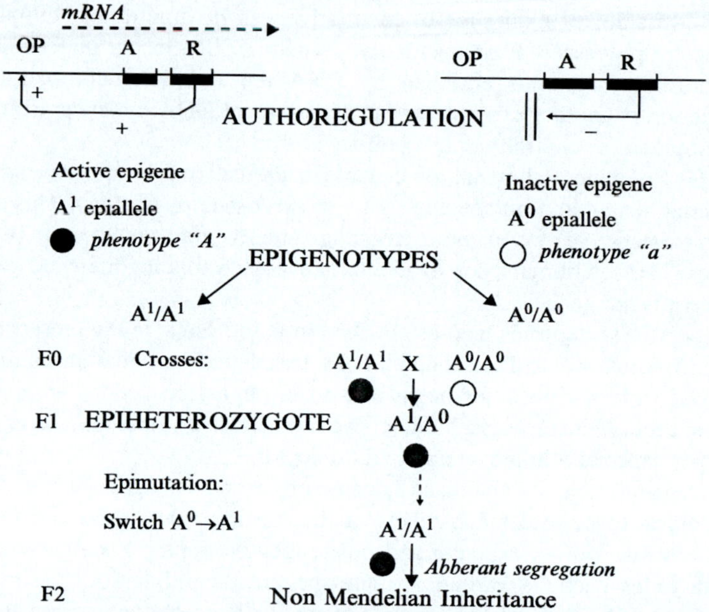

Fig. 22.2 A scheme of an epigene. Epigenes constitute autoregulatory feed back links via DNA binding proteins. The figure shows the positive auto-regulation at the transcription level. A^1 and A^0 designate active and non-active states of the epigene. Crosses may result in epimutation in the progeny of F1 epiheterozygotes

Figure 22.2 shows possibility of stochastic switching (trans-activation) from inactive to an active epigene state in cell epiheterozygotes. Such switching represent an epimutation with non-mendelian inheritance.

Noteworthy, if we have five independent epigens in the genome, the cell may have 64 potential variants of phenotype without any structural changes in DNA sequences.

22.4 Conclusion

The presented generalized approach to the genome organization and function posits a division of the genome on two subsystems OGE and FGE and assumes at least three types of heritable changes: mutations, variations and epigenetic alterations. Heritable changes of diverse facultative genome elements or variations occur simultaneously in many individuals. They might be site specific and are induced by various weak non-mutagenic environmental factors as temperature, interpopulation crosses or nutritional shifts ("genotrophs" in plants). The same is right for an epigenetic alterations.

Activation of mobile elements in nature results in mutation bursts: insertion – connected gene mutations and rearrangements. Insertion mutations in Drosophila embraces about 70% of spontaneous visible mutations. Facultativeness principle and generalized approach to the genome organization and function have an important implications. We indicate only on necessity a three-generation approach in demography and epidemiology [11, 12].

References

1. Berg RL (1982) Mutability changes in *Drosophila melanogaster* population of Europe, Asia and North America and probable mutability changes in human populations in the USSR. Jpn J Genet 57:171–183
2. Berg RL, Engels WR, Kreber RA (1980) Site-specific X-chromosome rearrangements from hybrid dysgenesis in Drosophila melanogaster. Science 210(4468):427–429
3. Berg RL, Engels WR, Kreber RA (1981) Site-specific X-chromosome rearrangements from hybrid dysgenesis in Drosophila melanogaster. Science 210:427–429
4. Blattner FR, Plnkett G et al (1997) The complete genome sequence of Escherichia coli K12. Science 227:1453–1462
5. Caporate LH (2000) Mutation is modulated: implication for evolution. Bioessays 22:388–395
6. Evgen'ev MB, Zelentsova H, Poluectova H, Lyozin GT, Velikovskaya VV, Pyatkov KL, Kidwell MG (2000) Mobile elements and chromosome evolution in the virilis group of Drosophila. Proc Natl Acad Sci USA 97:11337–11342
7. Golubovsky MD (1980) Mutation process and microevolution. Genetica 52(53):139–149
8. Golubovsky MD (1985) Organization of genotypes and types of hereditary variability in eukaryotes. Biol Bull Rev 100:323–329
9. Golubovsky MD (2000) A century of genetics: evolution of ideas and concepts. Borey, St. Petersburg
10. Golubovsky MD, Belyaeva ES (1985) Mutation outburst in nature and mobile genetic elements: study of a series of alleles in the singed bristle locus in *Drosophila melanogaster*. Genetika (Moscow) 21(10):1662–1670
11. Golubovsky MD, Manton KG (2005) A three-generation approach in biodemography is based on the developmental profile and the epigenetics of female gametes. Frontiers Biosci 10:187–191
12. Golubovsky MD, Manton KG (2005) Genome organization and three kinds of heritable changes: general description and stochastic factors (Review). Frontiers Biosci 10:335–344
13. Gregory TR (2001) Coincidence, coevolution, or causation? DNA content, cell size, and the C-value enigma. Biol Rev 76(1):65–101
14. Holliday R (1987) The inheritance of epigenetic defect. Science 238:163–170
15. Holliday R (1990) Mechanisms for the control of gene activity during development. Biol Rev 65:431–470
16. International Human Genome Sequencing Consortium (2001) Initial sequencing and analysis of the human genome. Nature 409:860–921
17. Jacob F, Wollman E (1961) Sexuality and the genetics of bacteria. Academic Press, New York
18. Kazazian H (2004) Mobile elements: drivers of genome evolution. Science 303:1626–1632
19. Khesin RB (1984) Genome Inconstancy. Nauka, Moscow
20. Khesin RB (2004) Genome inconstancy. Nauka, Moscow (in Russian)
21. King RC, Stansfield WD (2002) A dictionary of genetics, 6th edn. University Press, Oxford
22. Lim JK (1981) Site-specific intrachromosomal rearrangements in *Drosophila melanogaster*: cytogenetic evidence for transposable elements. Cold Spring Harb Symp Quant Biol 45:553–560

23. Mono J, Jacob F (1961) General conclusion: teleonomic mechanisms in a cellular metabolism, growth and differentiation. Cold Spring Harbor Symp Quant Biol 26:389–401
24. Nebert DW (2002) Transcription factors and cancer: an overview. Toxicology 181:131–141
25. Redon R, Ishikawa S, Fitch KR et al (2006) Global variation in copy number in the human genome. Nature 444:444–454
26. Shapiro JA (1999) Genome system architecture and natural genetic engineering in evolution. Annals NY Acad Sci 870:21–35
27. Shapiro JA (2002) Genome organization and reorganization in evolution: formatting for computation and function. Annals NY Acad Sci 981:11–134
28. Tchuraev RN, Galimzyanov AV (2009) Gene and epigene network: two levels of organizing of hereditary system. J Theor Biol 259(4):659–669
29. Tchuraev RN, Stupak V, Tropnia TS, Stupak E (2000) Epigenes: design and construction of new hereditary units. FEBS 486(3):200–202
30. Tyedmers J, Madariaga ML, Lindquist S (2008) Prion switching in response to environmental stress. PLoS Biol 6(11):2605–2614

Chapter 23
Recent Inserts of Transposable Elements Affect Structure and Functions of Human Genome

Anton Buzdin, Maria Suntsova, Olga Bantysh, Elena Aleksandrova, Anastasia Zabolotneva, Elena Gogvadze, and Nurshat Gaifullin

Abstract Transposable elements (TEs) are selfish fragments of DNA able to reproduce themselves into the host genomes. TEs typically occupy ~40–50% of the mammalian genomes. In our studies, we focus on evolutionary recent TE inserts that appeared in the DNA of human ancestor lineage after divergence with the chimpanzee ancestry, *i.e.* less than ~6 million years ago. These human specific elements (hsTEs) represent only a minor fraction of the whole TE cargo of the human genome. hsTEs are represented by the four families called HERV-K(HML-2), L1, Alu and SVA. The number of human specific copies for HERV-K(HML-2), L1, Alu and SVA families is approx. 150, 1,200, 5,500 and 860 copies per genome, respectively. Taken together, hsTEs shape ~6.4 megabases of human DNA, which is about 6-times lower than what is occupied by the human specific simple nucleotide polymorphisms, and 23-times smaller than the overall length of human specific deletions and duplications. However, although modest in terms of genomic proportion, hsTEs should be regarded as the perspective candidates for being molecular genetic agents of human speciation. Unlike most of random mutations and duplications, each novel insert of hsTE has provided to the recipient genomic locus a set of functional transcriptional factor binding sites positively selected during the TE evolution. For example, clusters of novel inserts of Alu elements may serve as CpG islets, SVA elements provide functional splice sites and polyadenylation signals, whereas L1 and HERV-K(HML-2) elements donate enhancers, promoters, splice sites and polyadenylation signals. Significant proportion of the human-specific genomic deletions,

A. Buzdin (✉) • M. Suntsova • O. Bantysh • E. Aleksandrova • A. Zabolotneva • E. Gogvadze
Shemyakin-Ovchinnikov Institute of Bioorganic Chemistry,
Miklukho-Maklaya 16/10, Moscow 117997, Russia
e-mail: bu3din@mail.ru

N. Gaifullin
Lomonosov Moscow State University, Lomonosovsky pr. 31-5,
Moscow 119192, Russia

C.E. Mothersill et al. (eds.), *Radiobiology and Environmental Security*,
NATO Science for Peace and Security Series C: Environmental Security,
DOI 10.1007/978-94-007-1939-2_23, © Springer Science+Business Media B.V. 2012

duplications and translocations has been also generated due to ectopic recombinations between the different individual TE inserts. Among the other, we report for the first time a detailed functional characteristics of the HERV-K(HML-2) hsTEs done at the genome-wide level. We have identified 65 active *in vivo* human specific promoters contributed by these elements. We also identified three cases of the hsTE-mediated human specific transcriptional regulation of functional protein-coding genes taking part in brain development during embryogenesis. We found ~180 human specific polyadenylation signals transferred by the SVA elements into the introns of known functional genes. Scaling of these data to the total number of the hsTEs predicts that hundreds of human genes are regulated by these elements. Finally, we discovered the first exclusively human specific TE family, represented by ~80 members formed by a combination of a part of a CpG islet of human gene *MAST2* ansd of the 3′-terminal part of an SVA retrotransposon. According to our estimates, this family, termed CpG-SVA, was far more active than the ancestral SVA family. Our data indicate that *MAST2* regulatory sequence was recruited during the evolution to provide effective CpG-SVA transcription in human testicular germ-line cells.

Keywords Human evolution • Genetic instability • Transposable elements • Human specific promoters • Antisense transcripts • Regulation of gene expression • Brain development • Hybrid family of retrotransposons

23.1 Introduction

23.1.1 Recent Evolution of the Human Genome

23.1.1.1 Major Genetic Differences Between Humans and Chimpanzees

Understanding of the genetic basis that accounts for the obvious differences in phenotypes of humans and their closest relatives, chimpanzees, is one of the most interesting tasks of modern life sciences. This task is also challenging, mostly due to strikingly high similarities in their genome structure and organization [25, 26, 106, 117]. Indeed, an average divergence between human and chimpanzee DNAs is about 1.24% [36], being as low as only 0.5% in protein coding regions [54]. Human and chimpanzee ancestor lineages diverged relatively recently in evolution, approximately 6 million years ago [57]. At present, we still don't know exactly what genetic traits make us humans, but a number of functionally important differences between human and great ape genomes have been identified. In general, they can be classified into the four major groups:

1. diverse chromosomal organization, including deletions, inversions, duplications and translocations
2. variations in copy number, genomic localization and functional status among the pre-existing common sequences

3. differences in protein coding regions
4. lineage-specific genomic insertions of transposable elements (TEs).

23.1.1.2 Non-TE Differences

Millions of mutated loci, mostly single nucleotide polymorphisms (SNPs), are known to be polymorphic in humans. However, they should not be considered here due to the lack of a functional role in human speciation for these sequences, which are presented only in a fraction of human population. Human and chimpanzee genomes, both approximately $3 \cdot 10^9$ base pairs in size, share ~98.8% sequence identity [36], thus making identification of functional human-specific sequences finding the needle in a haystack. Theoretically, recent success in human and chimpanzee DNA sequencing projects [25, 116] has provided an instrument for the direct comparison of genomes with the subsequent association of genomic changes with interspecies differences at the level of protein expression. However, in practice the chimpanzee genome draft, currently available in public databases, is not sufficiently accurate for such a comprehensive study [20], although many successful large-scale bioinformatical screenings have been performed [56, 81, 110, 111]. A plenty of lineage-specific substitutions, deletions, insertions, duplications, expressed pseudogenes, anonymous RNAs, transposable elements have been identified nowadays, but the omnibus study still remains to be done.

Cytogenetic differences. The comparison of human and African great ape karyotypes using fluorescent in situ hybridization has revealed the most important lineage-specific distinction, which is the fusion in human lineage of two ancestral chromosomes (human chromosome 2), corresponding to chimpanzee chromosomes 12 and 13 [141]. Other major points are numerous changes in centromeric and telomeric regions [62, 97, 102, 114] and lineage-specific rearrangements and amplifications of several gene families in non-recombining parts of Y chromosome [47]. Apart from translocations, insertions and deletions have together given rise to at least 150 Mb of genomic DNA sequence that is either present or absent in humans as compared to chimpanzees, according to the recent estimation by Kehrer-Sawatzki and Cooper [72]. Interestingly, mostly chromosome ends were the "hot spots" of recent genome evolution [68].

Emerging or inactivation of functional genes. Few functional genes are known to distinguish human and ape DNA. First of all, this is the functional deletion of an exon within the protein coding sequence of human gene *CMP* for syalic acid hydroxylase. Mutation caused by the human specific insertion of an Alu retroelement into 92 bp-long *CMP* exon, disrupted normal open reading frame for this enzyme and resulted in the lack of N-glycolyl neuraminic acid (Neu5Gc) on a surface of human cell membranes [22, 65]. Neu5Gc, thus, is replaced in humans by its precursor, N-acetyl neuraminic acid (Neu5Ac). This absence of Neu5Gc is the major biochemical distinction between human and chimpanzee, which, theoretically, may influence intercellular interactions and embryo development. Some other ancestor genes, mostly encoding olfactory receptors [45], have been lost or pseudogenised in the human lineage due to premature stop codon accumulations [56, 134]. On the contrary, transcribed human specific sequence termed *c1orf37-dup* gene, encoding for short transmembrane protein of unknown function, is selectively expressed in several human tissues including brain [140]. Also, a number of new copies of genes involved in immune response such as leukocyte receptors or antigens, have been acquired, lost or mutated during the recent human lineage evolution [43]. Finally, Pollard et al. [110] recently reported a novel human specific gene *HAR1F* for a putative regulatory RNA that is expressed specifically in the developing human neocortex from 7 to 19 gestational weeks, a crucial period for cortical neuron specification and migration.

Gene duplications. Gene duplications may influence cell physiology by providing additional copies of transcribed genes, thus escaping the original qualitative control of gene expression. For example, 7–11 copies of the olfactory receptor gene *OR-A* reside in human DNA, whereas the chimpanzee genome posesses only one copy of that gene. Different human copies are transcribed with different specificities, depending on their new genomic context [78, 127]. Similarly, eight genes for keratinocyte growth factor KGF were mapped in human DNA, in contrast to only five copies in the chimpanzee [143].

Lineage specific nucleotide substitutions. Millions of human specific single nucleotide substitutions, short deletions, duplications or microsatellite amplifications have been documented to the date [72, 117]. Many of them have been mapped in the regulatory genomic regions or in protein coding sequences. For example, chimpanzee dopamin receptor gene *D4* has 12 bp long deletion, as compared with its human ortholog [86]. However, the biological significance of these numerous changes accounting for a total of ~36 megabases in our DNA (mostly single nucleotide substitutions) is still unclear.

Differences in gene expression. Identifying differentially transcribed sequences may be a better solution for the direct finding of functional genes that might be involved in human speciation [37, 40]. For example, Nadezhdin et al. managed to identify differential transcription of a gene for transthyretin, the carrier of thyroid hormones, in the cerebella of humans and chimpanzees [101]. However, one has to compare samples from the same sex/physiological state groups of tissue donors. Due to an extremely limited number of the available chimpanzee tissue specimens, no reliable comparizon has been made so far, and the observed interspecies differences in gene expression remain frequently less in amplitude than the intraspecies ones [37].

23.1.1.3 Insertions of Transposable Elements

TEs are DNA fragments, capable of self-reproducing and changing their location into the host genome, *i.e.* to transpose. These selfish repetitive elements proliferate either directly via their DNA copies (DNA transposons), or through RNA intermediates (retroelements) utilizing the mechanism termed ‘reverse transcription’ and the RNA-dependant DNA polymerase enzyme, called reverse transcriptase (RT). The newly formed DNA copy of the element then integrates into the genome, using

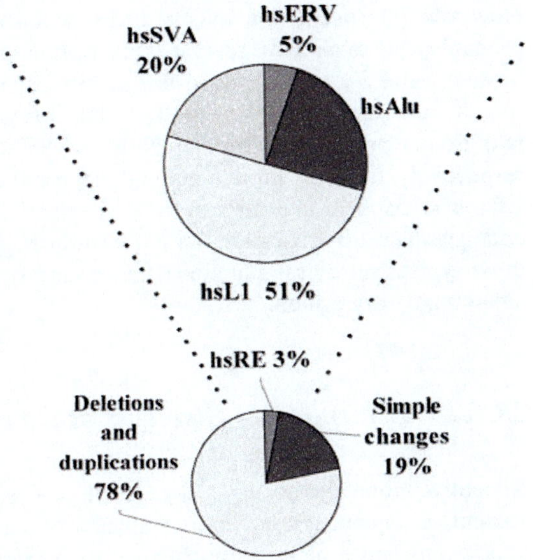

Fig. 23.1 Endogenous retroviruses occupy ~5% of the DNA shaped by human specific transposable elements, which, in turn, form only 3% of the total lineage specific DNA

a combination of host and self-encoded proteins, depending on the TE origin [11]. Retroelements, which constitute >42% of human DNA, are the only class of TEs, able to transpose in mammals [10, 124]. Four retroelement families (L1, Alu, SVA and HERV-K(HML-2)) were transpositionally active after the divergence of human and chimpanzee ancestries, thus forming relatively modest fraction of human-specific inserts (~7,800 copies [98], compared to a total of ~3 millions of human retroelements [77, 132]).

Together, human specific retroelements constitute approximately 6.4 megabases of the human DNA (Fig. 23.1), which is sixfold lower than that formed by short nucleotide substitutions, and 23-fold lower than human specific deletions/duplications. However, such a modest proportion is somewhat compensated by the active role of functional Genome Reshapers that is being played by human retrotransposons [10, 31, 123, 124, 136]. TEs are known to be recombination hot spots (e.g., human specific Alu-Alu recombinations resulted in deletion of at least 400 kb of human DNA [120]). It is known that retroelements can modify the activity of pre-existing human genes [9, 10, 31, 131]. At least one third of all human specific retroelements has been mapped within or close to genes [98]. Therefore, REs may well be one of the causative agents responsible for the phenotypic differences between *Homo sapiens* and its closest relatives, *Pan paniscus* and *Pan troglodytes* chimpanzees. These differences can be envisioned to arise not from the appearance of any new and/or disappearance of old genes but due to variations in the regulation of some genes common for the related species.

The first group, (~1,200 human specific members), is the L1 family of authonomous retrotransposons. The full-length primate L1s are about 6 kb long elements encoding two open reading frames, for RT/integrase and RNA binding protein.

However, L1 inserts are mostly 5′-truncated deficient copies originated, most probably, due to abortive reverse transcription [5]. The next two groups, Alu (~300 bp long) and SVA (~1.5 kb in size) retroposons, are non-authonomous TEs that recruit "heterologous" RT of the L1 origin for their own proliferation [133]. These two groups, presented in human DNA by ~5500 and ~860 lineage-specific copies, respectively, lack any protein coding genes and can be regarded as the parasites of L1 retrotranspositional machinery [10]. Finally, authonomous HERV-K (HML-2) endogenous retroviruses are the most complex group of human TEs. They harbor three typical retroviral functional genes and one additional gene encoding for a small regulatory protein.

23.1.2 *Transposable Elements as Genome Reshapers*

Repetitive sequences occupy a huge fraction of essentially every eukaryotic genome. Repetitive sequences cover more than 50% of mammalian genomic DNAs, whereas gene exons and protein coding sequences occupy only ~3% and 1%, respectively. Numerous genomic repeats include genes themselves. Those generally encode "selfish" proteins necessary for the proliferation of TEs in the host genome. The major part of evolutionary "older" TEs accumulated mutations over time and fails to encode functional proteins. However, repeats have important functions also on the RNA level [49]. Repetitive transcripts may serve as multifunctional RNAs by participating in the antisense regulation of gene activity and by competing with the host-encoded transcripts for cellular factors. Moreover, polymorphic intron-located L1 and Alu elements have been shown recently to decrease transcription of the corresponding alleles when compared to the expression of retroelement-free alleles [80, 128, 129]. In addition, genomic repeats include regulatory sequences like promoters, enhancers, splice sites, polyadenylation signals and insulators, which actively reshape cellular genome and transcriptome.

23.1.2.1 TEs as Transcriptional Promoters

Whole-genome analysis revealed that about 25% of all human promoters contain REs in their sequence [130]. Moreover, 7–10% of experimentally characterized transcription factor binding sites (TFBS) were shown to be derived from repetitive sequences including simple sequence repeats and transposable elements [109]. TFBS that originated from repeats evolve more rapidly than non-repetitive TFBS but still show signs of sequence conservation on functionally critical bases. Such rapidly evolving TFBS are likely to direct species-specific regulation of gene expression, thus participating in evolutionary process (Fig. 23.2).

In the majority of examples reported to date, REs act as alternative promoters, but may also represent the only known promoter for some human genes. For example, L1 and Alu sequences act as the unique promoter for *HYAL-4* gene, necessary for hyaluronan catabolism [130]. The application of novel high-throughput techniques

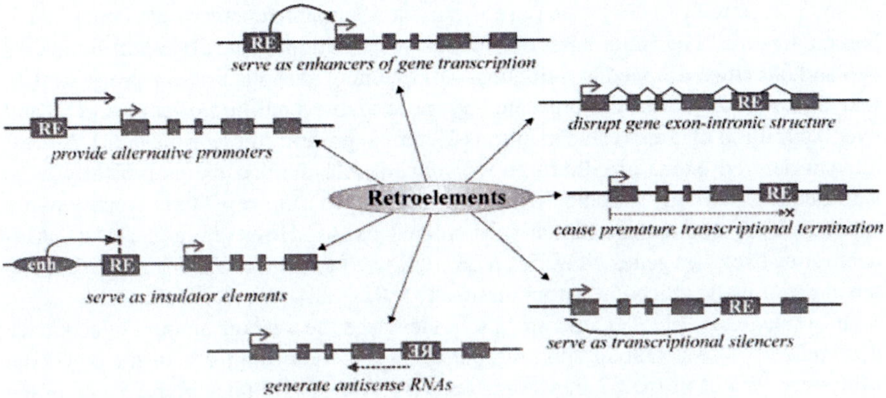

Fig. 23.2 Different mechanisms of RE influence on gene transcription

such as cap analysis of gene expression (CAGE) and paired-end ditag (PET) sequencing recently revealed 51,197 endogenous retrovirus (ERV)-derived promoter sequences. In 1,743 cases, ERVs were located in gene proximal or 5′ untranslated regions. 114 ERV-derived transcription start sites drive transcription of 97 human genes, producing chimeric transcripts initiated within LTR and read-through into known gene sequences [24].

23.1.2.2 TEs as Transcriptional Enhancers

There are many examples of TE enhancer activity in human tissues *in vivo*. For example, the ERV9 LTR element upstream of the DNase I hypersensitive site 5 (HS5) of the locus control region in the human β-globin cluster is responsible for controlling expression of this cluster in erythroid cells [87]. The enhancer of human apoliprotein A was shown to reside within LINE element [139]. Alu sequence is a part of enhancer element located in the last intron of the human CD8 alpha gene [58].

23.1.2.3 TEs as the Alternative Splice Sites

Apart from the modulation of transcription, TEs can also regulate splicing of pre-mRNA. In a genome wide comparison of the genomes of human and mouse, a total of 3,932,058 and 3,122,416 transposable elements have been identified in human and mouse, respectively. Interestingly, 60% of transposons in human are located in intronic sequences, whereas introns occupy only 24% of the genome [119]. All TE families in human can "exonize", *i.e.* be included in the exons of mature mRNA. TEs that are shared between human and mouse exhibit the same percentage of exonization in the two species, but the exonization level of a primate-specific retroelement Alu is far greater than that of other human transposons. This results in a higher overall level of transposon exonization in human than in mouse (1,824 exons compared with 506 exons, respectively) [119]. Alus are the most abundant repetitive elements in the human genome. The major burst of *Alu* retroposition took place 50–60 million years ago and has since dropped to a frequency of one new retroposition for every 20–125 new births [3, 28]. Alus are presented by more than 1.1 million of copies [21], and over 0.5 million of them reside in introns of human protein coding genes [84]. Almost all Alu-derived exons are alternatively spliced. Alu-derived exons typically have significantly weaker splicing signals compared to non-repetitive constitutively spliced exons and other alternatively spliced exons. However, at least six Alu-containing exons (in genes *FAM55C*, *NLRP1*, *ZNF611*, *ADAL*, *RPP38* and *RSPH10B*) are constitutively spliced in human tissues [85, 91, 122].

In some genes, Alu elements strikingly increased the average amount of sequence divergence between human and chimpanzee up to more than 2% in the 3′-UTRs. Moreover, 20 out of the 87 transcripts carrying Alu insert either in the 5′- or in the 3′-UTR contained more than 10% structural divergence in length. In particular, two-thirds of this structural divergence was found in the 3′-UTRs, and variable transcription start sites were conspicuous in the 5′-UTRs [118]. In both 5′- and 3′-UTR sequences, presence of an Alu element may be important for post transcriptional regulation of gene expression, for example by affecting protein translation, alternative splicing and mRNA stability [60]. Alu exonization might have played a certain role in human speciation. For example, there is a muscle-specific inclusion of an Alu-derived exon in mRNA of gene *SEPN1* (gene implicated in a form of congenital muscular dystrophy), which appeared due to a human-specific splicing change after the divergence of humans and chimpanzees [83]. The second example is the above mentioned functional deletion of an exon of human gene *CMP* for syalic acid hydroxylase.

Overall, the proportion of proteins with TE-encoded exons (approximately 0.1%), although probably underestimated, is much less than what the data at transcript level suggest (approximately 4%) [55].

23.1.2.4 TEs as Providers of Polyadenylation Signals

mRNA polyadenylation is an essential step for the maturation of almost all eukaryotic mRNAs, and is tightly coupled with termination of transcription in defining the 3′-end of genes. A polyadenylation signal (AAUAAA) nearby the 3′ end of pre-mRNA is required for poly(A) synthesis. The protein complex involved in the pre-mRNA polyadenylation is coupled with RNA polymerase II during the transcription of a gene, and only RNA polymerase II – products are terminally polyadenylated with the remarkable exception of two polyadenylated polymerase III – transcribed RNAs [7]. Autonomous retrotransposons encode proteins and utilize functional poly(A) signals at the 3′-termini of their genes. Therefore, insertions of these elements in genes in the sense orientation can influence the expression of neighboring genes by providing new poly(A) signals. For example, 5′ LTR of the retrovirus HERV-F may function as the alternative polyadenylation site for gene *ZNF195* [73]. Human genes *HHLA2* and *HHLA3* utilize HERV-H LTRs as the major polyadenylation signals [89]. Recently it was estimated that ~8% of all mammalian poly(A) sites are associated with TEs [82].

In general, there is a clearly seen strong negative selective pressure on the intron-located autonomous TE inserts oriented in the same transcriptional direction as the enclosing gene [12, 29, 131, 137, 142]. Indeed, all protein-coding intronic retroelements (including LINEs and LTR retrotransposons) oriented sense to gene transcription are underrepresented in all investigated genomes compared to statistically expected ratio of sense/antisense inserts. In contrast, non-autonomous retroelements like Alu don't employ polyadenylation of their transcripts and, thus, may have only casual AAUAAA sequences. However, such poly(A) signals are very weak and are highly affected by the surrounding sequence [113].

23.1.2.5 TEs as the Antisense Transcriptional Regulators

It has been demonstrated that TE inserts in gene introns are preferentially fixed in the antisense orientation relatively to enclosing gene transcriptional direction [96, 131]. Therefore, promoters of the intronic TEs may drive transcription of the RNAs that are complementary to gene introns and/or exons. Moreover, some retrotransposons are also known to possess bidirectional promoter [27, 34, 35, 39, 63, 94], and even downstream insertions of these elements relatively to genes may result in production of the antisense RNAs. Recently applied CAGE technology identified 48,718 human gene antisense transcriptional start sites within transposable elements [23].

One possible mechanism of the antisense regulation on the pre-mRNA level is connected with the generation of alternatively spliced mRNAs. It has been shown previously that antisense transcripts can inhibit splicing of pre-mRNA *in vitro* and *in vivo* [44]. The possible mechanism involves pairing of antisense transcript and a sense target RNA with the formation of double-stranded RNA that could induce the spliceosome to skip the paired region, thus forming an alternatively spliced transcript. This would result in the formation of non-functional RNAs containing multiple premature transcription termination codons. Normally, such RNAs are immediately degraded in the cytoplasm by nonsense-mediated decay machinery [38]. Alternatively, antisense transcript basepairing to the target RNA can lead to its rapid enzymatic degradation directly in the nucleus.

23.1.2.6 TEs as Recombination Agents

Recombination is a powerful factor of evolution that produces genetic variability by using reshuffling of already existing blocks of biological information [90]. Because of their high copy number and sequence similarity, TEs are the ideal substrates for illegitimate homologous recombination, also called ectopic recombination. The chance that an ectopic recombination will occur depends on the number of homologous sequences and on the length of the elements [6, 121]. Recombination causes genetic rearrangements that can be deleterious, advantageous or null. Alu-derived ectopic recombination generated 492 human-specific deletions, the distribution of which is biased towards gene-rich regions of the genome [120]. Finally, L1s were shown to join DNA breaks by inserting into the genome through endonuclease-independent pathway, thus participating in DNA double-strand breaks repair [100].

23.1.2.7 TE-Transduction of the Flanking Sequences

The ability to transduce 3′-flanking DNA to new genomic loci was firstly shown for the L1 elements [53, 99, 108]. L1s have a rather weak polyadenylation signal; therefore, RNA polymerase sometimes gets through it and terminates an RNA synthesis on any polyadenylation site located downstream. It was estimated that ~20% of all L1 inserts contain transduced DNA at the 3′ends. The length of these sequences varies from few bases to over 1 kb. Taken together, such transduced DNA makes up ~0.6–1% of the human genome. Therefore, L1-mediated transductions have the potential to shuffle exons and regulatory sequences to new genomic sites.

Recently it was shown that SVA elements are also able to transduce downstream sequence and it was estimated that about 10% of human SVA elements were involved in DNA transduction events [107, 133]. Moreover, SVA-mediated transduction can serve as a previously uncharacterized mechanism for gene duplication [138]. It the latter case new sequences may appear either on the 5′- or on the 3′ terminus of an SVA (5′ and 3′ SVA transduction, respectively). 3′ Transduction mechanism is similar to that proposed for L1 retrotransposon. The size of genomic sequence transferred in such a way may differ from several base pairs to over 1.500 bp. The most striking example is the transduction of a whole gene *AMAC* (acyl-malonyl condensing enzyme 1) in the great ape genomes [138]. Due to SVA 3′ transduction, human genome has two additional copies of *AMAC*.

Another kind of transduction results in attaching of new sequences to the 5′ end of an SVA. TE transcription initiation may proceed from any promoter located upstream in the genomic sequence. In this case termination of transcription and RNA processing usually occur using normal polyadenylation signal of a TE. This results in a mature RNA having on its 5′ end an additional copy of flanking genomic sequence and a copy of RE at its 3′ end. Subsequent reverse transcription and integration into the genome of a nascent cDNA result to a new RE genomic insert carrying 5′ transduced part [8].

23.2 Results

23.2.1 *Discovery of New Human TE Families*

23.2.1.1 RNA Recombination-Derived TEs

A typical LINE element encodes two proteins: ORF1p that is a RNA binding protein which likely helps reverse transcription as a nucleic acid chaperone [93], and ORF2p, the reverse transcriptase and the endonuclease [71]. Due to a ‘cis-preference’,

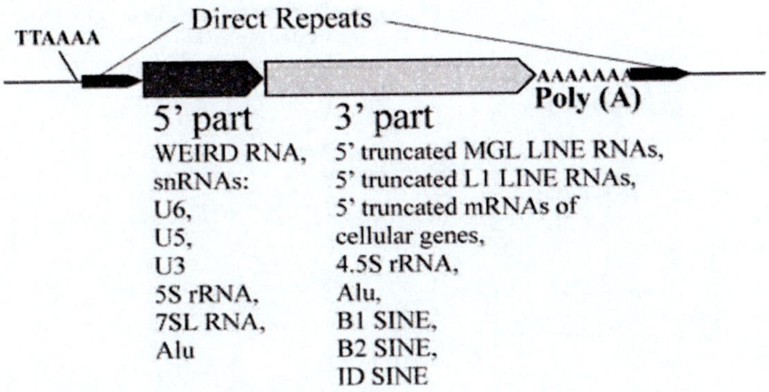

Fig. 23.3 Schematic representation of the bipartite chimeric retrogenes. Inserts are flanked by 10–20 bp long genomic direct repeats

the enzymatic machinery of a retrotransposition-competent LINE predominantly transposes its own copies [135]. However, LINEs are also able to mediate the transposition of other sequences, mostly non autonomous elements termed SINEs, but also cDNAs originating from different cellular RNAs, leading to the formation of processed pseudogenes [32]. Recently, we have shown that LINEs are involved in the formation of bi- and tripartite chimeric retrogenes during reverse transcription in many genomes including human and fungi [13, 14, 16, 41, 50]. Bipartite chimeric retrogenes with an unusual structure were identified in three mammalian and in one fungal genomes (Fig. 23.3).

A total of 82, 116, 66 and 31 elements were found in human, mouse, rat and rice blast fungus *Magnaporthe grisea* DNAs, respectively [13, 14, 41, 50]. These elements are composed of DNA copies from cellular transcripts either directly fused to each other or more frequently fused to the 3′ part of a LINE retrotransposon. The various cellular transcripts found in these chimeras correspond to messenger RNAs, ribosomal RNAs, small nuclear RNAs, and 7SL RNA.

The chimeras have the following common features: (1) 5′-parts are full-length copies of cellular RNAs; (2) 3′-parts are 5′-truncated copies of the corresponding RNAs (mostly LINEs); (3) sites of these truncations occur at random in the corresponding RNA; (4) both parts are directly joined with the same transcriptional orientation; (5) chimeras have a poly (A) tail at their 3′ end, and (6) chimeras are flanked by short direct repeats.

The last structural feature demonstrate that these elements were transposed as bipartite DNA copies. Indeed, mammalian chimeras carried at their 5′ ends a T_2A_4 hexanucleotide or its variants [13, 14, 50] that correspond to the T_2A_4 genomic site used by LINEs to initiate reverse transcription on oligo (A) motifs and separate newly inserted DNA by short tandem repeats [67]. The simultaneous integration of both parts of these chimeras was further supported by the data came from PCR-based evolutionary insertion polymorphism assay [13, 14].

This suggests that these bipartite elements are generated by a specific active mechanism. It frequently combines functional cellular transcripts that have nothing in common with transposable elements [19]. Many of the chimeras can be considered

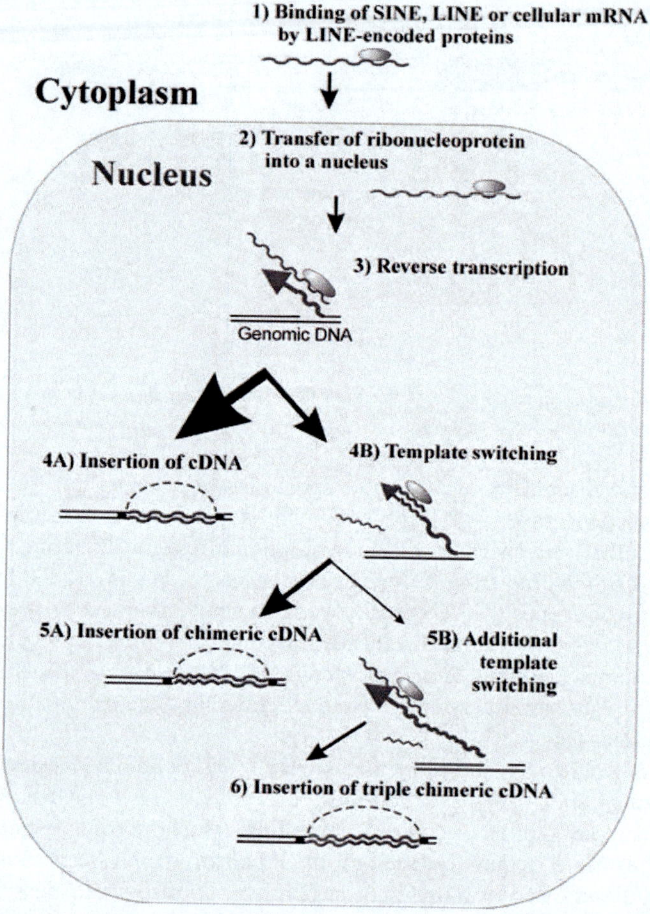

Fig. 23.4 Mechanism for the chimeras' formation using LINE enzymatic machinery. (*Step 1*) LINE pre-integration complex binds LINE, SINE or RNA in the cytoplasm. (*Step 2*) The resulting ribonucleoprotein is transferred to the nucleus. (*Step 3*) Reverse transcription of the bound RNA primed by a genomic DNA single-stranded break (target site primed reverse transcription). (*Step 4A*) Successful integration of the reverse transcribed cDNA copy into the genomic DNA. (*Step 4B*) Switch of templates on another RNA during the reverse transcription. (*Step 5A*) Integration of the chimera formed into genomic DNA. (*Step 5B*) The second template switch to another RNA with subsequent DNA reparation mediates formation of a tripartite chimeric retrogene insertion flanked by short direct repeats. The normal LINE integration pathway is: steps (*1*), (*2*), (*3*), (*4A*)

as new genes, as they were shown to be transcribed, some of them in a tissue-specific manner [14, 48, 50, 51]. Later on, in the mammalian and fungal genomes we found also the tripartite chimeras of a similar structure [51]. We further proposed that these chimeric retrogenes were generated through a mechanism involving RNA recombination during the reverse transcription of cellular RNAs (Fig. 23.4). This model includes a switch from the nascent cDNA serving as template for the reverse transcription of the 3′ part of the chimera to another RNA template corresponding to the 5′ part, followed by the chimera integration into the host genome [10].

Although RT main enzymatic activity is the continuous synthesis of the cDNA on RNA template, RT is able to switch templates during reverse transcription. For example, in retroviruses, RT jumps from one site of the RNA template to another site, are necessary for the synthesis of LTRs. Moreover, as retroviral particles usually contain two independent RNA molecules [126], the high template switch frequency significantly increases the retroviral diversity through recombination between these RNAs [69]. These recombination events most probably account for the mosaic structure of most retroviruses [66, 125].

This model for the chimera formation was further supported by results obtained with human L1 LINE element using an elegant experimental system of retrotransposition *in vitro* [46]. The authors managed to characterize 100 *de novo* retrotransposition events in HeLa cells. Importantly, one insert (1%) represented a newly formed chimera similar to those we identified in human genome, consisting of a full length U6 snRNA fused to a 5′ truncated L1. Similar results were obtained *in vivo* with a transgenic mouse model for L1 retrotransposition by Babushok and coauthors that characterized 33 novel retrotransposition events. 13% of these events likely result from template switching during reverse transcription [1]. Interestingly, it has been recently postulated that RT template jumps from LINE RNA to host genomic DNA might facilitate integration and, thus, could be normally required for successful LINE retrotransposition [1, 4].

Besides generating chimeric retrogenes, template switching events during LINE reverse transcription could give rise to chimeric SINE elements [104] and to mosaic rodent L1 structures, likely resulting from RNA recombination between L1 templates [8, 61]. Evolution of certain LINE families might also involve RNA-RNA recombination, resulting in the fusion of the 3′ part of a LINE to a new sequence at their 5′ end, as suggested by the observation that the 5′-untranslated regions of human, murine, rat and rabbit L1 families are not homologous to each other [42]. Interestingly, RT encoded by another member of LINE superfamily – R2 from arthropods, was documented to jump from one template to another *in vitro*, with R2-R2 chimeras being formed [4].

Furthermore, it is speculated that LTR-containing retrotransposons and SINEs themselves represent chimeric elements [75, 76, 92, 105]. A phylogenetic analysis of the ribonuclease H domain revealed that LTR-containing retroelements might have been formed as a fusion between DNA transposon and non-LTR retrotransposon [92]. tRNA-derived SINEs likely descended from retroviral strong-stop DNAs [105]. They consist of two regions: a conservative, including a tRNA promoter and a core domain, and a variable one similar to 3′-terminal sequence of different LINE families. The core domain of tRNA-like SINEs has conservative regions similar to fragments of lysine tRNA-primed retroviral LTRs. On the basis of these structural peculiarities it was suggested that tRNA-derived SINEs emerged due to the integration of retroviral strong-stop DNA into the LINE 3′-terminal part. The RE formed could be transcribed by RNA polymerase III and spread through the genome. Such a mechanism of SINE formation could also explain how these elements can transpose

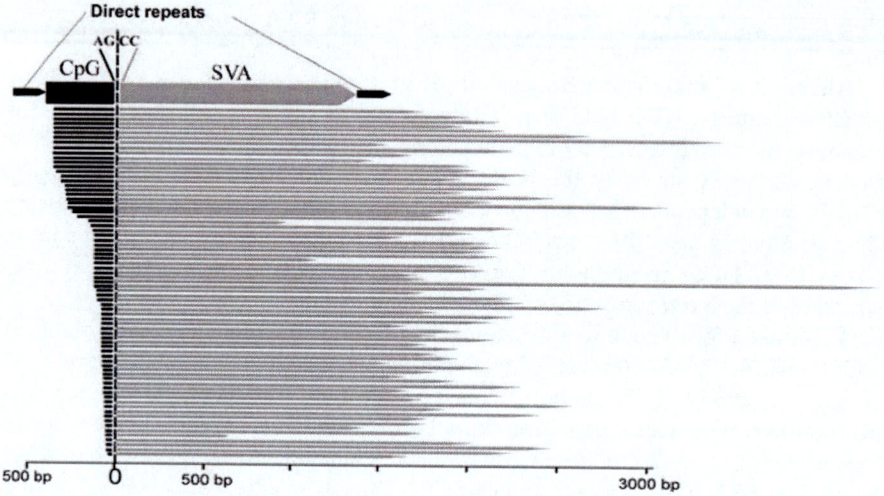

Fig. 23.5 Structure of chimeric CpG-SVA retrotransposons. CpG-SVA Inserts are flanked by direct repeats. Lengths of 5′ terminal (exonic) part vary from 35 to 383 bp, lengths of 3′ (SVA-derived) part vary from 662 to 4255 bp. 5′ Terminal parts are homologous to the first exon of *MAST2* gene, 3′ terminal parts – to SVA retrotransposon. Junction point between the two parts is identical in all CpG-SVA elements (canonical splice acceptor site AG from the side of exonic part and non-canonical splice donor site CC from the side of SVA). All SVA fragments start from the position 396 of the SVA consensus sequence

in the genome. Namely, it seems very likely that they recruited the enzymatic machinery from LINEs through a common "tail" sequence [105].

23.2.1.2 Human-Specific Hybrid Family CpG-SVA

Detailed structural analysis of the human specific SVA retrotransposons revealed 76 elements of an unusual structure. At the 5′ termini these elements carried copies of the first exon of *MAST2* gene, whereas at the 3′ end – SVA retrotransposon sequences. The border between exonic and SVA parts was located exactly between canonical acceptor splice site AG from exonic part and non-canonical donor splice-site CC from SVA-part (396 position in the SVA consensus sequence). Lengths of both parts of chimeric elements significantly varied: from 35 to 383 bp for the 5′-terminal part and from 662 to 4,255 bp for the 3′ terminal part. The border between the two parts was constant in all the chimeras (Fig. 23.5). On the 3′ terminus, the chimeras harbored a poly (A) sequence of variable length. These bipartite elements were flanked by 12–18 bp long direct repeats. In one case the length of direct repeats was unusually big (131 bp). Presence of the direct repeats surrounding chimeric inserts suggest implication of L1 retrotranspositional machinery in their formation, whereas poly (A) sequence indicates that retrotransposed RNA was transcribed by RNA polymerase II. The identified family of chimeric REs was called "CpG-SVA"

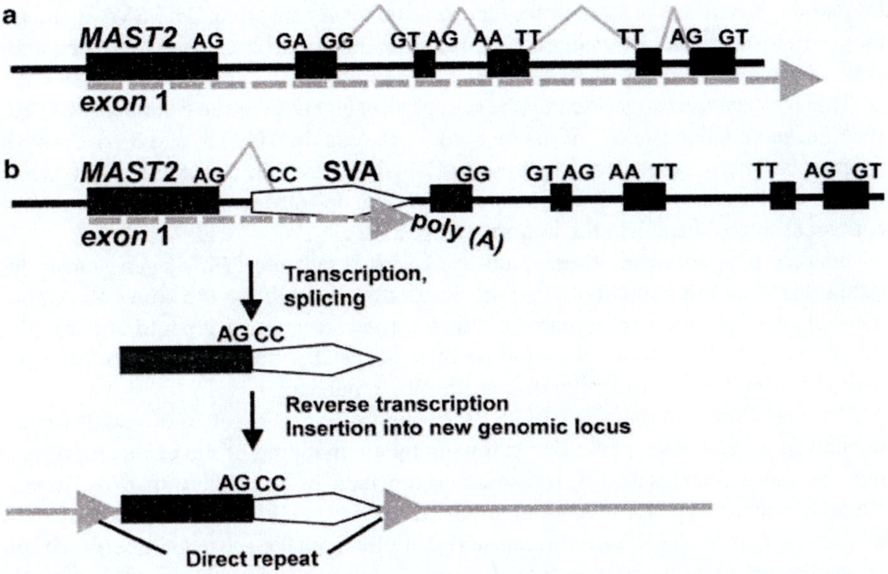

Fig. 23.6 Proposed mechanism of CpG-SVA family formation. **(a)** – schematic representation of genomic locus comprising human gene *MAST2*. *Dotted arrow* designates transcriptional direction, exons and splice sites are shown. **(b)** – Insert of an SVA retrotransposon in the sense orientation has changed gene exon-intronic structure and gave rise to aberrantly spliced mRNA polyadenylated at SVA sequence. Copy of this mRNA has inserted into a new locus of human genome and gave rise to CpG-SVA family that continued proliferation in human DNA. However, the ancestral allele of *MAST2* gene comprising SVA insert was lost due to the negative selection

because its 5′ terminal part complementary to the first exon of *MAST2* gene included a CpG island sequence. CpG-SVA elements were found only in human genomic DNA, whereas separately both SVA retrotransposons and *MAST2* exon sequence exist in the genomes of all great apes. Therefore, CpG-SVA may be regarded as a new human specific family of retrotransposons [2].

Two other papers describing the same family of hybrid retrotransposons (CpG-SVA) have been simultaneously published, where this family was termed either "MAST2-SVA" [59] or "SVA-F1" [30].

Basing on the structural features of the identified CpG-SVA family members, we purposed a mechanism for their formation (Fig. 23.6). At the first stage, SVA retrotransposon most probably has inserted into the first intron of *MAST2* gene in the sense orientation. After that there was formed an aberrant RNA driven by *MAST2* promoter and terminally processed using SVA polyadenylation signal. This RNA was further spliced which resulted in a fusion of the first exon of *MAST2* with a 3′-terminal fragment of an SVA (starting from 393 nucleotide of the SVA consensus sequence). This spliced chimeric RNA was then reverse transcribed by the L1 retrotranspositional machinery followed by integration of a nascent cDNA into the genome. This resulted to emerging of the master copy of CpG-SVA inserted into human DNA and flanked by direct repeats. The newly Inserted CpG-SVA element appeared to be transcriptionally active, possible due to the enclosed CpG-islet, and gave rise to a new family of REs.

This hypothesis is supported by the observation that there is the canonical *MAST2* gene acceptor splice site AG of on the border between the *MAST2*- and SVA-derived fragments. The putative donor splice site CC within an SVA is not canonical, what may be explained by the peculiarities of *MAST2* exon-intronic structure where non-canonical splice sites form the majority (Fig. 23.6).

Interestingly, at present there is no fixed SVA insert into *MAST2* gene intron in the human genome. Apparently, an ancestral allele containing the above SVA element in gene intron was eliminated by the negative selection as it could not provide functional *MAST2* mRNA formation because of the aberrant splicing of transcripts and/or preliminary polyadenilation on the SVA sequence.

We have found among the CpG-SVA elements several cases of 5′ and 3′ transduction of unrelated genomic DNA, proven by the mapping of the enclosing direct repeats. As in the classical 3′ transduction mechanism, it is likely that the downstream genomic fragments were captured due to "getting through" of SVA polyadenilation signals by the RNA polymerase II complex with the subsequent termination on any downstream sequence. In case of 5′ CpG-SVA transduction, there was apparently transcription of CpG-SVA elements initiated from upstream genomic promoters. Overall, we identified 18 and 11 cases of the 5′ and 3′ CpG-SVA transductions, respectively. The size of transferred genomic sequence differed from 8 to 854 bp for 5′- and from 141 to 734 bp for 3′ transduction events. Remarkably, four CpG-SVA elements contained both 5′ and 3′ transduced sequences. These four elements were highly identical and consisted of 364 bp long *MAST2* exon and 2,143–3,361 bp long SVA sequences. SVA length variations were caused by the Instability its internal satellite repeat modules. The double transducer CpG-SVAs were flanked by Alu sequence (member of evolutionally ancient AluSc family) at the 5′-termini and by the 400 bp long sequence including evolutionally ancient AluSp element at the 3′ ends. These structure similarities evidence common ancestry of these four elements from a single progenitor CpG-SVA element.

Once the exonic parts of the chimeras varied in length, but not in their primary structure, the SVA-derived parts had very different both lengths and primary structure. In the SVA parts there were different genetic changes like insertions, deletions, duplications, quantitative changes in tandem repeat composition and even insertions of retrotransposons. Together with the presence of transduced genomic sequences, this enabled us to construct phylogenetic tree for the members of CpG-SVA family to trace their reciprocal neighborhood. According to the primary structure similarity, CpG-SVA elements were grouped into three major branches (Fig. 23.3). Interestingly, although there was a kind of correlation between the size of "exonic" part and sequence localization on the tree, all three above brunches contained elements having exonic parts of very different lengths. There was also no connection between the position on a tree and lengths of the SVA parts. In several cases different tree brunches were including elements with the exactly same lengths of exonic part. For example, brunch 2 contained one CpG-SVA element with 364 bp long exonic part, whereas brunch 1 – five such elements. Exonic parts of seven elements from brunch 2 and of one element from brunch 3 were 148 bp long. There were also similar coincidences for the lengths 64, 76, 88 and 361 bp. These coincidences of exonic part sizes evidence that there were multiple independent events when CpG-SVA elements with identical exonic parts were formed.

The observed peculiarities of distribution of lengths of CpG-SVA exonic parts may be explained by the following factors: (1) there could be multiple functional transcription start sites within CpG-SVA, or (2) in some cases reverse transcription of the CpG-SVA RNA could terminate before the complete copying of the template has finished. The resulting shortened CpG-SVA inserts could, in turn, generate new elements having even shorter exonic parts, etc.

What are the functions of exonic part of CpG-SVA? Considering that (1) the first exon of *MAST2* gene includes CpG island, (2) CpG islands usually play major roles in gene transcriptional regulation, and (3) *MAST2* is strongly upregulated in testis, It can be hypothesized that the exonic part provides increased transcription of CpG-SVA family members in testis. This may be beneficial for the CpG-SVA family as it facilitates fixation of new inserts in the genome. To be fixed, RE insertion must occur into germ line cells, *e.g.* those localized in testis. Indeed, in terms of proliferation in the genome, the evolutionary young family CpG-SVA should be considered as very successful one: offsprings of only one among more than 1,000 SVA copies that resided in human DNA at that time (i.e. < 0.1%) have generated 76 new fixed inserts (~9% of all 860 human specified SVA elements) [2]. Experimental investigation of this hypothesis will be a matter of our further studies.

23.2.2 *Functional Characterization of a Family of Human Specific Endogenous Retroviruses HERV-K(HML-2)*

23.2.2.1 Identification of Human Specific Promoters

Promoter activity of human specific LTRs was investigated in both *in vitro* and *in vivo* assays. In transient transfection experiments with the luciferase or GFP reporter genes, the same human specific element from contig L47334 displayed very low promoter activity in three of the ten cell lines tested, moderate activity (10–20% of the SV40 early promoter) was observed in six cell lines and, finally, the maximal value of ~100% of SV40 promoter activity was obtained in Tera-1 cells, similarly to the above enhancer activity tests [34]. In the experiments by Lavie et al. [79], five human specific proviral 5′ LTRs have demonstrated the promoter strengths as high as 5–15% of the cytomegalovirus (CMV) promoter activity in Tera-1 cells (AP000776 – 15% of CMV promoter expression, AC025757 – 9%, AC072054 – 8%, AC025420 – 6% and AL590785 – 5%). The authors have demonstrated that the promoter activities of these elements directly depend on the methylation status of their CpG dinucleotides. Interestingly, the same five LTRs were strongly transcriptionally repressed in T47D cells [79].

In *in vivo* experiments, 5′ RACE (rapid amplification of cDNA ends) – based mapping of transcriptional start sites for five actively transcribed human specific LTRs provided evidence for the presence of two functional promoter regions within the LTR sequence [74]. Both promoters possess TATA box motif and other upstream regulatory sequences. The first promoter was the canonical element located in the LTR U3 region, whereas the second one was mapped in the very 3′ terminus of the LTR R region. Both promoters appeared to be active in solitary LTRs and in full-length proviruses. Surprisingly, this second non-canonical element was even more active than the classical U3-located retroviral promoter. Therefore, the R region is excluded from most transcripts initiated on LTRs, whereas a classical retroviral life cycle model implies that the transcription is driven from between the LTR U3 and R elements (first promoter), and the R transcript is a 5′-terminal component of the newly synthesized proviral RNA. Such a mode of proviral DNA transcription is a basis of the life cycle that provides the possibility of template jumps during proviral RNA reverse transcription. A shift of the transcriptional start site can be explained by the presence of at least two alternative promoters within the LTR, one of which is normally used for viral gene expression, and the other for transcription of retrotransposition-competent copies of the integrated provirus. The latter type of transcripts is supposed to be far less abundant, what basically corresponds to the above observations. It should be mentioned that alternative promoters with unknown functions were found earlier for many other retrotransposons [10, 31, 103].

Recently, we performed the comprehensive study of the expression of human specific LTRs *in vivo* in human germ-line tissue (testicular parenchyma) and in the corresponding tumor (seminoma) sampled from the same patient [17]. These were chosen because of markedly high endogenous retroviral transcriptional activity in germ-line cells, which is most probably needed to make *de novo* retroviral integrations inheritable [88, 112]. To this end, a new experimental technique that makes it possible to detect repetitive element own promoter activity has been developed [18]. This technique, termed GREM (genomic repeat expression monitor), combines the advantages of 5′-RACE and nucleic acid hybridization techniques. GREM is based on hybridization of total pools of cDNA 5′ terminal parts to genome wide pools of repetitive elements flanking DNA, followed by selective PCR amplification of the resulting hybrid cDNA-genome duplexes. A library of cDNA/genomic DNA hybrid molecules obtained in such a way can be used as a set of tags for individual transcriptionally active repetitive elements [18]. The method is both quantitative and qualitative, as the number of tags is proportional to the content of mRNA driven from the corresponding promoter active repetitive element. The GREM outcome was a set of amplified cDNA/genomic DNA heteroduplexes, below referred to as Expressed LTR Tags (ELTs), which were further cloned and sequenced. This study was the first detailed characterization of the functional promoters provided by a particular group of genomic repetitive elements. The data obtained in such a way suggest that at least 45% of human specific LTRs possessed promoter activity, and a total of 60 new human promoters have been identified. Individual LTRs were expressed at markedly different levels ranging from ~0.001% to ~3% of the housekeeping beta-actin gene transcript level. Although HS elements formed several subclusters on a phylogenetic tree [15, 95], no clear correlation between LTR primary structure and transcriptional activity was found. In contrast, the LTR status (solitary, 5′ or 3′ proviral) was an important factor affecting LTR activity: promoter strengths of solitary and 3′ proviral LTRs were almost identical in both tissues, whereas 5′ proviral LTRs displayed higher promoter activity (~2-fold and ~5-fold greater in testicular parenchyma and seminoma, respectively). These data suggest that a proviral sequence harbors some yet unknown downstream regulatory elements that provide significantly higher 5′ LTR expression, especially in seminoma [17, 18]. Another important factor affecting promoter activity was the LTR distance from genes: the relative content of promoter-active LTRs in gene-rich regions was significantly higher than in gene-poor genomic loci.

The data obtained suggest also a selective suppression of transcription in both tissues for proviral 3′ LTRs located in gene introns. Such a transcriptional suppression might be aimed at silencing of the proviral gene expression in gene-rich regions. In testicular parenchyma, the promoter strength of intronically located solitary LTRs was also significantly decreased. This may suggest yet unknown mechanism(s) for selective suppression of "extra" promoters generated due to mutations or viral integrations and located within gene introns or very closely to genes. Such a mechanism might minimize possible destructive effects of undesirable transcription. Many transcriptionally competent LTRs were mapped near known human genes, and as many as 86–90% of all genes located in close proximity to promoter active LTRs are known to be transcribed in testis. However, in general no clear-cut correlation was observed between transcriptional activities of genes and closely located LTRs [17]. Overall, LTRs provided at least 60 functional human specific promoters for host non-repetitive DNA, that are transcribed at different levels ranging from ~0.001% to ~3% of beta-actin transcript level.

23.2.2.2 Antisense Regulation of Functional Genes by the Human Specific HERV-K(HML-2) Elements

Later on, we reported the first evidence for the human specific antisense regulation of gene activity occurring due to promoter activity of HERV-K(HML-2) endogenous retroviral inserts [49, 52]. Human-specific LTRs located in the introns of genes *SLC4A8* (for sodium bicarbonate cotransporter) and *IFT172* (for intraflagellar transport protein 172) *in vivo* generate transcripts that are complementary to exons within the corresponding mRNAs in a variety of human tissues (Fig. 23.7). As shown by using 5′RACE technique (rapid amplification of cDNA ends), in both cases the LTR-promoted transcription starts within the same position of the LTR consensus sequence, which coincides with the previously reported HERV-K (HML-2) LTR transcriptional start site [74].

The effect of the antisense transcript overexpression on the mRNA level of the corresponding genes was investigated using quantitative real-time RT-PCR. Almost fourfold increase in *SLC-AS* expression led to 3.9-fold decrease of *SLC4A8* mRNA level, and overexpression of *IFT-AS* transcript 2.9-fold reduced the level of *IFT172* mRNA. In all cases the level of the antisense RNAs in the transfected cells was close to or lower than in many human tissues [52]. Similarly, intronically-located

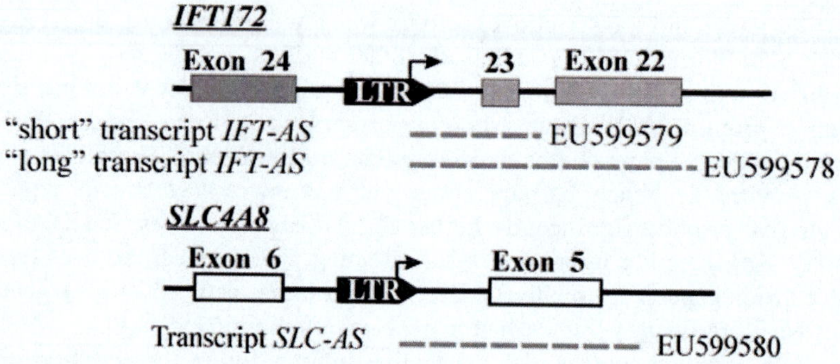

Fig. 23.7 Types of the antisense transcripts found and their corresponding accession numbers

representatives of an LTR retrotransposon family from rice genome called Dasheng likely regulate tissue-specific expression of several adjacent functional genes *via* antisense transcripts driven by the LTRs [70].

23.2.2.3 Enhancer Activity of Human Specific HERV-K(HML-2)

HERV-K (HML-2) LTR sequence harbors a complete set of regulatory elements required for regulation of the retroviral transcription *in vivo* and include a functional enhancer element including multiple transcription factor binding sites. HERV-K (HML-2) LTR enhancer activity was extensively studied *in vitro*, mostly for non-human specific members [33, 64, 115], with the only exception of the human specific solitary LTR from the genomic contig L47334 [115]. In transient transfection experiments on a panel of 10 mammalian cell lines, this LTR has demonstrated enhancer activity only in Tera-1 human testicular embryonal carcinoma cells (thus showing ~8-fold increase in luciferase expression, as compared to control plasmid lacking the enhancer element) [115].

In our recent studies, we found that ~ one-third of all human specific HERV-K (HML-2) LTRs are located in the close gene neighborhood. Nine such elements reside in the upstream regions of known human genes, close to transcriptional start site (*i.e.* at the distance less than 5 kb). In our experiments, three elements over nine have shown a strong enhancer effect in cell culture tests (up to ninefold increase in transcriptional activity). However, only one element, located upstream human gene *PRODH*, has demonstrated a correlation between the enhancer activities *in vitro* and *in vivo*. In the case of two other elements, the LTR inserts were deeply methylated in all the investigated tissues. In contrast, the LTR from *PRODH* region was mostly unmethylated in genomic DNAs of human brain and spinal cord. Our further studies revealed that the LTR enhancer activity is fully regulated by the methylation: the higher is the level of the methylation, the weaker is the LTR enhancer activity, and vice versa. Importantly, *PRODH* promoter is unmethylated in all the tissues, and this gene is transcribed predominantly in the central nervous system (CNS). In the experiments with the mouse brain progenitor cells, we have shown that *in vitro* the LTR insert has a strong enhancer activity on the *PRODH* promoter, thus fivefold increasing transcriptional level of a reporter gene. Furthermore, we have identified a family of transcriptional factors SRY/SOX, that are the most likely candidates for being the LTR activity mediators in germ cells and in the CNS.

PRODH encodes a CNS-specific isoform of the proline dehydrogenase. Unlike the liver isoform, *PRODH* is involved not only in the proline catabolism, but, mostly, in the synthesis of neuromediators like dopamine, GABA, aspartate and glutamate. The deficiency in PRODH activity causes first-type hyperprolinemia, that is frequently linked with severe cognitive disorders and CNS malfunctions, and, in several documented cases, with schizophrenia. Due to its important functions, *PRODH* expression is tightly regulated in human brains, and its expression profile in humans has little in common with the rodent ones. It should be noted, that *PRODH* promoter sequence itself is highly conserved among the mammalians, and the major structural distinction of the *PRODH* upstream region in human, rat, mouse and chimpanzee is the presence of the HERV-K (HML-2) LTR insert in human.

23.3 Concluding Remarks

Thus, the detailed analysis of a small fraction of human specific transposable elements revealed that they may regulate our genes by acting both in *cis* (as promoters and enhancers) and in *trans* (as antisense regulators and RNA recombination hotspots). At least three genes have been identified that are the subjects of a human-specific regulation by the TEs. Considering that only a relatively small portion of the human specific TEs was thoroughly analyzed to the date (~2% of all human specific TEs), one can expect that in the future the detailed genome-wide functional characterization of all human-specific TEs will make it possible to identify tens- or hundreds of genes having unique for human expression profiles. This knowledge, hopefully, at least partly will help us to answer the question “What makes us humans?” ... On the molecular level, of course.

Acknowledgments The authors were sponsored by the Russian Foundation for Basic Research grants 09-04-12302 & 10-04-00593-a, by the President of the Russian Federation grant MD-480.2010.4 and by the Program “Molecular and Cellular Biology” of the Presidium of the Russian Academy of Sciences. A.Buzdin presentation was supported by the NATO fellowship.

References

1. Babushok DV, Ostertag EM, Courtney CE, Choi JM, Kazazian HH Jr (2006) L1 integration in a transgenic mouse model. Genome Res 16:240–250
2. Bantysh OB, Buzdin AA (2009) Novel family of human transposable elements formed due to fusion of the first exon of gene MAST2 with retrotransposon SVA. Biochemistry (Mosc) 74:1393–1399

3. Batzer MA, Schmid CW, Deininger PL (1993) Evolutionary analyses of repetitive DNA sequences. Methods Enzymol 224:213–232
4. Bibillo A, Eickbush TH (2004) End-to-end template jumping by the reverse transcriptase encoded by the R2 retrotransposon. J Biol Chem 279:14945–14953
5. Boissinot S, Chevret P, Furano AV (2000) L1 (LINE-1) retrotransposon evolution and amplification in recent human history. Mol Biol Evol 17:915–928
6. Boissinot S, Davis J, Entezam A, Petrov D, Furano AV (2006) Fitness cost of LINE-1 (L1) activity in humans. Proc Natl Acad Sci USA 103:9590–9594
7. Borodulina OR, Kramerov DA (2008) Transcripts synthesized by RNA polymerase III can be polyadenylated in an AAUAAA-dependent manner. RNA 14:1865–1873
8. Brosius J (1999) Genomes were forged by massive bombardments with retroelements and retrosequences. Genetica 107:209–238
9. Brosius J (1999) RNAs from all categories generate retrosequences that may be exapted as novel genes or regulatory elements. Gene 238:115–134
10. Buzdin AA (2004) Retroelements and formation of chimeric retrogenes. Cell Mol Life Sci 61:2046–2059
11. Buzdin A (2006) Transposable elements and their use for target site specific gene delivery. Curr Pharmacogenomics 4:1–8
12. Buzdin A (2007) Human-specific endogenous retroviruses. Sci World J 7:1848–1868
13. Buzdin A, Ustyugova S, Gogvadze E, Vinogradova T, Lebedev Y, Sverdlov E (2002) A new family of chimeric retrotranscripts formed by a full copy of U6 small nuclear RNA fused to the 3' terminus of l1. Genomics 80:402–406
14. Buzdin A, Gogvadze E, Kovalskaya E, Volchkov P, Ustyugova S, Illarionova A, Fushan A, Vinogradova T, Sverdlov E (2003) The human genome contains many types of chimeric retrogenes generated through in vivo RNA recombination. Nucleic Acids Res 31: 4385–4390
15. Buzdin A, Ustyugova S, Khodosevich K, Mamedov I, Lebedev Y, Hunsmann G, Sverdlov E (2003) Human-specific subfamilies of HERV-K (HML-2) long terminal repeats: three master genes were active simultaneously during branching of hominoid lineages(small star, filled). Genomics 81:149–156
16. Buzdin A, Vinogradova T, Lebedev Y, Sverdlov E (2005) Genome-wide experimental identification and functional analysis of human specific retroelements. Cytogenet Genome Res 110:468–474
17. Buzdin A, Kovalskaya-Alexandrova E, Gogvadze E, Sverdlov E (2006) At least 50% of human-specific HERV-K (HML-2) long terminal repeats serve in vivo as active promoters for host nonrepetitive DNA transcription. J Virol 80:10752–10762
18. Buzdin A, Kovalskaya-Alexandrova E, Gogvadze E, Sverdlov E (2006) GREM, a technique for genome-wide isolation and quantitative analysis of promoter active repeats. Nucleic Acids Res 34:e67
19. Buzdin A, Gogvadze E, Lebrun MH (2007) Chimeric retrogenes suggest a role for the nucleolus in LINE amplification. FEBS Lett 581:2877–2882
20. Chen FC, Chen CJ, Li WH, Chuang TJ (2007) Human-specific insertions and deletions inferred from mammalian genome sequences. Genome Res 17:16–22
21. Chen C, Ara T, Gautheret D (2009) Using Alu elements as polyadenylation sites: a case of retroposon exaptation. Mol Biol Evol 26:327–334
22. Chou HH, Takematsu H, Diaz S, Iber J, Nickerson E, Wright KL, Muchmore EA, Nelson DL, Warren ST, Varki A (1998) A mutation in human CMP-sialic acid hydroxylase occurred after the Homo-Pan divergence. Proc Natl Acad Sci USA 95:11751–11756
23. Conley AB, Miller WJ, Jordan IK (2008) Human cis natural antisense transcripts initiated by transposable elements. Trends Genet 24:53–56
24. Conley AB, Piriyapongsa J, Jordan IK (2008) Retroviral promoters in the human genome. Bioinformatics 24:1563–1567
25. consortium, C. s. a. a (2005) Initial sequence of the chimpanzee genome and comparison with the human genome. Nature 437:69–87

26. Cooper GM, Brudno M, Green ED, Batzoglou S, Sidow A (2003) Quantitative estimates of sequence divergence for comparative analyses of mammalian genomes. Genome Res 13:813–820
27. Copeland CS, Mann VH, Brindley PJ (2007) Both sense and antisense strands of the LTR of the Schistosoma mansoni Pao-like retrotransposon Sinbad drive luciferase expression. Mol Genet Genomics 277:161–170
28. Cordaux R, Hedges DJ, Herke SW, Batzer MA (2006) Estimating the retrotransposition rate of human Alu elements. Gene 373:134–137
29. Cutter AD, Good JM, Pappas CT, Saunders MA, Starrett DM, Wheeler TJ (2005) Transposable element orientation bias in the Drosophila melanogaster genome. J Mol Evol 61:733–741
30. Damert A, Raiz J, Horn AV, Lower J, Wang H, Xing J, Batzer MA, Lower R, Schumann GG (2009) 5'-Transducing SVA retrotransposon groups spread efficiently throughout the human genome. Genome Res 19:1992–2008
31. Deininger PL, Moran JV, Batzer MA, Kazazian HH Jr (2003) Mobile elements and mammalian genome evolution. Curr Opin Genet Dev 13:651–658
32. Dewannieux M, Esnault C, Heidmann T (2003) LINE-mediated retrotransposition of marked Alu sequences. Nat Genet 35:41–48
33. Domanskii AN, Akopov SB, Lebedev Iu B, Nikolaev LG, Sverdlov ED (2002) Enhancer activity of solitary long terminal repeat of the human endogenous retrovirus of the HERV-K family. Bioorg Khim 28:341–345
34. Domansky AN, Kopantzev EP, Snezhkov EV, Lebedev YB, Leib-Mosch C, Sverdlov ED (2000) Solitary HERV-K LTRs possess bi-directional promoter activity and contain a negative regulatory element in the U5 region. FEBS Lett 472:191–195
35. Dunn CA, Romanish MT, Gutierrez LE, van de Lagemaat LN, Mager DL (2006) Transcription of two human genes from a bidirectional endogenous retrovirus promoter. Gene 366:335–342
36. Ebersberger I, Metzler D, Schwarz C, Paabo S (2002) Genomewide comparison of DNA sequences between humans and chimpanzees. Am J Hum Genet 70:1490–1497
37. Enard W, Khaitovich P, Klose J, Zollner S, Heissig F, Giavalisco P, Nieselt-Struwe K, Muchmore E, Varki A, Ravid R, Doxiadis GM, Bontrop RE, Paabo S (2002) Intra- and interspecific variation in primate gene expression patterns. Science 296:340–343
38. Fasken MB, Corbett AH (2005) Process or perish: quality control in mRNA biogenesis. Nat Struct Mol Biol 12:482–488
39. Feuchter A, Mager D (1990) Functional heterogeneity of a large family of human LTR-like promoters and enhancers. Nucleic Acids Res 18:1261–1270
40. Fu N, Drinnenberg I, Kelso J, Wu JR, Paabo S, Zeng R, Khaitovich P (2007) Comparison of protein and mRNA expression evolution in humans and chimpanzees. PLoS One 2:e216
41. Fudal I, Bohnert HU, Tharreau D, Lebrun MH (2005) Transposition of MINE, a composite retrotransposon, in the avirulence gene ACE1 of the rice blast fungus Magnaporthe grisea. Fungal Genet Biol 42:761–772
42. Furano AV (2000) The biological properties and evolutionary dynamics of mammalian LINE-1 retrotransposons. Prog Nucleic Acid Res Mol Biol 64:255–294
43. Gagneux P, Varki A (2001) Genetic differences between humans and great apes. Mol Phylogenet Evol 18:2–13
44. Galante PA, Vidal DO, de Souza JE, Camargo AA, de Souza SJ (2007) Sense-antisense pairs in mammals: functional and evolutionary considerations. Genome Biol 8:R40
45. Gilad Y, Man O, Paabo S, Lancet D (2003) Human specific loss of olfactory receptor genes. Proc Natl Acad Sci USA 100:3324–3327
46. Gilbert N, Lutz S, Morrish TA, Moran JV (2005) Multiple fates of L1 retrotransposition intermediates in cultured human cells. Mol Cell Biol 25:7780–7795
47. Glaser B, Yen PH, Schempp W (1998) Fibre-fluorescence in situ hybridization unravels apparently seven DAZ genes or pseudogenes clustered within a Y-chromosome region frequently deleted in azoospermic males. Chromosome Res 6:481–486
48. Gogvadze EV, Buzdin AA (2005) New mechanism of retrogene formation in mammalian genomes: in vivo recombination during RNA reverse transcription. Mol Biol (Mosk) 39:364–373

49. Gogvadze E, Buzdin A (2009) Retroelements and their impact on genome evolution and functioning. Cell Mol Life Sci 66:3727–3742
50. Gogvadze EV, Buzdin AA, Sverdlov ED (2005) Multiple template switches on LINE-directed reverse transcription: the most probable formation mechanism for the double and triple chimeric retroelements in mammals. Bioorg Khim 31:82–89
51. Gogvadze E, Barbisan C, Lebrun MH, Buzdin A (2007) Tripartite chimeric pseudogene from the genome of rice blast fungus Magnaporthe grisea suggests double template jumps during long interspersed nuclear element (LINE) reverse transcription. BMC Genomics 8:360
52. Gogvadze E, Stukacheva E, Buzdin A, Sverdlov E (2009) Human specific modulation of transcriptional activity provided by endogenous retroviral inserts. J Virol 83:6098–6105
53. Goodier JL, Ostertag EM, Kazazian HH Jr (2000) Transduction of 3'-flanking sequences is common in L1 retrotransposition. Hum Mol Genet 9:653–657
54. Goodman M (1999) The genomic record of Humankind's evolutionary roots. Am J Hum Genet 64:31–39
55. Gotea V, Makalowski W (2006) Do transposable elements really contribute to proteomes? Trends Genet 22:260–267
56. Hahn Y, Lee B (2006) Human-specific nonsense mutations identified by genome sequence comparisons. Hum Genet 119:169–178
57. Haile-Selassie Y (2001) Late Miocene hominids from the Middle Awash, Ethiopia. Nature 412:178–181
58. Hambor JE, Mennone J, Coon ME, Hanke JH, Kavathas P (1993) Identification and characterization of an Alu-containing, T-cell-specific enhancer located in the last intron of the human CD8 alpha gene. Mol Cell Biol 13:7056–7070
59. Hancks DC, Ewing AD, Chen JE, Tokunaga K, Kazazian HH Jr (2009) Exon-trapping mediated by the human retrotransposon SVA. Genome Res 19:1983–1991
60. Hasler J, Samuelsson T, Strub K (2007) Useful 'junk': Alu RNAs in the human transcriptome. Cell Mol Life Sci 64:1793–1800
61. Hayward BE, Zavanelli M, Furano AV (1997) Recombination creates novel L1 (LINE-1) elements in Rattus norvegicus. Genetics 146:641–654
62. Horvath JE, Viggiano L, Loftus BJ, Adams MD, Archidiacono N, Rocchi M, Eichler EE (2000) Molecular structure and evolution of an alpha satellite/non-alpha satellite junction at 16p11. Hum Mol Genet 9:113–123
63. Huh JW, Kim DS, Kang DW, Ha HS, Ahn K, Noh YN, Min DS, Chang KT, Kim HS (2008) Transcriptional regulation of GSDML gene by antisense-oriented HERV-H LTR element. Arch Virol 153:1201–1205
64. Illarionova AE, Vinogradova TV, Sverdlov ED (2007) Only those genes of the KIAA1245 gene subfamily that contain HERV(K) LTRs in their introns are transcriptionally active. Virology 358:39–47
65. Irie A, Koyama S, Kozutsumi Y, Kawasaki T, Suzuki A (1998) The molecular basis for the absence of N-glycolylneuraminic acid in humans. J Biol Chem 273:15866–15871
66. Jamain S, Girondot M, Leroy P, Clergue M, Quach H, Fellous M, Bourgeron T (2001) Transduction of the human gene FAM8A1 by endogenous retrovirus during primate evolution. Genomics 78:38–45
67. Jurka J (1997) Sequence patterns indicate an enzymatic involvement in integration of mammalian retroposons. Proc Natl Acad Sci USA 94:1872–1877
68. Kakuo S, Asaoka K, Ide T (1999) Human is a unique species among primates in terms of telomere length. Biochem Biophys Res Commun 263:308–314
69. Kandel ES, Nudler E (2002) Template switching by RNA polymerase II in vivo. Evidence and implications from a retroviral system. Mol Cell 10:1495–1502
70. Kashkush K, Khasdan V (2007) Large-scale survey of cytosine methylation of retrotransposons and the impact of readout transcription from long terminal repeats on expression of adjacent rice genes. Genetics 177:1975–1985
71. Kazazian HH Jr (2004) Mobile elements: drivers of genome evolution. Science 303:1626–1632

72. Kehrer-Sawatzki H, Cooper DN (2007) Structural divergence between the human and chimpanzee genomes. Hum Genet 120:759–778
73. Kjellman C, Sjogren HO, Salford LG, Widegren B (1999) HERV-F (XA34) is a full-length human endogenous retrovirus expressed in placental and fetal tissues. Gene 239:99–107
74. Kovalskaya E, Buzdin A, Gogvadze E, Vinogradova T, Sverdlov E (2006) Functional human endogenous retroviral LTR transcription start sites are located between the R and U5 regions. Virology 346:373–378
75. Kramerov DA, Vassetzky NS (2001) Structure and origin of a novel dimeric retroposon B1-diD. J Mol Evol 52:137–143
76. Kramerov DA, Vassetzky NS (2005) Short retroposons in eukaryotic genomes. Int Rev Cytol 247:165–221
77. Lander ES, Linton LM, Birren B, Nusbaum C, Zody MC, Baldwin J, Devon K, Dewar K, Doyle M, FitzHugh W, Funke R, Gage D, Harris K, Heaford A et al (2001) Initial sequencing and analysis of the human genome. Nature 409:860–921
78. Lane RP, Cutforth T, Young J, Athanasiou M, Friedman C, Rowen L, Evans G, Axel R, Hood L, Trask BJ (2001) Genomic analysis of orthologous mouse and human olfactory receptor loci. Proc Natl Acad Sci USA 98:7390–7395
79. Lavie L, Kitova M, Maldener E, Meese E, Mayer J (2005) CpG methylation directly regulates transcriptional activity of the human endogenous retrovirus family HERV-K(HML-2). J Virol 79:876–883
80. Lebedev YB, Amosova AL, Mamedov IZ, Fisunov GY, Sverdlov ED (2007) Most recent AluY insertions in human gene introns reduce the content of the primary transcripts in a cell type specific manner. Gene 390:122–129
81. Lee J, Cordaux R, Han K, Wang J, Hedges DJ, Liang P, Batzer MA (2007) Different evolutionary fates of recently integrated human and chimpanzee LINE-1 retrotransposons. Gene 390:18–27
82. Lee JY, Ji Z, Tian B (2008) Phylogenetic analysis of mRNA polyadenylation sites reveals a role of transposable elements in evolution of the 3'-end of genes. Nucleic Acids Res 36:5581–5590
83. Lev-Maor G, Ram O, Kim E, Sela N, Goren A, Levanon EY, Ast G (2008) Intronic Alus influence alternative splicing. PLoS Genet 4:e1000204
84. Levy A, Sela N, Ast G (2008) TranspoGene and microTranspoGene: transposed elements influence on the transcriptome of seven vertebrates and invertebrates. Nucleic Acids Res 36:D47–D52
85. Lin L, Shen S, Tye A, Cai JJ, Jiang P, Davidson BL, Xing Y (2008) Diverse splicing patterns of exonized Alu elements in human tissues. PLoS Genet 4:e1000225
86. Livak KJ, Rogers J, Lichter JB (1995) Variability of dopamine D4 receptor (DRD4) gene sequence within and among nonhuman primate species. Proc Natl Acad Sci USA 92:427–431
87. Long Q, Bengra C, Li C, Kutlar F, Tuan D (1998) A long terminal repeat of the human endogenous retrovirus ERV-9 is located in the 5' boundary area of the human beta- globin locus control region. Genomics 54:542–555
88. Lower R, Lower J, Kurth R (1996) The viruses in all of us: characteristics and biological significance of human endogenous retrovirus sequences. Proc Natl Acad Sci USA 93:5177–5184
89. Mager DL, Hunter DG, Schertzer M, Freeman JD (1999) Endogenous retroviruses provide the primary polyadenylation signal for two new human genes. Genomics 59:255–263
90. Makalowski W (2000) Genomic scrap yard: how genomes utilize all that junk. Gene 259:61–67
91. Makalowski W, Mitchell GA, Labuda D (1994) Alu sequences in the coding regions of mRNA: a source of protein variability. Trends Genet 10:188–193
92. Malik HS, Eickbush TH (2001) Phylogenetic analysis of ribonuclease H domains suggests a late, chimeric origin of LTR retrotransposable elements and retroviruses. Genome Res 11:1187–1197

93. Martin SL (2006) The ORF1 protein encoded by LINE-1: structure and function during L1 retrotransposition. J Biomed Biotechnol 2006:45621
94. Matlik K, Redik K, Speek M (2006) L1 antisense promoter drives tissue-specific transcription of human genes. J Biomed Biotechnol 2006:71753
95. Medstrand P, Mager DL (1998) Human-specific integrations of the HERV-K endogenous retrovirus family. J Virol 72:9782–9787
96. Medstrand P, van de Lagemaat LN, Mager DL (2002) Retroelement distributions in the human genome: variations associated with age and proximity to genes. Genome Res 12:1483–1495
97. Meneveri R, Agresti A, Rocchi M, Marozzi A, Ginelli E (1995) Analysis of GC-rich repetitive nucleotide sequences in great apes. J Mol Evol 40:405–412
98. Mills RE, Bennett EA, Iskow RC, Luttig CT, Tsui C, Pittard WS, Devine SE (2006) Recently mobilized transposons in the human and chimpanzee genomes. Am J Hum Genet 78:671–679
99. Moran JV, DeBerardinis RJ, Kazazian HH Jr (1999) Exon shuffling by L1 retrotransposition. Science 283:1530–1534
100. Morrish TA, Gilbert N, Myers JS, Vincent BJ, Stamato TD, Taccioli GE, Batzer MA, Moran JV (2002) DNA repair mediated by endonuclease-independent LINE-1 retrotransposition. Nat Genet 31:159–165
101. Nadezhdin EV, Vinogradova TV, Sverdlov ED (2001) Interspecies subtractive hybridization of cDNA from human and chimpanzee brains. Dokl Biochem Biophys 381:415–418
102. Nickerson E, Nelson DL (1998) Molecular definition of pericentric inversion breakpoints occurring during the evolution of humans and chimpanzees. Genomics 50:368–372
103. Nigumann P, Redik K, Matlik K, Speek M (2002) Many human genes are transcribed from the antisense promoter of L1 retrotransposon. Genomics 79:628–634
104. Nishihara H, Smit AF, Okada N (2006) Functional noncoding sequences derived from SINEs in the mammalian genome. Genome Res 16:864–874
105. Ohshima K, Hamada M, Terai Y, Okada N (1996) The 3' ends of tRNA-derived short interspersed repetitive elements are derived from the 3' ends of long interspersed repetitive elements. Mol Cell Biol 16:3756–3764
106. Osada N, Wu CI (2005) Inferring the mode of speciation from genomic data: a study of the great apes. Genetics 169:259–264
107. Ostertag EM, Goodier JL, Zhang Y, Kazazian HH Jr (2003) SVA elements are nonautonomous retrotransposons that cause disease in humans. Am J Hum Genet 73:1444–1451
108. Pickeral OK, Makalowski W, Boguski MS, Boeke JD (2000) Frequent human genomic DNA transduction driven by LINE-1 retrotransposition. Genome Res 10:411–415
109. Polavarapu N, Marino-Ramirez L, Landsman D, McDonald JF, Jordan IK (2008) Evolutionary rates and patterns for human transcription factor binding sites derived from repetitive DNA. BMC Genomics 9:226
110. Pollard KS, Salama SR, Lambert N, Lambot MA, Coppens S, Pedersen JS, Katzman S, King B, Onodera C, Siepel A, Kern AD, Dehay C, Igel H, Ares M Jr, Vanderhaeghen P, Haussler D (2006) An RNA gene expressed during cortical development evolved rapidly in humans. Nature 443:167–172
111. Prabhakar S, Noonan JP, Paabo S, Rubin EM (2006) Accelerated evolution of conserved noncoding sequences in humans. Science 314:786
112. Prudhomme S, Bonnaud B, Mallet F (2005) Endogenous retroviruses and animal reproduction. Cytogenet Genome Res 110:353–364
113. Roy-Engel AM, El-Sawy M, Farooq L, Odom GL, Perepelitsa-Belancio V, Bruch H, Oyeniran OO, Deininger PL (2005) Human retroelements may introduce intragenic polyadenylation signals. Cytogenet Genome Res 110:365–371
114. Royle NJ, Baird DM, Jeffreys AJ (1994) A subterminal satellite located adjacent to telomeres in chimpanzees is absent from the human genome. Nat Genet 6:52–56
115. Ruda VM, Akopov SB, Trubetskoy DO, Manuylov NL, Vetchinova AS, Zavalova LL, Nikolaev LG, Sverdlov ED (2004) Tissue specificity of enhancer and promoter activities of a HERV-K(HML-2) LTR. Virus Res 104:11–16

116. Ruvolo M (2004) Comparative primate genomics: the year of the chimpanzee. Curr Opin Genet Dev 14:650–656
117. Sakaki Y, Watanabe H, Taylor T, Hattori M, Fujiyama A, Toyoda A, Kuroki Y, Itoh T, Saitou N, Oota S, Kim CG, Kitano T, Lehrach H, Yaspo ML et al (2003) Human versus chimpanzee chromosome-wide sequence comparison and its evolutionary implication. Cold Spring Harb Symp Quant Biol 68:455–460
118. Sakate R, Suto Y, Imanishi T, Tanoue T, Hida M, Hayasaka I, Kusuda J, Gojobori T, Hashimoto K, Hirai M (2007) Mapping of chimpanzee full-length cDNAs onto the human genome unveils large potential divergence of the transcriptome. Gene 399:1–10
119. Sela N, Mersch B, Gal-Mark N, Lev-Maor G, Hotz-Wagenblatt A, Ast G (2007) Comparative analysis of transposed element insertion within human and mouse genomes reveals Alu's unique role in shaping the human transcriptome. Genome Biol 8:R127
120. Sen SK, Han K, Wang J, Lee J, Wang H, Callinan PA, Dyer M, Cordaux R, Liang P, Batzer MA (2006) Human genomic deletions mediated by recombination between Alu elements. Am J Hum Genet 79:41–53
121. Song M, Boissinot S (2007) Selection against LINE-1 retrotransposons results principally from their ability to mediate ectopic recombination. Gene 390:206–213
122. Sorek R, Ast G, Graur D (2002) Alu-containing exons are alternatively spliced. Genome Res 12:1060–1067
123. Sverdlov ED (1998) Perpetually mobile footprints of ancient infections in human genome. FEBS Lett 428:1–6
124. Sverdlov ED (2000) Retroviruses and primate evolution. Bioessays 22:161–171
125. Swanstrom R, Parker RC, Varmus HE, Bishop JM (1983) Transduction of a cellular oncogene: the genesis of Rous sarcoma virus. Proc Natl Acad Sci USA 80:2519–2523
126. Temin HM (1993) Retrovirus variation and reverse transcription: abnormal strand transfers result in retrovirus genetic variation. Proc Natl Acad Sci USA 90:6900–6903
127. Trask BJ, Friedman C, Martin-Gallardo A, Rowen L, Akinbami C, Blankenship J, Collins C, Giorgi D, Iadonato S, Johnson F, Kuo WL, Massa H, Morrish T, Naylor S, Nguyen OT, Rouquier S, Smith T, Wong DJ, Youngblom J, van den Engh G (1998) Members of the olfactory receptor gene family are contained in large blocks of DNA duplicated polymorphically near the ends of human chromosomes. Hum Mol Genet 7:13–26
128. Ustiugova SV, Amosova AL, Lebedev Iu B, Sverdlov ED (2006) A tissue-specific decrease in the pre-mRNA level of L1- and alu-containing alleles of human genes. Bioorg Khim 32:103–106
129. Ustyugova SV, Lebedev YB, Sverdlov ED (2006) Long L1 insertions in human gene introns specifically reduce the content of corresponding primary transcripts. Genetica 128:261–272
130. van de Lagemaat LN, Landry JR, Mager DL, Medstrand P (2003) Transposable elements in mammals promote regulatory variation and diversification of genes with specialized functions. Trends Genet 19:530–536
131. van de Lagemaat LN, Medstrand P, Mager DL (2006) Multiple effects govern endogenous retrovirus survival patterns in human gene introns. Genome Biol 7:R86
132. Venter JC, Adams MD, Myers EW, Li PW, Mural RJ, Sutton GG, Smith HO, Yandell M, Evans CA, Holt RA, Gocayne JD, Amanatides P, Ballew RM et al (2001) The sequence of the human genome. Science 291:1304–1351
133. Wang H, Xing J, Grover D, Hedges DJ, Han K, Walker JA, Batzer MA (2005) SVA elements: a hominid-specific retroposon family. J Mol Biol 354:994–1007
134. Wang X, Grus WE, Zhang J (2006) Gene losses during human origins. PLoS Biol 4:e52
135. Wei W, Gilbert N, Ooi SL, Lawler JF, Ostertag EM, Kazazian HH, Boeke JD, Moran JV (2001) Human L1 retrotransposition: cis preference versus trans complementation. Mol Cell Biol 21:1429–1439
136. Wessler SR (1998) Transposable elements and the evolution of gene expression. Symp Soc Exp Biol 51:115–122
137. Wheelan SJ, Aizawa Y, Han JS, Boeke JD (2005) Gene-breaking: a new paradigm for human retrotransposon-mediated gene evolution. Genome Res 15:1073–1078

138. Xing J, Wang H, Belancio VP, Cordaux R, Deininger PL, Batzer MA (2006) Emergence of primate genes by retrotransposon-mediated sequence transduction. Proc Natl Acad Sci USA 103:17608–17613
139. Yang Z, Boffelli D, Boonmark N, Schwartz K, Lawn R (1998) Apolipoprotein(a) gene enhancer resides within a LINE element. J Biol Chem 273:891–897
140. Yu H, Jiang H, Zhou Q, Yang J, Cun Y, Su B, Xiao C, Wang W (2006) Origination and evolution of a human-specific transmembrane protein gene, c1orf37-dup. Hum Mol Genet 15:1870–1875
141. Yunis JJ, Prakash O (1982) The origin of man: a chromosomal pictorial legacy. Science 215:1525–1530
142. Zemojtel T, Penzkofer T, Schultz J, Dandekar T, Badge R, Vingron M (2007) Exonization of active mouse L1s: a driver of transcriptome evolution? BMC Genomics 8:392
143. Zimonjic DB, Kelley MJ, Rubin JS, Aaronson SA, Popescu NC (1997) Fluorescence in situ hybridization analysis of keratinocyte growth factor gene amplification and dispersion in evolution of great apes and humans. Proc Natl Acad Sci USA 94:11461–11465

Chapter 24
Ecological Risk Assessment to Benthic Biocenoses

V.G. Tsytsugina

Abstract On the basis of experimental data and cytogenetic studies of marine and freshwater invertebrates natural populations it was shown that different species had approximately equal part (%) of full value posterity at the same average level of chromosome mutagenesis for populations, independing on deleterious factors and females fecundity. This phenomenon made it possible to calculate specimens reproductive contribution for species with different females fecundity and to assess expected reduction of population adaptive potential and the increase of ecological risk at different levels of population damage. It is suggested that the adaptation to pollution will be more effective in populations with high female fecundity. Adaptive possibilities of populations and the resistance of posterity were considered as a alternative properties of populations Ranges of chromosome mutagenesis critical levels for different hydrobionts taxons were determined. An algorithm of ecological risk assessment to benthic communities was proposed.

Keywords Benthic communities • Pollution • Fecundity • Adaptation • Ecological risk assessment

24.1 Introduction

Approaches to ecological risk assessment are based mainly on "dose (concentration) – effect" dependencies [4]. But the most constructive and adequate approach to ecological risk assessment is the study of adaptive potential of natural populations. It is known that genetic variability and adaptive possibility are connected with effective

V.G. Tsytsugina (✉)
The A.O. Kovalevsky Institute of Biology of the Southern Seas, NAS,
Nakhimovv prospect, 2, Sevastopol 99011, Ukraine
e-mail: vtsy@ibss.iuf.net

C.E. Mothersill et al. (eds.), *Radiobiology and Environmental Security*,
NATO Science for Peace and Security Series C: Environmental Security,
DOI 10.1007/978-94-007-1939-2_24, © Springer Science+Business Media B.V. 2012

population size (i.e. the number of specimens which breed and determine genetic structure of next posterities). Effective population size depends on the dispersion of reproductive contribution of specimens. Increasing of reproductive contribution dispersion leads to decreasing of effective population size and adaptive potential [1]. For the assessment of specimens reproductive contribution a number (%) of posterity with spontaneous mutagenesis (up to 2% cells with chromosome aberrations [8]) may be used as the criterion of full value posterity because one with higher number of cells with chromosome aberrations is less viable [10]. Proceeding from data on the number (%) of full value posterity in population and the mean fecundity of females it is possible to calculate the number of full value posterity per a female (i. e. reproductive contribution). Obviously the increase of ecological risk may be expected if there is less than one full value embryo (or larva) per a female. The average levels of chromosome mutagenesis in populations inducing such effect may be considered as critical ones.

24.2 Adaptive Populations Possibilities

24.2.1 Adaptive Potential of Hydrobionts Populations with Different Females Fecundity

24.2.1.1 Experimental Data

Experimental data on separate and combined effect of ionizing radiation and chemical mutagens on the posterity of *Chaetogammarus olivii* (Amphipoda) and *Idothea baltica* (Isopoda) females with different fecundity (5, 20 and 40 eggs) have shown that approximately equal part (%) of full value embryos at the same average level of chromosome mutagenesis for populations was observed, independently of the eggs number and a type of injure factors (Table 24.1) [10].

24.2.1.2 Natural Populations

This data were compared with the results of cytogenetic studies of embryos and larvas in natural invertebrates populations (20 marine and freshwater species from 11 taxons of coelenterates, annelids, mollusks, arthropods) from different biotopes in 1975–2003 [7, 8, 11] (Table 24.2). These species have the fecundity from several eggs to hundreds ones. In Table 24.2 it can be seen that (such as in experiments) different species had approximately equal part (%) of full value posterity at the same average level of chromosome mutagenesis for populations.

The observed phenomenon makes it possible to calculate reproductive contribution of specimens for species with different fecundity as well as to assess expected reduction of population adaptive potential and the increase of ecological risk at different levels of environmental pollution.

Table 24.1 Part (%) of full value crustacean posterity at the same average level of chromosome mutagenesis (Experimental data)

Mutagen	Mean number of cells with chromosome aberrations, %	Species	Mean fecundity of females (number of eggs)	Number of embryos with spontaneous mutagen-esis, (%)
^{90}Sr		*Chaetogammarus olivii*	40	19
^{137}Cs		*Chaetogammarus olivii*	5	20
$Pb^{(2+)}$		*Chaetogammarus olivii*	40	28
$Pb^{(2+)}$	5.0±0.6	*Chaetogammarus olivii*	5	18
Chlorphene		*Chaetogammarus olivii*	40	30
$Pb^{(2+)}$+, +chlorphene		*Chaetogammarus olivii*	40	30
^{137}Cs+, + chlorphene		*Chaetogammarus olivii*	5	22
^{90}Sr		*Idothea baltica*	20	20
^{90}Sr		*Chaetogammarus olivii*	40	5
^{90}Sr		*Chaetogammarus olivii*	5	5
^{90}Sr + ^{137}Cs+, $Pb^{(2+)}$+, +chlorphene	8.0±08	*Chaetogammarus olivii*	40	5
^{90}Sr		*Idothea baltica*	40	8

Table 24.3 presents the results of such calculations for females with different fecundity (5, 20, 50, 100 eggs) at the mean number cells with chromosome aberrations from 1.5% to 8.5%.

It can be seen that the reduction of adaptive possibility may be expected in populations with low fecundity of females (5 eggs) already at environmental pollution inducing in the mean 6% cells with chromosome aberrations (less than one full value embryo per a female). Obviously the adaptation to pollution will be more effective in populations with higher female fecundity (20–100 eggs) even at higher mean level of damage.

24.3 Adaptive Populations Possibilities and Posterity Resistance

On the other hand, it was shown on *Chaetogammarus olivii* (as a example) that the posterity of females with high fecundity is less resistant to deleterious effect [9]. Populations of the Black Sea Amphipoda produce 3–4 generations a year [3]. The 1st and the most numerous generation appears in spring, the 2nd – in summer, the 3rd – in autumn and the 4th one – at the end of a year. The life span of these crustaceans is 6–10 months. Crustaceans of the 1st and the 2nd generations breed in

Table 24.2 Part (%) of full value posterity in natural hydrobionts populations

Species	Habitat	Mean number of cells with chromosome aberrations, %	Number of embryos (larvas) with spontaneous chromosome mutagenesis,%
Polydora ciliata	Black Sea		95
Glaucus sp.	Indian Ocean		91
Mytilus galloprovincialis	Black Sea		88
Chaetogammarus olivii	Black Sea		76
Idothea baltica	Black Sea	1.5±0.5	82
Idothea metallica	Atlantic Ocean		80
Melita palmata	Aegean Sea		85
Mysidacea gen. sp.	Black Sea		83
Decapoda gen. sp.	Atlantic Ocean		85
Pterocuma pectinata	Black Sea		65
Dikerogammarus haemobaphes	Dnieper River, Kakhovsky reservoir		60
Pontogammarus crassus	Dnieper River, Kiev reservoir	2.5±0.5	56
Pontogammarus robustoides	Dnieper River, Kremenchug reservoir		56
Decapoda gen. sp.	Atlantic Ocean		65
Glaucus sp.	Indian Ocean		40
Melaraphe neritoides	Black Sea		43
Anchylomera Brossevilei	Indian Ocean	4.0±0.6	50
Chaetogammarus olivii	Black Sea		44
Idothea baltica	Black Sea		45
Pontogammarus crassus	Dnieper River, Kakhovsky reservoir		40
Pontogammarus crassus	Dnieper, Kiev reservoir		45
Pontogammarus crassus	Dnieper River, Kremenchug reservoir		18
Pontogammarus robustoides	Dnieper River, Kakhovsky reservoir	6.0±0.8	13
Pontogammarus robustoides	Black Sea		11
Podon polyphaemoides			15
Velella velella	Pacific Ocean		5
Monodacna caspia	Black Sea	8.3±1.3	6
Gammarus lacustris	10-km Ch NPP zone		3

Table 24.3 Reproductive contribution calculated for females with different fecundity

Mean level of chromosome mutagenesis, %	Number of posterity with spontaneous mutagenesis,%	Mean fecundity of females (number of eggs)	Number of posterity with spontaneous mutagenesis per a female
		5	4
		20	16
1.5±0.5	80	50	40
		100	80
		5	3
2.5±0.5	60	20	12
		50	30
		100	60
		5	2
4.0±0.6	40	20	8
		50	20
		100	40
		5	0.75
6.0±0.8	15	20	3
		50	7.5
		100	15.0
8.5±0.5	5	5	0.25
		20	1.0
		50	2.5
		100	5.0
>10	no		

summer and autumn, the 3rd, the 4th and partly the 2nd generations – in winter and spring. Depending on water temperature the time needed for maturation varies from 1.5 to 3 months. Maturation of these crustaceans is faster in summer and autumn; therefore the size of females during reproduction is always smaller than in winter and spring. Big individuals of the 3rd and 4th generations with the length of 6–10 mm dominate in the population in winter and spring months. They are the main reproduction fund for the spring breeding. They are going to die in May after finishing reproduction. Radical change of age and size composition of the population happens in this period. Females of 4–5 mm length dominate in the population of the crustaceans at the end of May and the beginning of June. They are crustaceans of the 1st generation of the current year. Already bigger females of the 2nd and the 3rd generations of 5–6 mm size dominate in late autumn.

Experiments with acute γ-irradiation were carried out in different seasons: in spring (April) when females producing large amount of eggs breed and in summer (June) when in population there are only females with low fecundity [9]. In Table 24.4 it can be seen that the posterity of females with high fecundity demonstrates less resistance to the irradiation. It is possible that age physiological peculiarities of parents contribute to definite extent into this difference.

However, the main cause apparently is as follows. As it was shown in the work [3]., fecundity of females depends on their size. The relation is expressed by the power

Table 24.4 Cytogenetic effect of acute γ-irradiation (5 Gy) on posterity of *Chaetogammarus olivii* females with different fecundity

Mean fecundity of females (number of eggs)	Mean number of cells with chromosome aberrations, %	Lim
40	35.7±2.0	20.0–53.6
5	15.5±1.0	3.6–30.0

Table 24.5 Reproductive contribution of *Chaetogammarus olivii* specimens under the action of ^{90}Sr

Mean fecundity of females (number of eggs)	Mean number of cells with chromosome aberrations, %	Number of embryos with spontaneous mutagenesis, %	Number of embryos with spontaneous mutagenesis per a female
5	8.7±1.0	15	0.75
40	9.2±1.2	10	4.0

function with the index of extent, equaled to 3, and the constant to 0.0409. Small females with the size of 3–4 mm produce averagely three eggs, bigger females (5–6 mm) – 12 eggs and the biggest females (7–10 mm) – 40–50 eggs. The relation between sizes of females and their fecundity is observed for other groups of Crustacea, specifically, for Cumacea, Mysidacea and some others. Radioresistance of posterity depends, apparently, on fecundity of females and is decreased significantly for big females, producing large amounts of eggs. It is obvious, that fecundity of females and resistance of posterity compensate each other. Maximum adaptation of each generation is reached in such a way. The alternative (number of embryos – embryos' resistance) is solved differently in different seasons. In spring., in a decisive time for a population, when a radical change of its composition occurs, the winter big crustaceans of older age groups, making up the base of the population and its main reproductive fund for spring reproduction, completely realize their biotic potential before their mass death, producing a numerous, though less resistant posterity. In May-June young females of the 1st generation produce a small number of eggs, significantly more resistant ones. The alternative is solved in this case for the benefit of greater life ability of a small number of individuals.

From the data obtained it might be inferred that obviously higher resistance posterity in populations with low females fecundity and more wide adaptive possibility of populations with high females fecundity can be considered as a alternative properties of populations with different females fecundity.

As a model of such alternative the analysis of cytogenetic study of *Ch. olivii* population in the sewage region may be discussed. The adaptation of this population to pollution is accompanied by growing of non-specific resistance to high damage by incorporated ^{90}Sr (22 MBq/l) [9].

In order to assess adaptive possibility of this population in different seasons when females with high or low fecundity breed calculations of specimen reproductive contribution were made (Table 24.5). The results show that the posterity of females with high fecundity (40 eggs) is less resistant but amount of full value embryos per a female is 5 times more than this in females with low fecundity.

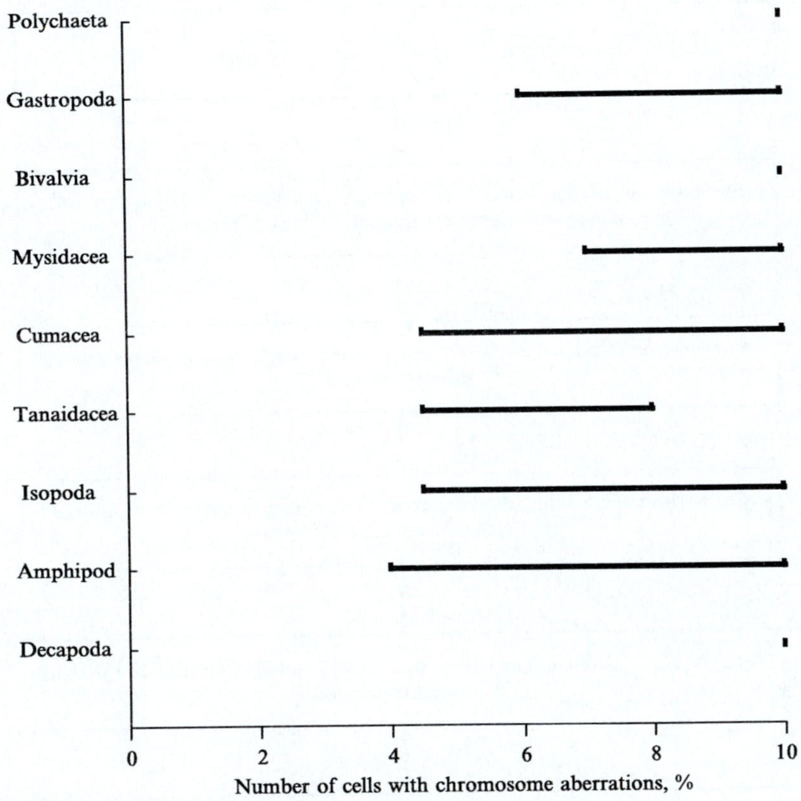

Fig. 24.1 Ranges of chromosome mutagenesis critical levels for different taxons

This example clearly demonstrates that though the resistance of the posterity in populations with low fecundity of females may be high but chances on the adaptation in populations with high females fecundity will be obviously greater.

24.4 An Algorithm of Ecological Risk Assessment to Biocenoses (Communities)

Of the basis of the benthic invertebrates fecundity [2, 5, 6, 12, 13] as well as of the calculations of specimens reproductive contribution at the different average numbers of chromosome aberrations for populations we determined ranges of chromosome mutagenesis critical levels for the basic macrozoobenthic taxons (Fig. 24.1). It can be seen that variations of critical levels are different. These ranges are more wide in taxons with greater species difference in females fecundity. Lower levels are determined for small crustaceans.

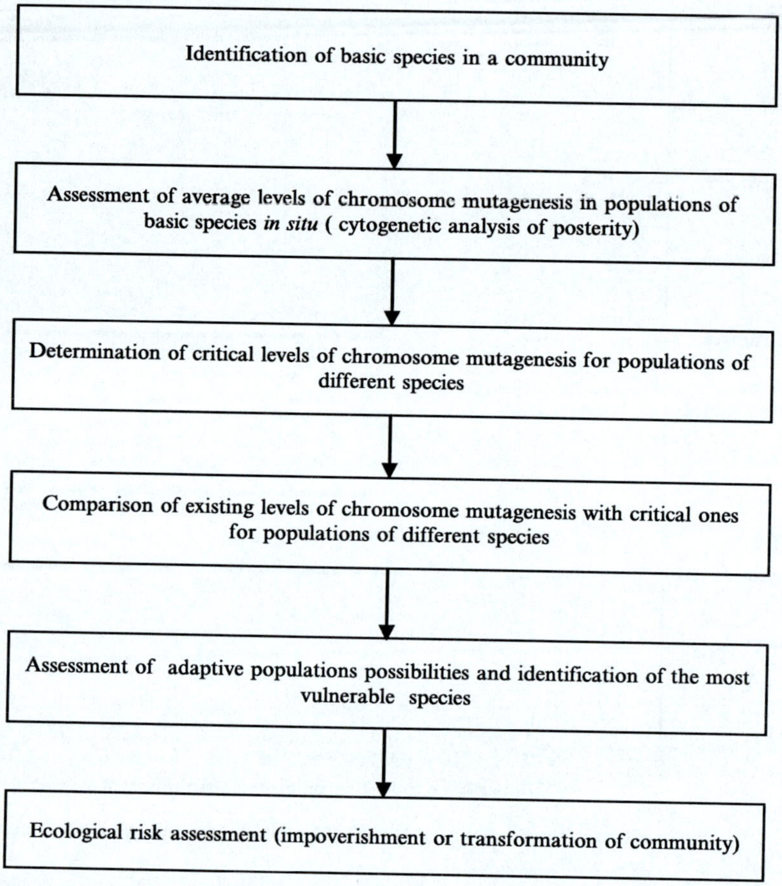

Fig. 24.2 An algorithm of ecological risk assessment to benthic communities

The results obtained allow an algorithm of ecological risk assessment to biocenoses (communities) to be proposed (Fig. 24.2).

24.5 Conclusion

On the basis of experimental data and the results of cytogenetic studies of marine and freshwater invertebrates populations (coelenterates, annelids, mollusks, crustaceans) approaches to ecological risk assessment to the benthic biocenoses (communities) were development.

References

1. Crow JF, Kimura M (1970) An introduction to populations genetic theory. Harper and Row, New York, 430 p
2. Grassle JF, Grassle JP (1986) Analysis of existing information on susceptability of deep-sea animals/Contaminant levels and relative sensitivities to contaminants in the deep-sea communities. NOVA Technical Memorandum NOS OMA 26, Rockville
3. Greze II (1977) Amphipods of the Black sea and their biology. Naukova Dumka, Kiev, 150 p (in Russian)
4. Jones D, Domotor S, Higley K, Kocher D, Bilyard G (2003) Principles and issues in radiological ecological risk assessment. J Environ Radioact 60:19–39
5. Makkaveeva EB (1977) Invertebrates of the Black Sea macropnytes. Naukova Dumka, Kiev, 150 p
6. Reish D (1957) The life history of the polychaetous annelid Neanthes caudate (delle Chiaje) including a summary of development in the Family Nereidae. Pac Sci 11:216–228
7. Tsytsugina VG (1979) Chromosome variability in marine organisms under natural conditions (on the problem of biological effect of contaminants). In: Proceeding of the international symposium interaction between water and living matter, vol II, 6–10 Oct 1979. Nauka, Moscow, pp 94–98
8. Tsytsugina VG (1990) Chromosome mutagenesis in population of aquatic biota in the Black Sea, Aegean Sea and Danube and Dnieper Rivers, 1986–1989 In: Proceeding of seminar on comparative assessment of the environmental impact of radionuclides released during three major nuclear accidents: Kyshtym, Windscale, Chernobyl, vol II, Luxembourg, 1–5 Oct 1990, pp 895–904
9. Tsytsugina VG, Polikarpov GG (1998) Pecularities of generative structure of hydrobionts populations and their radiosensitivity. In: Proceeding of topical meeting. Mol – Belgium, 01–05 June. IUR. pp 3–11
10. Tsytsugina VG, Polikarpov GG (2006) Assessment of expected reduction of crustacean populations adaptive possibilities at different levels of deleterious action. Mar Ecol J 5(1):92–98, in Russian
11. Tsytsugina VG, Polikarpov GG (2006) The estimation of the Chernobyl accident influence on the natural populations of the Black Sea basin hydrobionts by cytogenetic criteria. In: Burlakova EB, NaidichVI (eds) 20 years of the Chernobyl accident: past, present and future. Nova Science Publication Inc, New York, Chapter 16, pp 287–383
12. Vinogradova ZA (1950) Biology of the Black Sea mollusks. Trudy Karadag. biolog. stancii., 9. 100–159 (in Russian)
13. Vinogradova ZA (1951) Fecundity of the Beack Sea Decapodia, Trudy Karadag. biolog. stancii., 11. 69–91, (in Russian)

Chapter 25
Molecular Changes in Radiation Induced Thyroid Carcinomas in Mice

O. Klymenko, K.-J. Heiliger, I. Gonzalez Vasconcellos, C. Dalke, M.J. Atkinson, and M. Rosemann

Abstract Thyroid carcinomas arising from follicular epithelial cells are the most common endocrine malignancy in man. During studies performed on the population of the Marshall-Islands and after the accident at the Chernobyl nuclear power plant in 1986 a large increase in benign thyroid nodules and thyroid cancer, especially among children was shown. Thyroid follicular carcinomas are categorised into 3 histotypes: papillary, follicular, and undifferentiated (anaplastic). Papillary thyroid carcinomas are the predominant type of the thyroid cancer in patients exposed to external radiation, particularly in children. The classic oncogenic genetic alterations commonly seen in thyroid cancer include RET/PTC rearrangements, Ras point-mutations, PAX8-peroxisome proliferator-activated receptor γ (PPARg) fusion oncogene and BRAF mutation. For some families that share the FNMTC syndrome (familiar follicular nonmedullary thyroid carcinoma) a predisposition to thyroid tumors has been described. The genes responsible for FNMTC have been identified through linkage analyses of the affected families, TCO (thyroid tumors with cell oxyphilia), PRN1, and NHTC1. This study investigates gene changes in radiation induced follicular thyroid carcinoma. Following low dose exposure of the thyroid by off-targeted irradiation of the alpha-emitter ^{227}Thorium, 5 cases of Follicular thyroid carcinomas (FTC) and 1 case of thyroid hyperplasia developed in a highly susceptible mouse strain. In such cases, Comparative Genomic Hybridization (CGH) for numerical changes was performed on the whole genome. Copy number loss affecting the entire chromosome 14 in 3 out of 6 cases was observed. A similar pattern of chromosome 14 deletions was already reported in thyroid tumours of other mouse strains following high-dose exposure or oncogene activation and therefore

O. Klymenko (✉) • K.-J. Heiliger • I.G. Vasconcellos • C. Dalke
• M.J. Atkinson • M. Rosemann
Institute for Radiation Biology Helmholtz Centre Munich,
Neuherberg, Germany
e-mail: olena.klymenko@helmholtz-muenchen.de

C.E. Mothersill et al. (eds.), *Radiobiology and Environmental Security*,
NATO Science for Peace and Security Series C: Environmental Security,
DOI 10.1007/978-94-007-1939-2_25, © Springer Science+Business Media B.V. 2012

suggests that this deletion is not associated with genetic predisposition in different mouse strains to thyroid tumourigenesis.

Keywords Mice • Radiation induced follicular thyroid carcinoma

25.1 Introduction

25.1.1 *Epidemiology*

Malignant late effects in the thyroid gland following irradiation currently attract much attention and have been widely investigated for various aspects. The first investigations were performed on the Marshall Islands population contaminated from nuclear-weapon test explosions [22]. In this study, a clear increase of benign thyroid nodules was seen in all age classes exposed to the nuclear fallout [23, 24]. However malignant thyroid carcinoma appeared mainly in patients that were exposed as children.

Thyroid gland late effect studies showed great importance after the Chernobyl nuclear power plant accident in 1986 when a significant increase in the incidence of thyroid cancer among children in Ukraine and Belarus was observed. The regions affected by the fallout from the Chernobyl NPP accident were heavily contaminated with different radioisotopes, with ^{137}Cs and ^{131}I being the most pronounced. The incidence of thyroid carcinomas in children in affected regions increased from less than 1 per million to more than 80 per million. In this category of patients papillary thyroid carcinoma was the predominant type of thyroid cancer [1, 14, 19]. The relative risk for thyroid cancer has been found to be significantly higher for patients who were exposed during childhood as compared to those exposed at an older age [11]. At present there still remains an increasing trend of thyroid malignancy among population in the aforementioned regions [21].

Another study involved investigation of an Israeli population, mainly childhood, treated for tinea capitis in 1950 showed increase in long term relative risk of thyroid cancer formation [16]. Now a new study investigates population of Serbia that was also treated for tinea capitis with ionising radiation in 50th. Late health consequences can be expected in treated people [18].

25.1.2 *Classification of Thyroid Tumours*

On a global scale thyroid carcinomas are the most common endocrine neoplasm with an incidence of roughly 1%. Approximately 95% of all thyroid tumours arise from thyroid follicular epithelial cells whereas 5% of thyroid tumours are medullar thyroid carcinomas of parafollicular C-cell origin. Thyroid follicular cells are responsible for iodide uptake and thyroid hormone synthesis. They may undergo neoplastic transformation into carcinoma of 3 histotypes: papillary, follicular, and undifferentiated (anaplastic).

Papillary thyroid carcinomas are the most common thyroid tumour in countries with sufficient iodine diet and comprise up to 80% of all thyroid malignancies. Follicular thyroid carcinomas are more prevalent in regions with insufficient iodine intake and represent approximately 10–20% of all thyroid malignancies. Up to 7% of all thyroid tumours are represented by the familial form of nonmedullary thyroid carcinoma (FNMTC). FNMTC is defined by the presence of differentiated thyroid cancer of two or more first-degree relatives. Thus diagnosis of FNMTC may be reduced due to the presence of coexisting syndromes such as familial adenomatous polyposis (FAP) or Gardner's syndrome. In addition it is also difficult to specifically attribute malignancy to FNMTC when exposure to environmental factors such as ionising radiation is known to increase the risk of developing thyroid carcinomas [2, 9, 13, 20].

Anaplastic thyroid carcinomas are refractory to iodine uptake and characterised by a poor prognosis. Medullary carcinomas or C-cell carcinomas arise from calcitonin-secreting cells and in 20–25% are associated with inherited syndromes (multiple endocrine neoplasia) and familial medullary thyroid carcinomas. In the remaining 75% cases, medullary carcinomas are sporadic.

25.1.3 Gene Changes in Thyroid Carcinomas

The aetiology of most thyroid cancers is not clear. Some might develop as a late effect following exposure to ionising radiation. However mutations of DNA repair genes could also lead to gene rearrangements and development of the diseases.

The classic oncogenic genetic alterations commonly seen in thyroid cancer include RET/PTC rearrangements (cytogenetically detectable as an intra-chromosome 10 translocation), Ras point-mutations, PAX8-peroxisome proliferator-activated receptor γ (PPARg) fusion oncogene and BRAF mutation.

External radiation is a well-known exogenous agent that can cause thyroid carcinomas, specifically papillary thyroid carcinomas. The predominant molecular alteration in these tumours are chromosomal rearrangements affecting the RET proto-oncogene. This results in the synthesis of chimeric proteins, composed of the catalytic Tyrosine-Kinase domain of the RET proto-oncogene and variable C-terminal domain fusion to different genes. There are three major forms of rearrangement with RET/PTC1, RET/PTC2 and RET/PTC3 being the most frequent variants. RET/PTC1 is most often found in patients who were exposed to external radiation. On the other hand, RET/PTC3 is most often found in papillary thyroid carcinomas that developed within the first decade after exposure to the Chernobyl accident fall-out. RET/PTC3 is often associated with a solid variant of papillary thyroid carcinoma, exhibiting a more aggressive clinical appearance while RET/PTC1 does not. RET rearrangements are detected in 15–25% of papillary thyroid carcinomas [7, 8, 12, 17].

Follicular carcinomas are characterised by somatic rearrangements in the gene encoding the nuclear receptor PPARγ (the peroxisome proliferator-activated receptor gamma). PPARγ rearrangements identify in approximately 25–30% of follicular carcinomas. PAX8-PPARγ is a thyroid-specific mutation and one of the members of a family of PPARγ rearrangements in follicular carcinomas [5].

Somatic point mutation in the BRAF gene has also been identified in thyroid cancer [10]. BRAF gene produces the protein BRAF, which is involved in cell signalling and in cell growth. BRAF gene is located on chromosome 7 and a mutation that alerts valine 599 to glutamine acid (V559E) in the BRAF kinase domain has been identified in 35–45% of papillary thyroid [25].

Activation of Ras mutations are found in benign and malignant follicular neoplasms, and rarely in papillary thyroid carcinomas. Mutations in the Ras genes (H-Ras, K-Ras and N-Ras) arise from single base substitutions at codons 12, 13 or 61. These mutations have been found in 10–15% of all human cancers and in up to 50% of follicular thyroid malignancies [3]. In addition the p53 tumour suppressor gene also appears to play a role in the genesis of anaplastic thyroid cancers [4].

A predisposition to thyroid tumors has been described for some families that share the FNMTC syndrome (familiar follicular nonmedullary thyroid carcinoma). The genes responsible for FNMTC have been identified through linkage analyses of the affected families, TCO (thyroid tumors with cell oxyphilia), PRN1, and NHTC1 [9, 13].

In addition to these point mutations and chromosomal translocations, little is known about the contribution of DNA copy number changes in thyroid cancerogenesis after radiation exposure.

25.1.4 *Aim of the Study*

To use an experimental animal model of radiation-induced thyroid carcinoma in a genome-wide screen of DNA alteration using array CGH. Comparison of gene alteration in the genome of mouse strains with different sensitivity to irradiation after induction of the thyroid tumours.

25.2 Material and Methods

25.2.1 *Tissues and Genomic DNA Extraction*

Follicular thyroid carcinomas (FTC) for CGH studies developed in FVB/N mice following low-dose exposure to the thyroid by off-targeted radiation of bone-seeking alpha-emitter ^{227}Th. Following necroscopy, they were fixed in 5% formalin, embedded in paraffin and histologically analysed in the Institute for Pathology (Fig. 25.1). DNA from those tumours were extracted using the Qiagen DNAease kit (Qiagen Inc). Case, diagnosis and sex of mice are shown in Table 25.1.

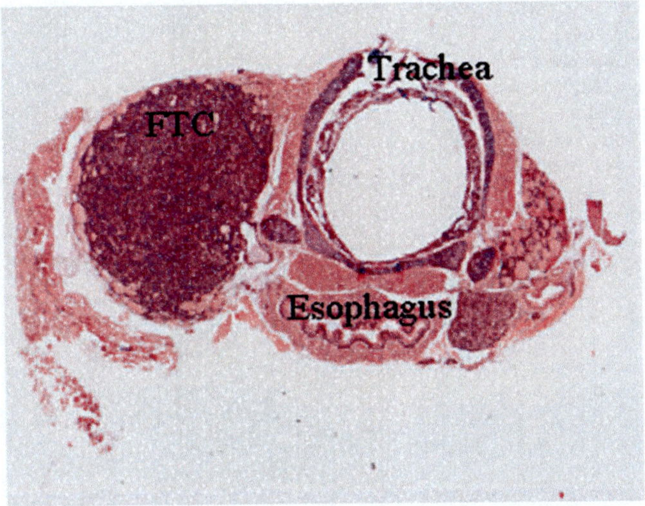

Fig. 25.1 FTC developed in mouse of FVB/N mouse strain exposed to Th-227

Table 25.1 Case, diagnosis and sex of mice

Case number	Tissue	Diagnosis	Sex
032	Thyroid gland	FTC	Male
457	Thyroid gland	FTC	Male
909	Thyroid gland	FTC	Male
910	Thyroid gland	FTC	Female
1026	Thyroid gland	FTC	Male
1375	Thyroid gland	FTC	Male

Reference DNA from mice tails was obtained by digestion of a tail tip in a 750 µl of lysis buffer overnight. The buffer was composed of 50 mM Tris, pH 8 100 mM ethylene diaminotetraacetic acid (EDTA), 100 mM NaCl and 1% sodium dodecyl sulfate (SDS) with 0.5 mg/ml proteinase K at 55°C. After adding to the mixture 250 µl of 6 M NaCl and centrifuging for 10 min at 10,000 rpm, supernatant was transferred to a fresh tube and DNA precipitated by adding one volume of isopropanol. DNA was pelleted, washed twice for 15 min with 75% ethanol, air-dried, and resuspended in 200 µl of H_2O (Ampuwa).

25.2.2 Whole Genome Amplification

Whole genome amplification of the thyroid tumor DNA was done using GenomPlex® Complex Whole Genome Amplification (WGA) kit (Sigma) following the protocol of the manufacturer. The quality of the WGA DNA was than qualitatively determined by loading 10 µl of the amplified DNA onto 1.5% agarose gel.

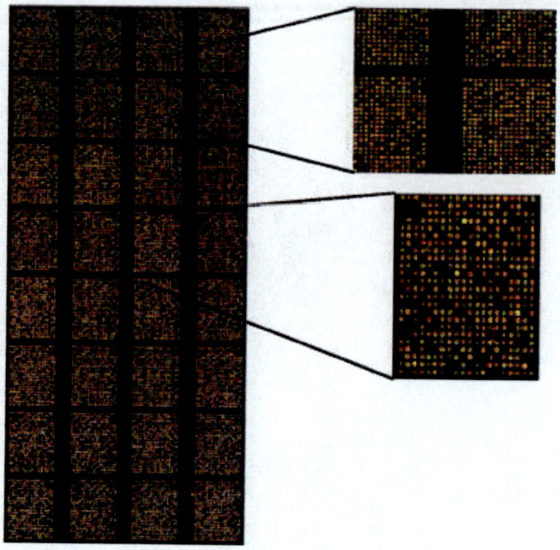

Fig. 25.2 Scan of the 1 Mb CGH array slide with measured intensity ration for each BAC clone. All BAC clones are repeated in quadruplets. *Green* channel: Tumour DNA; *Red* channel: reference DNA

25.2.3 Comparative Genomic Hybridization (CGH) on BAC Arrays

CGH is a technique that can identify and map DNA copy number changes in tumours relative to a normal tissue in a single hybridisation experiment [15]. 1 Mb CGH arrays containing approximately 3,000 BAC probes that cover the entire genome (Codelink DNA array slides). For each experiment, 450 ng of either female or male reference DNA and 450 ng of tumour DNA were labelled with Gy3-dCTP and Cy5-dCTP (Perkin Elmer Life Science), respectively, in a random-primed labelling reaction. Fluorescent labelled tumour and reference DNA of the opposite sex were co-hybridised with a presence of mouse cot-1 DNA, (to block repetitive sequences) washed and dried in an automated TECAN HS400 hybridisation station. Co-hybridization was performed 'sex-mismatch', meaning that male tumour DNA was hybridized with female reference DNA and opposite and the expected difference in number of X-chromosomes could serve as an internal control for the experiment. The slides were scanned with a TECAN Laser Scanner and the red-to-green intensity ratio determined for each BAC clone with an array analysis software (GenePix Pro 6.0, Axon Laboratories) (Fig. 25.2).

For every single BAC clone on the array, a shift in the intensity ratios of the tumour DNA fluorescence signal to the reference DNA fluorescence signal is indicative of DNA copy number changes in the tumour sample. For all genomic regions where the tumour DNA carries the normal two copies, the related BACs on the array show an equal intensity of green (i.e. tumour DNA) and red (i.e. reference DNA) fluorescence intensity. In case of a region of chromosome or genes being lost or deleted, the tumour DNA hybridizes less intense and hence causing an overall colour-shift to red. In contrary, if the tumour cells acquired a DNA-copy number in a certain genomic region, this would result in an increased hybridisation signal of the tumour DNA onto the related BAC clone, equivalent to a colour-shift towards green.

Further analysis of the data was performed with CAPweb array CGH evaluation platform, which was accessible via Internet server.

25.3 Results

25.3.1 *Array CGH Analyses*

In three out of six cases (032, 910 and 1026) a deletion of the entire chromosome 14 was observed. In addition, case number 910 also suggested a deletion of the whole chromosome 7. The remaining three cases (457, 909 and 1375) did not show significant chromosome copy number changes. The total number of BAC spots showing a useful hybridisation signal varied between 75% and 97%. The expected 'sex-mismatch'-signal could be observed only in case numbers 032, 457 and 910, whereas cases 909, 1026 and 1375 failed to pass this qualitative quality parameter. For case 1026 a 'sex-mismatch'-signal could be found in a repeated experiment after using whole-genome preamplification where deletion of the chromosome 14 was confirmed. This hybridisation produced 53% detectable BAC clones. All cases exhibit significant noise, which probably masked the detection of smaller regions of copy number changes on other chromosomes. One of the reasons that could lead to the appearance of huge noise and loss of a number of BAC clones could be that DNA samples consisted of small DNA fragments (partially degraded DNA). A PCR reaction with different sets of primers was performed to determine the length-range of DNA fragments presented in investigated samples. The results in Fig. 25.3 clearly demonstrate that DNA samples do not contain fragments much longer than ~200 bp. All array CGH profiles are shown in Fig. 25.4.

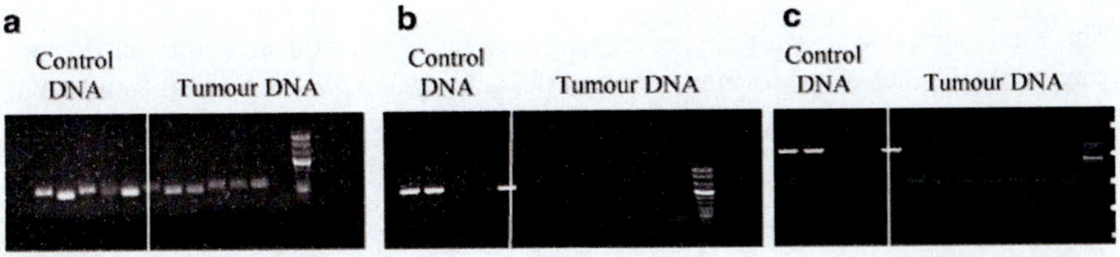

Fig. 25.3 Gel-electrophoresis of PCR products: **(a)** show fragments of 180 bp range; **(b)** show fragments of 530 bp range; **(c)** shows fragments of 650 bp range. Tumours DNA were loaded with next order of cases 457, 909, 910, 1026 and 032 respectively

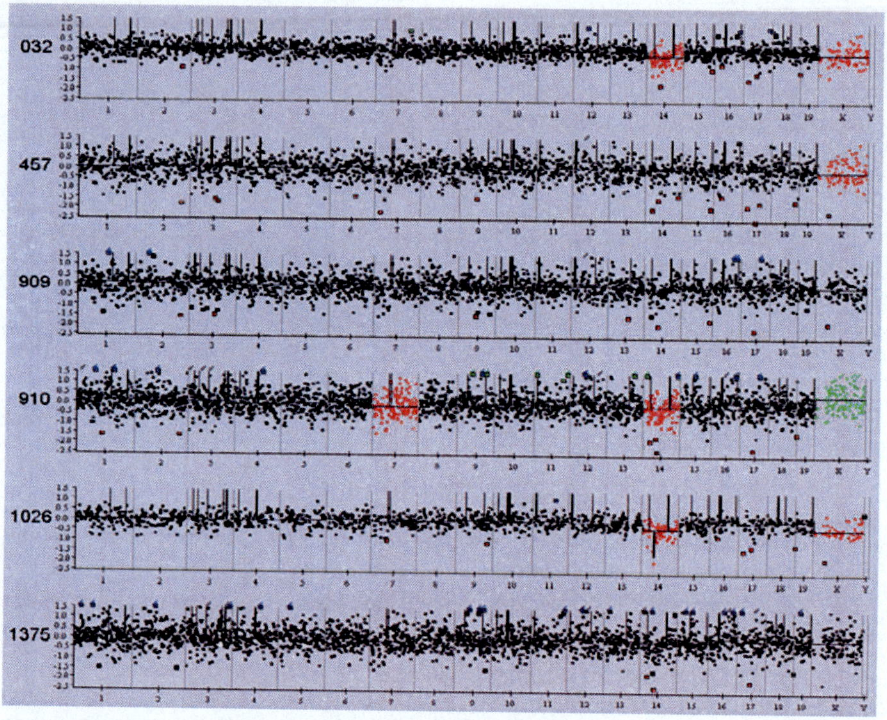

Fig. 25.4 Result of array CGH analysis. Numbers along abscissa relate to chromosome, from *left* to *right* is the distance from p-arm telomere to q-arm telomere. Numbers on the ordinate axis give relative copy number changes in tumour DNA as a logarithm to base 2. From *top* to *bottom* case 032, 457, 909, 910, 1026 and 1375. Each single point represents hybridisation of tumour DNA relative to normal DNA on one distinct genomic BAC probe. *Dots in black* are classified as unchanged in tumours copy numbers, *dots in red* classified as reduced copy numbers and dots in *green* as increase in copy numbers. BAC clones taken into *squares* are outliers

25.4 Discussion

This study aimed to investigate gene changes in radiation induced follicular thyroid carcinoma. CGH was performed for numerical changes within the whole genome in 5 cases of follicular thyroid carcinomas and 1 case of thyroid hyperplasia that developed following low dose exposure of the thyroid by off-targeted radiation of the alpha-emitter ^{227}Th in strain FVB/N. Using array CGH we could show deletions of the entire chromosome 14 in 3 out of 6 cases. Despite the fact that DNA obtained from formalin fixed paraffin embedded tissue was not of high quality and one case failed to show internal control signal, after performing whole genome amplification, deletion of chromosome 14 in that particular case was confirmed. These observations correspond to data from recent studies where follicular thyroid carcinomas were induced by ^{131}I in backcrossed mice (G. Hölzwimmer, GSF Pathology) [6]. Therein, 5% of mice (F1-hybrids and backcrosses of C3H/He and C57BL/6), treated with relatively high doses of ^{131}I (activity 111 kBq), developed follicular thyroid carcinomas after exposure. In 38% of these tumours deletion of the chromosome 14 was also observed. Moreover in 75% cases of the irradiated mice G. Hölzwimmer detected loss of heterozygosity (LOH) on chromosome 14. Because injection of the alpha-emitter ^{227}Th used in the present study targets predominately the mouse skeleton, the gamma-dose from ^{227}Th daughter products (mainly ^{233}Ra) that could expose the thyroid gland must be estimated to be low (personal communication Dr. M. Rosemann). The observation that follicular thyroid carcinoma in both high doses ^{131}I treated C3H x C57BL/6 mice and low dose ^{227}Th treated FVB/N mice share the same histological type and the same molecular alteration implies that the congenital predisposition of FVB/N mice is caused by very early steps of the carcinogenic process of thyrocytes. In contrast, both the gross-histomorphology of a tumour as well as large chromosomal alterations are probably determined in the later steps of the malignant progression. The observation in recent studies suggests that there is no correlation between deletion of chromosome 14 and genetic predisposition in different mouse strains to thyroid tumourigenesis.

25.5 Summary

Our observations suggest that congenital predisposition of FVB/N mice is caused by very early steps of the malignant transformation of thyrocytes, without any association to deletion of chromosome 14 in tumours itself.

Acknowledgments I would like to thank Head of the Department of Radiation Cytogenetics Helmholtz Center Munich Prof. Dr. H. Zitzelsberger and to the Head of the Master course in Radiation Biology University College London Prof. Claus R. Trott.

References

1. Baverstock K, Egloff B, Pinchera A, Ruchti C, Williams D (1992) Thyroid cancer after Chernobyl. Nature 359:21–22
2. Fagin JA (1997) Familial nonmedullary thyroid carcinoma–the case for genetic susceptibility. J Clin Endocrinol Metab 82:342–344
3. Fagin JA (2002) Minireview: branded from the start-distinct oncogenic initiating events may determine tumor fate in the thyroid. Mol Endocrinol 16:903–911
4. Fagin JA, Matsuo K, Karmakar A, Chen DL, Tang SH, Koeffler HP (1993) High prevalence of mutations of the p53 gene in poorly differentiated human thyroid carcinomas. J Clin Invest 91:179–184
5. Farid NR (2004) Molecular basis of thyroid cancer, pp 87–90
6. Hölzwimmer GC (2007) Histologische und molekulargenetische Untersuchungen von strahleninduzierten Schilddrüsentumoren im Mausmodelled^eds. Ludwig-Maximilians-Universität, München

7. Komminoth P (1997) The RET proto-oncogene in medullary and papillary thyroid carcinoma. Molecular features, pathophysiology and clinical implications. Virchows Arch 431:1–9
8. Lin X, Finkelstein SD, Zhu B, Silverman JF (2008) Molecular analysis of multifocal papillary thyroid carcinoma. J Mol Endocrinol 41:195–203
9. McKay JD, Thompson D, Lesueur F, Stankov K, Pastore A, Watfah C, Strolz S, Riccabona G, Moncayo R, Romeo G, Goldgar DE (2004) Evidence for interaction between the TCO and NMTC1 loci in familial non-medullary thyroid cancer. J Med Genet 41:407–412
10. Namba H, Nakashima M, Hayashi T, Hayashida N, Maeda S, Rogounovitch TI, Ohtsuru A, Saenko VA, Kanematsu T, Yamashita S (2003) Clinical implication of hot spot BRAF mutation, V599E, in papillary thyroid cancers. J Clin Endocrinol Metab 88:4393–4397
11. Nikiforov Y, Gnepp DR, Fagin JA (1996) Thyroid lesions in children and adolescents after the Chernobyl disaster: implications for the study of radiation tumorigenesis. J Clin Endocrinol Metab 81:9–14
12. Nikiforov YE, Rowland JM, Bove KE, Monforte-Munoz H, Fagin JA (1997) Distinct pattern of ret oncogene rearrangements in morphological variants of radiation-induced and sporadic thyroid papillary carcinomas in children. Cancer Res 57:1690–1694
13. Pal T, Vogl FD, Chappuis PO, Tsang R, Brierley J, Renard H, Sanders K, Kantemiroff T, Bagha S, Goldgar DE, Narod SA, Foulkes WD (2001) Increased risk for nonmedullary thyroid cancer in the first degree relatives of prevalent cases of nonmedullary thyroid cancer: a hospital-based study. J Clin Endocrinol Metab 86:5307–5312
14. Richter H, Braselmann H, Hieber L, Thomas G, Bogdanova T, Tronko N, Zitzelsberger H (2004) Chromosomal imbalances in post-chernobyl thyroid tumors. Thyroid 14:1061–1064
15. Ried T, Liyanage M, du Manoir S, Heselmeyer K, Auer G, Macville M, Schrock E (1997) Tumor cytogenetics revisited: comparative genomic hybridization and spectral karyotyping. J Mol Med 75:801–814
16. Sadetzki S, Chetrit A, Lubina A, Stovall M, Novikov I (2006) Risk of thyroid cancer after childhood exposure to ionizing radiation for tinea capitis. J Clin Endocrinol Metab 91:4798–4804
17. Santoro M, Melillo RM, Fusco A (2006) RET/PTC activation in papillary thyroid carcinoma: European Journal of Endocrinology Prize Lecture. Eur J Endocrinol 155:645–653
18. Shvarts S, Sevo G, Tasic M, Shani M, Sadetzki S (2010) The tinea capitis campaign in Serbia in the 1950s. Lancet Infect Dis 10(8):571–576
19. Stsjazhko VA, Tsyb AF, Tronko ND, Souchkevitch G, Baverstock KF (1995) Childhood thyroid cancer since accident at Chernobyl. BMJ 310:801
20. Sturgeon C, Clark OH (2005) Familial nonmedullary thyroid cancer. Thyroid 15:588–593
21. Sumner D (2007) Health effects resulting from the Chernobyl accident. Med Confl Surviv 23:31–45
22. Takahashi T, Trott KR, Fujimori K, Simon SL, Ohtomo H, Nakashima N, Takaya K, Kimura N, Satomi S, Schoemaker MJ (1997) An investigation into the prevalence of thyroid disease on Kwajalein Atoll Marshall Islands. Health Phys 73:199–213
23. Takahashi T, Simon SL, Trott KR, Fujimori K, Nakashima N, Arisawa K, Schoemaker MJ (1999) A progress report of the Marshall Islands nationwide thyroid study: an international cooperative scientific study. Tohoku J Exp Med 187:363–375
24. Takahashi T, Fujimori K, Simon SL, Bechtner G, Edwards R, Trott KR (1999) Thyroid nodules, thyroid function and dietary iodine in the Marshall islands. Int J Epidemiol 28:742–749
25. Xing M (2005) BRAF mutation in thyroid cancer. Endocr Relat Cancer 12:245–262

Chapter 26
Study of *Arabidopsis thaliana* Patterns of Experimental Mutational Variability

Olga Usmanova and Timur Usmanov

Abstract A summary of results of prolonged genetic and physiological research initiated by N.V. Timofeev-Ressovsky and carried out based on the unique genetic collection of the mutants of the classical model object – *Arabidopsis thaliana* – under the leadership and immediate participation of P.D. Usmanov in the Laboratory of Physiological Genetics of the Genetic Department of the Institute of Plant Physiology and Genetics of the Academy of Sciences of the Republic of Tajikistan.

The article contains the following: the summary of development of the Arabidopsis genetic collection; the results of the study of the Arabidopsis mutation variability; creation of the wide spectrum of mutant forms differing by their phenotype, the study of their genetic nature and physiological and biochemical peculiarities; the analysis of the genotypic variability of the photosynthetic apparatus characteristics in order to ascertain the mechanism of the photosynthesis genetic control and inter-relations between the cell organelle (karyon, chloroplasts, mitochondrion).

Based on the complex genetic, physiological and biochemical study of the mutant lines from the mentioned Arabidopsis collection, there have been established experimental model systems for ecological and genetic research, and the systems for evaluation of the impact of extreme factors and anthropogenic influence on the gene pool of plant populations and analysis of genetic mechanisms of plants adaptation.

Keywords Arabidopsis • Mutants • Phenotype • Genotype • Markers • Experimental models • Test-systems

---

O. Usmanova (✉) • T. Usmanov
Institute of Plant Physiology and Genetics Academy of Sciences of the Republic of Tajikistan, Aini Str., 299/2, Dushanbe 734063, Republic of Tajikistan
e-mail: rustim.us@mail.ru

C.E. Mothersill et al. (eds.), *Radiobiology and Environmental Security*,
NATO Science for Peace and Security Series C: Environmental Security,
DOI 10.1007/978-94-007-1939-2_26, © Springer Science+Business Media B.V. 2012

26.1 Introduction

At the end of 50th of the previous century special attention has been paid to the creation of collections of such plant species which offered special advantage as objects of scientific research in different divisions of biology. *Arabidopsis thaliana (L) Heynh* – the "plant drosophila" – can be attributed to this category with confidence. Good fruit inception at self-pollination and controlled cross-pollination; relative simplicity of cultivation both in soil and on growth medium with agar in test tubes, Koch and Petri dishes with all possible regulation of root nutrition, light and temperature regimes. Due to the abovementioned characteristics Arabidopsis has stared being used as a model object in genetic, physiological and biochemical research [4, 6, 18].

The Central Seed Bank with hundreds of samples of populations and genetically pure lines of Arabidopsis [5] has been established in Frankfurt-am-Main (Germany). Publication of the special annual journal "Arabidopsis Information Service" has been initiated there. Arabidopsis researchers from many countries joined around it. A perfect model object – "plant drosophila" – Arabidopsis first emerged in Tajikistan at the beginning of 60th of the previous century following the example of N. V. Timofeev-Ressovsky. Being in Tajikistan, N. V. Timofeev-Ressovky in his reports, lectures and speeches had repeatedly reiterated that a successful choice of the initial material for an experimental research had the decisive importance for the successful solving of the correctly formulated scientific task. In his article "Genetics and plant physiology" [13] he proved the importance of a joint approach of genetics and physiology to the solution of the fundamental practical as well as the most important practical problems of the modern biology. N. V. Timofeev-Ressovky was one of the first who emphatically urged to use Arabidopsis mutants to study the genetic fundamentals of the photosynthesis.

This period is supposed to be the beginning of the unique genetic collection of Arabidopsis for genetic and physiological research (including mutagenesis) under the leadership and the immediate participation of P. D. Usmanov, who was a worthy pupil and, later on, a worthy friend of the unforgettable N. V. Timofeev-Ressovsky.

It is known that the *Arabidopsis* genus includes 14 species, some of which are available in the Central Asia. Some species are available in the North Africa, the Mediterranean, Europe and Siberia. "The Flora of the USSR" (1939) includes 5 species. The mentioned number will grow up to 8 with the inclusion of *A. Wallichii* (Hook.f.et Thom) N. Busch, *A. bursifolia* (D.C.) Botsch, isolated recently from other genuses, and a new species *A. Korshinski* Botsch. Seven species of the abovementioned eight are available in the territory of the Central Asian republics and five in Tajikistan [38]. N. I. Vavilov has rightly pointed out that the unique climatic, ecologic and geographic conditions of Tajikistan stipulate its rich morpho-physiological diversity. Tajikistan is an acting evolutionary arena, one of the loci of the intensive form-building processes. Five species of the *Arabidopsis* genus growing in Tajikistan represent a natural polyploid set: *A. thaliana*, 2n = 10; *A. Wallichii*, 2n = 16; *A. mollissima*, 2n = 26; *A. pumila*, 2n = 32; *A. korshinski*, 2n = 48

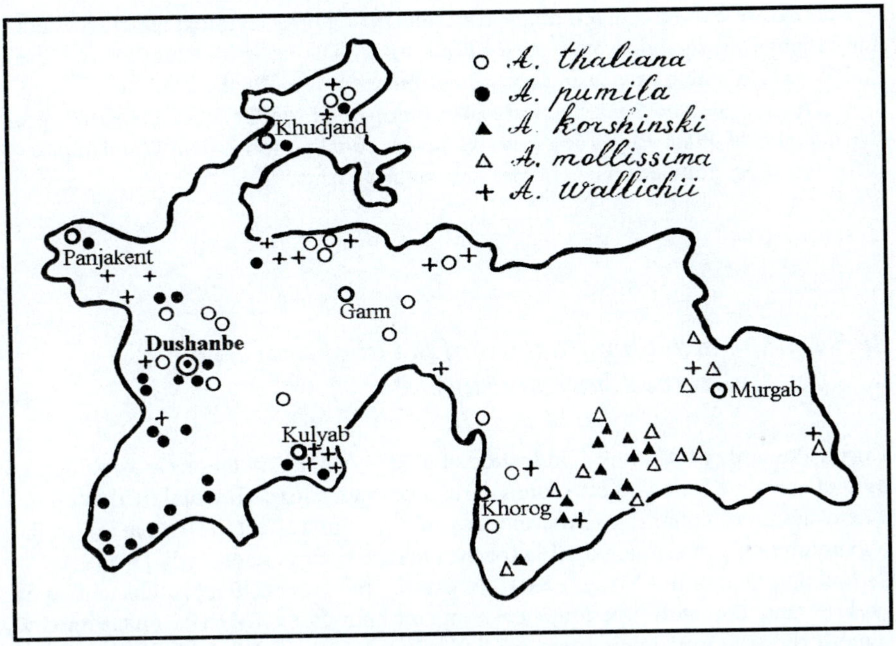

Fig. 26.1 Map of Tajikistan with the marked locations of growth of the Arabidopsis species

(Fig. 26.1), as well as 8 ecotypes of Arabidopsis: Anzob, Vakhsh, Tajik, Kurama, Shugnan, Rengentau, Turkestan, Kharangon, and populations at different altitudes: Kondara, 1,100 m above the sea level; Khodja-Obi-Gharm, 1,800 m; Sorbo, 2,200 m; and Shokhdora, 3,400 m.

26.2 Materials and Methods

Mainly, the work on the cultivation of the Arabidopsis mutant forms and their collection, study and experiments have been carried out at the Biological station "Siekukh" of the Institute of Plant Physiology and Genetics of the Academy of Sciences of the Republic of Tajikistan. It is located on the southern slope of the Ghissar range at the altitude of 2,300 m above the sea level, 73 km north of Dushanbe. Plants were grown on special experimental plots annually starting from June to September-October.

Arabidopsis was also grown in the greenhouse of the Genetics Department (Dushanbe, 800 m above the sea level) as well as in laboratories on agar in test tubes and Koch dishes (aseptic culture) under the light installation in strictly control condition [4, 37].

Methods of experimental mutagenesis, genetic analysis, cytological, biochemical and biophysical methods were used in this work. The experimental data obtained underwent statistical processing according to the commonly accepted methods.

Currently, the genetic collection of Arabidopsis of the Genetics Department of the Institute of Plant Physiology and Genetics contains more than 400 samples – different races, populations, ecotypes and mutant forms.

26.3 Results

26.3.1 The Mutation Analysis of the Photosynthetic Apparatus Characteristics

During the study of patterns of induction of photosynthetic mutations in Arabidopsis by mutagens we followed the common radiobiological experimental methodology, where the main condition is the plotting of "the dose-effect" (DE) curves in the maximum range of doses, possible for the characteristics under study [14].

Ionizing radiation – X-rays (XR) and gamma radiation (GR); protons and alpha rays (P and α); and supermutagens: nitrozomethylurea (MNU), nitrozomethyl biureth (NMB), nitrozoethylurea (NEU), ethylene imine (EI) and ethyl methane sulphonat (EMS) have been used as mutagens.

It has been established for the most of the genes, determining the characteristics of the photosynthetic apparatus that they react in a different manner to the mutagen impact. To compare the genetic impact caused by different mutagens in Arabidopsis, the index of "the degree of sterility" was used as a standard curve [9]. Comparison of the frequency of chlorophyll mutations caused by NMB, NMU and XR at the different degrees of sterility has shown that at a low level of sterility (up to 8%) mutation yield curves of the mentioned mutagens are practically congruent. This indicates the identical impact of the compared mutagens on a cell, i.e. that the lethal and mutagen effects have common mechanisms in this case. There is the most clear-cut distinction between mutagens at the interval of sterility from 10% to 30%. The maximum number of mutations is noted at the use of NMB, then NMU and XR, though the comparison of the genetic impact at higher levels of sterility leads to unclear results. The abovementioned allows making a very important methodical conclusion: evaluate the relative genetic efficiency of the mutagens under study, it is necessary to plot DE curves on many experimental points and compare them with each other. Otherwise, the conclusions will be erroneous.

The phenotypic spectra have been plotted (Fig. 26.2) for the evaluation of the specificity of the mutagen impact in induction of different levels of mutations. In this figure, the mutants are arranged according to the increasing frequency of mutations of the *viridis* class. Therefore, according to the efficiency of causing of the mutation frequency of this class, the mutagens arrange as follows: NMU, NMB, P, XR and GR, alpha rays, EMS, NEU and EE. According to mutations of the *xantha* type this order looks as follows: EI, α-particles, EMS, P, NEU, XR and GR, NMB and NMU.

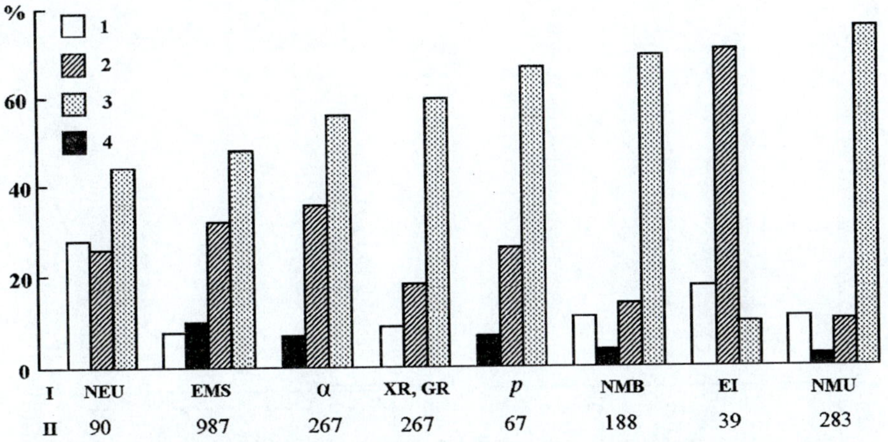

Fig. 26.2 Phenotypic spectra of chlorophyll mutations induced by different mutagens. I – mutagens; II – number of the mutant families. *1* – albina; *2* – xantha; *3* – viridis; *4* – others

The Fig. 26.2 shows that the heavy charged particles in the spectrum do not have mutations of the *albina* type. Possibly, it is indicative of the specificity of the P and α-particles impact on the genetic apparatus [16].

It should be pointed out that the share of the chlorophyll mutations is significant among the experimentally caused mutations [3, 4, 11, 12, 16, 34]. Therefore, the chlorophyll mutations are widely used in the mutation research as a handy test-system for the quantitative characteristic of the mutation variability of plants.

In order to obtain the point (chlorophyll and morphological) and the genome (polyploidy) mutations of Arabidopsis its seeds and pollen grains of the race *Dijon, Enkheim* and *Columbia* were exposed to X-rays and processed by EMS and colchisine. For the X-rays (Fig. 26.3) the linear correlation of the yield of the chlorophyll mutations according to the dose of radiation is established at the maximum of 32 kg, while for EMS the exponential correlation of the yield of mutations at the maximum of 24 mM [21, 33].

The impact of the UV ($\lambda = 254$ nm) on the pollen, and the impact of the natural UV radiation on the pollen and plants of Arabidopsis has been studied. Trial experiments have shown that the UV radiation ($\lambda = 254$ nm) in the interval of 60–120 min caused inactivation of the pollen grains. The results of the basic experiments have shown that the frequency of the dominant embryonic lethal mutations increased naturally according to the radiation dose.

UV dose, min	0	8	16	24	32
Embryos under study	2,026	924	1,132	1,054	1,202
Lethal among them, %	4.1	14.5	33.5	52.8	68.4
Average number of embryos in a seedpod	48	51	52	53	48

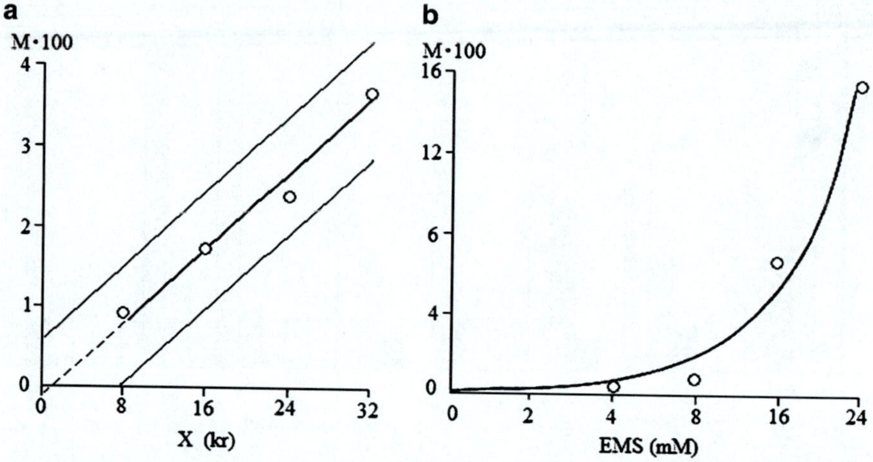

Fig. 26.3 Dependency of the recessive chlorophyll mutations frequency in the A.thaliana (**a**, Dijon race; **b**, Columbia race) on doses of the X-ray irradiation of the pollen and the EMS concentration (processing of seeds). On the X-axis: dose is in kr; concentration is in mM. On the Y-axis: average number of mutations (M) per 100 diploid cells

During the special experiments the lack of photo-reactivating impact of the light and the dark repair during or shortly after the UV irradiation of the pollen has been proved. The results obtained during the study of the mutation variability of Arabidopsis and the variability of the genotypic structure of the model populations of Arabidopsis and wheat allow to assert that the UV-B is the factor of selection impacting the direction of micro-evolutionary processes in the populations of the higher plants. Most probably, this is the role of the high mountain UV radiation in the process of the morphogenesis of the higher plants [26, 28, 30].

26.3.2 Genetic Analysis and Evaluation of the Number of Genes Determining the Photosynthetic Apparatus Characteristics

The mutational and hybridologic methods of analysis of the genotype structure allowed carrying out the inventory of karyogenes determining the photosynthetic apparatus characters. To receive the objective evaluation of the number of genes determining the character "the chlorophyll insufficiency", two successive stages of the experimental work should be carried out. At the first stage the frequency of mutations is determined by any mutagen, and the phenotypic spectrum of mutations is plotted. At the second stage determine the number of genes for a definite phenotypic character (phene) by the cross-breeding on the allelism. The second stage of the experiments is very labour-intensive, as is known. And, it is advisable that one should use the method without that stage, or bring it to the minimum. This idea was assumed as a basis of our method. To evaluate the number of genes determining any character, one can limit him/herself by the first stage of the experiment, and instead of the second stage use the following formula:

$ni = \frac{Pij}{Poj}$, where *ni* is the number of genes determining *Ai* – the character of the plant (*i*=0,1 ….,*N*); *Pij* is the frequency of mutations *Pij* character under the influence of *Bj* – mutagene (*j*=1,2,….*M*). The values of *n* and *Poj* are determined from the reference experiment over the *Ao* character (in our case *Ao* is the *fuska* character), and *Pij* is ascertained during the experiment on determination of the phenotypic spectra of mutations. In order to calculate the number of genes we accepted n=12 (adopted from Muller's work) [10] by the *fuska* character, and $Pij : Poi = 40$ (ascertained in our experiments independently of Muller). Then, it follows from the offered formula (as a result of calculation), that $ni \sim 500$, i.e. the mentioned number of the genes of the Arabidopsis genotype determines the characters of the chlorophyll genesis and the lamella system of chloroplasts [16].

One of the main results of the genetic analysis was a significant expansion of the available collection by new genetically pure lines of Arabidopsis, whose chromosomes are marked with multiple signal genes. These and other genetically pure lines of Arabidopsis essentially alleviate the labour-intensive genetic analyses and are used by scientific research groups.

During some special experiments some photosynthetic mutants of Arabidopsis were cross-bred with the tester lines *vc'er, gl'an* and *lu'co* in order to determine the coupling groups. For the random sampling, including 34 genes, determining the photosynthetic apparatus characters, the following distribution of the mutant genes on the chromosomal coupling groups was obtained. In the first group: *atroc, atvi 1, atvicos 1 and 2, cif 1, cif 3, flavi 4, vimac 2*; in the second group: *sid, xa 1–9, viluts 1*; in the third group: *almac 2, bf, clavi 17, lu 1, tr 2*; in the fourth group: *as 3,ch 13, xa 1–5, xas 1–1, 1–2, 1–3, 1–4, 1–5, vimac 6*; in the fifth group: *cla, chloti, flavi 1–1, 1–2, 1–3, 1–4, 1–5, flavici, tr 1*. It follows from the above, that the genes determining the photosynthetic apparatus characters are located in all five chromosomal coupling groups of Arabidopsis [16].

It is significant that in the overwhelming majority of cases the photosynthetic mutations induced by ionizing radiation and chemical mutagens in Arabidopsis are the monofactorial disjoining recessives, i.e. they have a nature of the point (gene) mutations. Along with this, some cases showed that the phenes of the chlorophyll insufficiency can be stipulated by both chromosome mutations located in the first and the third chromosomes, and by mutations of the plastid genes.

26.3.3 Phene Analysis and Phene Genesis of the Photosynthetic Apparatus

The research of patterns of the genotypic variability of plants has found that the genotypic differences (the genotypic environment) of plants stipulated by the point and chromosome mutations have a significant impact on the character of the interactions and interrelations of the cell organelle. The received mutant forms of

Arabidopsis differed by the big phenotypic variability and this phenomenon allowed to study the phenotypic display and photosynthetic mutations for determination of the adaptive potentials of the mutant genotypes.

Variation of the composition of the agar nutritive medium had a significant impact on the Arabidopsis mutants. As a result, some lethal mutants of Arabidopsis reached the phase of fruitification (without nutritive additions lethal seedlings usually die at the stage of cotyledonary leaves within 2–18 days from the date of emergence of seedlings). The cultivated plants were typical dwarfs evenly coloured yellow (*79 xa, 127 xa*) and whitish (*9al*), i.e. the normalization did not take place by the character of the colour of plant [24, 25]. Analysis of the ultra-structural organization of the cell components has shown that the plastid completely lacked the lamella structures; however, the genetic blockage of the development of the plastids caused the growth of the number of mitochondrion:

Lines	The number of plastids	The number of mitochondrion
En	8±0.14	9±0.13
79 xa	3±0.13	15±0.20
127 xa	3±0.14	16±0.17

It is interesting to point out that the growth of the number of mitochondrion was accompanied with the good development compared to the norm of their membranous organization. This fact speaks well of the idea on the functional interchangeability of the energy transforming organelle of the plant cell [2].

The results show that the energy needed for the biosynthesis ensuring growth and development of the mutant plants is delivered exclusively by mitochondrion which works with the increased functional load in the lack of the normal chloroplasts.

During the study of the biochemical mutant *virido-albina 40/3* the new evidence of the availability of competitiveness between chloroplasts for the product of metabolism, amino acid leusine, has been received. The leusine deficit caused by the impact of the mutant gene leads to creation of the undeveloped plastids and the normal chloroplasts inside the same cell (the so-called, pseudo-mixed cells, in contrast to the real, characteristic of the cytoplasmic chloroplast mutants). Addition of leusine into the nutritive medium (D,L – 200 mg/l) led to the finalization of the chloroplast genesis and complete recovery of the chloroplasts ultrastructure, and extinction of the pseudo-mixed cells in the mutant. The concrete mechanism of the formation of hetero-plastids, i.e. the pseudo-mixed cells, in contrast to the real mixed cells, formed under the influence of the plastid mutations, has been found [1].

Based on the experimental data [24, 25] and the results of many other authors, the patterns of the phylogenetic, the ontogenetic, the paratypic and the mutational variability of the photosynthetic membranes systems of the representatives of different systematic plant groups have been analyzed. It allowed to establish the homology of the variability of the photosynthetic membranes stipulated by the

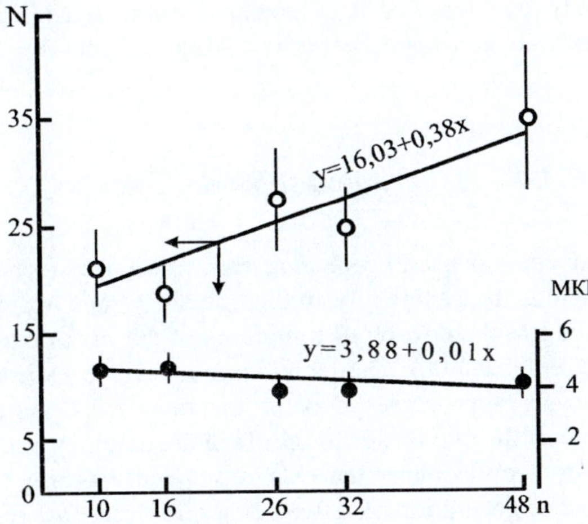

Fig. 26.4 Interrelation between number of chromosomes (n), chloroplasts diameter (*dark circles*) and average number of chloroplasts (*light circles* – N) in cells of cancellous mesophyll in five species of the Arabidopsis genus

mutation events both the genome and the plastome, and by the environmental conditions, in the cells of prokaryotes and eukaryotes.

The mutational variability of the chloroplasts of the higher plants depicts, practically, all types of structures represented in the scheme on the lower floors of the evolutionary spiral. This testifies to the fact that the chloroplasts of the higher plants preserve the historic chronicle of emergence and improvement of the photosynthetic membranes in their genotype.

It is necessary to draw attention to the point mutations of the genetic systems of the cell, which almost always lead to the emergence of the types of the membrane systems located on the lower floors of the evolutionary spiral.

To study the genetic control of the dimensions and the number of chloroplasts both the gene and the genome mutations were used. The dimensions and the number of the chloroplasts in the cells of the cancellous mesophyll of the first pair of the real leaves were studied on the five species of Arabidopsis genus growing in Tajikistan, and forming the polyploidy row (2n=10, 16, 24 and 48); (Fig. 26.4).

The results of the regression analysis have shown that the increase of the number of the chromosomes correlates positively with the number of the chloroplasts. For the coefficient of the regression, which characterizes the angle of the curve slope according to the dimensions of the chloroplasts, a reliable difference from a zero was not found ($b=0.014\pm0.013$; $P=0.38$). This testifies to the lack of correlation between the alteration of the number of chromosomes, on the one hand, and the dimensions of the chloroplasts, on the other hand [15, 20]. Therefore, the control of the number of the chloroplasts in the cell is carried out irrespective of the control of their dimensions, i.e. the number of the chloroplasts in the cell is controlled by the chromosome apparatus, but the dimensions are determined by the plastid genes. The established patterns were confirmed during experiments on the artificial polyploid series of Arabidopsis of *Dijon* and *Columbia* races (2n=10; 3n=15 and 4n=20), and sugar beet (2n=18; 3n=27 and 4n=36).

26.3.4 Experimental Model Systems

For ecological and genetic research several test-systems have been elaborated. They include the genetically studied heterozygous Arabidopsis plants split by the recessive lethal chlorophyll mutations of the *al, xa* and *ch* types (e.g., 104 *xa*, 127 *xa*, P 10/53 *xa*, etc.). The mentioned lines have a specific character – their splitting by the genotype corresponds to the ratio 1:2:1. As a result of the lethal mutations impact the mutant seedlings die in the heterozygous status in the phase of formation of the cotyledonary leaves. Consequently, during the period of fruitification in the second generation only two genotypic classes are registered: one part of the homozygous plants of the wild type and two parts of the heterozygous on the lethal factor plants, which, phenotypically, do not differ from the norm, but which are accurately indentified by the embryo-test method [9, 18]. These lines are elementary by their structure. They represent "the micro-populations", which being reproduced give plants attributable to two genotypic classes at the ratio 1:2. Shifts in this ratio under the influence of the natural factors allow analysis of the structural variability patterns impacted by the agents under study [16, 30].

The study of impact of the lethal mutations in heterozygous state on physiological processes revealed some mutant heterozygotes of Arabidopsis (*9 al, 127 xa, P20/31 xa, 19 ch, 29 ch*), which statistically exceed the initial race *Enkheim* in the growth, the development and the seed productivity. This indicates the effect of the monohybrid heterosis. The phenotypic display of the monohybrid heterosis depends both on the impact of the mutant genes in the heterosis state, and on the environmental conditions, even if the plants have been grown under the same conditions, but in different seasons [23]. This experimental model can be used as a sample of the scientific approach for elaboration of effective genetic and selection methods of increase of the crop-productive power of agricultural plants.

A highly sensitive complex test-system was designed based on the semi-dominant chlorophyll mutation of the chlorine type with the recessive lethal impact. It allows the evaluation of the frequency of different types of mutations caused by physical agents or anthropogenic environmental pollutants with the minimal labour and time costs [22, 31].

The complex test-system allows, after a single processing of air-dry seeds of Arabidopsis, taking of proper account of the mutations frequency on the induction of:

- The somatic mutations registered on the surface of the plants' leaves (the phase of four true leaves 14 days after seeding) attributable to different genotypic classes (+/+, chl/+, chl/chl);

- The recessive, embryonic and chlorophyll lethal mutations being established by the embryonic (germinal) test in the seed generations of plants M1, the generation grown from the seeds exposed to the mutagenic impact;
- The lethal mutations emerging in the fourth chromosome coupling group, taken into consideration in M2 generation [7, 8].

Among the mutations impacting the structure and the function of the photosynthetic apparatus, allelic mutations are of the biggest interest. Their establishment through the genetic analysis serves as a telling argument in favour of the fact that the structure of the same gene is the unit of variation.

As a result of the cross-breeding of eight chlorophyll mutants (random sampling), attributable to *chlorina* class, three independent genes have been established (*ch3, flavi 1–1* and *ch5*), which irrespective of one another determine yellow-green coloration of plants. Based on one gene *flavi 1* six allelomorphic conditions compiling a series of multiple alleles have been determined: *flavi 1–1, flavi1–2, flavi 1–3, flavi 1–4, flavi 1–5, flavi 1–6*, located in the fifth chromosome coupling group. Mutations in different sub-units of the same gene (*flavi 1*) determining the chlorophyll insufficiency, impacting in a different manner all the parameters of the meso-structure and the work of the photosynthetic apparatus, lead, as a rule, to the visible decrease of the index of photosynthesis [29, 35, 36]. The key role in suppression of the function of the photosynthetic apparatus in allelic mutants plays the patterns of the ultra-structural organization of the system of the photosynthetic membranes of chloroplasts [34].

The obtained results allowed offer the experimental model based on the use of allelic mutations of Arabidopsis for the study of mechanisms of sustainability and adaptation of the photosynthetic apparatus to the adverse environmental factors.

Special attention deserve the unique experimental models – "the gene – stem fasciation", based on the coupling of the gene *cla'* (the morphologic mutant, fascinated seed pod) with the tester line *lu'co* (the gene *lu* stipulates the altering yellow-green coloration of plants in the ontogenesis, the gene *co* determines a very delayed development), and the allelic mutations of the *flavi* series determining yellow-green coloration of plants and delayed development. All the signal genes *cla', lu', co* and *flavi*, as was established, are located in the same fifth chromosomal group of coupling [19]. For the first time it is shown that the gene *cla* in combination with the different genes *lu' co* and the genes from a series of multiple alleles (flavi 1–1 to flavi 1–5) on the chlorophyll insufficiency leads to the emergence of the new mutant forms with strongly fasciated stem (up to 25 mm). The experimental model based on the fasciation character can be used for the study of the genetic control of the photosynthetic productivity, in general, and the hereditary heterosis, in particular.

For the first time an experimental model system consisting of 32 differently marked signal genes of the genetically pure lines of Arabidopsis has been elaborated. They were received due to multiple cross breedings between themselves and the initial *Columbia* race with the marker lines *tr', vc', er','gl* and *an'*, and

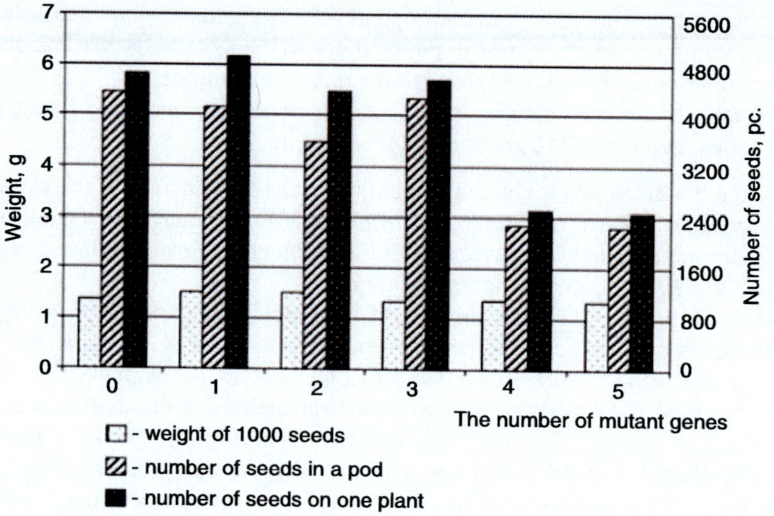

Fig. 26.5 Variability of the weight of seeds from one plant depending on number of mutant genes introduced into the Arabidopsis genotype

allowed the fruitful study of the phylogenetics of the complex physiological processes of the impact of the mutant gene in different genotypical media on the growth, the development and the productivity of plants. It is shown that different genotypical media has a different impact on the display of the genes *tr'*, *vc'*, *er'*, *gl'* and *an'*, if evaluate their actions according to the fertility character (weight of 1,000 seeds, the number of seeds in a pod and on one plant). A regular decrease of values of these indices was found out depending on the number of the mutant genes, introduced into the genotype. Maximal values of the weight of seeds, the number of the seeds in a pod and on one plant were registered when solitary genes *tr'*, *vc'*, *er'*, *gl'* and *an'* were introduced into the Arabidopsis genotype, while minimal values were identified following the introduction of four and five mutant genes in the genotype of Arabidopsis. It is established that the variability of the weight of seeds and the number of seeds, formed on a single plant, depends on different combinations of the signal genes. The weight of seeds from a single Arabidopsis plant is a constant value on condition of introduction of one to three genes. At the introduction of four and five genes this index diminishes drastically [17, 23, 32] (Fig. 26.5).

The "test-system" for determination of the impact of the factors of a space flight on the life activity of the higher plants including some Arabidopsis lines (*En, tr'*, 310) was approved under the conditions of imponderability on orbital space stations "SALUT (Author's Certificate № 919174. Published in the official bulletin of the State Committee of the USSR on inventions and findings, 1982, № 3, p. 254; [27]).

References

1. Abdullaev KA, Usmanov PD, Nasyrov YS (1972) Heteroplastids in the cells of the mutant *40/3 of Arabidopsis thaliana (L.) Heynh.* Reports of the Academy of Science of the Republic of Tajikistan, Dushanbe "Donish", XV(8): pp 48–50
2. Gofmann PO (1971) On interrelations between chloroplasts and mitochondrion. In: Photosynthesis Biochemistry and Biophysics. Irkutsk, Dushanbe "Donish", SB AS USSR: Publisher, pp 94–98
3. Gostimsky SA (1981) Genetic control of photosynthesis in the higher plants. Abstract of Doctor of Biology thesis, 30 p
4. Ivanov VI (1974) Radiobiology and radiogenetics of Arabidopsis. In: Problems of Space Biology. M.: Nauka, Dushanbe "Donish", 191 p
5. Kranz AR (1978) Demonstration of new and additional population samples and mutant lines of the AIS seed bank. Arabid Inf Serv 15:118–139
6. Kvitko KV, Muller A (1961) *Arabidopsis thaliana (L.) Heynh.* – A new object for genetic research. In: Genetic Research. L.: LGU Publisher, Dushanbe "Donish", pp 79–91
7. Mednik IG, Usmanov PD (1982a) About an interesting case of incomplete dominance in *Arabidopsis.* In: Arabid Inf Serv, Frankfurt/Main 19:87–92
8. Mednik IG, Usmanov PD (1982b) The structure of chimeric plants and estimation of frequencies induced mutations in Arabidopsis thaliana (L.) Heynh. In: Arabid Inf Serv, Frankfurt/Main 19:79–86
9. Muller AJ (1963) Embrionen test sum Hachweis reseasiven Letal faktoren bei *Arabidopsis thaliana.* Biol Zbl 82:133–163
10. Muller AJ (1972) Mutageni tatsprufung von Chemikalien bei *Arabidposis thaliana.* Biol Zbl B 91(1):S31–S48
11. Robbelen G (1963) The effects of two endogenous factors on artificial mutagenesis in Arabidopsis thaliana. In: Induction of mutations and mutation processes Prague, pp 42–45
12. Sidorova KK (1981) Book genetics of the Pea mutants. Nauka, Novosibirsk, 168 p
13. Timofeeff-Resovssky NV (1968) Genetics and plant physiology. In: Genetic research in memory of Dontso Kostoff. Sofia, Bulg. Acad. of Scien.: Dushanbe "Donish", Publisher, pp 97–99
14. Timofeev-Ressovsky NV, Ivanov VI, Korogodin VI (1968) Use of the hitting principle in radiobiology. M.: Atomizdat, 228 p
15. Usmanov PD (1977) To the question of genetic control of chloroplasts functioning. Agric Biol M.: Dushanbe "Donish", Kolos, XII(5):769–778
16. Usmanov PD (1984) Genotypic variability of characters of the photosynthetic apparatus of the higher plants, Abstract of Doctor of Biology thesis, M., 45 p
17. Usmanov TP (2007) Impact of mutant genes in different genotypic media on physiological functions of *Arabidopsis thaliana (L.) Heynh.* Abstract of the Candidate of Biology thesis, Dushanbe, 23 p
18. Usmanov PD, Muller A (1970) Use of embryo-test for analysis of the embryo lethals induced by irradiation of pollen seeds of *Arabidopsis thaliana (L.) Heynh.* Genetics 6(7):50–60
19. Usmanov PD, Startsev GA (1979) The appearance of stem fasciation in *Arabidopsis thaliana* F_2 by crossing the mutant *90* with *lu'co.* Arabid Inf Serv 16:99–102
20. Usmanov PD, Usmanova OV (1973) On the genetic control of the chloroplast size and number in mesophyll cells of species of *Arabidopsis thaliana.* Arabid Inf Serv 10:20
21. Usmanov PD, Usmanova OV (1980) Frequency of haplophase and dominant lethal mutations induced by UV-rays in pollen seeds of *Arabidopsis thaliana (L.) Heynh.* under conditions of Alpine Hissar Mountain ranges. Arabid Inf Serv 17:142–144
22. Usmanov PD, Usmanova OV (1990) Use of Arabidopsis as a test-system for evaluation of the mutagene factors impact, in: Genetic aftermaths of the environmental pollution by the mutagene factors, All-Union Coordination Conference, 8–10 Oct, Agenda, M. – Samarkand, pp 170–180

23. Usmanov TP, Usmanova OV (2007) Impact of mutant genes and their different combinations on fertility of *Arabidopsis thaliana (L.) Heynh.* In: Proceedings of the Academy of Sciences of the Republic of Tajikistan, Department of Biology and Medical Sciences 3(156):36–42
24. Usmanov PD, Abdullaev KA, Tagejeva SV (1978a) Die Variabilitat des photosynthetischen Membransystems und die Chloroplasten-evolution. In: Wissenschaftliche Zeitschrift der Humboldt-Univers. zu Berlin, Math.-Natl Res B 27(5):S505–S515
25. Usmanov PD, Abdullaev KA, Tulakin VG (1978b) Normalization and study of recessive chlorophyll mutants of *Arabidopsis thaliana (L.) Heynh.* In: The plants cell culture. Kiev, pp 300–304
26. Usmanov PD, Mednik IG, Tulakin VG, Muzyka VI (1980) The possible role of natural UV-rays as a factor of selection in populations of *Arabidopsis*. Arabid Inf Serv 17:145–148
27. Usmanov PD, Mednik IG, Tulakin VG, Muzyka VI (1984) Impact of the space flight factors on the dormant seeds of different genetic lines of *Arabidopsis thaliana (L.) Heynh.* In: Biologic research on orbital space stations "Salut", the USSR Academy of Science, Department of Physiology, M., pp 64–68
28. Usmanov PD, Mednik IG, Lipkind BI, Giller YuE (1987) Genetic patterns of plants reaction on the UV-radiation. Plant Physiol 34(4):720–729
29. Usmanova OV, Veselova TV, Veselovsky VA, Usmanov PD (1991) Functioning of the photosynthetic apparatus of the allelic mutants of *Arabidopsis thaliana (L.) Heynh.* Plant Physiol 38(4):649–654
30. Usmanov PD, Mednik IG, Usmanova OV (1994) Genetic and evolutionary aspects of the middle-waves UV-radiation impact on the higher plants. In: Proceedings of the Academy of Sciences of the Republic of Tajikistan, Department of Biology 1:41–50
31. Usmanov PD, Makhmadbekov TS, Nazarova KT (1997) Semi-dominant chlorophyll mutation with recessive lethal impact in *Arabidopsis thaliana (L.) Heynh.* In: Proceedings of the Academy of Sciences of the Republic of Tajikistan. Department of Biology, 1–2, pp 74–80
32. Usmanov PD, Usmanova OV, Usmanov TP (2000) Phene genetic study of fertility of *Arabidopsis thaliana (L.) Heynh.* In: International conference devoted to the 100th anniversary of N.V.Timofeev-Ressovsky, Minsk, pp 116–118
33. Usmanova OV (1982) On the character of interrelation between embryo lethals and chlorophyll mutations induced by the X-ray radiation of pollen seeds of Arabidopsis. In: Sensitivity of organisms to mutagene factors and emergence of mutations, Papers of the All-Union Conference, Vilnius, pp 48–49
34. Usmanova OV (1990) Mutational variability of mesostructure of the photosynthetic apparatus of the higher plants. Abstract of the Candidate of Biology thesis, Dushanbe, 24 p
35. Usmanova OV, Usmanov PD (1983) Allelic mutations of *Arabidopsis thaliana (L.) Heynh.* Arabid Inf Serv 20:35–40
36. Usmanova OV, Usmanov PD (1989) Impact of different allelic states of the gene *flavi-1* on the variability of some morpho-physiological characters and ultra-structural organization of chloroplasts in *Arabidopsis thaliana (L.) Heynh.* Armenian Biol J 25:1335–1337
37. Veleminsky J, Gichner T (1964) Sterile culture of *Arabidopsis* on agar medium. Arabid Inf Serv 1:34–35
38. Yunusov SY, Kasyanenko AG, Usmanov PD (1969) Distribution and fruiting time of Arabidopsis species in Tajikistan. Arabid Inf Serv Gottingen 6:3

Chapter 27
Methods for Predicting ^{137}Cs Contamination Levels of Soil Suitable to Obtain Plant and Fodder Products in Compliance with the Adopted Standards

Alexey V. Panov, Rudolf M. Alexakhin, and Anna A. Muzalevskaya

Abstract Using southern districts in the Kaluga region affected by the Chernobyl accident as an example, a comparative analysis has been made of estimated levels of ^{137}Cs contamination of agricultural lands with data of real measurements. The estimated data which consider only the radionuclide decay are quite reliable, however for better accuracy scales of agrotechnical countermeasures need to be considered.

Keywords Radio nuclides • Radioactive contamination • Contamination density • Protective measures

27.1 Introduction

At all stages after a radiation accident the basis to assess the radioecological situation in the agricultural complex and planning of agricultural production in accordance with the adopted standards are data on radionuclide contamination of soils [6]. Following the Chernobyl accident the results of radiological survey of the agricultural land have become the basic information for decision making on the affected land remediation [10]. It should be noted that measures of the large-scale radiological survey of agricultural lands are rather expensive, laborious, and are usually held in rounds, once every 4–5 years, on the same territory [9]. With this in mind, methods for predicting radionuclide contamination of agricultural lands have recently come into use when planning farming on the Chernobyl affected lands [8].

A.V. Panov (✉) • R.M. Alexakhin • A.A. Muzalevskaya
Russian Institute of Agricultural Radiology and Agroecology (RIARAE),
Kievskoe shosse, 109 km, Obninsk 249032, Russia
e-mail: riar@mail.ru

C.E. Mothersill et al. (eds.), *Radiobiology and Environmental Security*,
NATO Science for Peace and Security Series C: Environmental Security,
DOI 10.1007/978-94-007-1939-2_27, © Springer Science+Business Media B.V. 2012

In 2009, Atlas of the present and predicted (till 2056) aspects of impacts of the Chernobyl NPP accident for Russia and Belarus was released, where with some settlements as an example, principles are justified for prediction of ^{137}Cs and ^{90}Sr contamination levels of territories considering only decay of these radionuclides [7]. This approach is quite adequate for mapping, but how much accurate it is for predicting radionuclide contamination of a specific site (arable land, meadow, pasture) is still a question.

As was apparent after investigations following the Chernobyl accident, reduction in the contamination density of farm lands with time is influenced by a number of factors other than radionuclide decay:

- radionuclide migration in soil [3];
- protective agrotechnical measures such as plowing of agricultural lands which results in radionuclide redistribution in the soil profile [2];
- part of radionuclides over a long period of time can be removed with plant and fodder products [5].

Estimation of the impact of all these disregarded factors on the prediction accuracy of radionuclide contamination of farm lands is essential for both optimization of the radiological situation and planning of agricultural production in the affected regions.

The objective of the present paper is a comparative analysis of the predicted values of ^{137}Cs contamination density of farm lands estimated for different periods after the Chernobyl NPP accident and real measurements, as well as evaluation of the role of factors influencing the accuracy of these predictions.

27.2 Methodology

In 2007, in the context of Atlas preparation [7], specialists from the RIARAE and Kaluga center "Agrochemradiology" carried out a radiological survey of agricultural lands on 36 farms in three southern districts in the Kaluga region affected by the Chernobyl accident. On these farms 155 plots (135 arable land and 20 pastures) were monitored. The total surveyed area was 7.9 thousand hectares, with 90% of it being an arable land (Table 27.1). The area of plots varied from 2 to 250 ha, averaging some 52 ha. The criterion for selection of plots was relatively uniform level of ^{137}Cs contamination.

Table 27.1 Characteristics of the study districts in the Kaluga region

District	Number of farms	Arable land		Hayland and pastures	
		Number of plots	Area, ha	Number of plots	Area, ha
Zhizdra	11	41	1,933	3	317
Ulianovsk	15	53	2,320	11	341
Khvastovichi	10	41	2,753	6	219

On the each study plot soil samples were collected in accordance with the methodology described in [1] and ^{137}Cs concentrations were measures in the samples. Based on this information, the density of plot contamination in 2007 was estimated by the formula:

$$Q_i = A \times h \times d \times 10^{-3},\ \mathrm{kBq/m^2}, \tag{27.1}$$

where Q_i is the soil contamination density (the radionuclide inventory in the arable 0–20 cm layer per 1 m^2), kBq/m^2; A is the radionuclide concentration in soil, Bq/kg; h is the depth of arable horizon, cm; d is the specific mass of soil, g/cm^3; 10^{-3} is the coefficient to convert Bq/m^2 to kBq/m^2.

Previously, in various periods after the Chernobyl accident (1992–2005), the same plots were also subject to radiological survey which was accompanied by the study of soil properties (estimation of agrochemical parameters and mechanical composition) and collection of data on the application of protective and remedial measures on these farm lands [4]. Based on the collected information during the initial radiological survey, forecasts were made of soil contamination levels by ^{137}Cs (Q_r) in 2007. In this case an expression was used that takes into account only decay of the given radionuclide:

$$Q_r(t) = Q_{t_1} \cdot \exp\left(\frac{(-0.693 \cdot \Delta t)}{T_{\frac{1}{2}}}\right),\quad \mathrm{kBq/m^2}, \tag{27.2}$$

where Q_{t_1} is the density of ^{137}Cs contamination of farm lands in the year of the first radiological survey, kBq/m^2; Δt is the time (number of years) from the first radiological survey till 2007; $T_{1/2}$ is the ^{137}Cs half-life equal to 30.17 years.

The estimated for 2007 data on ^{137}Cs contamination of agricultural lands were then compared with the data of real measurements carried out in 2007.

27.3 Results and Discussion

The comparative analysis has demonstrated that estimated levels of ^{137}Cs contamination of the study plots in 2007 differed from the real (measured) ones by 12% towards larger values (Fig. 27.1). The correlation between the estimated and actual data was rather high (r=0.87). Since the predicted values which took into account only ^{137}Cs decay were overestimated, it was suggested that in real conditions measurement of this parameter could be also influenced by other factors (^{137}Cs migration in the soil and trophic chains, specific features of the agricultural practice, application of countermeasures). To find out the level of influence of these factors between pairs of estimated (Q_r) and measured (Q_i) values for ^{137}Cs contamination of plots in 2007, the ratio (Q_r/Q_i) was determined, the resulting data were divided into groups for each study parameter (factor) and data samplings were statistically processed.

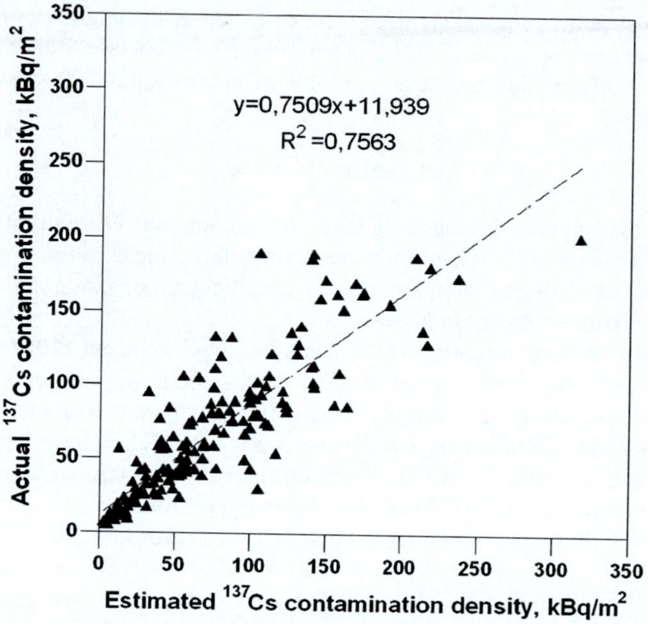

Fig. 27.1 Field of regression between estimated and actual ^{137}Cs contamination density of farm lands

Evidently, the closer to unity is the ratio (Q_r/Q_i), the more accurate is the prediction based only on the consideration of ^{137}Cs decay. If the ratio (Q_r/Q_i) differs from unity, the measurement of ^{137}Cs contamination density is influenced by another factor in addition to the radionuclide decay.

Classification of (Q_r/Q_i) data by the dates of the first radiological survey of the plots has identified two groups of results obtained in different periods after the Chernobyl accident: 1992–1994 and 1999–2005 (Fig. 27.2). It is suggested that with time part of radionuclides are removed from the soil with the agricultural produce. Then the ratio (Q_r/Q_i) for 1992–1994 will differ from unity more than that for 1999–2005. However, as seen from Fig. 27.2, the time factor (annual removal of the radionuclide with products) did not produce significant effect on the accuracy of estimations. Except for the 1992 survey data, the other results from comparison showed a high correlation coefficient (0.72–0.95) and insignificant scattering in (Q_r/Q_i). The mean estimated to actual values for ^{137}Cs contamination (except for 1992 and 2001) were close to 1 and ranged between 0.96 and 1.1. Significant differences in mean values in 1992 and 2001 are explained by small samplings. In these years only 7 plots were surveyed, whereas in the other time their number varied from 11 to 35. In 1992, considering large data scatter from the (Q_r/Q_i) mean, methods of soil sampling and ^{137}Cs measurement were probably violated. Hence, it may be concluded that ^{137}Cs is annually removed from the soil with phytomass in negligible amounts,

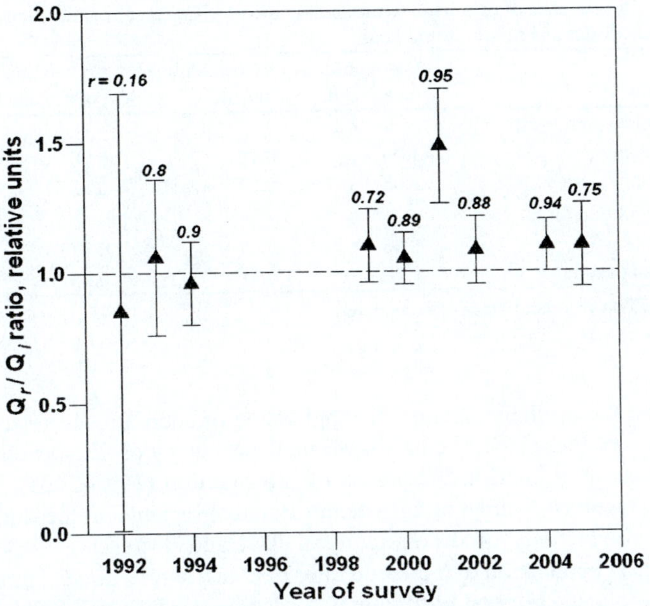

Fig. 27.2 Dependence of the ratio of estimated to actual ^{137}Cs contamination density of farm lands on the time of initial survey (numerical values – correlation coefficient *r*)

thereby producing no significant effect on the prediction accuracy of cesium contamination of the agricultural land within both small and large time spans after radioactive fallout (15 years and more).

Classification of the study plots by soil mechanical composition did not reveal any influence of this factor on the accuracy of predictions (Table 27.2). The ratio of estimated to measured cesium contamination densities of farm lands differed from unity by 5–6% for both soil groups typical for the given region which is well within the standard system of measurement errors. For a long time after radioactive fallout on both sandy and loamy soils the bulk ^{137}Cs is in the upper 20 cm layer, migration of this radionuclide has a local pattern and it does not leave the top horizon of the soil cover.

Division of the study plots into the types of agricultural land has demonstrated that on arable land predictions proved to be more accurate than on meadow-pasture lands. For arable land, the ratio of estimated to measured data on ^{137}Cs contamination was nearest to unity at a small confidence interval (Table 27.2). On meadow-pasture lands this ratio was somewhat higher. On the one hand, it could be associated with greater removal of ^{137}Cs radionuclides with fodder compared with plant products produced on arable land. On the other hand, active application of agroameliorative measures on meadow-pasture lands could be a potential reason responsible for the difference between the estimated and actual ^{137}Cs depositions.

Table 27.2 Dependence of estimated to measured ratio of ^{137}Cs contamination density of plots on soil properties and type of agricultural land

Classification	Geometric mean, Q_r/Q_i	95% confidence interval	Correlation coefficient	Number of plots
Soil mechanical composition*				
Sandy, sandy loam	1.06	0.15	0.90	46
Light-, medium loam	1.05	0.07	0.85	108
Type of agricultural land				
Arable	1.06	0.06	0.87	135
Haylands and pastures	1.08	0.31	0.85	20

*1 plot of 155 on peat land was not considered

Based on the available data on the application of countermeasures on the study plots, effects of these have been assessed on the accuracy of ^{137}Cs levels estimation for the agricultural land. In the period of investigation (1992–2005), 30% of the study plots received a range of agrochemical and agrotechnical measures aimed at increasing productivity of the agricultural land and decreasing ^{137}Cs uptake by plants, namely application of increased rates of potassium fertilizers, liming and use of rock phosphates. Mineral fertilizing was accompanied by soil replowing, deeper than in the usual practice. This caused redistribution of radionuclides in the top soil. Classification of plots into two groups, with and without countermeasures, showed that the ratio (Q_r/Q_i) amounted to 1.03 in the latter case and 1.15 with countermeasure application.

A better insight into the effects of countermeasures on the estimation accuracy of ^{137}Cs contamination of the agricultural land can be gained from the soil agrochemical properties changeable in response to countermeasures. Thus, on the study plots located in the southern districts in the Kaluga region in the absence of countermeasures average pH varies between 4.4 and 5.1 and the mean weighted content of mobile potassium ranges from 7.6 to 9.3 mg/100 g soil [4]. Agrochemical measures reduce soil acidity and increase soil content of mobile potassium. Since mineral fertilizing of the agricultural land was accompanied by agrotechnical works, the dynamics of soil agrotechnical parameters can be used as an indicator of scales and effectiveness of countermeasures.

All the study plots were divided into 3 groups by soil agrochemical parameters influenced by countermeasures. Treating soil acidity as one of these parameters it should be noted that in the absence of countermeasures (pH 4.0–5.2) the accuracy of prediction proved to be the highest (Q_r/Q_i=0.97). With the decrease in soil acidity caused by liming the reliability of predictions dropped since the (Q_r/Q_i) value grew to 1.11–1.14 (Fig. 27.3a).

The same situation is reported for the parameter describing soil content of mobile potassium. The application of increased rates of potassium fertilizers which caused increment in the fraction of mobile potassium to 17–35 mg/100 g soil the prediction accuracy fell by 17% (Fig. 27.3b). Naturally, the estimation accuracy was influenced by the scales of agrotechnical works (deep replowing) which have resulted in

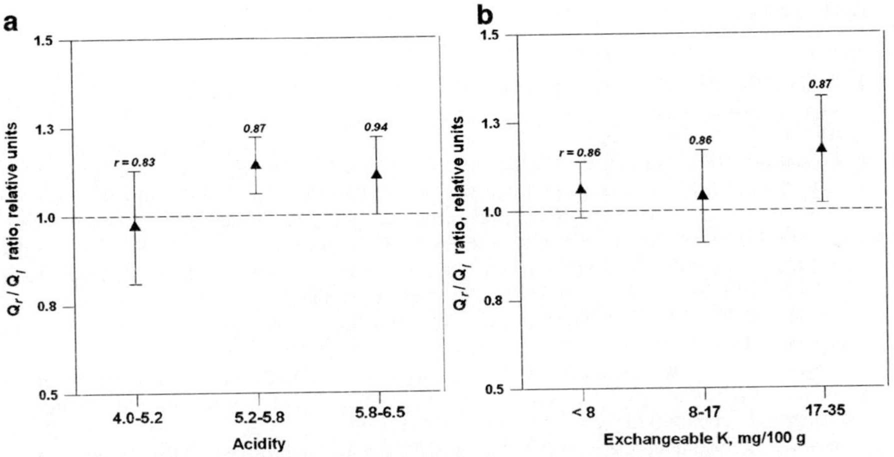

Fig. 27.3 Dependence of the ratio of estimated to actual ^{137}Cs contamination density of farm lands on soil agrochemical properties: **a**, acidity and **b**, content of exchangeable potassium (numerical values – correlation coefficient *r*)

redistribution of radionuclides to the lower (25–40 cm) soil layers rather than mineral fertilizing. This in turn contributed to the reduction in ^{137}Cs contamination density on the plots.

It should be concluded that the farm lands in the southern districts of the Kaluga region affected by the Chernobyl accident saw smaller scales of countermeasures compared to the Bryansk region [3, 4]. It cannot be ruled out that at larger scales of countermeasures the reported regularities in reducing accuracy of predictions of ^{137}Cs contamination of farm lands could have been more pronounced.

Consequently, it is concluded that long-term prediction of ^{137}Cs levels on farm lands can be reasonably accurate based only on the radionuclide decay. In this case the estimation error is within 5–8%. Despite the fact that on meadow-pasture lands ^{137}Cs removal with fodder products is slightly higher than with plant stuffs produced on arable land, transfer of this radionuclide with phytomass do not produce noticeable effect on the reliability of predicted values. With time after radioactive fallout ^{137}Cs is neither removed from the soil with agricultural produce nor migrate into the deeper horizons. The bulk of the radionuclide is deposited in the top 20 cm soil independent of the mechanical composition. At the same time estimated values are always higher than the actual data. In addition to radionuclide decay, predictions of ^{137}Cs levels on farm lands must take into account a factor of countermeasure application. On plots which saw agrotechnical works part of radionuclides are transferred to the deeper soil layers, thereby resulting in reduced contamination density of the agricultural land. Therefore, on these plots estimated values for ^{137}Cs contamination need to be reduced by 15–20% dependent on the scale of agrotechnical measures.

References

1. (2007) Methodology to survey agricultural lands affected by radioactive contamination. RIARAE, Obninsk, 28 p
2. Alexakhin RM, Korneev NA (eds) (1992) Agricultural radioecology. Ekologia, 400 p
3. Aleksakhin RM, Sanzharova NI, Fesenko SV, Geras'kin SA, Spiridonov SI, Panov AV, Isamov-Jr. NN, Nizhebovsky AV, Prudnikov PV, Pasternak AD (2006) Results of investigations performed in the chernobyl accident zone and practical results of the accident consequences liquidation in the agricultural sector of the russian federation. P. 195–231. In the Book: 20 Years After the Chernobyl Accident: Past, Present and Future. Ed. by EB Burlakova & VI Naidich. Nova Science Publishers, Inc. New York, 358 p
4. Dubovaya VG (2001) Analysis of factors responsible for contamination levels of farm products and justification of countermeasures in the long term after the Chernobyl accident (illustrated by contaminated districts of the Kaluga region) Candidate thesis – Obninsk, 28 p
5. Gulyakin IV, Yudintseva EV, Gorina LI (1975) Accumulation of 137Cs in yield depending on species characteristics of plants, Agrokhimiya 7:12–19
6. Instruktsii i metodicheskie ukazaniya po otsenke radiatsionnoi obstanovki na zagryaznennoi territorii (Manuals and Methodological Instructions on Assessing the Radiation Situation on Contaminated Territory), Moscow: Goskomgidromet SSSR, 1989
7. Izrael Yu A, Bogdevich IM (eds) (2009) Atlas of the present and predicted aspects of the Chernobyl accident impacts in the affected areas of Russia and belarus. Foundation "Infosfera"–NIA-Priroda, Moscow-Minsk, 138 p
8. Panov AV, Muzalevskaya AA, Aleksakhin RM, Prudnikov PV, Vlasenko EV (2008) On the problem of the production of normatively clean fodder production and plant industry products under 137Cs contaminated soil conditions, Radiatsionnaya Gigiena 1(2):4–13
9. Prudnikov PV, Karpechenko SV, Novikov AA et al (2007) Agrochemical and agroecological condition of soils in the Bryansk region. Klintsy Municipal Printing House, Bryansk, 608 p
10. Vorobiov GT, Guchanov DE, Markina ZN et al (1994) Radioactive contamination of soils in the Bryansk region. Grani, Bryansk, 148 p

Chapter 28
Effects of Chronic Irradiation in Plant Populations

Stanislav Geras'kin, Alla Oudalova, Nina Dikareva, Elena Chernonog, Julia Prytkova, Vladimir Dikarev, and Tatiana Novikova

Abstract An assessment of the state of plant and animal populations inhabiting polluted territories and the analysis of mechanisms of their adaptation to adverse environmental conditions undoubtedly have general biological importance. Consequently, studies that examine biological effects on non-human biota in natural settings provide a unique opportunity for obtaining information about the potential biological hazard associated with radioactive contamination. The results of long-term field studies in the Bryansk region of Russia affected by the Chernobyl accident and in the Semipalatinsk Test Site, Kazakhstan are presented. Although radionuclides cause primary damage at the molecular level, there are emergent effects at the level of populations, non-predictable solely from knowledge of elementary mechanisms of the pollutants' influence. Plant populations growing in areas with relatively low levels of pollution are characterized by the increased level of both cytogenetic disturbances and genetic diversity. Radioactive contamination of the plant environment activates biological mechanisms, changing a population's resistance to exposure. However, there are radioecological situations where enhanced radioresistance has not evolved or has not persisted.

Keywords Radioactive contamination • Plant populations • Absorbed doses • Genetic effects • Reproductive ability • Radio-adaptation

S. Geras'kin (✉) • A. Oudalova • N. Dikareva • E. Chernonog • J. Prytkova • V. Dikarev • T. Novikova
Russian Institute of Agricultural Radiology and Agroecology, Kievskoe shosse, 109 km, Obninsk 249020, Russia
e-mail: stgeraskin@gmail.com

C.E. Mothersill et al. (eds.), *Radiobiology and Environmental Security*, NATO Science for Peace and Security Series C: Environmental Security,
DOI 10.1007/978-94-007-1939-2_28, © Springer Science+Business Media B.V. 2012

28.1 Introduction

One of the major difficulties in the implementation of an ecological risk assessment is a lack of knowledge about the effect from chronic, low-level exposures to contaminants. The acute phytotoxic effects of environmental factors are well known, but the effects of long-term chronic exposure to low pollutant concentrations is neither well understood nor adequately included in risk assessments. There is growing evidence [27, 34] that subletal effects of chronic exposure to environmental contaminants lead to population-level consequences. Although low dose rates may be obtained in the laboratory by protraction, these may not adequately represent true environmental radiation exposures. To properly understand the effect of normal, real-world contaminant exposures, one must consider actual field conditions. Only in the field we can see what actually happened. However, for most wildlife groups and endpoints there are very few studies that link accumulation, chronic exposure and biological effects in natural settings [6]. These data gaps imply that the protection of the environment from ionizing radiation will require more experimental data related to effects of chronic low-level exposure to radionuclides at the population level. In this connection, several questions arise:

- what do we know about mutagenic effect of chronic low dose rate radiation exposure?
- what do we know about the fate of induced by radiation mutations in altered ecological conditions?
- can chronic low dose rate radiation exposure be regarded as ecological factor changing the genetic make-up of a population?

To answer these questions, a well-directed accumulation of knowledge is needed and appropriate experimental data analysis is necessary. Previously completed and ongoing field studies that have been carried out in Laboratory of Plant Ecotoxicology, RIARAE with different species of wild and agricultural plants are briefly summarized in Table 28.1. A wide range of radioecological situation and climatic zones have been covered in frames of this work. To illustrate the main findings, two field studies are discussed here in more details.

28.2 Material and Methods

In 2005–2007 seeds of crested hairgrass (*Koeleria gracilis* Pers.) were collected from four locations of the Semipalatinsk Test Site (Kazakhstan). Radiation background at the sites and activity concentrations of the most dose-forming radionuclides in the soil samples were measured. Absorbed doses to crested hairgrass were calculated. Squashed slides for cytogenetic analysis were prepared of coleoptiles (2–5 mm of length) of germinated seeds. In every slide, all ana-telophase cells (4,800–11,900 ana-telophases in 30–90 slides) were scored to calculate frequency of aberrant cells. Detailed description of methods used is given in [10].

Table 28.1 Field studies on wild and agricultural plants

Species	Site and time	Assay and/or endpoints
Winter rye and wheat, spring barley and oats	10-km ChNPP zone (12–454 MBq/m²), Ukraine, 1986–1989	Morphological indices of seeds viability, cytogenetic alterations in intercalary and seedling root meristem [7]
Scots pine, coach-grass	30-km ChNPP zone (2.5–27 μGy/h), Ukraine, 1995	Cytogenetic alterations in seedling root meristem [8]
Scots pine	Radioactive waste storage facility, Leningrad Region, Russia, 1997–2002	Cytogenetic alterations in needles intercalary and seedling root meristems [9]
Scots pine	Briansk Region radioactively contaminated in the Chernobyl accident (7–130 mGy/y), Russia, 2003–2009	Cytogenetic alterations in seedling root meristem, enzymatic loci polymorphism, abortive seeds [11, 12]
Crested hairgrass	Semipalatinsk Test Site (0.7–36 μGy/h), Kazakhstan, 2005–2008	Cytogenetic alterations in coleoptiles of germinated seeds [10]

To study biological effects in chronically exposed Scots pine (*Pinus sylvestris* L.) populations six test sites were chosen in the Bryansk Region of Russia radioactively contaminated as a result of the Chernobyl accident. Pine cones were collected in autumns of 2003–2008. Activity concentrations of radionuclides in soil samples were measured, and doses to the pine trees' generative organs were estimated. Aberrant cells were scored in root meristem of germinated seeds in ana-telophases of the first mitoses. The method of isozymic analysis of megagametophytes was used for an estimation of genetic variability in Scots pine populations. Five enzymatic loci (GDH, LAP, MDH, DIA, and 6-PGD) were studied in endosperms of the seeds collected in 2005. Detailed description of materials and methods used can be found in [11, 12].

28.3 Results and Discussion

The Semipalatinsk Test Site (STS) used to be the main site for testing nuclear weapon in the former Soviet Union. Between 1949 and 1989, 459 nuclear tests were conducted at the STS. Among them, 116 explosions were above ground. As a result, large-scale, heterogeneous radioactive contaminations occurred at the STS and adjusted territory. The STS is a unique place to study effects of chronic low dose rate exposure on non-human species over several generations. The wide range of plots different in levels and spectrum of radioactive contamination, an availability of plots with dominating contribution of particular types of radiation (α-, β-, and γ-radiations) to dose absorbed by plants and animals as well as specific climatic conditions provide a unique opportunity for studying long-term biological effects in

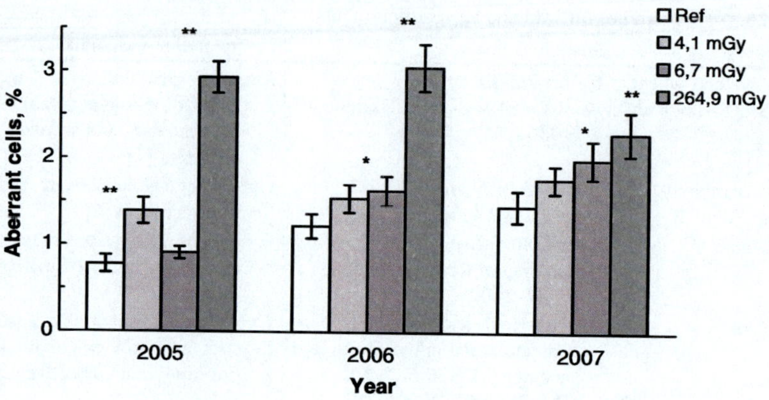

Fig. 28.1 Aberrant cells frequency in coleoptiles of germinated seeds of crested hairgrass, %. Significant difference from the reference level: * p<0.10; ** p<0.05

chronically exposed ecosystems against the background of extreme environmental conditions. A study of crested hairgrass populations, a typical wild cereal for Kazakh steppe, showed that the frequency of cytogenetic alterations in coleoptiles of germinated seeds increases proportionally to the dose absorbed by plants (Fig. 28.1). Severe alterations of single and double bridges as well as laggard chromosomes contribute mainly to the observed cytogenetic effect [10]. The agreement between findings from 3 years of study (2005–2007), different in weather conditions, suggests the leading role of radioactive contamination in an occurrence of cytogenetic effects.

Dose rate in the epicenter of nuclear tests exceeds 30 μGy/h, which is more than 3 fold of the predicted no-effect dose rate of 10 μGy/h derived in the EC ERICA project [6]. It is, however, well below the threshold for statistically significant effects (100 μGy/h) derived at the FASSET Radiation Effects Database analysis [30]. It is not surprising, than, that in our study there are found significant cytogenetic effects in crested hairgrass populations but no pronounced morphological alterations. Thus, the findings obtained are in agreement with the benchmark values proposed in the FASSET and ERICA projects to restrict radiation impacts to biota.

Investigations over the last several decades [22, 23, 25] have demonstrated that forest ecosystems are vulnerable to human-induced environmental stresses, resulting in a progressive loss of biodiversity and general declines in forest health. Adverse impact to the forest ecosystems lead to concomitant changes in the quality and extent of wildlife habitat [5]. Forest trees have gained much attention in recent years as nonclassical model eukaryotes for population, evolutionary and ecological studies [13]. Because of their potential to affect many other species, any responses to selection pressures that are exerted on such keystone species as forest trees are especially important to quantify. The low domestication, large open-pollinated native populations, and high sensitivity to environmental exposure make conifers almost an ideal species for the study of environmental effects of radioactive contamination.

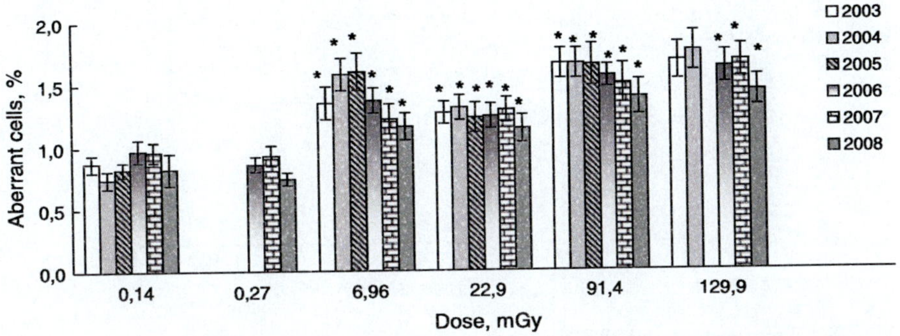

Fig. 28.2 Aberrant cells in root meristem of germinated seeds (%) from Scots pine populations collected in the Bryansk Region of Russia, 2003–2008. Ref and Ref 1 are the reference populations. Significant difference from the reference populations: * – $p<0.05$

The Chernobyl accident caused dramatic and long-term increases in ambient radiation doses to many forest environments. Sites still exist in the Bryansk Region of Russia, 20 years after the Chernobyl accident, where radioactive contamination significantly exceeds background. In the study reported herein, cytogenetic effects in Scots pine populations growing in the Bryansk Region have been investigated for 6 years, and the findings are presented in Fig. 28.2. There were no significant differences in frequencies of cytogenetic abnormalities, observed in the same study site from year to year. Thus, the effects observed can be regarded as quite robust and replicable over time. Aberrant cell frequency in root meristem of germinated seeds collected from these populations significantly exceeds the reference level and shows correlation with the dose absorbed. Combined with data from other our studies [7, 9], these findings indicate that an increased level of cytogenetic alterations is a typical phenomenon for plant populations growing in areas with relatively low levels of pollution.

Absorbed doses in generative organs of pine trees were assessed with an original dosimetric model [12]. In 22 years after the ChNPP accident, the annual doses are about 30 times lower than the 0.4 mGy/h guideline proposed by IAEA as safe for terrestrial plants [17]. On the other hand, dose rate for two most contaminated sites exceeded the ERICA generic predicted no-effect value of 10 μGy/h, applicable as an incremental dose rate in addition to the background for generic ecosystems [6]. These comparisons suggest that radiation levels at the study sites were high enough to induce cytogenetic rather than morphologic abnormalities in the exposed populations. Indeed, the aberrant cell frequency, as well as the occurrence of severe types of cytogenetic alterations in root meristem of germinated seeds collected from the impacted populations, exceeded the reference level during all 6 years of study (Fig. 28.2). This supports the conclusion that exposure to low levels of radiation resulted in cytogenetic effects in the pine tree populations contaminated from the Chernobyl accident. It should be noted that, in the STS study, a wide range of doses from 4 to 265 mGy absorbed by the plants was studied, and a regular dependence of

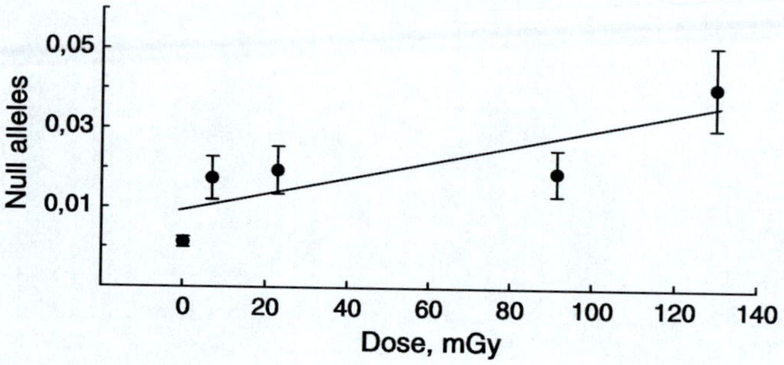

Fig. 28.3 Frequency of null alleles in enzymatic loci of endosperms (2005) in dependence on annual dose absorbed by generative organs of pine trees

cytogenetic effects on dose was revealed. On the contrary, in the Bryansk Region, the range of doses absorbed by the pine trees at the study sites is much narrower; this could be the reason for an absence of statistically significant increase of biological effect with the dose absorbed in some years of observations.

It is becoming increasingly clear that cytogenetic alterations detected in our studies might only be tip of an iceberg, reflecting global structural and functional rearrangements induced in exposed populations. An increase in mutation rate can affect the population genetic structure by producing new alleles or genotypes, and thereby has ecologically relevant effect. Alterations in the genetic make-up of populations are of primary concern because somatic changes, even if they lead to a loss of some individuals, will not be critical in populations with a large reproductive surplus. To analyze whether an exposure to radionuclides causes changes in population genetic structure, we evaluated frequencies of three different types of mutations (null allele, duplication and changing in electrophoretic mobility) of enzymatic loci in endosperm and embryos of pine trees from the studied populations. It is found that chronic radiation exposure results in the significant increase of total occurrence of enzymatic loci mutations. In particular, frequencies of mutations for loss of enzymatic activity increase with the dose absorbed by generative organs of pine trees (Fig. 28.3).

There are plenty of theoretical interpretations of evolution, but what is important is to see what happens in practice. Mutations in plant or animals are not necessary bad events when they do not adversely affect the population fitness. Mutation is one of the mechanisms that maintains genetic variation within a natural population and thus enables that population to cope with an adversely changing environment. Indeed, phenotypic variability in the exposed pine tree populations, estimated via the Zhivotovsky index [38], significantly exceeds the reference level and increases with dose absorbed by generative organs of pine trees (Fig. 28.4).

Populations living under unfavorable conditions (harsh climate, border of the species areal, man-caused pollution) exhibit not only greater polymorphism but also higher heterozygosity than those in optimal conditions. A decrease in heterozygosity within individuals has been associated [34] with decreased resistance to diseases,

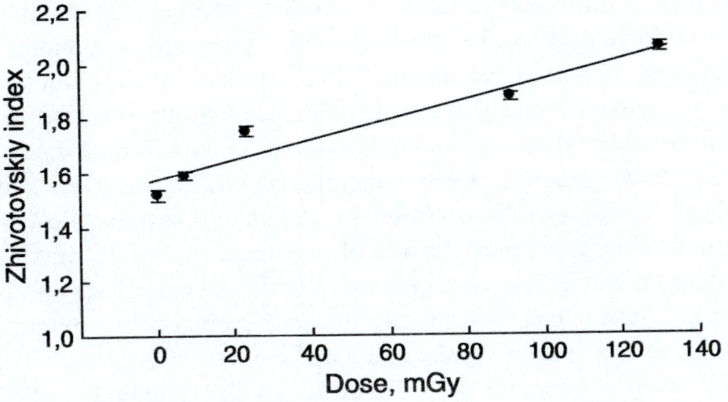

Fig. 28.4 Phenotypic variability estimated via the Zhivotovsky index (2005) in dependence on annual dose absorbed by generative organs of pine trees

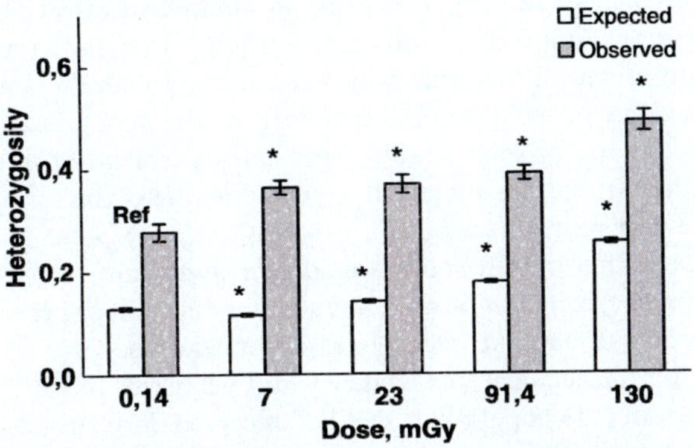

Fig. 28.5 Heterozygosity in endosperms of Scots pines in dependence on annual dose absorbed by generative organs of pine trees. Significant difference from the Ref population: * $p<0.01$

decreased growth rates, and decreased fertility. This would suggest that variations in individual heterozygosity may affect population growth and recruitment. The observed heterozygosity in pine tree populations at the radioactively contaminated sites is essentially higher than the expected one and increases with dose absorbed by generative organs of pine trees (Fig. 28.5).

From the data presented we can conclude that the relationship between radioactive contamination and genetic variability provides evidence of adaptation which optimizes the physiological response of a population to environmental changes. Keeping in mind all the data mentioned, it could be concluded that a high level of mutation occurrence is intrinsic for descendants of pine trees in the investigated populations, and genetic diversity in the populations is essentially conditioned by radiation exposure. So, in spite of their low values, dose rates observed can be considered as a factor able to modify genetic structure of populations. Furthermore, an increased genetic diversity within the population of keystone species is likely to be positively correlated with increased species diversity of the depended community [37].

Although great progress has been made in understanding the nature of mutations, too little is yet known about the way in which mutations can lead to observable effects in life traits of organisms and populations. The traditional theory of population and quantitative genetics mostly concerned evolution based on standing variation and largely ignored evolution based on *de novo* mutations. However, a recent theory has increasingly explored the role of mutations [4, 26]. A high level of mutational variability could lead to an acceleration of microevolutionary processes in the populations. Because plants lack a germ line, epigenetic and genotypic changes are readily generated and selected in the soma and can be transmitted to progeny [36]. In particular, this fundamental property underlies the remarkable phenotypic and genotypic plasticity of plants, and their elasticity to rapidly changing environments, including high levels of man-made pollution [24].

The effect of severe stress on populations is often thought to eliminate the most susceptible individuals [28, 29]. However, an alternative effect is to change the number of offspring produced by individuals [35]. It is true that a much larger number of seeds are produced than that which develop into adult plants, and that the changes in frequency of the different genotypes are due to a greater death of some genotypes than others. In such a way, a greater percentage of tolerant offspring would be expected from trees subjected to pollutants. This is a form of response to selection, and a very powerful one.

In the first year after the Chernobyl accident a significant decrease in reproductive ability of pines (reduction of seed mass and their number per cone, as well as increase in portion of abortive seeds) was observed at doses over 1 Gy [3, 18]. Eleven years after the accident this tendency still persisted. In 1997, the portion of abortive seeds from pine populations that had received doses of 10–20 Gy in 1986 significantly exceeded the correspondent reference level [3]. The effect of radioactive contamination on reproductive ability of pine trees was also observed at the South Urals radioactive trail. Chronic exposure of pine trees at dose rates of 4.2–6.3 μGy/h resulted in a significant decline of seed mass, as well as an increase in the fraction of abortive seeds [20]. At a lower dose rate of 0.8 μGy/h, the enhancement in percent of abortive seeds was not observed. In 2000–2001, decrease in pollen viability as well as increase the number of anomalous pollen grains in Scots pine populations from Bryansk region at dose rates of 1.8–5.4 μGy/h have been detected [31]. In contrast to the results mentioned above, we failed to find any clear linkage between reproductive ability and doses absorbed by generative organs of pine trees (Fig. 28.6). So, the high mutation rate found in our study (Figs. 28.2 and 28.3) had no effect on the reproductive ability of the exposed populations.

An appearance of some standing factors (either of natural origin or man-made) in the plants' environment may activate genetic mechanisms, changing a population resistance to a particular stress. Laboratory studies of repair inhibitions, dose-effect relationships for low- and high-LET radiations, measurements of unscheduled DNA synthesis and an efficacy of the single strand breaks recovery [32] suggest that the divergence of populations in terms of resistance is connected with a selection for

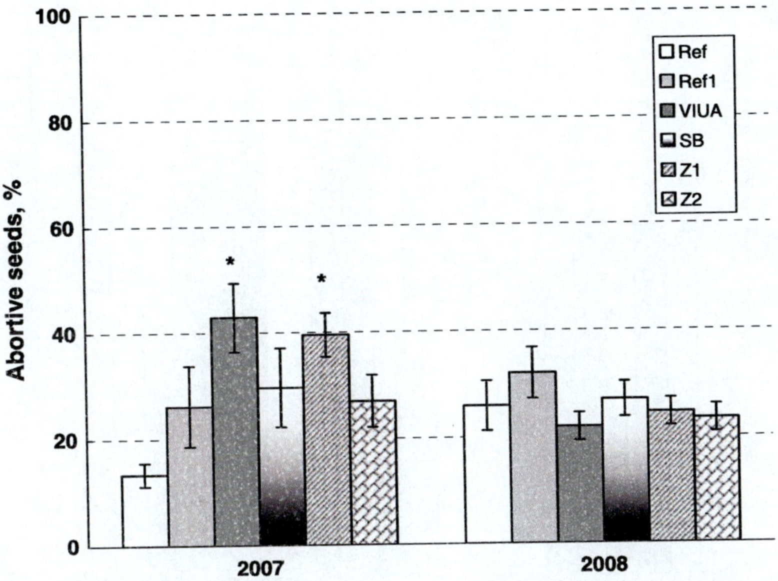

Fig. 28.6 Proportion of abortive seeds in the Scots pine populations at the study sites. Difference from the Ref population is significant: * $p<0.05$

changes in the effectiveness of repair systems. Another study on the possible mechanisms of adaptations to radioactive contamination [21] showed extremely low (more than 10-fold) recombination levels, and a higher level of global genome methylation in chronically irradiated plants that may have prevented extensive genome rearrangements. Although our understanding of mechanisms of adaptation is far from complete, these studies give good evidence that the response of biological systems to stress have been produced by normal evolutionary processes in relation to the environments in which the populations occur. This is, of course, a wasteful system since metabolic energy has to be used to produce the adaptive trait even when it is not required. Actually, increased fitness in unfavorable environments is associated [15] with decreased fitness in favorable environments. As a result, there are situations [16] when resistance to environmental changes has not evolved or has not persisted. Moreover, adaptation is often observed in one species but not found in others, despite an equivalent opportunity and exposure conditions [2]. Indeed, at the South Urals radioactive trace, radioresistance increased 3–4 times in radiosensitive plants, but remain practically unchanged in radioresistant species [32]. It seems that if we examine radio-adaptation in nature, we find as much evidence of it is not occurring, as we do of it is existing. It quickly becomes obvious that the role of microevolutionary processes in a population's response to low-level chronic exposure is still not clearly understood.

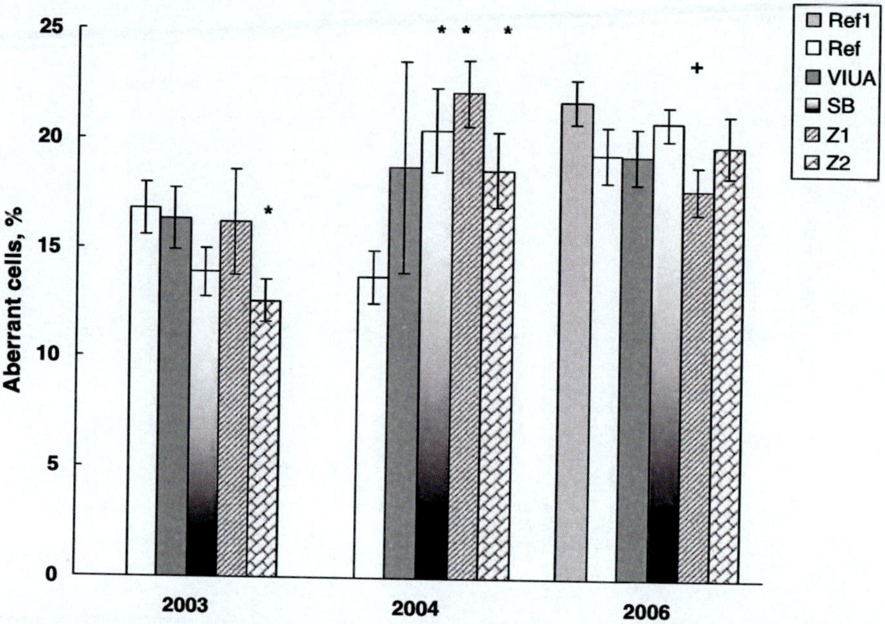

Fig. 28.7 Aberrant cell frequency in root meristems of Scots pine seedlings after acute exposure of seeds to 15 Gy of γ-rays. * -difference from the Ref site is significant, $p<0.05$; + – difference from the Ref1 site is significant, $p<0.05$

The response of a population exposed to low dose rate irradiation depends on both the type of organism and the biophysical properties of radiation (relative biological effectiveness, linear energy transfer, dose rates etc.). Contrary to the increased radioresistance of seeds from plant populations inhabiting radioactively contaminated territories described in Shevchenko et al. [32] and Kalchenko and Fedotov [19], no significant difference in resistance to subsequent γ-ray exposure between seeds collected from the reference and exposed Scots pine populations was found in our study (Fig. 28.7). Similarly, the seeds from the crested hairgrass populations that have been experiencing radiation exposure for more than a half century and are bearing the elevated levels of cytogenetic abnormalities do not show any reliable increase in resistance to the additional acute γ-ray exposure (Fig. 28.8). Moreover, the results remain qualitatively the same at considerable variation in the dose rate of acute irradiation (more than 70 times, from 2,970 Gy/h to 39 Gy/h). An improved DNA repair capacity and ability to germinate under abiotic stress (salinity and accelerated ageing) was shown in seeds embryos of evening primrose growing near the Chernobyl NPP on sites contaminated with γ- and β-emitters, while on the α-, β- and γ-contaminated site such an improvement was not found [1]. This is in line with data on the successful adaptation of wild vetch populations on sites most highly

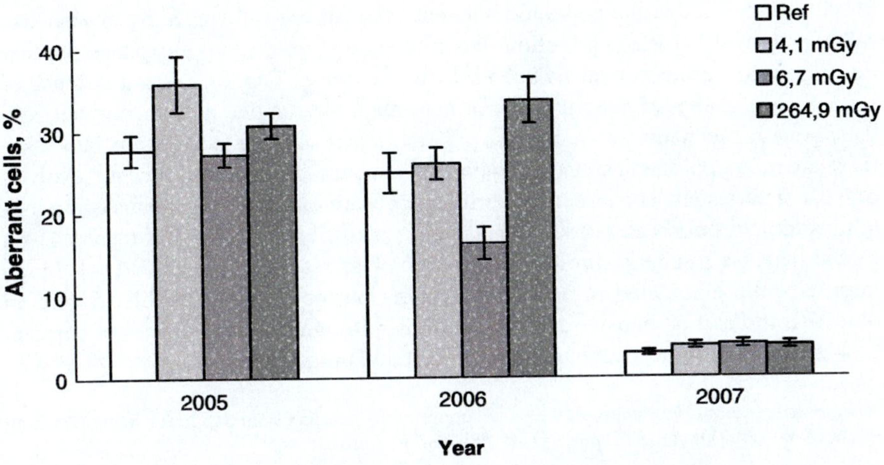

Fig. 28.8 Aberrant cell frequency in coleoptile of crested hairgrass after dry seeds exposure to an acute γ-ray dose of 68.8 Gy (dose rate of 2970 Gy/h) in 2005 and 2006 and dose of 50 Gy (dose rate 39 Gy/h) in 2007. * – difference from the Ref site is significant, $p<0.05$

contaminated by β-emitters, but not by α-emitters [33]. Consequently, there are good theoretical and practical reasons for more attention being paid to the mechanisms by which populations become more radioresistant, and to those situations where radio-adaptation appears not to be taking place.

28.4 Conclusions

A basic level of concern within a newly developing system for radiological protection of the environment is a population. Of specific importance in this context are studies on plant and animal populations inhabiting sites with contrasting levels and spectra of radioactive contamination. Special attention should be paid to population-level effects such as radio-adaptation, changes in sexual, age and genetic structure of populations, since knowledge of elementary mechanisms of the radionuclides' impact is insufficient to predict them. Corresponding studies are likely to increase in importance as the rate at which we change the environment worldwide continues to accelerate. It is of great value for both justification of radiation exposure risk assessments to humans and the environment as well as for understanding the complicated processes in the biosphere caused by increasing levels of man-made impact.

The findings presented here clearly indicate that plant populations growing in areas with relatively low levels of pollution are characterized by an increased level of both cytogenetic alterations and genetic diversity. Concordant responses between changes in population genetic structure and elevated levels of cytogenetic damage provide evidence that the population genetic changes are influenced by exposure to radionuclides. Man-made pollution may influence an evolution of exposed populations through a contaminant-induced selection process. The long-term existence of some factors (either of natural origin or man-made) in the plants' environment activates genetic mechanisms, changing a population's resistance to exposure. However, there are radioecological situations where enhanced radioresistance has not evolved or has not persisted. The effects of chronic exposure on living organisms and populations remain poorly explored, and represent a much needed field of research as a global increase in energy production from nuclear reactors is forecast [14]. Much more is to be elucidated in our understanding before we will be able to give an objective and comprehensive assessment of the biological consequences of chronic, low-level radiation exposures to natural plant and animal populations.

Acknowledgements Works presented were supported by Russian Foundation for Basic Research (grant 08-04-00631) and ISTC projects № 3003 and K-1328.

References

1. Boubriak II, Grodzinsky DM, Polischuk VP, Naumenko VD, Gushcha NP, Micheev AN, McCready SJ, Osborne DJ (2008) Adaptation and impairment of DNA repair function in pollen of *Betula verrucosa* and seeds of *Oenothera biennis* from differently radionuclide-contaminated sites of Chernobyl. Ann Bot 101:267–276
2. Bradshaw AD (1991) Genostasis and the limits to evolution. Phil Trans R Soc Lond B 333:289–305
3. Fedotov IS, Kalchenko VA, Igonina EV, Rubanovich AV (2006) Radiation and genetic consequences of ionizing irradiation on population of *Pinus sylvestris* L. within the zone of the Chernobyl NPP. Radiat Biol Radioecol 46:268–278, in Russian
4. Futuyma DJ (2010) Evolutionary constraint and ecological consequences. Evolution 64:1865–1884
5. Galbraith H, LeJeune K, Lipton J (1995) Metal and arsenic impacts to soils, vegetation communities and wildlife habitat in Southwest Montana uplands contaminated by smelter emissions: I. Field evaluation. Environ Toxicol Chem 11:1895–1903
6. Garnier-Laplace J, Della-Vedova C, Andersson P, Copplestone D, Cailes C, Beresford NA, Howard BJ, Howe P, Whitehouse P (2010) A multi-criteria weight of evidence approach for deriving ecological benchmarks for radioactive substances. J Radiol Prot 30:215–233
7. Geras'kin SA, Zimina LM, Dikarev VG, Dikareva NS, Zimin VL, Vasiliyev DV, Oudalova AA, Blinova LD, Alexakhin RM (2003) Bioindication of the anthropogenic effects on micropopulations of Pinus sylvestris L. in the vicinity of a plant for the storage and processing of radioactive waste and in the Chernobyl NPP zone. J Environ Radioact 66:171–180
8. Geras'kin SA, Dikarev VG, Zyablitskaya YY, Oudalova AA, Spirin YV, Alexakhin RM (2003) Genetic consequences of radioactive contamination by the Chernobyl fallout to agricultural crops. J Environ Radioact 66:155–169
9. Geras'kin SA, Kim JK, Oudalova AA, Vasiliyev DV, Dikareva NS, Zimin VL, Dikarev VG (2005) Bio-monitoring the genotoxicity of populations of Scots pine in the vicinity of a radioactive waste storage facility. Mutat Res 583:55–66
10. Geras'kin SA, Mozolin EM, Dikarev VG, Oudalova AA, Dikareva NS, Spiridonov SI, Tetenkin VL (2009) Cytogenetic effects in *Koeleria gracilis* Pers. populations from the Semipalatinsk Test Site (Kazakhstan). Radiat Biol Radioecol 49:147–157, in Russian

11. Geras'kin SA, Vanina JC, Dikarev VG, Novikova TA, Oudalova AA, Spiridonov SI (2010) Genetic variability in Scotch pine populations of the Bryansk Region radioactively contaminated in the Chernobyl accident. Biophysics 55:324–331
12. Geras'kin SA, Oudalova AA, Dikareva NS, Spiridonov SI, Hinton T, Chernonog EV, Garnier-Laplace J (2011) Long-term observations on Scots pine populations affected by the Chernobyl accident. Ecotoxicology 20:1195–1208
13. Gonzalez-Martinez SC, Krutovsky KV, Neale DB (2006) Forest-tree population genomics and adaptive evolution. New Phytol 170:227–238
14. Grimes RW, Nuttall WJ (2010) Generating the option of a two-stage nuclear renaissance. Science 329:799–803
15. Hickey DA, McNeilly T (1975) Competition between metal tolerant and normal plant populations; a field experiment on normal soil. Evolution 29:458–464
16. Hoffmann AA, Hercus MJ (2000) Environmental stress as an evolutionary force. Bioscience 50:217–226
17. IAEA (1992) Effects of ionizing radiation on plants and animals at levels implied by current radiation protection standards, Technical reports series N 332. International Atomic Energy Agency, Vienna
18. Ipatyev V, Bulavik I, Braginsky V, Goncharenko G, Dvornik A (1999) Forest and Chernobyl: forest ecosystems after the Chernobyl nuclear power plant accident: 1986–1994. J Environ Radioact 42:9–38
19. Kalchenko VA, Fedotov IS (2001) Genetic effects of acute and chronic ionizing radiation on *Pinus sylvestris* L. inhabiting the Chernobyl meltdown area. Russ J Genet 37:427–447
20. Kalchenko VA, Spirin DA (1989) Genetic effects revealed in populations of *Pinus sylvestris* L. growing under exposure to small doses of chronic irradiation. Russ J Genet 25:1059–1064
21. Kovalchuk I, Abramov V, Pogribny I, Kovalchuk O (2004) Molecular aspects of plant adaptation to life in the Chernobyl zone. Plant Physiol 135:357–363
22. Kozlov MV, Zvereva EL (2007) Industrial barrens: extreme habitats created by non-ferrous metallurgy. Rev Environ Sci Biotechnol 6:231–259
23. Kozlowski TT (2000) Responses of woody plants to human-induced environmental stresses: issues, problems, and strategies for alleviating stress. Crit Rev Plant Sci 19:91–170
24. Macnair M (1993) The genetics of metal tolerance in vascular plants. New Phytol 124:541–559
25. Mankovska B, Seinnes E (1995) Effects of pollutants from an aluminum reduction plant on forest ecosystems. Sci Total Environ 163:11–23
26. Orr HA, Unckless RL (2008) Population extinction and the genetics of adaptation. Am Nat 172:160–169
27. Peterson CH, Rice CD, Short JW, Esler D, Bodkin JL, Ballachey BE, Irons DB (2003) Long-term ecosystem response to the Exxon Valdez oil spill. Science 302:2082–2086
28. Pitelka LF (1988) Evolutionary responses of plants to anthropogenic pollutants. Trends Evol Ecol 3:233–236
29. Prus-Glowacki W, Wojnjcka-Poltorak A, Oleksyn J, Reich PB (1999) Industrial pollutants tend to increase genetic diversity: evidence from field-grown European Scots pine. Water Air Soil Pollut 116:395–402
30. Real A, Sundell-Bergman S, Knowles JF, Woodhead D, Zinger I (2004) Effects of ionizing radiation exposure on plants, fish and mammals: relevant data for environmental protection. J Radiol Prot 24:A123–A137
31. Scock AV, Glasoun IN, Samoshkin EN (2005) Influence of radioactive contamination on pollen viability and anomaly in Scots pine from Bryansk region. Forest J 5:7–11
32. Shevchenko VA, Pechkurenkov VL, Abramov VI (1992) Radiation genetics of natural populations: genetic consequences of the Kyshtym accident. Nauka Publishers, Moscow, 221 p, in Russian
33. Syomov AB, Ptitsyna SN, Sergeeva SA (1992) Analysis of DNA strand break induction and repair in plants from the vicinity of Chernobyl. Sci Total Environ 112:1–8

34. Theodorakis CW (2001) Integration of genotoxic and population genetic endpoints in biomonitoring and risk assessment. Ecotoxicology 10:245–256
35. Valladares F, Gianoli E, Gomez JM (2007) Ecological limits to plant phenotypic plasticity. New Phytol 176:749–763
36. Walbot V (1985) On the life strategies of plants and animals. Trends Genet 1:165–170
37. Whitham TG, Bailey JK, Schweitzer JA, Shuster SM, Bangert RK, LeRoy CJ, Lansdorf EV, Allan GJ, DiFazio SP, Potts BM, Fischer DG, Gehring CA, Lindroth RL, Marks JC, Hart SC, Wimp GM, Wooley SC (2006) A framework for community and ecosystem genetics: from genes to ecosystems. Nat Rev Genet 7:510–523
38. Zhivotovsky L (1980) An index of intrapopulation diversity. J Gen Biol 41:828–836 (in Russian)

Chapter 29
Reduced Rb1 Expression Causes Impaired Genome Stability in Bone-Cells and Predispose for Radiation-Induced Osteosarcoma

Michael Rosemann, Iria González-Vasconcellos, Tanja Domke, Michaela Nathrath, and Mike J. Atkinson

Abstract The risk of cancer after exposure to ionizing radiation is currently defined only as a function of the received dose. Genetic factors that modify individual susceptibility to radiation-induced cancer are excluded from the risk assessment. We report the mapping of QTLs that confer increased susceptibility to radiation-induced osteosarcoma in the mouse. The strongest candidate locus, on chromosome 14, contains a functional polymorphism weakening the efficiency of the Rb1 promoter. The Rb1 allele associated with increased susceptibility is preferentially retained during allelic loss at the Rb1 gene in radiation-induced tumors. In combination with allelic losses of CDKN2a/P16, an upstream regulator of Rb1, 100% of all analyzed tumors exhibit a defect affecting this pathway. Alpha-irradiation of knockout mice with a bone-specific expression reduction of Rb1 or P16 confirmes that these genes can alter the susceptibility for osteosarcoma, either by increasing the tumor-risk or by shortening their latency. These results suggest that common germ-line polymorphisms causing impaired expression of known tumor-suppressor genes can modify individual susceptibility to radiation-induced cancer.

Keywords Radiation-induced cancer • Osteosarcoma • Mouse model • Genetic modifier • Inherited susceptibility • Alpha-emitter • Incorporated isotopes • Instability • Radiation-risk • Individual predisposition

M. Rosemann (✉) • I. González-Vasconcellos • T. Domke • M. Nathrath • M.J. Atkinson
Helmholtz Center Munich, German Research Center for Environmental Health,
Institutes of Radiation Biology and Clinical Cooperation Group Osteosarcoma,
Munich, Germany
e-mail: ROSEMANN@GSF.DE

C.E. Mothersill et al. (eds.), *Radiobiology and Environmental Security*,
NATO Science for Peace and Security Series C: Environmental Security,
DOI 10.1007/978-94-007-1939-2_29, © Springer Science+Business Media B.V. 2012

29.1 Introduction

Ionising radiation has been identified as a carcinogenic agent through epidemiological studies of cohorts exposed to relatively high doses, such as A-bomb survivors and those receiving therapeutic or accidental irradiation. The dose response data from these groups is used in extrapolating the risk of radiation-induced cancer to low-exposure groups [19]. Such population-based extrapolations carry with them a degree of uncertainty, and cannot accurately predict outcome in individual cases. This is especially true in those individuals carrying a germ-line mutation that predisposes them to the carcinogenic effects of ionising radiation e.g. P53, Rb1, and ATM [2, 10]. However, common gene polymorphisms also may influence individual susceptibility to the carcinogenic effects of ionising radiation, as they do other malignancies [12, 15, 21].

Uncertainties of tumor causality preclude studies in man that are designed to identify genetic effects on radiation carcinogenesis. However, mouse models show both direct causality and strain-dependent differences in radiation-induced tumor incidences, suggesting that modifiers of susceptibility can be identified using the mouse [9, 18].

Skeletal deposition of bone-seeking alpha particle-emitting radionuclides (e.g. Ra, Pu, Am), results in highly localized irradiation of the bone-forming cell compartment. The incidence of osteosarcoma in man and mouse is increased by a number of orders of magnitude following such exposures [6, 13]. Inbred BALB/c mice show a high incidence of osteosarcoma following injection of ^{227}Th (4 Gy skeletal dose), whilst at the same dose CBA strain animals are relatively resistant (Fig. 29.1). The polygenic basis of the strain difference was established in irradiated (BALB/c x CBA) F1 hybrid animals. These show an intermediate sensitivity as compared to their parental strains, indicating the action of multiple low-penetrance modifier genes.

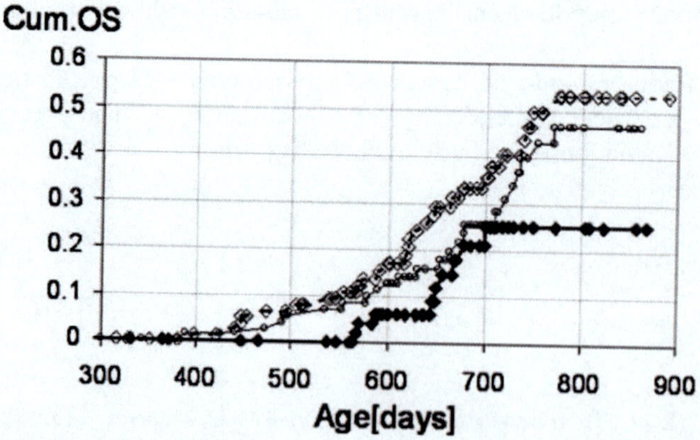

Fig. 29.1 Cumulative incidences of osteosarcoma development following injection with 185 Bq/g 227Th at the age of 100 days showing greater sensitivity of BALB/c strain (*open diamonds*) compared to CBA/CA strain (*closed diamonds*). The F1 hybrid animals (*open circles*) are of intermediate susceptibility

The variable penetrance of osteosarcoma in certain familial cancer syndromes [14] also suggests the existence of yet unknown susceptibility genes in man. We describe in this report mapping, identification and first functional studies of such susceptibility genes in radiation-induced osteosarcoma in mice.

29.2 Genome-Wide Mapping of Murine Susceptibility Genes

We have previously employed genetic mapping techniques in backcrossed mice strains to identify loci in the genome that harbour potential susceptibility genes for alpha-radiation induced osteosarcoma [16, 17]. These genome-wide screens for quantitative trait loci (QTL) established a total of six distinct chromosomal site that modify osteosarcoma predisposition in the mouse strains BALB/c, CBA/CA, C3H/H and 102/Nhg (Fig. 29.2).

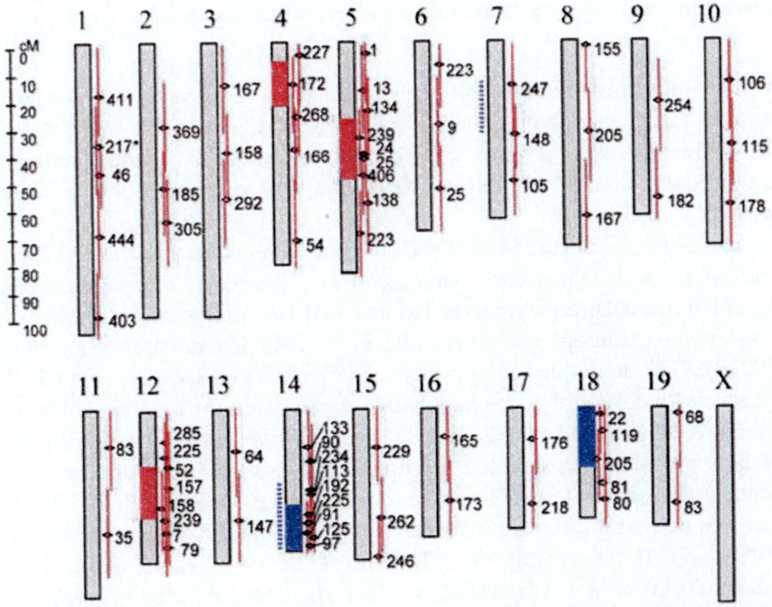

Fig. 29.2 Approximate positions (derived from MGD database) of BALB/c/CBA polymorphic microsatellite markers are given as *black diamonds*. Numbers are related to the name of each marker in the MIT panel (* exception is 217, that was originally D3Mit217, but has now been allocated to chromosome 1). Red lines indicate the sweep-radius of each marker, i.e. the interval for which linkage to susceptibility genes could be detected. Chromosomal regions showing linkage of osteosarcoma susceptibility with BALB/c alleles (*red*) and CBA/Ca alleles (*blue*) are shown together with two osteosarcoma susceptibility loci mapped in preceding study (hatched blue bars) (Reproduced from [17])

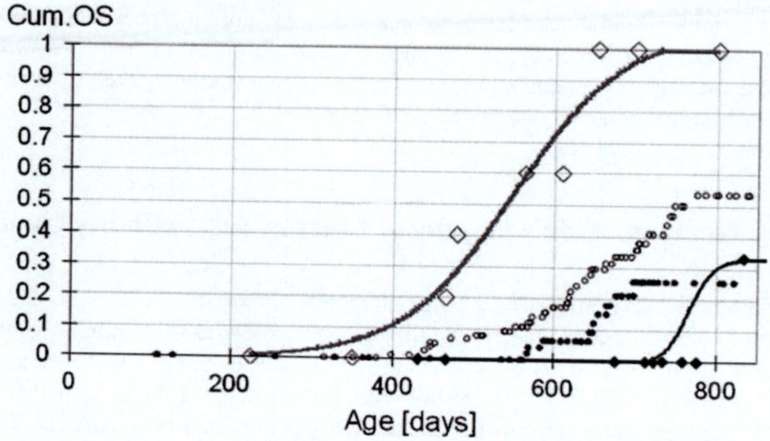

Fig. 29.3 Additive effect of five susceptibility loci on tumor incidence in backcross F2 animals grouped according to their genotypes. Those inheriting high-risk genotypes at all five susceptibility loci (*open diamonds*) show higher sensitivity than parental BALB/c (*open circles*) or CBA/CA mice (*closed circles*). F2 animals inheriting low-risk genotypes at all five loci exhibiting lowest osteosarcoma frequency (1/11) with the longest latency time (*closed diamonds*)

Animals inheriting the high-risk genotypes at five of these loci showed a 100% osteosarcoma incidence, with an average latency of 430 days. In contrast, animals inheriting low-risk gene variants at these five loci exhibit a marked tumor resistance, with only 1 out of 11 animals developing osteosarcoma with a latency of 680 days (Fig. 29.3).

Markers between D14Mit234 and D14Mit97 showed linkage with the BALB/c-allele being associated with an increased osteosarcoma resistance. Of 24 animals developing tumors all but 5 were heterozygous (CBA and BALB/c allelotypes) at these markers. This is in contrast to tumor-free animals, where only 10 of 36 animals were heterozygous ($p=0.0002$). QTL mapping using a normalized, censored latency time t(L)C as a trait parameter indicated that the likelihood ratio statistic for a QTL in this interval exceeds 20, equivalent to LOD=4.39 [17]. Significantly, this region on chromosome 14 overlaps with the osteosarcoma susceptibility locus we previously mapped in a different mouse cross [16], giving a compound LOD score of 5.44. Haplotype analysis was carried out for all 169 mice using additional markers placed between D14Mit234 and D14Mit97 (Fig. 29.4). The lowest number of recombinants was found for markers D14Mit192 and D14Mit225, which span only a 3 Mbp interval [5].

29.2.1 *Somatic Losses-of-Heterozygosity in Tumors*

To further narrow down the genomic position of an tumor susceptibility gene we analysed a set of radiation-induced osteosarcoma in (BALB/c x CBA) F1 hybrid mice for the presence of somatic gene alterations (loss-of-heterozygosity, LOH).

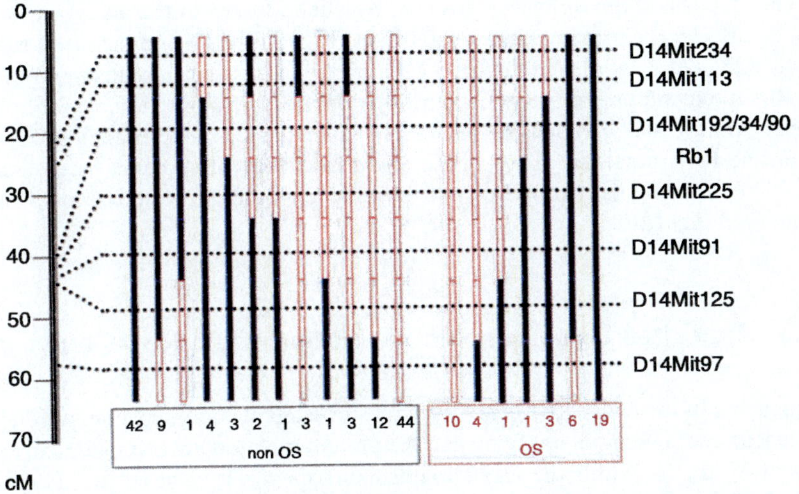

Fig. 29.4 Haplotype analysis of radiation-osteosarcoma susceptibility with additional markers to narrow down the principal susceptibility locus on chromosome 14. Germline haplotypes in 44 mice with and 125 mice without tumor are shown here with tracks of BALB/CBA heterozygote genotypes (*dark bars*) and BALB/BALB homozygote genotypes (*open bars*)

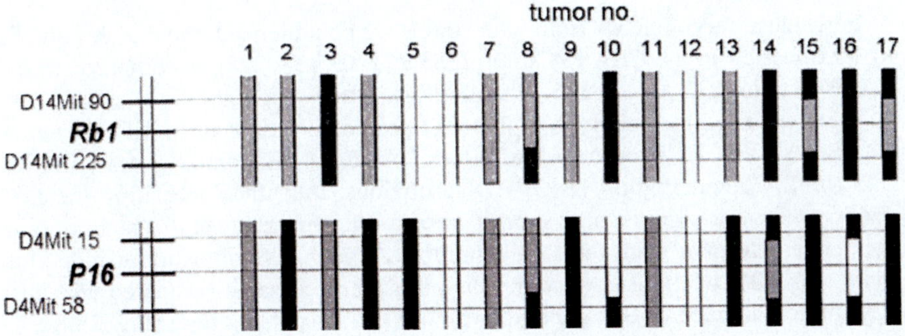

Fig. 29.5 Pattern of allelic-loss in 17 radiation-induced osteosarcoma found in BALB/c x CBA/CA F1 hybrid mice. Two loci in the mouse genome, encompassing RB1 on chromosome 14 and encompassing P16 on chromosome 4 are shown. Allelotyping was done for intra-genic SNPs of the two genes and for microsatellite-markers flanking them proximal and distal (Reproduced from [8]). *Black*: retention of heterozygosity; *grey*: loss of maternal BALB/c allele; *white*: loss of paternal CBA/CA allele

In the interval originally mapped by F2 linkage analysis, we now also found LOH in 13 out of 17 tumors (Fig. 29.5). Interestingly, there was a preferential loss of the BALB/c-allele, with 10 of the 17 tumors (58.8%) showing loss of the BALB/c allele, compared to only 3 of 17 (17.6%) tumors showing loss of the CBA-allele [8]. This preferential loss of the BALB/c allele in the tumors suggests that the BALB/c gene variant may have a tumor suppressor function in these animals.

The smallest common region affected by allelic losses in the analyzed tumors was defined by the proximal marker D14Mit192 (72.3 Mbp) and the distal marker D14Mit225 (75.3 Mbp). Within this 3 Mpb region on murine chromosome 14 maps the Rb1 tumor suppressor gene (73.6 Mbp), which is a strong candidate for both an osteosarcoma- and a radiation sensitivity predisposing gene. Patients carrying germ-line Rb1 mutations show an increased incidence of sporadic osteosarcoma [3, 7, 11], as well as a propensity to develop post-therapy osteosarcoma in the radiation field [4].

29.3 Inherited Hypomorphic Variations of the Rb1-Gene

Sequencing of the entire Rb1 gene transcript revealed no strain-specific differences in the Rb1 coding region, implying that the protein function is not impaired in either of the two strains. Within the 1,190 bp upstream of exon 1, however we identified a TCGCCC hexanucleotide duplication that is present only in the BALB/c allele, positioned between nt 1053 and nt 1084 (Fig. 29.6, base numbering according Genebank M86180). This duplication lies 177 bp upstream from the first exon, and is 5′ to the predicted binding sites for Sp1, ATF and E2. This alteration has therefore the potential to change the Rb1-transcription in the BALB/c strain relative to the CBA/CA strain.

In whole embryo mRNA from both strains we could indeed find a 50% reduced Rb1 expression in the CBA/CA strain (183 AU, CI 138–228) as compared to the BALB/c strain (232 AU, CI 165–299, p=0.038, *t*-test) (Fig. 29.7).

When we used the genotype of this *Rb1* promoter variant to stratify the entire BALB/c x CBA F2 backcross into either BALB/BALB homozygotes (B/B) or BALB/CBA heterozygotes (B/C) it was obvious, that mice inheriting the B/C genotype have a significantly shorter latency and higher overall osteosarcoma incidence than mice with the B/B genotype. At 500 days after tumor induction more than 40% of the B/C mice developed osteosarcoma as compared with just 12% tumors in mice of B/B genotype (p=0.0007, Log-Rank Test) (Fig. 29.8).

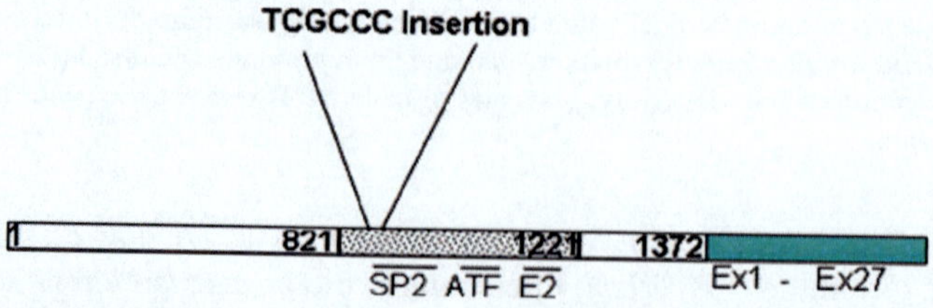

Fig. 29.6 Sequence of the *Rb1* promoter region between nt1053 and nt1084 determined for strains BALB/c and CBA/CA , showing position of the BALB/c specific hexanucleotide insertion relative to the SP1, ATF and E2 core binding sites (base numbering according to Genebank accession no. M86180)

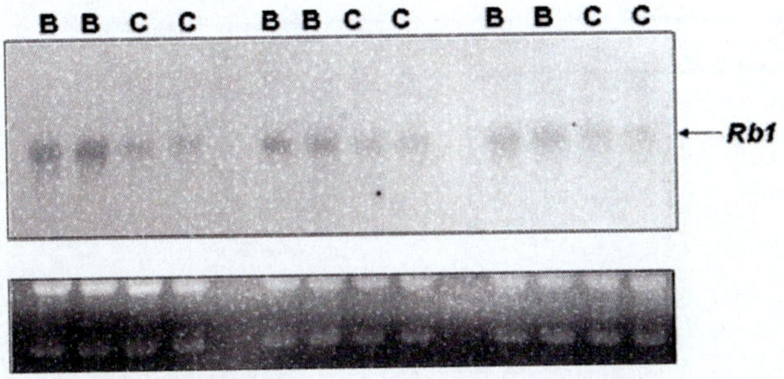

Fig. 29.7 *Rb1* expression in BALB/c and CBA/CA embryo RNA by northern blot analysis. *Lower row*: 18S and 28S RNA as loading control

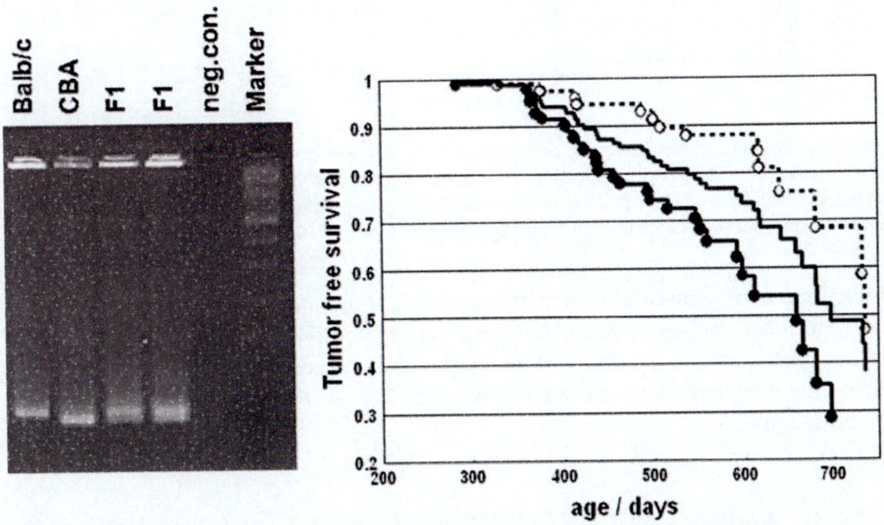

Fig. 29.8 Kaplan-Meier plot of osteosarcoma development in BALB × CBA backcrossed mice grouped according to their genotypes at the *Rb1* promoter. Difference between the B/B (*dashed line, open circles*) and the B/C genotypes (*filled line, black circles*) is highly significant (p=0.0007, Log Rank test). Intermediate line (*without symbols*) represent the pooled cohort of both genotypes

29.4 The Rb1 Pathway and Other Affected Elements

The Rb1 protein acts as a growth regulator and tumor suppressor by controlling transition of the G1/S cell cycle checkpoint in conjunction with the p16 tumor suppressor [20]. Temporary arrest at this checkpoint is an essential event in the post-irradiation DNA damage repair process, allowing efficient repair before entry into S-phase [1]. The lower level of transcription of Rb1 from the CBA promoter may be expected to lead to an inefficient arrest, and hence the entry of cells with unrepaired damage to enter S phase. A similar mechanism has been proposed for the increased

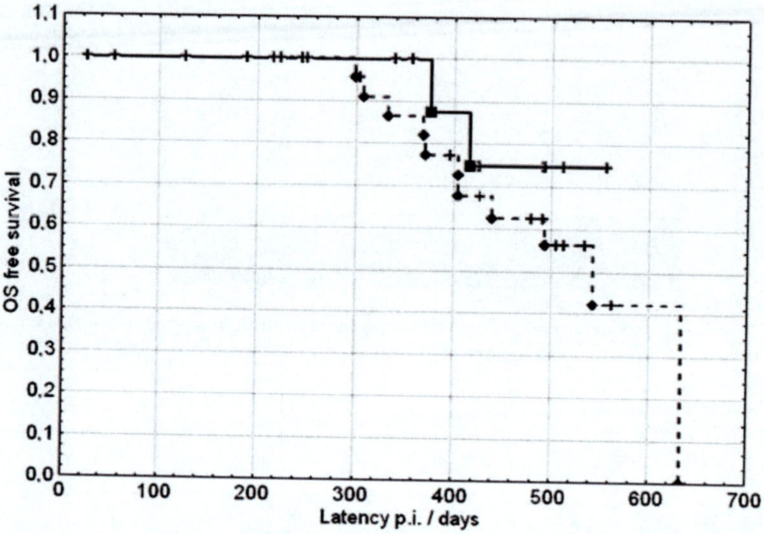

Fig. 29.9 Kaplan-Meier curve for development of osteosarcoma in female FVB/N-RB1LoxP x CreCol mice after 227Thorium injection. All mice carried heterozygote or homozygote alleles of the conditional RB1-LoxP allele [8]. *Dashed line*: Mice inheriting Cre-Recombinase -->RB1-deletion in normal bone; *full line*: Mice not inheriting Cre-Recombinase -->RB1 functionally normal.

incidence of pristane-induced plasmacytoma is associated with inheritance of a p16 promoter variant in BALB/c mice [21]. It should be noted that the p16 locus did not show any association with increased tumor susceptibility in our study nor is there any evidence of an association between human P16 germline mutations and osteosarcoma.

29.4.1 Reduced Rb1 and P16 Gene Expression Show Different Effects onto Osteosarcoma Development

To test how Rb1 and P16 germline variants differ in their effects onto radiation osteosarcomagenesis, we used heterozygote P16 knock-out and Rb1 knock-out mice and induced tumors again by Th-227 incorporation [8]. Both mouse lines had a pre-existing defect in one copy of either the P16- or the Rb1 gene in all bone cells, and were compared in this experiment with their wildtype littermates (i.e. with a set of mice on the same genetic background, but with the normal state of both target genes).

Out of 42 animals derived from conditional Rb1 knockout mating, 13 developed osteosarcoma with a median latency time of 402 days. Of these 13 tumors, 11 were found in 24 mice that had already a single copy of the *Rb1* gene lost in every normal bone cell. Of the remaining 18 mice without a pre-existing *Rb1* defect in bone, only 2 were diagnosed with an osteosarcoma (median latency 392 days). As shown in Fig. 29.9, the time course of osteosarcoma induction is similar in heterozygote Rb1

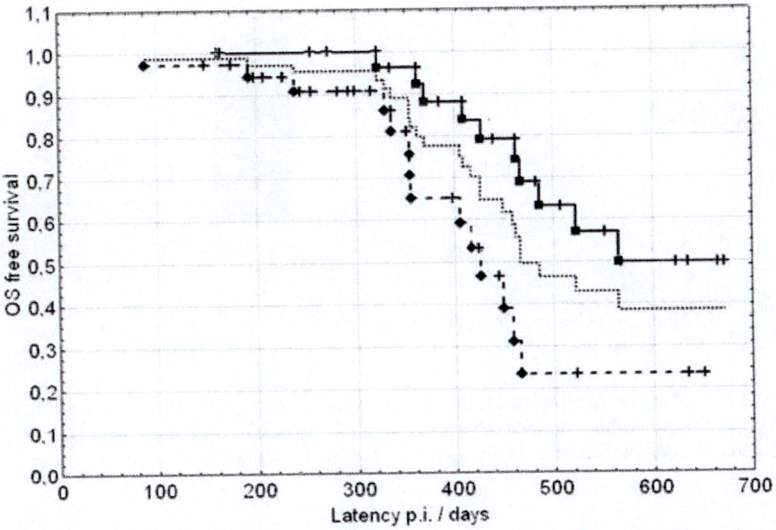

Fig. 29.10 Kaplan-Meier curve for development of osteosarcoma in female C57/BL6-P16 k.o. mice after 227Thorium injection. Littermate mice are either P16 +/– heterozygote or P16 wildtype (Reproduced from [8]). *Dashed line*: Mice inheriting P16 +/– germline defect; *full line*: Mice inheriting wildtype P16; *dotted line*: pooled cohort of both genotypes

knockout and in wildtype mice, but the tumor incidence is significantly different ($p=4 \times 10^{-5}$ Fisher-Yates exact test).

Of a total of 70 animals from the P16 knockout breeding, 24 developed osteosarcoma (Fig. 29.10, median latency time 413 days). Of those 24 tumors, 14 were found among 36 mice with a preexisting defect in one copy of the P16 gene (median latency 355 days). Of the remaining 34 mice without a pre-existing P16 germline defect, 10 were diagnosed with an osteosarcoma (median latency 445 days). The relative numbers of tumors in the heterozygote P16 knockout and in the wildtype mice were statistically not different ($p=0.28$, Fisher's exact test), but they arise much earlier in P16+/– mice as compared to their wildtype littermates ($p=0.018$, Mann–Whitney Test).

We can therefore conclude that a germline defect of the P16 gene, leading to a 50% reduced expression, changes the tumor growth kinetics, but not the tumor incidence rate. In contrast, a reduced Rb1 expression due to an congenital variant has the capacity to increase osteosarcoma risk, but not the latency time or tumor kinetics.

29.5 Rb1 Increases Genome Instability Following Radiation-Exposure

Whereas Rb1 inactivating germline mutations are know to predispose for radiation-induced osteosarcoma in man [4], such an association is not yet known for hypomorphic Rb1 alleles (i.e. alleles that result only in a lower gene-expression). To understand

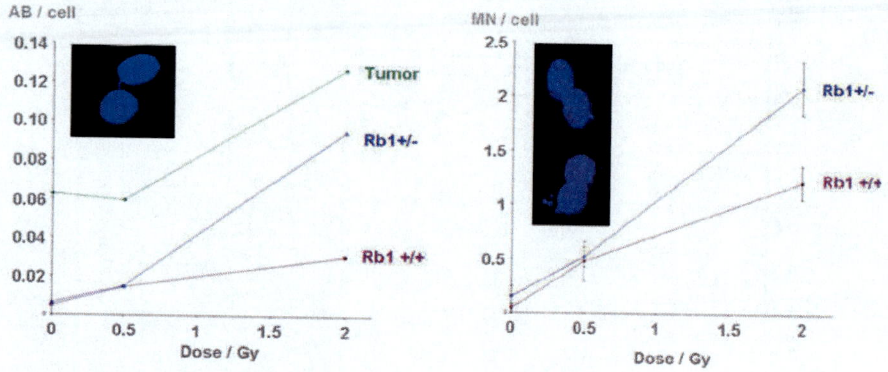

Fig. 29.11 Anaphase bridges (*left*) and micronuclei (*right*) in Rb1+/– and Rb1+/+ osteoblast cultures in-vitro after gamma irradiation with 0.5 and 2 Gy. For Anaphase bridges, data for an tumor cell line established from one of the radiation-induced osteosarcoma is shown as comparison

why a 50% reduced Rb1 expression in normal osteoblast cells is sufficient to confer an increased tumor risk, we investigated the acute effects of a gamma-irradiation in in-vitro cultures of osteoblasts from the above described heterozygote Rb1 knockout mice and their wildtype littermates. Phenotypically normal osteoblasts with a monoallelic Rb1 loss (i.e. with a 50% reduced gene expression) after 2 Gy gamma-irradiation exhibit a significantly increased number of post-mitotic genome defects, such as anaphase bridges (Fig. 29.11a) and micronuclei (Fig. 29.11b). Both are indicative of an increased genomic instability in the heterozygote Rb1 knockout osteoblasts.

29.6 Conclusion

We have shown here that a reduced expression of 50% of an important tumor susceptibility gene (Rb1) is sufficient to increase the risk for radiation induced osteosarcoma from 11% to 46%. This reduced Rb1 expression causes an increase of radiation-induced genomic instability in phenotypically normal osteoblasts. Both observations taken together suggest, that in normal wildtype osteoblasts the radiation damage can be tolerated up to a certain level whilst in cells with reduced Rb1-expression the induced genomic instability quickly exceeds a threshold and hence causes an overproportional rise in tumor incidence.

Genetic mapping of additional modifier loci in the outbred BALB/c x CBA strains could also show that, in addition to a principal susceptibility gene such as Rb1 other loci in the genome exist which, in an additive manner, can further contribute to an individual tumor risk. Whether susceptibility genes exist at these loci, or if they rather harbour regulatory elements remains elusive. Such regulatory elements could control expression of susceptibility gene *in-trans*, or could even modulate entire cellular pathways. Albeit a deeper functional knowledge of such heritable susceptibility factors in man is still missing, such factors could in principle by used already today to predict an individual tumor risk following occupational, medical or accidental radiation exposure.

Acknowledgement This work was supported by a grants from the European Commission (EURATOM) FIGH-CT99- 00001 and from the German ministry BMBF (KVSF 03NUK007).

References

1. Brugarolas J, Moberg K, Boyd SD, Taya Y, Jacks T, Lees JA (1999) Inhibition of cyclin-dependent kinase 2 by p21 is necessary for retinoblastoma protein-mediated G1 arrest after gamma-irradiation. Proc Natl Acad Sci USA 96:1002–1007
2. Cox R (1994) Molecular mechanisms of radiation oncogenesis. Int J Radiat Biol 65:57–64
3. Draper GJ, Sanders BM, Kingston JE (1986) Second primary neoplasms in patients with retinoblastoma. Br J Cancer 53:661–671
4. Eng C, Li FP, Abramson DH, Ellsworth RM, Wong FL, Goldman MB, Seddon J, Tarbell N, Boice JD Jr (1993) Mortality from second tumors among long-term survivors of retinoblastoma. J Natl Cancer Inst 85:1121–1128
5. Ensembl mouse genome sequence (2010) http://www.ensembl.org/Mus_musculus/
6. Fry SA (1986) Studies of U.S. radium dial workers: an epidemiological classic. Radiat Res 150(Suppl):S21–S29
7. Gilman PA, Wang N, Fan SF, Reede J, Khan A, Leventhal BG (1986) Familial osteosarcoma associated with 13;14 chromosomal rearrangement. Cancer Genet Cytogenet 17:123–132
8. Gonzalez Vasconcellos I, Domke T, Kuosaite V, Esposito I, Sanli-Bonazzi B, Nathrath M, Atkinson MJ, Rosemann M (2011) Differential effects of genes of the Rb1-signaling pathway on osteosarcoma incidence and latency in alpha-particle irradiated mice. Radiat Environ Biophys 50(1):135–141
9. Groch KM, Khan MA, Brooks AL, Saffer JD (1997) Lung cancer response following inhaled radon in the A/J and C57BL/6J mouse. Int J Radiat Biol 71:301–308
10. Hahn H, Wojnowski L, Zimmer AM, Hall J, Miller G, Zimmer A (1998) Rhabdomyosarcomas and radiation hypersensitivity in a mouse model of Gorlin syndrome. Nat Med 4:619–622
11. Hansen MF, Koufos A, Gallie BL, Phillips RA, Fodstad O, Brøgger A, Gedde-Dahl T, Cavanee WK (1985) Osteosarcomas and retinoblastomas: a shared chromosomal mechanism revealing recessive predisposition. Proc Natl Acad Sci USA 82:6216–6220
12. Hemminki K, Mutanen P (2001) Genetic epidemiology of multistage carcinogenesis. Mutat Res 473:11–21
13. Luz A, Müller WA, Linzner U, Strauss PG, Schmidt J, Müller K, Atkinson MJ, Murray AB, Gossner W, Erfle V, Hofler H (1991) Bone tumor induction after incorporation of short-lived radionuclides. Radiat Environ Biophys 30:225–227
14. Meisner LF, Gilbert E, Ris HW, Haverty G (1979) Genetic mechanisms in cancer predisposition: report of a cancer family. Cancer 43:679–689
15. Ponder BA (1990) Inherited predisposition to cancer. Trends Genet 6:213–228
16. Rosemann M, Lintrop M, Favor J, Atkinson MJ (2002) Bone tumorigenesis induced by alpha-particle radiation: mapping of genetic loci influencing predisposition in mice. Radiat Res 157:426–434
17. Rosemann M, Kuosaite V, Kremer M, Favor J, Quintanilla-Martinez L, Atkinson MJ (2006) Multilocus inheritance determines predisposition to alpha-radiation induced bone tumourigenesis in mice. Int J Cancer 118:2132–2138

18. Santos J, Herranz M, Perez DCI, Pellicer A, Fernandez-Piqueras J (1998) A new candidate site for a tumor suppressor gene involved in mouse thymic lymphomagenesis is located on the distal part of chromosome 4. Oncogene 17:925–929
19. UNSCEAR (2000) Sources and effects of ionizing radiation, annex I. United Nations Publications, New York, pp 298–450
20. Zacksenhaus E, Bremner R, Jiang Z, Gill RM, Muncaster M, Sopta M, Phillips RA, Gallie BL (1993) Unraveling the function of the retinoblastoma gene. Adv Cancer Res 61:115–141
21. Zhang S, Qian X, Redman C, Bliskovski V, Ramsay ES, Lowy DR, Mock BA (2003) p16 INK4a gene promoter variation and differential binding of a repressor, the ras-responsive zinc-finger transcription factor, RREB. Oncogene 22:2285–2295

Chapter 30
Biophysics of Heavy Ions

Marco Durante

Abstract Densely ionizing radiation, such as heavy ions, produce biological damage which is different from that normally produced by sparsely ionizing radiation, such as X- or γ-rays which are a large component of the natural radiation background. In fact, as a result of the different spatial distribution of the energy deposited, along the core and penumbra of the track, DNA lesions are exquisitely complex, and difficult to repair. RBE factors are normally used to scale from X-ray to heavy ion damage, but it should be kept in mind that RBE depends on several factors (dose, dose rate, endpoint, particle energy and charge, etc.) and sometimes heavy ions produce special damages that just cannot be scaled from X-ray damage. The special characteristics of heavy ions can be used to treat tumors efficiently, as it is currently done in Japan and Germany.

Keywords Heavy ions • Radiobiology • High LET • Densely ionizing radiation • Particle therapy

30.1 Introduction

The biological effectiveness of ionizing radiation strongly depends on the linear energy transfer, or LET, and it is well known that it is, for many endpoints, higher than sparsely ionizing radiation for LET values between 50 and 200 keV/μm in water. This different biological effectiveness os normally attributed to the different spatial distribution of lesion density in the DNA. While physicists know very well,

M. Durante (✉)
GSI Helmholtzentrum für Schwerionenforschung Biophysics Department, and Technische Universität Darmstadt, Planckstraße 1, 64291 Darmstadt, Germany
e-mail: M.Durante@gsi.de

C.E. Mothersill et al. (eds.), *Radiobiology and Environmental Security*,
NATO Science for Peace and Security Series C: Environmental Security,
DOI 10.1007/978-94-007-1939-2_30, © Springer Science+Business Media B.V. 2012

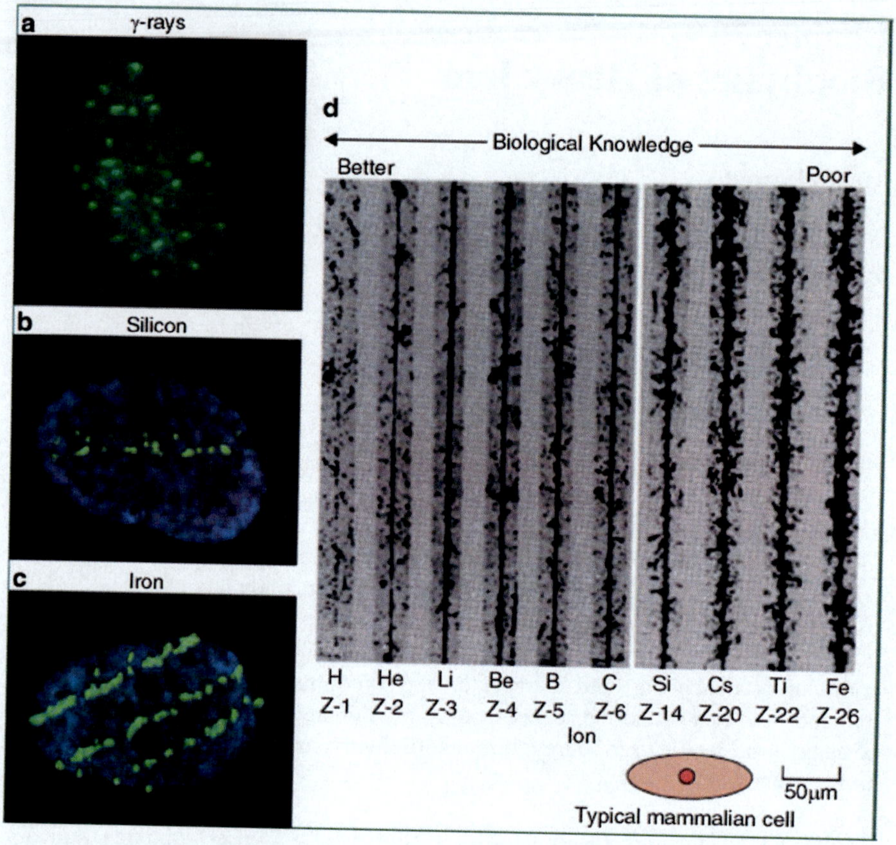

Fig. 30.1 Three nuclei of human fibroblasts exposed to (**a**) *γ-rays*, (**b**) *silicon ions*, or (**c**) *iron ions*; and immunostained for detection of γ-H2AX. Every green focus corresponds to a DNA double-strand break. In the cell exposed to sparsely ionising γ-rays (**a**), H2AX foci are uniformly distributed in the nucleus. Cells exposed to heavy ions show DNA damage along tracks – one silicon (**b**) and three iron (**c**) particles, respectively. Spacing between DNA double strand breaks is reduced at very high-LET. (**d**) Tracks of different ions, from protons to iron, in nuclear emulsions, show increasing LET as charge, Z, increases (From ref. [1])

form nuclear emulsions, how different a track of a heavy ion is compared to photons, more recently this could be visualized directly in mammalian cells, exploiting markers of DNA lesions such as phosphorilated histone γH2AX (Fig. 30.1) [1] or the accumulation of GFP-tagged repair proteins, such as 53BP1 (Fig. 30.2) [2]. Clearly, heavy ions produce "streaks" of DNA lesions in the cell nucleus, and the density of lesions increase with LET (Fig. 30.1), and they can be hardly repaired or moved following exposure (Fig. 30.2). This observation begs the question of whether the damage induced by heavy ions is different from that produced by X-rays. The answer is unfortunately not simple: even if the DNA damage is more difficult to repair, this

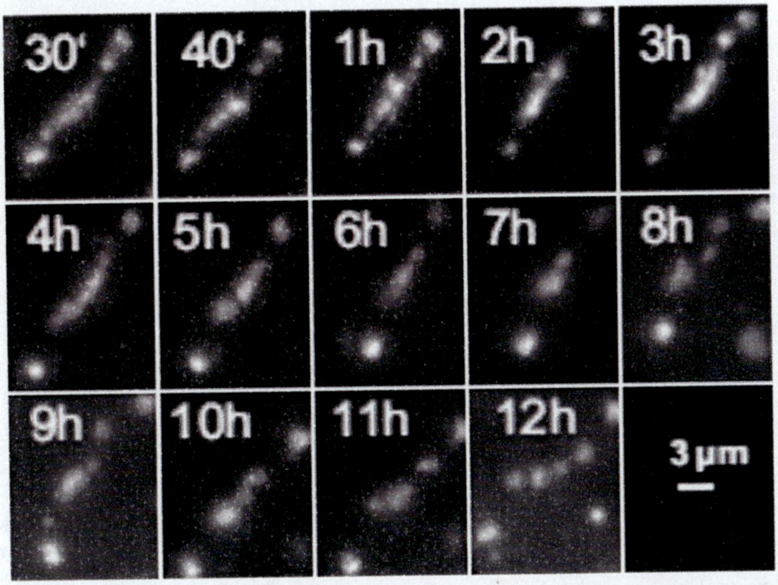

Fig. 30.2 Quantitative analysis of the motion of DNA double-strand breaks (DSBs) after high LET irradiation. Time-dependent changes of a single Ni-ion-induced 53BP1-GFP streak in a human tumor cell showing the typical motional behavior of individual proteins along the trajectory over the time course of 12 h after irradiation. Compared to Fig. 30.1, these pictures show the evolution of the damage in living cells, exploiting GFP-tagged proteins expressed in the cell, instead of fixing and staining the samples (From ref. [2])

may lead to an increased cell killing, but not necessarily to increased late risk: a dead cell cannot represent a risk, although the bystander effect may play a dominant role in explaining the effectiveness of high-LET radiation for late effects.

30.2 Heavy Ion Radiobiology

Because heavy ions are not present on Earth, their study is not relevant for radiation protection, and neither it has been for radiation therapy for many years. However, heavy ions are now often used in therapy [3] and they represent a major risk for human space exploration [4] (Fig. 30.3).

30.2.1 Particle Therapy

The rationale of oncological particle therapy is simply based on the different energy deposition of charged particles (the Bragg curve) and photons (exponential attenuation). Figure 30.4 immediately suggests that charged particles have a better energy

Fig. 30.3 Interest for radiobiology of heavy ions is linked to two main applications: cancer therapy (*left*) and protection of astronauts in long term space missions (*right*). In one case, we are interested to exploit the ability of heavy ions to kill cells; in the second, to protect the crews from long-term late effects. Although the exposure conditions are very different (high dose, fractionated, localized irradiation in therapy; low dose, chronic, whole body in space), the two topics share several research topics, such as studies on stochastic risk of heavy ions, or on radioprotectors (Table 30.1)

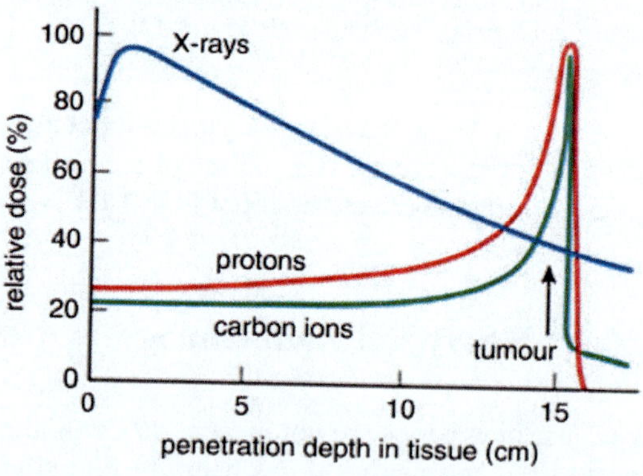

Fig. 30.4 Rationale of using charged particles for cancer therapy. While X-rays deposit more energy on the surface than in the tumor, the opposite holds for charged particles, such as protons and carbon ions

deposition pattern than X-rays for therapy, as recognized by Wilson already in 1946. Protontherapy is today widely spread in the world, and is considered a cutting-edge technology, with clinical results at least comparable to X-ray IMRT. However, apart from the favorable dose distribution, protons donor really add biological advantages, as their RBE is close to 1. On the other hand, heavy ions combine an increased biological effectiveness to a high RBE, and reduced oxygen enhancement ratio (OER), in the Bragg peak. Carbon ions are for instance low-LET (about 10 keV/μm) in the entrance channel, but high-LET (up to 80 keV/μm) in the Bragg peak, thus

Table 30.1 Some research topics relevant for both hadrontherapy and space radiation protection

	Hadrontherapy	Space radiation protection
Particles	H and C	All ions from H to Ni
Maximum energy	~400 MeV/n	~10 GeV/n
Dose	60–80 Gy-eq. in the target volume. Dose to the normal tissue depend on the treatment plan	50–150 mSv on the space station, up to 1 Sv for the Mars mission
Exposure conditions	Partial-body, fractionation (2 Gy-eq./day in the target volume)	Total-body, low dose-rate (1–2 mSv/day)
Individual radiosensitivity	Patient selection, personalized treatment planning	Personalized medical surveillance of the crewmembers
Mixed radiation fields	Effects of primary particles and fragments for tumor cell killing and side effects	Cosmic radiation is a mixed field. Effects of shielding
Late stochastic effects of heavy ions	Risk of secondary cancers in patients	Risk of cancer in astronauts
Normal tissue deterministic effects	Early and late morbidity	Cataracts, CNS damage, other late degenerative effects
Radioprotectors	Protection of the normal tissue, but not of the tumor. Drugs	Protection from heavy ions at low doses and protons at high doses (solar particle event). Dietary supplements
Biomarkers	Predicting risk of secondary cancers or late morbidity	Reducing uncertainties in risk estimates
Bystander effect	Role in tumor cell killing	Role in stochastic risk at very low fluence

providing sparing of the normal tissue and high effectiveness in the tumor. The clinical results, so far based on a fairly limited number of cancer patients (about 5,000) are indeed very good, and after the clinical trials in NIRS (Japan) and GSI (Germany), several new centers are under constructions in Europe and Asia.

30.2.2 Radiation Protection in Space

Although protons are by far the most common particle in space radiation, heavy ions play a major role because energy deposition increase with z^2, and the RBE increases with LET. Therefore, heavy ions are nowadays acknowledged by space agencies as a major barrier to human space exploration. Cancer risk is of course the main concern, because it is well documented that radiation can induce cancer, but the RBE of heavy ions is not known, due to the lack of epidemiological studies and the only limited animal studies, performed at particle accelerators [4]. In addition to cancer, several others late effects cause concern, including damage to the central nervous system, cataracts, risk of cardiovascular diseases, and hereditary effects. Both NASA and ESA support large experimental campaigns to study these effects, considering that space agencies are now shifting their programs to exploration.

The NASA program is based at the Brookhaven National Laboratory (Upton, NY), whereas the European program is based at GSI (Darmstadt, Germany).

Notwithstanding the large differences in exposure conditions (high dose, fractionated acute partial-body exposure in therapy; low dose, chronic whole-body exposure in space), the two topics share several research topics as summarized in Table 30.1.

30.3 Conclusions

Biological effects of densely ionizing radiation are becoming a key topic in radiobiology, because of the interest for heavy ions coming from space radiation protection and particle therapy. Large experimental campaigns are currently under way at accelerators, and it is likely that they will lead to a reduction of the uncertainty on the late risk of heavy ions.

References

1. Cucinotta FA, Durante M (2006) Cancer risk from exposure to galactic cosmic rays: implications for space exploration by human beings. Lancet Oncol 7:431–435
2. Jakob B, Splinter J, Durante M, Taucher-Scholz G (2009) Live cell microscopy analysis of radiation-induced DNA double-strand break motion. Proc Natl Acad Sci USA 106:3172–3177
3. Durante M, Loeffler JS (2010) Charged particles in radiation oncology. Nat Rev Clin Oncol 7:37–43
4. Durante M, Cucinotta FA (2008) Heavy ion carcinogenesis and human space exploration. Nat Rev Cancer 8:465–472

Chapter 31
De Novo Mutations in Y-Chromosome STR Loci Revealed in Paternal Lineages of Siberian Tundra Nentsi Population

Ludmila Osipova, Tatiana Karafet, and Michael Hammer

Abstract Y-chromosome short tandem repeats (STR) markers are widely used in human population genetic studies and forensic applications. Estimation of Y-STR mutation rate has a key role for dating the origin of Y chromosome lineages and for paternity tests. Previous studies demonstrated significant interlocus difference in mutation rate and the positive correlation to STR repeat length. The different ethnic groups and various Y-chromosome haplogroups defined by single-nucleotide polymorphisms were characterized by different Y-STR mutation rates. To date no Y-STR pedigree mutation studies were reported in native Siberian populations. Siberian Tundra Nentsi population possesses many unique genetic features and represents a good model for genealogical studies because of large family sizes, available ancestry information and relatively isolated life style in the extreme north environment. We thoroughly selected 50 paternal lineages with deep genealogical depth from 2 to 6 generations with the total number of 330 males. The number of descendants varied from 2 to 25 per paternal lineage, in average seven males per lineage. We also included in our analysis six Komi, three Russian and one Khant families who lived in the same villages. Totally 34 STR loci from non-recombining part of Y-chromosome were studied: DYS19, DYS390, DYS391, DYS393, DYS385A, DYS385B, DYS426, DYS388, DYS392, DYS439, DYS389-1, DYS389-2, DYS458, DYS447, DYS449, DYS459, DYS454, DYS464, DYS455, DYS457, DYS448, H4, DYS607, CDY-1, CDY-2, DYS460, YCA-1, YCA-2, DYS576, DYS570, DYS438, DYS456, DYS442, and C4. Eleven males (4%) had different alleles in six or seven STR loci compared to their paternal ancestors. We believe that those men were born out-of-wedlock or

L. Osipova (✉)
Institute of Cytology and Genetics SB RAS, Lavrentjeva Ave. 10, Novosibirsk 630090, Russia
e-mail: ludos84@mail.ru

T. Karafet • M. Hammer
Genomic Analysis and Technology Core University of Arizona, Tucson, USA

C.E. Mothersill et al. (eds.), *Radiobiology and Environmental Security*,
NATO Science for Peace and Security Series C: Environmental Security,
DOI 10.1007/978-94-007-1939-2_31, © Springer Science+Business Media B.V. 2012

were adopted. Traditionally native Siberian people adopt and raise all children who lost their parents. Mutations were observed in 21 out of 34 Y-STR loci. We found relatively higher number of mutation gains (60%) versus losses. Our study has thus shown that the chosen Y-STR loci represent a powerful tool to estimate mutation rates for forensic and population genetic purposes.

Keywords Y chromosome • Short tandem repeats • De novo mutation • Tundra nentsi

31.1 Introduction

The past two decades witnessed an explosion in data from the Y chromosomes in human populations. The inability of the Y chromosome to undergo recombination over most of its length means that the genotype of any individual is traceable with certainty to a single lineage (consisting of the father, the paternal grandfather, one paternal great-grandfather, etc.). Because Y chromosomes have unisexual transmission, migration and genetic drift will have quite different effects on the population genetic structure of Y chromosomes when compared to autosomes, particularly due to the difference in effective population size. Compared to the mitochondrial genome containing 16.5 thousand base pairs, Y chromosome numbers approximately 60 million base pairs, and provides the researchers with a potentially more powerful instrument [1].

The low level of polymorphism on the on the NRY hindered research for many years. By the end of 1990s a novel mutation detection method and direct sequences have led to discovery of many new Y-specific single nucleotide polymorphisms (SNPs). More than 200 SNP's on the NRY were identified by 2001 [2, 3]. Almost 600 mutations were incorporated into the Y chromosome binary haplogroup tree [4].

Y-STRs are a short tandem repeat (STR) on the Y chromosome (Fig. 31.1). Due to their diversity, simple structure and the unique feature of being inherited by the paternal lineage, Y chromosome STRs (Y-STRs) have become a valuable tool in evolution studies, forensic medicine, medical genetics and genealogy analysis [5, 6, 21]. Distinct Y chromosomes identified by STRs are designated "haplotypes," as suggested by de Knijff [7]. Comparing haplotypes of two men with known mutation rates for Y-STRs, one can assess the number of generations to their closest ancestor by paternal lineage.

To date, about 500 polymorphic Y-STRs have been discovered. The most noted, well-described and widely used are DYS19, DYS385, DYS389I, DYS389II, DYS390, DYS391, DYS392, DYS393 и YCAII [8]. (Table 31.1). Less used are DYS434, DYS435, DYS436, DYS437, DYS438, DYS439, DYS460, DYS461, GATA A10, GATA C4, GATA H4 [9, 10]. Many Y-STRs have been found recently, and even more still are to be discovered [11].

Microsatellites show high levels of polymorphism. For example, the quadruplet TAGA (DYS19) may be repeated from 10 to 19 times, the triplet ATA

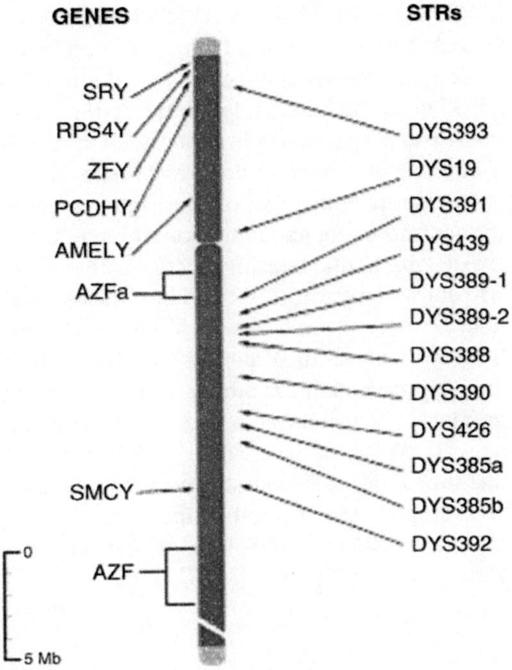

Fig. 31.1 A schematic diagram of human Y chromosome

Table 31.1 Structure of some Y-STRs, corresponding GeneBank accession numbers and sizes [8]

Marker	GeneBank	Sequence	Size
DYS19 (DYS394)	X77751	$(TAGA)_3TAGG(TAGA)_n$	143–179
DYS385	Z93950	$(AAGG)_{6\text{–}7}(GAAA)_n$	240–312, 324
DYS389 I	G09600	$(TCTG)_3(TCTA)_n$	235–267
DYS389 II	G09600	$(TCTG)_n(TCTA)_nN_{28}(TCTG)_3(TCTA)_n$	343–387
DYS390	G09611	$(TCTG)_n(TCTA)_n(TCTG)_{0\text{–}1}(TCTA)_{0\text{–}4}$	188–232
DYS391	G09613	$(TCTG)_3(TCTA)_n$	128–160
DYS392	G09867	$(TAT)_n$	181–217
DYS393	G09601	$(AGAT)_n$	104–140
DYS437	AC002992	$(TCTA)_m(TCTG)_n(TCTA)_4$	180–196
DYS438	AC002531	$(TTTTC)_1(TTTTA)_{0,1}(TTTTC)_n$	201–241
DYS439	AC002992	$(GATA)_n$	232–260
DYS460 (GATA A7.1)	G42675	$(ATAG)_n$	105–133
DYS461 (GATA A7.2)	G42671	$(TAGA)_n(CAGA)_1$	144–172
GATA A10	G42674	$(TCCA)_2(TATC)_n$	150–178
GATA C4	G42673	$(TCTA)_2[(TCTA)_2(TGTA)_2]_{2,3}(TCTA)_n$	238, 246–274
GATA H4	G42676	$(AGAT)_4CTAT(AGAT)_2(AGGT)_3$ $(AGAT)_nN_{24}$ $(ATAG)_4(ATAC)_1(ATAG)_2$	268–292

(DYS388) – 10–16 times, DYS385 contains from 7 to 28 repeated quadruplets GAAA etc. The most informative Y–STRs have been chosen as genealogical markers. At present, standard genealogic DNA tests are conducted using 12, 17, 25, 37 or 67 markers, although six markers is often already enough to attribute the given haplotype, depending on the combination of repeat numbers, to one of the Y-chromosome haplogroups, from A to R.

For the effective use of these microsatellite markers, it is necessary, besides further improving and reduction in price of typing methods, to create comprehensive databases that take into account regional and ethnic attributes of people, which will allow comparing STR variants and estimating its variation in different populations. It is also important to refine on a convenient and simple classification of these repeats, as for some Y-STR with a simple repeat structure, it is easy to find the consensus, and for others, characterized by a complex structure, it is often a problem. For instance, some obliged changes were made in the classification to include motives of the Y-STRs DYS19 and DYS390 (the first ones to be described), that turned out to be variable as new data had emerged. To prevent such changes, according to the general guidelines of the International Society for Forensic Genetics (ISFG), "alleles should be named considering the variable and non-variable repeats" [12, 25].

31.2 Materials and Methods

A Tundra Nentsi population of the size of about 2,000 people from Pur district of Yamalo-Nenets Autonomous Okrug has been under study since 1992. A comprehensive gene pool survey including the reproductive and demographic indices, health evaluation, and the genetic analysis employing the state-of the-art methods and molecular genetic markers has been performed. The population has been extensively studied including the classic blood group marker polymorphism rates as well as the mtDNA and Y-chromosome variabilities [13–16].

We attempted to achieve two goals – first, to study Y-chromosome microsatellite haplotype diversity in the Siberian population of Tundra Nentsi, and, second, to evaluate the Y-chromosome microsatellite mutation level associated with the adaptation of the Nentsi population to the technogenic environment.

In the present study, we are primarily concerned with the second aspect, which appears important due to the prevalence of chemical and radiation mutagenic factors in the human habitats. A multidisciplinary radioecology and genetic study by the Siberian Branch of the Russian Academy of Sciences was carried out over the period of 1997–2007 [17–20]. It has been shown that the areas inhabited by the Nentsi are saturated with the longlived isotopes emerged from the nuclear tests at the Novaya Zemlya Test Site and deposited into a prolonged food chain "lichen-reindeer-human". The current absolute contamination values (according to the Radiation safety standards of Russian Federation, 1999), however, are not hazardous to health. The present-day mean caesium-137 content in 237 samples of lichen was 153 Bq/kg (varying from

7 to 685 Bq/kg), while in 17 samples of reindeer meat the mean value was 118 Bq/kg (17 samples). The circulating caesium-137 values were measured in the urine of aboriginal people of Pur district (1–44 Bq/kg in 65 samples), as well as in the placentas of parturient women (1–27 Bq/kg in 64 samples). The aboriginal people appear to be affected by the low and ultra-low doses of internal ionizing radiation. The mechanism of action of those doses, especially, in the susceptible cohorts (including pregnant women) is significantly different than that of the short-term external irradiation (as it is shown by Burlakova E.B. with co-authors as well as by other researchers). During gestation, and, particularly, at critical stages of embryogenesis, teratogenic or embryotoxic exposure may occur [19].

The results of our 5-year long cytogenetic study confirm the negative effects of exposure to hazardous technogenic factors on human genome. The study was performed by a routine classical method of differential staining. The group under study was 369 people from the exposed region, including 250 of aboriginal people and 119 locals of European descent. As follows from the results presented in the Table 31.2, the total chromosome aberration frequency in all groups is significantly higher than the control level. The frequencies of dicentric and ring chromosomes indicative of radiation were significantly higher than the control level in all groups except the aboriginal children.

In the group of adult aboriginal population alone (N = 178), a significant increase in the total chromosome aberration rate (3.71 ± 0.27) as well as the frequencies of rings and dicentric chromosomes (0.44 ± 0.10) is observed.

The classical cytogenetic approach, thus, indicates at a stochastic influence of mutagenic factors on the cellular genetic mechanisms of the aboriginal population, possibly, causing an elevated level of disabilities in the local population.

FISH analysis with rRNA probe was performed unfortunately on a single local patient, year of birth 1957. (Grafodatsky, Rubtsov, personal communication). It demonstrated the presence of a marker chromosome carrying ribosomal genes in 50% of as much as 300 metaphases analysed. The patient was clinically infertile and died at the age of 50.

Recently, an elevated level of different kinds of pathologies with underlying genetic causes, including the congenital heart and vessel disfunctions, eye pathologies, allergic diseases and neuropsychic abnormalities has been observed among the children in the population under study.

Over the period of 1992–2008, when the population monitoring of the Samburg group of Tundra Nentsi has been carried out, we collected blood samples from about 80% of the total population together with all individual information required. All survey participants gave informed consent. From the overall pool of participants we constructed 50 paternal genealogical lineages ("extended families") totaling 330 people including the family founders and their descendants. Those lineages differed in the temporal depth of information available (from 2 to 7 generations) and the number of descendants (from 2 to 25 descendants, with a mean value of about seven descendants in a family). About 70% of male family founders in this sample were Tundra Nentsi. The others were represented by the Forest Nentsi (n = 6), Komi (n = 6), Russians (n = 3), and Khanty (1), who resided in the same area and used to marry the

Table 31.2 Results of cytogenetic analysis of aboriginal and non-aboriginal population of Yamalo-Nenets Autonomous Okrug

		Number of people	Number of cells (total)	Total cells with aberrations (N) / Dispersion% / Mean% ± m	Dicentrics and ring chromosomes (N) / Dispersion% / Mean% ± m
Aboriginal population	**Total**	250	26,631	860 0–23.8 ***3.23 ± 0.21	92 0–5.5 ***0.35 ± 0.07
	Adult	178	18,642	691 0–23.8 ***3.71 ± 0.27	82 0–5.5 ***0.44 ± 0.10
	Children	72	7,989	169 0–18 ***2.12 ± 0.32	10 0–1.3 0.13 ± 0.08
Control		105	10,472	162 0–5 1.55 ± 0.24	12 0.11 ± 0.06

Significance levels: significantly higher ***(p<0.001); significantly higher **(p<0.01); significantly higher *(p<0.05); and (without asterisk) – non-significantly higher (p>0.05) than the control level

Tundra Nentsi women. Since the peak exposure to the short-lived radionuclides has been in its highest values in the period of 1955–1963, we specifically focused on that time frame, tentatively considering September, 1955 to be the starting date of the technogenic exposure in the form of ionizing radiation (IR).

In our work, we used the following 34 Y-chromosome STR loci as the markers: DYS19, DYS390, DYS391, DYS393, DYS385A, DYS385B, DYS426, DYS388, DYS392, DYS439, DYS389-1, DYS389-2, DYS458, DYS447, DYS449, DYS459, DYS454, DYS464, DYS455, DYS457, DYS448, H4, DYS607, CDY-1, CDY-2, DYS460, YCA-1, YCA-2, DYS576, DYS570, DYS438, DYS456, DYS442, and C4. We genotyped these loci in 5 two multiplex reactions according to Redd et al. [21] with slight modifications.

Experimental study of Y-chromosome microsattelite polymorphisms was carried out in 2004–2005, in the period of work stay of L.P. Osipova at the University of Arizona, Tucson (in the laboratory of Prof. Michael Hammer).

31.3 Results and Discussion

Analysis of our results, first, discovered 11 nonpaternity cases (4% of the participants under study). The minority of those were the children born out of wedlock, from the fathers, who arrived to the area to work at oil and gas fields. The rest were either unrelated adopted orphans or the adopted relatives, who changed their family name to the adoptive father's family name. The traditional ethical rules in the

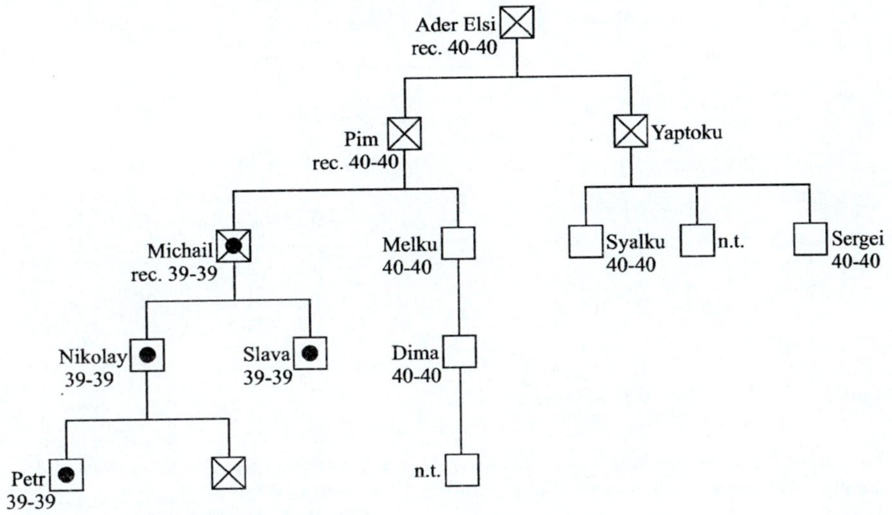

Fig. 31.2 Genealogy of Tundra Nentsi, de novo mutation in CDY locus. *Squares* with *black dots* designate individuals with mutation. "Rec" stands for "reconstructed"

aboriginal people populations were aimed at the avoidance of orphan hood. Typically, the children, whose parents have died, were adopted either by the families of their close relatives or by the childless families. Although genealogical information has been collected with maximal scrutiny, the high resolution of STR markers did not allow to unambiguously attribute those 11 individuals to the putative parental lineages. It was important to distinguish nonpaternity from mutation. We defined nonpaternity as a mismatch between fathers and sons at≥2 Y STR loci. In 11 cases of nonpaternity, a mismatch occurred in at least 6–7 STR markers.

It is relevant to present here two paternal genealogical trees, in which the de novo mutations were identified unambiguously.

In the first family tree presented in Fig. 31.2 (paternal line #5, family founder Ader Elsi, who was a Tundra Nenets), we identified a two de novo mutation in CDY locus. It is apparent that the mutation event occurred in the germ line of Michail, the member of Ader family born in 1940 and departed in 1970. It is from him that the "marker" branch of paternal lineage stemmed, within which the CDY 40–40 variant is replaced by CDY 39–39 variant, which is further inherited. That mutation may be considered as induced by technogenic exposure, since his elder son, Nikolai, was born in 1961 (the younger son, Vyacheslav was born in 1966). This genealogy includes five generations of descendants, for whom the reliable information about their relation was available.

The founder of the second family was the Forest Nenets Aivasedo Lelyu, who settled among the Samburg Tundra Nentsi and married the Tundra Nentsi woman. The de novo mutation in CDY-2 locus was shown to occur in his grandson, Nakoti (Fig. 31.3).

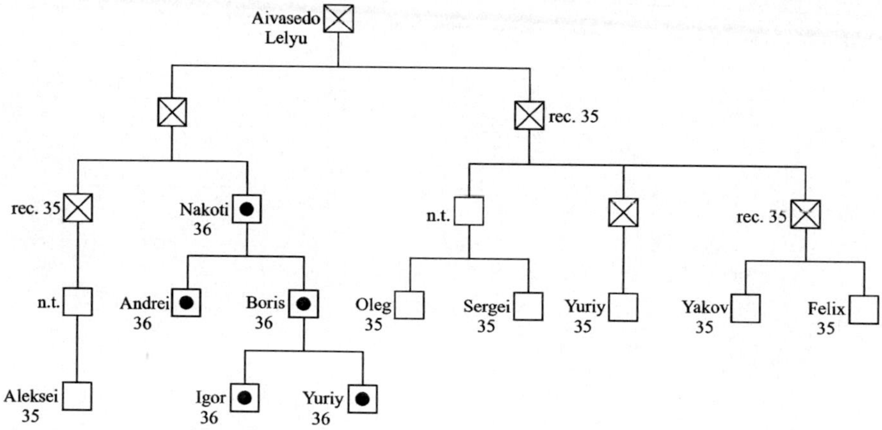

Fig. 31.3 Genealogy of Forest Nentsi, de novo mutation in CDY-2 locus. *Squares* with *black dots* designate individuals with mutation. "Rec" stands for "reconstructed"

In our sample, we discovered 59 "de novo" mutations, which, typically, involved a single locus. The mutations directed to an increase of repeat number prevailed (60% vs. 40%).

The "de novo" Y-chromosome STR loci mutations were discovered in 28 pedigrees out of 50, i.e., in more than a half of cases. Out of 34 Y-STR loci employed, as much as 21 loci were involved in mutation process, which confirms the high adequacy of the selected markers for the study of mutation component (among other microevolutionary factors) in effect in human populations. The highest mutability was identified in CDY locus, which was responsible for one third of all STR mutations. This locus provides increased male relative differentiation. Mutations in DYS449 and DYS464 loci occurred three times less frequent.

31.4 Conclusions

1. The selected markers are highly suitable for the study of mutation component (among other microevolutionary factors) in effect in human populations, for forensic identification purposes, and for evolutionary studies. For example, the men born out of wedlock often turn to the geneticists with a request to try to identify the ethnic origin of their biological father.
2. High frequency of the discovered "de novo" Y-STR mutations, in fact, put more questions, than they provide the answers. First, the selected populations resides at high latitudes with a severe climate. The possibility that the Y-microsatellite mutation rate in the High North populations is increased cannot be ruled out. In order to investigate the effect of high latitudes on the observed mutation rates of Y-STR loci, similar studies in the other populations within the same climatic zone

of residence should be carried out. Our male database includes such aboriginal populations of Siberia and Altai under demographic monitoring.

3. We are currently unable to make an unambiguous and mathematically accurate statement about a direct relation between the level of microsatellite mutability and the technogenic ionizing radiation. This stays to be a working hypothesis, though.
4. Currently, the obtained results with the consideration of the number of meiotic events are carefully analyzed with the use of different statistical methods, since the different mutation rates for the particular Y-STR loci has been published. We find important to obtain the correct estimates of those rates for a selected population, and we are planning to expand our male population sample and the number of paternal lineages to this end.

References

1. Underhill PA, Kivisild T (2007) Use of y chromosome and mitochondrial DNA population structure in tracing human migrations. Annu Rev Genet 41:539–564
2. Underhill PA, Shen P, Lin AA, Jin L, Passarino G, Yang WH, Kauffman E, Bonne-Tamir B, Bertranpetit J, Francalacci P et al (2000) Y chromosome sequence variation and the history of human populations. Nat Genet 26:358–361
3. Hammer MF, Karafet TM, Redd AJ, Jarjanazi H, Santachiara-Benerecetti S, Soodyall H, Zegura SL (2001) Hierarchical patterns of global human Y-chromosome diversity. Mol Biol Evol 18:1189–1203
4. Karafet TM, Mendez FL, Meilerman MB, Underhill PA, Zegura SL, Hammer MF (2008) New binary polymorphisms reshape and increase resolution of the human Y chromosomal haplogroup tree. Genome Res 18(5):830–838
5. Kayser M, Kittler R, Erler A, Hedman M, Lee AC, Mohyuddin A, Mehdi SQ, Rosser Z, Stoneking M, Jobling MA et al (2004) A comprehensive survey of human Y-chromosomal microsatellites. Am J Hum Genet 74:1183–1197
6. Kayser M (2007) Uni-parental markers in human identity testing including forensic DNA analysis. Biotechniques 43:3042
7. de Knijff P (2000) Messages through bottlenecks: on the combined use of slow and fast evolving polymorphic markers on the human Y chromosome. Am J Hum Genet 67:1055–1061
8. Gusmao L, Alves C (2005) Y chromosome STR typing. Methods Mol Biol 297:67–82
9. White PS, Tatum OL, Deaven LL, Longmire JL (1999) New, male-specific microsatellite markers from the human Y chromosome. Genomics 57(3):433–437
10. Ayub Q, Mohyuddin A, Qamar R, Mazhar K, Zerjal T, Mehdi SQ, Tyler-Smith C (2000) Identification and characterisation of novel human Y-chromosomal microsatellites from sequence database information. Nucleic Acids Res 28(2):e8
11. Ballantyne KN, Goedbloed M, Fang R, Schaap O, Lao O, Wollstein A, Choi Y, van Duijn K, Vermeulen M, Brauer S, Decorte R, Poetsch M, von Wurmb-Schwark N, de Knijff P, Labuda D, Vézina H, Knoblauch H, Lessig R, Roewer L, Ploski R, Dobosz T, Henke L, Henke J, Furtado MR, Kayser M (2010) Mutability of Y-chromosomal microsatellites: rates, characteristics, molecular bases, and forensic implications. Am J Hum Genet. doi:10.1016/j.ajhg.2010.08.006
12. Gill P, Brenner C, Brinkmann B, Budowle B, Carracedo A, Jobling MA, de Knijff P, Kayser M, Krawczak M, Mayr WR, Morling N, Olaisen B, Pascali V, Prinz M, Roewer L, Schneider PM, Sajantila A, Tyler-Smith C (2001) DNA commission of the international society of

forensic genetics: recommendations on forensic analysis using Y-chromosome STRs. Forensic Sci Int 124(1):5–10

13. Osipova LP, Posukh OL, Ivakin EA, Kryukov YuA, Karaphet TM (1996) Gene pool of native inhabitants of the Samburg Tundra Russ J Genet 32(6):722–728. Translated from Genetika (1996) 32(6):830–836
14. Duzhak TG, Osipova LP, Posukh OL, Chasovnikova OB, Ostashevskii VA, Lyakhovich VV (2001) Genetic polimorpisms of CYP1A1, GSTM1 and P53 genes in a unique Siberian population of Tundra Nentsi and its farmacogenetic importance. Int J Circumpolar Health 60:228–234
15. Karafet TM, Osipova LP, Gubina MA, Posukh OL, Zegura SL, Hammer MF (2002) High levels of Y-chromosome differentiation among native Siberian populations and the genetic signature of a boreal hunter-gatherer way of life. Hum Biol 74(6):761–789
16. Tambets K, Rootsi S, Kivisild T, Help H, Serk P, Loogvali EL, Tolk HV, Reidla M, Metspalu E, Pliss L, Balanovsky O, Pshenichnov A, Balanovska E, Gubina M, Zhadanov S, Osipova L et al (2004) The western and eastern roots of the Saami–the story of genetic "Outliers" told by mitochondrial DNA and Y chromosomes. Am J Hum Genet 74(4):661–682
17. Osipova LP, Posukh OL, Ponomareva AV, Matveeva V.G, Shcherbov BL, Strakhovenko VD, Sukhorukov FV (1999b) The Tundra Nentsy: health and environmental radiation on the territory they inhabit. In: Armannsson H (ed) The proceedings of the 5th international symposium on geochemistry of the earth's surface, Reykjavik, 16–20 Aug 1999, pp 215–218
18. Osipova LP, Posukh OL, Koutzenogii KP, Sukhorukov FV, Matveeva VG, Grafodatskii AS, Konovalova NA, Sukhovey YG, Petrov SA, Lefranc G, Lefranc M-P (1999a). Epidemiological Studies for the Assessment of Risks from Environmental Radiation on Tundra Nentsi Population. In: Baumstark-Khan C et al (eds) NATO science series. Series 2: Environmental security. Proceedings of the NATO advanced research workshop "Fundamentals for the assessment of risks from environmental radiation", vol 55, Brno, Czech republic, 6–10 Oct 1997, pp 35–42
19. Osipova LP, Koutsenogii KP, Shcherbov BL, Strakhovenko VD, Petrov SA (2002) Environmental radioactivity for risk assessment of health status in nature and human population of Northern Siberia. In:The 5-th international conference on environmental radioactivity in the Arctic and Antarctic, St. Petersburg, 16–20 June 2002, pp 124–127
20. Korovkina ES, Tuzikov FV, Tuzikova NA, Osipova LP, Buneva VN, Nevinsky GA (2004) The evaluation of biological health in the population of Tundra Nentsy (North-Western Siberia) from the data on autoantibodies and lipoproteins of the donors sera. Nucleosides Nucleotides Nucleic Acids 23(6/7):1009–1013
21. Redd AJ, Agellon AB, Kearney VA, Contreras VA, Karafet T, Park H, de Knijff P, Butler JM, Hammer MF (2002) Forensic value of 14 novel STRs on the human Y chromosome. Forensic Sci Int 130(2–3):97–111
22. Shen P, Wang F, Underhill PA, Franco C, Yang WH, Roxas A, Sung R, Lin AA, Hyman RW, Vollrath D et al (2000) Population genetic implications from sequence variation in four Y chromosome genes. Proc Natl Acad Sci 97:7354–7359
23. Tilford CA, Kuroda-Kawaguchi T, Skaletsky H, Rozen S, Brown LG, Rosenberg M, McPherson JD, Wylie K, Sekhon M, Kucaba TA, Waterston RH, Page DC (2001) A physical map of the human Y chromosome. Nature 409(6822):943–945
24. Underhill PA, Jin L, Lin AA, Mehdi SQ, Jenkins T, Vollrath D, Davis RW, Cavalli-Sforza LL, Oefner PJ (1997) Detection of numerous Y chromosome biallelic polymorphisms by denaturing high-performance liquid chromatography. Genome Res 7:996–1005
25. Y Chromosome, Consortium (2002) A nomenclature system for the tree of human Y-chromosomal binary haplogroups. Genome Res 12(2):339–348

Chapter 32
Effects of Chronic Gamma Irradiation on Reproduction in the Earthworm *Eisenia fetida*

Deborah H. Oughton, Turid Hertel-Aas, Alicja Jaworska, and Gunnar Brunborg

Abstract In recent years ecological impacts of ionizing radiation have emerged as an important research field. But there are still considerable knowledge-gaps regarding biological effects of chronic irradiation in wildlife, particularly for endpoints related to reproduction. Reproduction is considered to be one of the most sensitive radiation-associated endpoints, and determines not only the fate of the single organism, but may also influence population dynamics and the balance of higher ecological units. Based on their radioecological properties and their important role in the soil ecosystem, earthworms have been identified by ICRP as one of the reference animals and plants (RAPs) to be used in environmental radiation protection. This paper will present results of a series of studies carried out on the effects of ionising radiation on earthworms, covering long term-effects on reproduction, as well as studies on biological processes such as recovery, acclimatization and adaptive response, using a number of molecular biomarkers. The aim is to show how the studies can improve predictions of the way individuals and populations respond to chronic exposures of ionising radiation.

Keywords Environmental effects • Ionising radiation • Earthworm • Reproduction • Chronic exposure

---

D.H. Oughton (✉) • T. Hertel-Aas
Department of Plant and Environmental Sciences, Norwegian University of Life Sciences, 1432 Aas, Norway
e-mail: deborah.oughton@umb.no

A. Jaworska
Norwegian Radiation Protection Authority, Osteraas, Norway

G. Brunborg
Norwegian Institute of Public Health, Oslo, Norway

C.E. Mothersill et al. (eds.), *Radiobiology and Environmental Security*, NATO Science for Peace and Security Series C: Environmental Security, DOI 10.1007/978-94-007-1939-2_32, © Springer Science+Business Media B.V. 2012

32.1 Introduction

There are still considerable knowledge-gaps regarding biological effects of chronic irradiation in many wildlife species, particularly for endpoints related to reproduction [1]. In ecological risk assessment, impacts on populations of species are considered the most relevant for protection of ecosystems as a whole [2, 3]. Since the reproduction capacity of adults and the sexual maturation rate of offspring are of great importance for population dynamics, it follows that information on the effects of chronic exposures on these endpoints is required [4, 5]. Based on their radioecological properties and their important role in the soil ecosystem, earthworms have been identified as one of the reference animals and plants (RAPs) by ICRP [6], and they are also included as reference organisms in the framework for environmental risk assessment developed during the EU ERICA project [7, 8]. Invertebrates are generally known to be relatively radioresistant but as for most species, reproduction is known to be far more sensitive than mortality [9]. After nuclear accidents the local radioactivity levels can be very high, but will show a rapid decline due to the disintegration of short-lived radionuclides, i.e., after the Chernobyl accident, approximately 80% of the accumulated doses were absorbed within 3 months of the accident [9]. The long term effect of such high exposures on the population dynamics depends partly on whether the radiation induced damage is irreversible or if the organisms have the ability to recover after the exposure levels have declined. Mechanisms leading to acclimatization (physiological adjustments during the exposure history of individuals) or adaptation (inheritable traits acquired during the exposure history of the population), might also influence the long term effects of exposure. This paper summarises results from two studies. The main purpose of the first experiment [10] was to establish dose (rate) response relationships for reproduction endpoints in the earthworm Eisenia fetida in two subsequently exposed generations (F0 and F1). In the second experiment worms were exposed at a high dose rate and the ability to recover after a complete radiation induced inhibition of the reproductive capacity were studied.

32.2 Exposure Details

In both experiments the experimental set up was based on standardised OECD chemical ecotoxicology tests for earthworm reproduction [11]. Briefly, ten worms were placed in each replicate container applied with 677 g moist artificial OECD soil [11], and moist horse manure was spread on the soil surface (1.5 g wet w/worm/week). The endpoints were assessed after defined exposure periods, after which the cocoons were transferred to small boxes, the soil replaced and hatchlings removed after registration. More details on the experimental set up and dosimetry can be found in Hertel-Aas et al. [10].

32.2.1 *Dose-Response Relationships for Chronic Exposure*

In the first study [10], E. fetida was exposed to external ^{60}Co gamma irradiation during two generations (F0 and F1). Adult F0 reproduction capacity (i.e., number of cocoons produced, cocoon hatchability and number of F1 hatchlings) in controls and at five absorbed dose rates (0.18, 1.7, 4, 11 and 43 mGy/h) was measured over a 13 week exposure period. Survival, growth and sexual maturation of F1 hatchlings were examined for 11 weeks. F1 adults were exposed for a further 13 weeks for registration of reproduction capacity.

32.2.2 *Recovery*

In the second experiment, adult F0 worms were exposed at 17 mGy/h for 10 weeks. The dose rate and exposure time were chosen to achieve an accumulated dose, causing a complete inhibition of hatchability, based on results from the first experiment. Thereafter the worms were transferred to a climate chamber at 21±0.5°C for 16 weeks.

32.3 Results and Discussion

32.3.1 *Dose-Response Relationships*

Accumulated doses are shown in Table 32.1 and impacts on reproduction endpoints in Fig. 32.1.

There was no radiation-induced effect on the viability, growth, sexual maturation or cocoon production rates in F0 or F1. For F0, hatchability of cocoons produced during the first 4 weeks was reduced to 60% at 43 mGy/h (98% in controls), and none of the cocoons produced at 5–13 weeks hatched. At 11 mGy/h a pronounced effect on cocoon hatchability was not observed until 9–13 weeks, when hatchability was reduced to 25%. Also, the number of hatchlings per hatched cocoon was reduced at the two highest dose rates, and the total number of F1 hatchlings per adult F0 produced during the 13 weeks exposure period was reduced to 17% and 57% compared to the controls at 43 mGy/h and 11 mGy/h, respectively. The total number of F1 hatchlings was reduced also at 4 mGy/h, but the effect was of borderline significance. At the end of adult F0 exposure, dissection and microscopic examination revealed atrophic male reproductive organs in worms exposed at 43 mGy/h. For adult F1, the hatchability of cocoons produced at 11 mGy/h was reduced to 45–69% during the 13-week exposure period, and the total number of F2 hatchlings

Table 32.1 Dose rates and accumulated doses in F0 and F1 worms, are given as ranges due to the decay of ^{60}Co. Values are means ± SD (including 15% uncertainty)

Mean dose rate (mGy/h)	Accumulated dose (Gy)			
	F0			
	Week 4	Week 8	Week 13	Week 11
0.18–0.19	0.11 ± 0.02	0.23 ± 0.03	0.37 ± 0.05	0.39 ± 0.06
1.7–1.8	1.1 ± 0.2	2.2 ± 0.3	3.6 ± 0.5	3.7 ± 0.6
4.2–4.3	2.7 ± 0.4	5.4 ± 0.8	8.6 ± 1.3	9.0 ± 1
11.2–11.5	7.1 ± 0.9	14 ± 2	23 ± 3	24 ± 4
42.4–42.7	26 ± 4	53 ± 8	85 ± 13	–
	F1			
	Week 16	Week 20	Week 24	
0.18–0.19	0.53 ± 0.08	0.63 ± 0.09	0.73 ± 0.11	
1.7–1.8	5.0 ± 0.8	6.1 ± 0.9	7.0 ± 1.1	
4.2–4.3	12 ± 2	15 ± 2	17 ± 3	
11.2–11.5	33 ± 5	39 ± 6	45 ± 7	
42.4–42.7	–	–	–	

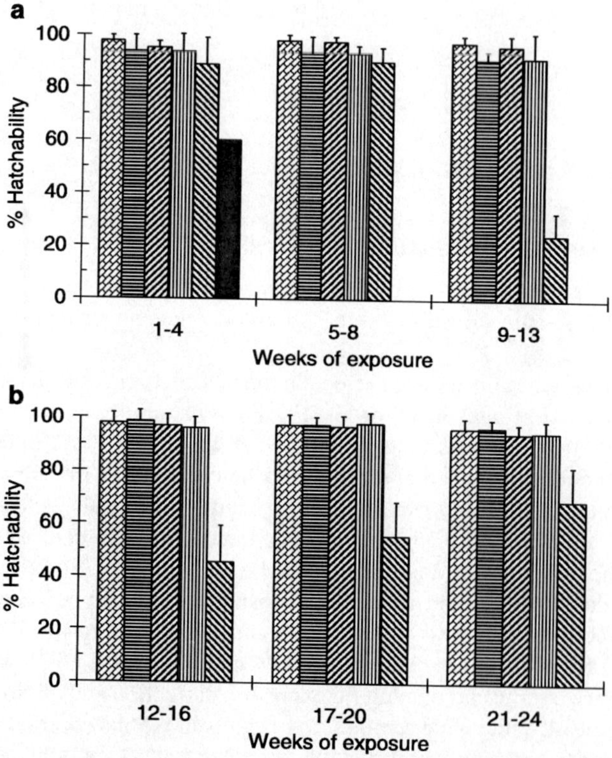

Fig. 32.1 Hatchability of cocoons produced during different exposure periods, shown as accumulated percentages 8 weeks after first counting. (**a**) F0 exposure, (**b**) F1 exposure. Means ± SD of replicates

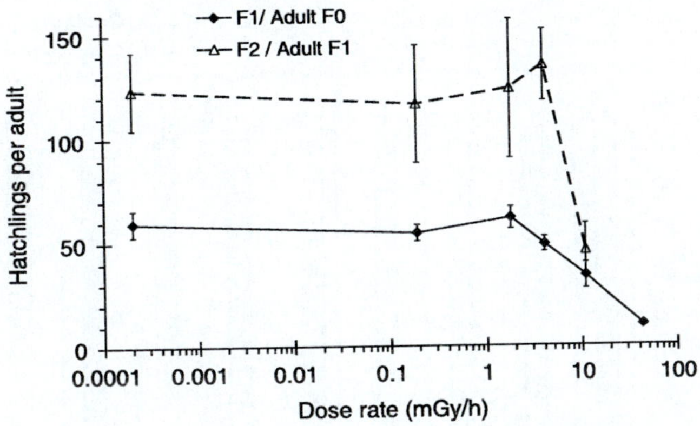

Fig. 32.2 The total number of F1 hatchlings per adult F0, and F2 hatchlings per adult F1, produced during the 13 weeks reproduction periods. External dose to controls was 0.2 μGy/h. Values are means ± SD for replicate boxes. Control: n = 12; 0.19–11 mGy/h, n = 4; 43 mGy/h, n = 1

produced per adult F1 was reduced to 37% compared to the controls. However, and in contrast to the results observed for F0, hatchability increased with time, suggesting a possible acclimatization or adaptation of the F1 individuals. This finding might be confounded by the increase in temperature during some of the F1 exposure periods (Fig 32.2).

32.3.2 *Recovery*

At 17 mGy/h, the hatchability of cocoons produced during week 7 and 8 was reduced to 6%, at which time the dose absorbed by the adult worms had reached 20 ± 3 Gy. None of the cocoons produced during week 9–10 hatched (absorbed dose 25 ± 4 Gy) (Fig. 32.3a). This was in good agreement with the results from the first experiment where there seemed to be a critical absorbed dose in F0 of around 20 Gy. Also the number of hatchlings per hatched cocoon and the numbers of F1 hatchlings produced per adult per week was significantly reduced during week 7–8 (Fig. 32.3b). At the end of the 10 week exposure period, the total number of F1 hatchlings produced per adult F0 was reduced to 51% compared to the controls. After transfer to the control area, none of the cocoons produced during the first 2 weeks by worms pre-exposed at 17 mGy/h area hatched, whereas after 3–4 weeks the hatchability increased slightly. A further increase was observed for cocoons produced during week 5–8, and during the next period the recovery seemed complete as the hatchability and the number of F1 hatchlings produced per adult F0 per week reached control levels. Furthermore, the number of F1 produced per adult F0 worm derived from the cocoons produced during week 13–16 was higher than for the controls.

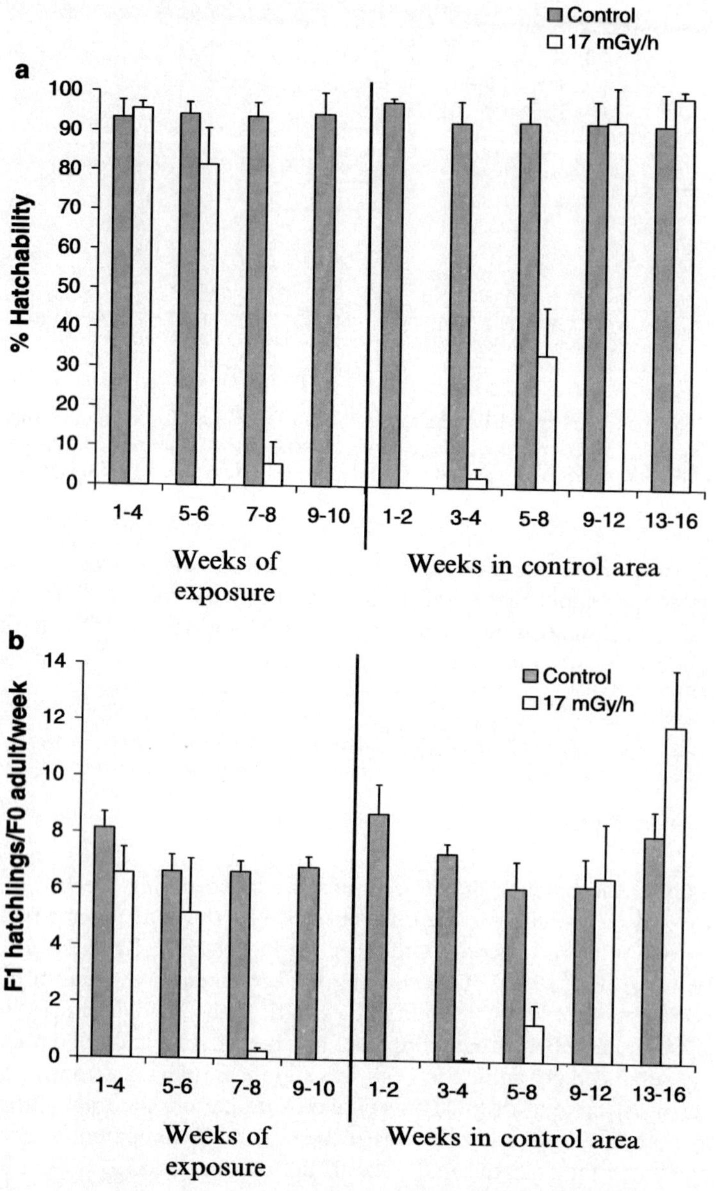

Fig. 32.3 **(a)** Hatchability of cocoons produced during different periods, shown as accumulated percentages 8 weeks after first counting. **(b)** Number of F1 hatchlings produced per adult F0 per week derived from cocoons produced during the same periods. Means ± SD for replicate boxes. Controls, n = 3; 17 mGy/h n = 3

32.4 Conclusion

In conclusion, the most sensitive endpoints examined in the current studies were the hatchability of cocoons and the number of hatchlings per cocoon. But relatively extensive exposure periods were needed for these effects to be expressed at lower dose rates. This illustrates the problems in extrapolating from acute to chronic studies, and the need for more ecologically relevant studies for understanding the potential impact of ionising radiation on non-human species. Short exposure times can lead to an underestimation of the potential effects on such endpoints, and studies on reproduction in any species must be designed to avoid such errors.

The likely mechanism behind the reduction in hatchability is that damage is induced in male germ cells (i.e. spermatogenic cells) directly, and/or that damage accumulates in the testes or seminal vesicles resulting in reduced sperm production or infertile sperm. In the F0 generation it appeared that inhibition of reproduction was related to a critical total accumulated dose of approximately 20 Gy. Although the lowest observed effect dose rates (4 mGy/h in F0 and 11 mGy/h in F1) were relatively high, they are comparable to those that may occur after major nuclear accidents. While earthworms showed the ability to regain their reproductive capacity after a complete radiation induced inhibition of cocoon hatchability, more studies are required to better understand the underlying mechanisms.

Acknowledgment This work was supported by the EU ERICA project (Contract: FI6R-CT-2004-508847).

References

1. Garnier-Laplace J, Copplestone D, Gilbin R, Alonzo F, Ciffroy P, Gilek M, Agüero A, Björk M, Oughton DH, Jaworska A, Larsson C-M, Hingston JL (2008) Issues and practices in the use of effects data from FREDERICA in the ERICA Integrated Approach. J Environ Radioact 99(9):1474–1483
2. Andersson P, Beaugelin-Seiller K, Beresford NA, Copplestone D, Della Vedova C, Garnier-Laplace J, Howard BJ, Howe P, Oughton DH, Wells C, Whitehouse P (2008) Deliverable 5: numerical benchmarks for protecting biota from radiation in the environment: proposed levels, underlying reasoning and recommendations. Report for the PROTECT project. EC Contract Number:036425 (FI6R). CEH-Lancaster, Lancaster. Available from: http://www.ceh.ac.uk/protect/
3. Andersson P, Garnier-Laplace J, Beresford NA, Copplestone D, Howard BJ, Howe P, Oughton DH, Whitehouse P (2009) Protection of the environment from ionising radiation in a regulatory context (PROTECT): proposed numerical benchmark values. J Environ Radioact 100:1100–1108
4. Stark JD, Banks JE, Vargas R (2004) How risky is risk assessment: the role that life history strategies play in susceptibility of species to stress. Proc Natl Acad Sci USA 101:732–736
5. Alonzo F, Hertel-Aas T, Gilek M, Gilbin R, Oughton D, Garnier-Laplace J (2008) Modelling the propagation of effects of chronic exposure to ionising radiation from individuals to populations. J Environ Radioact 99(9):1464–1473

6. ICRP (2008) Environmental protection: the concept and use of reference animals and plants. Publication 108. Annals of the ICRP 38(4–6): 3–242
7. Beresford N, Brown J, Copplestone D, Garnier-Laplace J, Howard B, Larsson C-M, Oughton D, Pröhl G, Zinger I (2007) An integrated approach to the assessment and management of environmental risks from ionising radiation. D-ERICA. EC 6th Framework, Contract N°FI6R-CT-2004-508847. https://wiki.ceh.ac.uk/display/rpemain/ERICA+reports
8. Larsson C-M (2008) An overview of the ERICA integrated approach to the assessment and management of environmental risks from ionising contaminants. J Environ Radioact 99(9): 1364–1370
9. UNSCEAR (1996) Sources and effects of ionizing radiation. Report to the general assembly, with scientific annex. United Nations, New York, pp 1–86
10. Hertel-Aas T, Oughton DH, Jaworska A, Bjerke H, Salbu B, Brunborg G (2007) Effects of chronic gamma irradiation on reproduction in the earthworm Eisenia fetida (Oligochaeta). Radiat Res 168:515–526
11. OECD Guideline for the testing of chemicals (2004) 222 earthworm reproduction test (*Eisenia fetida/Eisenia andrei*). Organisation for Economic Co-operation and Development, Paris